Current Chinese Economic Report Series

For further volumes:
http://www.springer.com/series/11028

Xiaoxi Li · Jiancheng Pan
Editors

China Green Development Index Report 2011

BEIJING NORMAL UNIVERSITY PUBLISHING GROUP
BEIJING NORMAL UNIVERSITY PRESS

Springer

Editors
Xiaoxi Li
Beijing Normal University
Beijing
People's Republic of China

Jiancheng Pan
National Bureau of Statistics of China
Beijing
People's Republic of China

ISSN 2194-7937 ISSN 2194-7945
ISBN 978-3-642-31596-1 ISBN 978-3-642-31597-8 (eBook)
DOI 10.1007/978-3-642-31597-8
Springer Heidelberg New York Dordrecht London

Library of Congress Control Number: 2012942031

Printed on acid-free paper

Springer is part of Springer Science+Business Media (www.springer.com)

Research Group Leaders

Li Xiaoxi (Professor, Vice-chairman of Academic Committee of Beijing Normal University, Academic Consultant of Southwestern University of Finance and Economics)
Pan Jiancheng (PhD, Deputy Director-General of China Economic Monitoring and Analysis Center, National Bureau of Statistics of China)

Leaders of Cooperating Institutions

Lai Desheng (Professor, Dean of School of Economics and Business Administration, Beijing Normal University)
Tang Renwu (Professor, Executing Dean of School of Management, Beijing Normal University)
Hu Biliang (Professor, Dean of School of Economics and Resource Management, Beijing Normal University)
Yang Zhifeng (Professor, Dean of School of Environment, Beijing Normal University)
Gan Li (Professor, Dean of the Research Institute of Economics and Management, Southwestern University of Finance and Economics)
Liu Fangjian (Professor, Executing Dean of School of Economics, Southwestern University of Finance and Economics)

Collaborative and Supporting Institutions

China Economic Monitoring and Analysis Center and Relevant Research Institutes, National Bureau of Statistics of China
Center for Forecasting Science, Chinese Academy of Sciences
Policy Research Center for Environment and Economy, Ministry of Environmental Protection
College of Resources Science and Technology, Beijing Normal University
School of Geography, Beijing Normal University
School of Environment, Beijing Normal University

School of Life Sciences, Beijing Normal University
College of Water Sciences, Beijing Normal University
Institute of National Accounts, Beijing Normal University
Philanthropy Research Institute, Beijing Normal University
School of Public Administration, Southwestern University of Finance and Economics
Research Institute of Energy Economics, Southwestern University of Finance and Economics
Experimental Economics Laboratory, Southwestern University of Finance and Economics

Coordinators

Zhao Junli (Director of Information Analysis Division, China Economic Monitoring and Analysis Center)
Zhang Jiangxue (PhD, School of Economics and Resource Management)
Zhao Zheng (PhD, School of Economics and Resource Management)

Contact Persons: Cong Yajing, Song Tao

Commenting Experts

Wu Jinglian (Professor, Development Research Center of the State Council)
Li Yining (Professor, Honorary Dean of Guanghua School of Management, Peking University)
Zhang Zhuoyuan (Professor, Member of Academic Committee, Chinese Academy of Social Sciences)
Wei Liqun (Professor, Executive Vice President, Chinese Academy of Governance)
Chen Xiwen (Professor, Deputy Commissioner of General Office of the Central Finance Leading Group of China)
Liu Shijin (Professor, Vice President of Development Research Center of the State Council)
Lu Zhongyuan (Professor, Vice President of Development Research Center of the State Council)
Gu Shengzu (Professor, Vice-chairman of Internal and Judicial Affairs Committee, the 11th National People's Congress of the People's Republic of China)
Zhou Hongren (Professor, Standing Vice-chairman of Advisory Committee for State Informatization)
Pan Yue (Professor, Vice Minister of Ministry of Environmental Protection)
Gan Zangchun (Professor, Member of the Leading Party Group, Deputy Chief Inspector of National Land, Ministry of Land and Resources)
Fan Gang (Professor, Vice President of China Society of Economic Reform)
Liu Wei (Professor, Vice President of Peking University)
Ge Jianping (Professor, Vice President of Beijing Normal University)
Bian Huimin (Professor, Vice President of Southwestern University of Finance and Economics)
Xu Xianchun (Professor, Deputy Commissioner of National Bureau of Statistics of China)
Zhang Xinshi (Professor, Member of the Chinese Academy of Sciences)
Niu Wenyuan (Professor, Chinese Academy of Sciences, Counselor of the State Council)

Wei Jie (Professor, Director of the National Center for Economic Research, Tsinghua University)
Hu Angang (Professor, Director of the Research Center for Contemporary China, Tsinghua University and Chinese Academy of Sciences)
Pan Jiahua (Professor, Director of Research Centre for Sustainable Development, Chinese Academy of Social Sciences)
Fan Hengshan (Professor, Director-General of Department of Regional Economy, National Development and Reform Commission)
Xia Guang (Professor, Director of Policy Research Center for Environment and Economy, Ministry of Environmental Protection)
Su Wei (Professor, Director-General of Department of Climate Change, National Development and Reform Commission)
Chen Dongqi (Professor, Acting Dean of Academy of Macroeconomics Research, National Development and Reform Commission)
Jia Kang (Professor, Director of the Institute for Fiscal Science Research, Ministry of Finance)
Zhang Jianhua (Professor, Director-General of Research Bureau, the People's Bank of China)
Lu Mai (Professor, Secretary General, China Development Research Foundation)
Qiu Dong ("Cheung Kong Scholars Chair Professor", Beijing Normal University)
Chen Zongsheng (Professor, Vice Secretary-General of Tianjin Municipal People's Government)

Expert Comments[1]

Wu Jinglian: China's economic growth has long been powered by resources and productive factors input, and this rapid growth is realized at the cost of resources and environment. It has raised public concern on how to make a transition from extensive to efficiency-based growth model. China Green Development Index, led by Li Xiaoxi and Pan Jiancheng, was compiled by Beijing Normal University and other authoritative institutions in 2010. It measures and compares the province-based green economic growth on a trial basis, filling the void and providing a perspective for China's transformation in economic growth model. Based on the research results in 2010, Li Xiaoxi and his fellow researchers improved China Green Development Index system and its measuring method according to the suggestions from the academic circle using the most accessible and essential data. In the report, they measured and compared the province- and city-based green economic development in China, with a prominence being given to the progress and challenges of green development during the 11th Five-Year Period (2006–2010). The evidence-loaded report provides information on China's economic development transformation, and provides local governments with new ideas for green development. I expect to see more fruits and government support in this field.

Li Yining: The book gives me three inspirations. First, the research is a pioneer as it is the first to compile China's green development. The index consists of first-grade, second-grade, and third-grade indicators, which respectively reflect the production and resource usage efficiency, the situation of environment and resources protection and pollutants emission, and government's related investment and management. Second, for the first time the report ranks and compares 30 provinces and cities according to their performance in striking a balance between

[1] We received many valuable comments from 28 famous experts. We greatly appreciate all the reviewers for their insightful comments and recommendations. Since the comments are as much as 24,000 words altogether, we herein only quote those of Professor Wu Jinglian and Li Yining. Comments from other experts can be found in the Chinese version.

economic expansion and environmental protection. We can see that some developed provinces and cities with smaller carrying capacity potential of natural resources and environment rank higher in green degree of economic growth. There must have been some local policies launched in favor of green development such as related investment, energy saving, and emission reduction and tertiary industry performance policies. This finding is quite encouraging which shows the importance of support of government policies in green development. Third, the regularly published report will boost the green development in China. The report clearly shows that some provinces in middle China are yet to be improved in green development. If the provinces ranking low in the report can pay more attention to green development and government support, they can still enjoy green development in the future, as illustrated by some provinces and cities in eastern area. In short, this report is quite inspiring for some local governments who regard GDP as the only index to measure economic growth. This newly developed research still needs improvements in index choosing and system building. I hope that the specialists in the research group will achieve further success.

Preface 1

With the rapid growth of global industrialization, there has been substantial consumption of fossil fuels such as coal, petroleum, and natural gas along with growing carbon dioxide emissions. Unprecedented environmental and ecological crisis has clouded the world. Fortunately, the Climate Conference in Copenhagen signaled hope amid the sluggish global economic recovery. Countries worldwide have been braced for developing their scientific and industrial strategies in the era of the post financial crisis with a green and low-carbon philosophy. In recent years, China has witnessed rapid growth in its economic growth, attracting worldwide attention. China's economic growth has long been powered by resources and productive factors' input, and this rapid growth is realized at the cost of resources and environment. It has raised public concern on how to make a transition from extensive to efficiency-based growth model. The Plan for the 12th Five-Year Period enunciated the goal of building a resource-saving and environment-friendly society through green development, and appealed for increased crisis awareness and ideas of low-carbon development from the public. A green economy is emerging. As a developing country with rapid industrialization, urbanization, and modernization, China especially requires green development to win the future world competitions.

The Plan for the 12th Five-Year Period brought up 12 indicators for resource conservation and environment protection, among which 11 indicators are binding indicators. This shows our determination towards green development. Confidence and wisdom is not enough, we also need to establish performance evaluation index systems and mechanisms as scientific guidance to promote green development. The China Green Development Index Report 2011, completed by Beijing Normal University, Southwest University of Finance and Economics, leading thinkers in various fields together with the help of China Economic Monitoring and Analysis Center (CEMA), is such a guidance report which plays a significant role to help us promote green development.

Nature has bequeathed us valuable resources. If we want to go further, we should adapt the law of nature and follow a green development path. Green economy has become a new driving force in the world economy after information and technology revolution, ushering our way into a green era. Under the guidance of the Chinese government, we are confident of making a bright future.

Ma Jiantang

Preface 2

The year 2011 marks the beginning of China's economic and social development in the 12th Five-Year Period. As the plan for the period suggests, it is necessary for China to strengthen the sense of crisis and promote the green and low-carbon idea against the ever-restrained resource and environment, and meanwhile to facilitate energy-saving and environment-friendly production and consumption.

The implementation of the plan means a lot for China in improving its sustainable development and eco-efficiency. To help in this progress, the Beijing Normal University, Southwestern University of Finance and Economics, and the National Bureau of Statistics of China have worked together and produced the China Green Development Index Report in the consecutive 2 years.

The 2010 report caught the spotlight of over 30 influential domestic and foreign news media such as the News Channel of CCTV, Xinhua News Agency and People's Daily, and network media like the Xinhua Net and Sina. It was highly appreciated by the public who also showed great concern.

The 2011 report saw more progress. Based on the advice from experts and institutions, the research team improved the indicators of China Green Development Index System, added in the City Green Development Index Measurement, and finally finished the China Green Development Index Report 2011: Regional Comparison.

This report is forward-looking, objective, and innovative. It starts with a summary of the green development achievements during the 11th Five-Year Period and looks into the future in the 12th Five-Year Period. It analyzes China's green development in recent years from perspectives such as past achievements, existing problems, regional comparison, and future opportunities and challenges.

It is to be noted that the rankings of provinces and cities are not fixed, and this is how the report helps in stimulating them to improve their rankings.

As an important force of scientific and technological innovation in China, Beijing Normal University, with abundant expert resources, will continue its exemplary role and work with others to contribute to the national economic and social development.

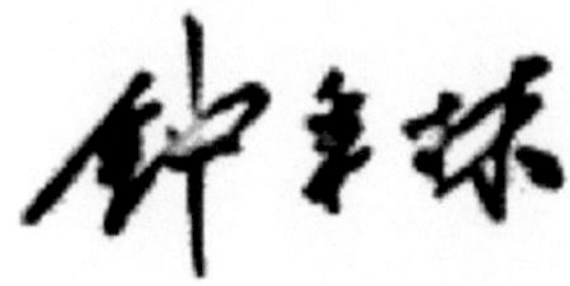

Zhong Binglin

Preface 3

The concept of green development has gradually become the common sense of each country. In 2008, the UN unveiled a plan for green politics and green economy, which was well-received and carried out by countries worldwide. China's green development obtained remarkable achievements. The concept of green development enjoyed popular support and green actions were advanced in China. The pace of developing an energy-efficient and environment-friendly society with circular economy is constantly being speeded up.

2011 commences China's 12th Five-Year Period (2011–2015). At this historical point, we should not only summarize the achievements in green development during the 11th Five-Year Period (2006–2010), but also explicitly realize the current shortage of natural resources and environmental constraints. This will help us to put forward constructive recommendations to the green development during and after the 12th Five-Year Period.

The China Green Development Index Report 2011, launched by Professor Li Xiaoxi, deputy director Pan Jiancheng, and the research team, did fruitful research into green development. This report measures the green development level of 30 provinces, municipalities, and autonomous regions as well as 34 large- and medium-sized cities in China. The city-based measurement is first introduced into the report.

For one green earth and one green dream, we need to take green actions. The academic circle needs to actively explore and research the general rule of green development, and thus provide intellectual support to the green development of the economic society. The report was completed by Prof. Li Xiaoxi, with the dedication of other leading thinkers in economics, management, environment, and resources together with the help of the China Economic Monitoring and Analysis Center (CEMA).

As a national key university directly under the Ministry of Education and one of the key constructions of the "211 Project", Southwestern University of Finance and Economics will continue to push forward green development research and actively explore the transitional law of economic development pattern. We will contribute our strength to economic development and social progress in China. We also believe that with the collective "green action" of the government, industry, and academic circle, our future will become greener.

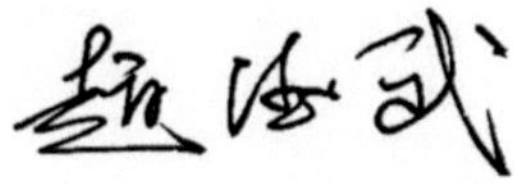

Zhao Dewu

Acknowledgments

After more than a year of hard work since November 2010, the study of "China green development index report—regional comparison" has made great progress. We were very excited when its Chinese version was published, and the English version has been completed and submitted to the press. Here, we want to express our acknowledgments for the cordial care from our leaders, the sincere cooperation of the experts, and the selfless dedication of the teachers and students.

First, we would like to thank the leaders and experts for their support and promotion for the successful completion of this report. Our deepest gratitude goes to the Director of the State Statistical Bureau, Professor Ma Jiantang, President of Beijing Normal University, Professor Zhong Binlin, and President of Southwestern University of Finance and Economics, Professor Zhao Dewu. They are all pleased to preface the book without any hesitation, from which we can feel their encouragement and trust. We are also deeply indebted to the Secretary of Beijing Normal University, Professor Liu Chuansheng. She has reaffirmed the green index report several times, and encouraged this research to be carried out continuously.

Thirty evaluation experts reviewed the report insightfully in spite of their busy schedule, which benefits the report greatly. As one of the most famous economists in China, Professor Wu Jinglian gave his valuable review to this report after his tight schedule of the organization of the 16th World Congress of the International Economic Association. We also thank deeply the famous economist, Professor Li Yining. He gave a fair and rigorous comment to the report soon after he came back from a fatiguing field trip. Sincere thanks also go to Professor Wei Liqun, executive vice president of the Chinese Academy of Governance, Professor Chen Xiwen, deputy director of Central Finance Leading Group Office, Professor Gu Shengzu, deputy director of Internal and Judicial Affairs Committee, Professor Liu Shijin and Professor Lu Zhongyuan, deputy directors of Development Research Center of the State Council, and officials from the Ministry of Environmental Protection, for their guiding ideas and practical recommendations which greatly benefitted the report. We also invited a number of experts in the field of resources, environment, and ecology to give their comments on the report to combine the ideas both from natural and social sciences. The deputy president of Beijing

Normal University, Professor Ge Jianping and deputy president of Southwestern University of Finance and Economics, Professor Bian Huimin, have always been concerned about the progress of our project and have given great support to this report.

The experts from the National Bureau of Statistics and other departments have carefully examined and identified the supplements and amendments of the indicators, thereby playing an irreplaceable role. Statistical experts work with high levels of professional standards and commitment with consciousness. We have a very pleasing and efficient cooperation with China Statistical Information Services Center, National Bureau of Statistics. In the process of discussion of the index system and the finalization of the report, the conscientiousness of the deputy director, Pan Jiancheng, and the seriousness of Zhao Junli director in the coordination process, and the professionalism of deputy chief Wang You Juan, director Shi Faqi, director Jiang Mingqing, and preciseness of director Mao Yuru, all left a deep and beautiful impression on me.

The close cooperation and mutual support of the school of economics and business administration, school of management, and school of economics and resource management contributed to the successful completion of this report. We also appreciate the support from the leader of Southwestern University of Finance and Economics and its departments. Professor Lai Desheng, the director of school of economics and business administration, Professor Tang Renwu, the executive director of management school, Professor Hu Biliang and Professor Zhang Qi, the director and secretary of the school of economics and resource management, Professor Wang Zhenyao, the director of the philanthropy research institute, Professor Liu Xuemin, the secretary of the college of resources science & technology and Professor Liu Fangjian, the director of school of economics in Southwestern University of Finance and Economics, all participate the discussion of the project and involve in the writhing of the book.

A number of professors in the related fields are responsible for the writing of the major chapters and features. They gave the report many views and analyses of scientific value, which demonstrates the advantages of the research team with the combination of arts and science. In science, there are Professor Liang Jinshe in the school of geography, Professor Mao Xianqiang in the school of environment, Professor Jiang Yuan in the college of resources science and technology, and Professor Wang Hongrui in the college of water sciences. In the social sciences, there are Professor Tang Fangfang in Peking University, Professor Li Baoyuan, Professor Yang Guanqiong, Professor Wang Luozhong, and Dr. Zhang Shengling, Dr. Lin Weibin, Dr. Lin Yongsheng, Dr. Zhao Zheng, and Ms. Fan Lina in Beijing Normal University, and Professor Liu Jiansheng, Dr. Liu Jinshi, Dr. Li Ding in Southwestern University of Finance and Economics. Dr. Hou Wanjun in the Research Office of the State Council, Dr. Yang Xiaoming in the Ministry of Environmental Protection, Dr. Wang Yuanlong in the Bank of China, and Dr. Zeng Xuewen in the Agricultural Bank of China all completed the writing of features, and played their respective advantages on linking theory with practice. The explanation of the percentile method of index by Professor Wang Shouyang and Li Qisai, Yang Cuihong, Pang Yong, Zhu Kunshen in center for forecasting science, Chinese Academy of Sciences gave us great inspiration.

Here, I would like to express my appreciation to the teachers and students for their wisdom and labor contributed to the project. They devoted all their efforts and did efficient work to complete the project.

Thanks are also given to the Green Index Group (GIG). They did a lot of work including comprehensive information collection and collating, and data entry. They also put forward the preliminary proposals for the indicator amendment based on the review experts' opinions. They also completed the preliminary calculation of the provincial and urban green development index, which provides a good basis for the selection and determination of the calculation scheme and index system. Particular thanks are given to Dr. Zhang Jiangxue. As the coordinator of this project, she worked hard and played an important role in the whole process of helping me to organize the project. As contact person of the project, Dr. Cong Yajing and Dr. Song Tao undertook respectively the city and regional green development index data entry and trial calculation work, and did a good job. Dr. Jiang Xin fully participated the indicator selection and provincial green development index calculation, and successfully completed the related work in the project. Dr. Sun Xiaopeng in the school of life sciences and Dr. Song Peng in the school of environment carefully completed the writing task. Mr. Cai Ning played an active role in the green development index selection and calculation. He also helped to contact the review experts and organization work such as typesetting which was greatly valued by everyone. Hou Rui, Wang Xiwei, Rong Tingting, Jin Jingjing and Wan Qian actively participated in the project enthusiastically and carefully. Zhou Jiancong, Zhu Lei, Yang Fei, Zhang Liang, Liu Yang and Xu Ligang, all the GIG members actively participated in the data collating and trial calculation of the green development index, and completed the relevant sections of the writing task. In the translation process of the manuscript, Dr. Zhao Zheng very carefully and seriously assisted me in the organization and coordination work. Many teachers and students in the translation quality work group, including Shi Faqiang, Ren Ran, Mao Xianqiang, Zheng Yanting, Liu Yimeng, Zhou Wei, Wu Min, Long Fei and Xiao Yihuan put forward many constructive amendments and recommendations on the translation of the manuscript. A number of off-campus friends helped in the translation, and played an important role, such as Professor Zhang Tianwei in the University of Colorado. English terminology collection group, including Zhu Lei, Zhang Liang and Xu Ligang initiatively collected and collated a lot of professional terms for the translation reference. Long Fei, Jin Jingjing, Qu Ge and Liu Shiyao typeset the whole translation draft arduously. Here, I would like to express my sincere thanks to all the teachers and students participating in the project for their hard and fruitful work.

Besides, I wish to express my profound appreciation for the translators, such as Ren Ran, Yang Jin, Bai Ruixue, Zhang Tianwei, Wu Min, Cheng Yang, Zheng Yanting, Liu Qian, Sun Xiaopeng, Liu Yimeng, Liu Xueye, Qian Cheng, Long Fei, Zhou Wei, Jiang Yuanqun, Zhang Ai, Zhao Guoqin, Xiao Yihuan, Tian Yaling, and so on. Through all their efforts and efficient work, the English version was completed and submitted to the press on time. Logistics are also very important for the project. Here, I would like to thank Wang Ying and Yan Ling, for their management of the funds use and reimbursement. I also appreciate the teachers and students who participated in the conference services several times and in particular those who were responsible for the contact of review experts and manuscript submission. They worked really hard and obtained praise from inside and outside experts.

Last but not the least, we appreciate the strong support from the Beijing Normal University Press. The Press president, Professor Yang Geng, took great care in the progress of the project. The editor-in-chief, Ye Zi arranged personally the editing and publishing of the book. As the planning editor of this book, Mr. Ma Hongli involved in the whole process of the discussion and writing in the research team, and did a lot work of guidance and assistance to complete on time the manuscript according to the quality requirements. Ms. Chen Jingsi worked tirelessly, and completed the book editing with high quality.

We have determined to continuously promote the green development index research under the support of the projects of "985" and "Classic China International Publishing Project", and improve this study to a new level along with the exchanges and cooperation with international friends.

Li Xiaoxi

Contents

Columns

Tables

Figures

Chapter 1
Introduction

Xiaoxi Li and Jiancheng Pan

On November 22, 2010, the Global Green Think-Tank Summit (GGTTS) Beijing made declaration to rewrite the crisis of human survival and achieve global economic green development. With severer global resources depletion and environmental degradation, the global economy is facing unprecedented challenges in climate change and natural disasters. Resources major economies rely on are depleted. Environmental degradation threatens human health worldwide. Meanwhile, economic competition becomes fiercer in rivalry, threatening the Earth's limit on resources and environment. And the rapid economic growth and severe environmental degradation make the world smaller and affect each person more deeply than before. Striking a balance between development and environment is our only choice. The summit called for the implementation of green new deal and the development of green markets and green trades worldwide especially in the developed countries first. It called for leading experts in all circles to build a think-tank for a sustained global growth mode. Furthermore, it called for people worldwide especially business elite to take actions as well. In the end, it suggested more focus be on negotiating a settlement of sustained economic and environmental development rather than on a debate over sharing the responsibility for environmental protection at the United Nations summit in Cancun, Mexico.

In 2011 Tohoku Earthquake and Tsunami which caused a number of nuclear accidents proved that the warnings are this true. It taught us a lesson that we can't afford to neglect the power of nature and that we must adapt to its law rather than mess it up. It also gives sense to our green talks: it is our responsibility to protect the environment for the sake of our children.

We will introduce what progress we made on green development index study in this part.

X. Li and J. Pan (eds.), *China Green Development Index Report 2011*,
Current Chinese Economic Report Series, DOI: 10.1007/978-3-642-31597-8_1,

1.1 Influences of China Green Development Index Report

In the introduction of the China Green Development Index Annual Report 2010: Inter-provincial Comparison, we introduced the background information and significance of the report. We should solve the contradictions between economic growth and resources and environment, reduce fossil fuels consumption and carbon dioxide emissions in response of climate change before achieving sustained economic growth. And governments should implement green new deal in response of financial crisis. The report offers a new development mode for governments, investors and businesses and raises public awareness of environmental protection. We also introduced the ideas, structure, indicators, measurement approach and results of the index system. Based on extensive comparison, we create a China-specific system including three first-class indicators, nine second-class indicators and 55 third-class indicators. The system focuses on green development, the government's role and the green productivity. According to the measurement results, we released the Green Development Index Ranking of 30 countries, municipalities and autonomous regions (exclusive of Tibet) in 2008. Since its official release on November, 2010, the China Green Development Index Annual Report 2010: Inter-provincial Comparison has attracted worldwide attention.

1.1.1 In-Depth Media Coverage Home and Abroad

The Medias including CCTV, Xinhua News Agency, People's Daily, Guangming Daily, China News Service and Reuter's News Agency gave it in-depth coverage. According to Reuters, "The index, which in part revives an earlier attempt to put a price on pollution, is a small example of how China is gradually changing its course after years of pursuing fast growth with little regard for the environment. It ranks cities and provinces according to their performance in striking a balance between economic expansion and environmental protection."

There were 2.43 and 1.21 million web pages in Chinese and in English respectively listed and even a listing of top key word query on Baidu. There were 7.42 and 1.03 million web pages in Chinese and English respectively on Google.

No negative report or negative evaluation was given on the orientation of the green development indicators system.

You can find a wide range of coverage in 50,000 words in another work of us.

1.1.2 Consensus on the Inter-Provincial Ranking of the Green Development Index

There are always debates over the ranking of green performance. Unsurprisingly, those who ranked low will question the ranking in many cases. However, it didn't happen this time. For example, knowing that Shanxi, a major GDP contributor esp. for its coal output, ranked the lowest, its government didn't frown upon the raking, instead, they showed us how hard they had been working on the green development by demonstrating its Overall Reform Pilot Zone for Resource-based Economic Transformation and other programs.

Local websites and newspapers also pay close attention to their rankings. Some local governments even tried to analyze why they ranker high without valuing the ranking only.

1.1.3 Tremendous Multi-Disciplinary Review Comments

Thirty senior experts were invited to give in-depth comments on China Green Development Index Report 2011: Regional Comparison for its multi-disciplinary nature. The experts include preeminent economists such as Wu Jinglian, Li Yining and Zhang Zhuoyuan, some specialists with government agencies such as Wei Liqun, Liu Shijin, Lu Zhongyuan, Gu Shengzu, Pan Yue and Su Wei and some researchers of environmental protection and sustainable development such as Zhang Xinshi, Niu Wenyuan and Pan Jiahua. Their suggestions make remarkable contributions to the report's improvement.

1.1.4 Green Development Index Attracting Global Attention for its Ideas and Methodology

China Council for International Cooperation on Environment and Development (CCICED) pays close attention to our research. In July 2010, CCICED launched China Environment and Development Outlook Feasibility Study and compared 21 reports in China, including Human Development Report 2009/2010, Report on Ecological Footprint in China, Chinese Environment and Development: The Century Challenge And The Strategy Choice, The Energy Development Report of China 2008, Environmental Policy Research Series. In January 2011, CCICED selected five representative reports which are Human Development Report 2009/2010, China Green Development Index Report 2011: Inter-provincial Comparison, China Sustainable Development Strategy Report, Green Book and China Development Report respectively. CCICED particularly spoke highly of the China Green Development Index Report 2011: Inter-provincial Comparison for it is the first system to monitor and measure green development index in China with in-depth analysis and accurate data. CCICED also invited leaders in charge of green development index study to participate in China Environment and Development Outlook Feasibility Study.

In May 2011, China Green Development Index Report 2011 was to be published in English sponsored by the Classic China—International Publishing Project. This project is launched by GAPP and aims to let the world know more about China through publishing books in foreign languages. The supported projects should be high-grade books selected from Chinese academic classics or Chinese literary classics.

1.2 The Latest Trends in Green Development

In recent years, green development has become a rising star in global economy. The data in last year's BP Statistical Review of World Energy told that due to economic recession, the total global energy consumption dropped 1.1 % for the first time since 1973, of which nuclear energy, nature gas and oil consumption reduced and coal consumption remained flat. Also, the CO_2 emissions experienced its first drop since 1998. This is encouraging but we expect that the energy consumption reduction comes from the improvement in energy efficiency and green economy, rather than economy recession.

1.2.1 China's New Thoughts on Green Development

Some economic concepts like low carbon economy, green economy and green growth have been discussed worldwide. To tackle the global financial crisis, the United Nations Environment Program (UNEP) has proposed two concepts—Green New Deal and Green Economy—to arouse the interest of governments and scholars. Recently, green development, among other frequently mentioned terms, has drawn more and more attention in China.

Most researches on green development, carried out by international organizations, governments and academic institutions, focus on environment and resources protection other than economic development. They emphasize green GDP accounting, net economic welfare and the expansion of wealth when accounting national economy; they stress green indexes when evaluating the quality and sustainability of environment; they underline the global alternative energy indexes when assessing resources and energy; and they value green consumption behaviors when it comes to life quality. Researches on economic development largely depend on the sustainable development evaluation index system, which concerns all aspects of lives.

Green development values both economic and environmental performance, which means we should lead a sustainable economic development model that can cost less, benefit more and produce fewer emissions. In short, developing countries like China need a green future where quality and performance of goods are balanced and where resources efficiency and environment protection are well developed; where the living standard of people is largely improved and where the economic and environmental performance are evened.

However, the concept of low carbon economy, proposed by developed countries, did not mention much about economic development, a vital part in developing countries who are struggling with industrialization and urbanization. So it is no accident that developing countries hold the principle of Common But Differentiated Responsibility and prefer green development. According to International Energy Agency (IEA), we will expect the fourth industrial revolution, the green industrial revolution, after entering the twenty first century. As green development is the driving force of the revolution, China will play a leading role in it.

1.2.2 Unfolding Green Development Forums and Activities

In recent years, there are many forums and activities related to green development in the news. For instance, the first annual meeting of World Green Forum was held in Hong Kong from May 26 to 27, 2011. At the forum, politicians and scholars from countries and regions worldwide have discussed issues like global green economy development model, the building of international green economy exchange platforms, experiences in this field, green policies, green innovations and green financing plan.

On November 18, 2011, the High-level Forum on Green Development and S&T Innovation was organized in Beijing to promote S&T innovation and green development by China's six departments including Ministry of Science and Technology (MOST), National Development and Reform Commission, Ministry of Foreign Affairs, Ministry of Industry and Information Technology, Ministry of Environmental Protection, and China Meteorology Administration (CMA). The central government showed great concern on the forum with important officials attending the opening ceremony and addressing at the forum.

On June, 27th, 2010, the International Green IT Service Cooperation Forum and 2010 China–Europe CIO Summit was jointly held by the China Center for Information Industry Development (CCID) and other institutions with its theme as "Green IT service is influencing global CIO's minds". On the forum, it was put forward that green development has become a new engine for global economy in this post-crisis era. According to related forecasting data from international institutions, the global green IT service market may surge to 5 billion dollars by 2015, and 2 billion dollars may come from the Asian-Pacific region, among which China, Japan, and Australia will take over a large portion.

Annual Summit of China Green Companies 2011, hosted by China Entrepreneur Club, Daonong Center for Enterprise, and Green Herald Magazine, was held in Qingdao from April 21 to 22, 2011. The China Greentech Initiative (CGTI) published China Greentech Report 2011. The report cites five specific trends that are driving China's emergence as a global greentech leader. It includes market updates of six greentech sectors, and summaries of 19 prioritized greentech market opportunities. These sectors include Cleaner Conventional Energy, Electric Power Infrastructure, Green Building, Cleaner Transportation, Renewable Energy, and Clean Water.

"Green Honor" ceremony was held in World Trade Center Association (WTCA) pavilion in Shanghai World Expo on June 7, 2011. Haier's green development achievement caught global attention because of its environmental protection. WTCA presented Global Sustainable Development Outstanding Achievement Award to Haier. WTCA is a not-for-profit, non-political association dedicated to the establishment and effective operation of World Trade Centers (WTCs) as instruments for trade expansion representing 216 members in 91 countries. It encourages the expansion of world trade and promotes international business relationships. It has more than 300 city members and more than 750,000 corporate members around more than 100 countries and regions. It has profound international influence.

Tibet Autonomous Region strives to develop green economy to face global warming. It is "origin of rivers" and "origin of zoology" of South Asia and Southeast Asia, and is also "staring motor" and "adjustment region" of the climate of China and even the Eastern Hemisphere. In order to face global warming, the central government of China increased ecological environmental protection and development of Tibet. Tibet would continue to pursue the road towards "Chinese characteristic and Tibetan specialty". In January 2008, Tibet drafted and published Tibet Autonomous Region Comprehensive Energy Saving and Emission Reduction Work Plan. Tibet boosted Grain for Green Project, cover the land for forestry and pasture. Tibet Ecological Safety Barrier Protection and Construction, total investment of 15.5 billion yuan, will be completed in 2030. By 2009, forest living woods stock was 2,273 billion cubic meter, total forest living beings was 1.898 billion ton, and total carbon stock was 0.953 billion ton. And these three indexes all rank first in the country.

To sum up, people in all fields are paying attention to green development and conduct green activities.

1.2.3 International Exchanges on Green Development

On June 10th, 2010, the International Energy Agency (IEA) said that the Gulf of Mexico oil spill was a potential "game changer" for crude oil supply, which might give rise to restrictions on subsea crude oil development and supply in the future. Meanwhile Barack Obama, the president of the United States, also said that American economy would suffer substantial ongoing effects from the spill. Obviously, the oil industry calls for green development, and so does the international community.

At the end of 2010, UN Climate Change Conference in Cancún delivered a balanced package of decisions to set all governments more firmly on the path towards a low-emissions future and encourage enhanced actions on climate change in the developing world. One of the summit outcomes was an agreement adopted by the states' parties that called for a large "Green Climate Fund", proposed to be worth $100 billion a year by 2020, to assist poorer countries in financing emission reductions and adaptation. In the twenty first century, the greatest challenge is global climate change and the biggest opportunity is green development.

The fast-growing industrial society has brought us both unprecedented opportunities together with resource and environmental problems. Now all parties have recognized that energy conservation, emission reduction, resource efficiency promotion and environmental protection can only be achieved through green development. Development is irreversible and we must cooperate to make it environmentally friendly.

China needs green development, which from a global perspective, concerns its long-term development and international image. Since energy, resource and environment issues have become diplomatic and political, China has to form a green development strategy in an international scope.

1.2.4 The 12th Five-Year Plan, China's First Green Development Plan

The Plan for the 12th Five-year Period stressed on green development, making it the first green development plan in China. It enunciated the goal of building a resource-saving and environment-friendly society through green development, and appealed for increased crisis awareness and ideas of low carbon development from the public. We need to focus on energy conservation and emission reduction and improve simulative and binding mechanisms, so as to facilitate an environment-friendly production and consumption model and enhance sustainable development.

The green indicators take more proportion in the plan for the 12th Five-year Period: eight resource and environment indicators taking up 33.3 %, greater than the seven ones taking up 27.2 % in the 11th Five-year Period plan. With one indicator in service industry and four in education, science and technology, the indicators on green development add up to 17, taking up 60.7 % of all indicators.

Also, China aims to cut the carbon dioxide emission per unit of GDP by 17 % from the level of 2010 by 2015, which is the first time China included such a goal for combating global climate change in its five-year economic and social development plan. Forest coverage and timber stock objectives are also set in the plan. Third, stimulative and binding mechanisms are set up to intensify the price reform for resource-based products and environmental protection. The plan also calls for evaluating the implementation of energy saving and emission reduction objectives and curbing the total energy consumption.

In the 12th Five-year Period, the work of green development will focus on six aspects, which are, climate change, resource management, circular economy development, environment protection, ecological restoration, and water conservancy and disaster prevention and mitigation. For example, in terms of combating climate change, the main objectives by 2015 are clearly stated in the plan, such as reducing CO_2 emissions per unit of GDP, increasing non-fossil energy consumption, forest coverage, timber stock and carbon fixation.

In a word, the 12th Five-Year Plan opens China's green development era, and marks China's participation in the global green revolution.

1.3 Improvements of China Green Development Index

The 2011 report bears two aims: first, to improve China Province Green Development Index System; second, to build China City Green Development Index System.

1.3.1 Improvements of China Province Green Development Index System

When releasing the 2010 report, we said that it was on a trial basis and there was still room for improvement. We endeavored to improve the 2011 report after soliciting suggestions from academic circles.

Since the end of 2010, the research group has held several meetings to discuss how to revise and compile the 2011 report. The following section will elaborate on the improvements in index system and measurement system.

Based on the 2010 report, the 2011 version makes no change in first-class indicators and revises second-class indicators and third-class indicators.

First-class indicators of the 2011 report remains the same as those of the 2010 report, and they are green degree of economic growth, carrying capacity potential of natural resources and environment and support degree of government policies.

The nine second-class indicators in the 2011 report are respectively green growth efficiency indicators, primary industry indicators, secondary industry indicators, tertiary industry indicators, resource abundance and ecological conservation indicators, environmental pressure and climate change indicators, green investment indicators, infrastructure indicators and environmental management indicators. Three of them are revised. First, the original resource and ecological conservation indicators are revised to resource abundance and ecological conservation indicators because the research focuses on the abundance of resources. Second, environment and climate change indicators are revised to environmental pressure and climate change indicators, for the research focuses on the pressure posed by emissions to the environment and reflect carrying capacity potential of natural resources and environment. Third, infrastructure and environment management indicators are revised to infrastructure indicators because there is an independent city-based research in the 2011 version.

For detail information of second-class and third-class indicators, please refer to Chaps. 6, 11, 16 and the appendix.

Now we will focus on the revision of third-class indicators.

The 2011 report includes 60 third-class indicators. The 60 indicators are more balanced among different fields, 22 indicators for green degree of economic growth, 19 for carrying capacity potential of natural resources and environment and 19 for support degree of government policies.

See framework of China Province Green Development Index System in Table 1.1.

Now we need to answer this question: How the 60 third-grade indicators in the table are determined?

In 2011, the research group revaluated the original 55 indicators in the 2010 China Green Development Index Annual Report: Inter-provincial Comparison, and with proper revisions and changes, 60 third-grade indicators were finally determined.

First, 50 more indicators have been supplemented based on related domestic and international materials such as statistical yearbooks, literatures and reports. Also, experts' opinions on the 2010 report are fully embodied in these 50 indicators.

Second, 10 out of the 50 indicators have been selected after several consultations. They are:

1. Electricity consumption per capita in urban areas,
2. Water-saving irrigation rate,
3. Proportion of effectively irrigated area in area of cultivated land,
4. Proportion of area of wetlands in total area of a region,
5. Total standing stock volume per capita,
6. Nitrogen oxides emissions from road communication per capita,
7. Per capita length of public transportation in operation in urban areas,
8. Ratio of rural population benefiting from water improvement projects to total rural population,
9. Ratio of green covered area to completed area in city,
10. Sudden accidents effecting environment.

Third, 5 of the indicators in 2010 report have been removed according to related statistics studies and experts' advice. They are:

1. Coal consumption for thermal power generation (the No. 18 indicator in 2010 report)
2. Industrial solid wastes discharge per unit of GDP (the No. 9 indicator in 2010 report)
3. Industrial solid wastes discharge per unit of GRP (the No. 36 indicator in 2010 report)
4. Per capita industrial solid wastes discharge (the No. 37 indicator in 2010 report)
5. Ecological restoration rate in ore districts (the No. 50 indicator in 2010 report).

Fourth, 8 indicators in the 2010 report have been revised. They are:

1. Energy consumption per unit of industrial added value by above-scale industrial enterprises (the No. 14 indicator in 2010 report—Energy consumption of industrial added value by above-scale industrial enterprises)

Table 1.1 China province green development index system

First-class indicators	Second-class indicators	Third-class indicators	
Green degree of economic growth	Green growth efficiency indicators	1. GDP per capita	6. COD emissions per unit of GRP
		2. Energy Consumption per unit of GDP	7. NOx emissions per unit of GRP
		3. Ratio of non-fossil energy consumption to total energy consumption	8. Ammonia nitrogen emissions per unit of GRP
		4. CO_2 emissions per unit of GRP	9. Electricity consumption per capita in urban areas
		5. SO_2 emissions per unit of GRP	
	Primary industry indicators	10. Labour productivity of primary sector	13. Proportion of water-saving irrigated area in effectively irrigated area
		11. Output ratio of land	
		12. Water-saving irrigation rate	
	Secondary industry indicators	14. Labour productivity of secondary sector	17. Ratio of industrial solid wastes utilized
		15. Water consumption per unit of industrial value added	18. Reuse rate of industrial water
		16. Energy consumption per unit of industrial value added at a cut-off level	19. Ratio of output value of six high energy-bearing industrial sectors to gross industrial output value
	Tertiary industry indicators	20. Labour productivity of tertiary sector	22. Proportion of employees of tertiary sector in total employees
		21. Proportion of value added of tertiary sector in GRP	

(continued)

Table 1.1 (continued)

First-class indicators	Second-class indicators	Third-class indicators	
Carrying capacity potential of natural resources and environment	Resource abundance and ecological conservation indicators	23. Water resources per capita	27. Proportion of area of wetlands in total area of a region
		24. Forest area per capita	28. Total standing stock volume per capita
		25. Forest coverage rate	
		26. Proportion of area of natural reserves in total area of a region	
	Environmental pressure and climate change indicators	29. CO_2 emissions per unit of land area	37. Ammonia nitrogen emissions per unit of land area
		30. CO_2 emissions per capita	38. Ammonia nitrogen emissions per capita
		31. SO_2 emissions per unit of land area	39. Consumption of chemical fertilizers per unit of area of cultivated land
		32. SO_2 emissions per capita	40. Consumption of pesticides per unit of area of cultivated land
		33. COD emissions per unit of land area	41. Nitrogen oxides emissions from road communication per capita
		34. COD emissions per capita	
		35. Nitrogen oxides emissions per unit of land area	
		36. Nitrogen oxides emissions per capita	

(continued)

Table 1.1 (continued)

First-class indicators	Second-class indicators	Third-class indicators	
Support degree of government policies	Green investment indicators	42. Ratio of environmental protection expenditures to government expenditures	45. Investment in converting cultivated land into forests and grassland per unit of area of cultivated land
		43. Ratio of investment in the treatment of environmental pollution to GRP	46. Ratio of expenditures for science and technology, education, culture, and medical and health care to government expenditures
		44. Per capita investment of water sanitation and toilet improvement in rural areas	
	Infrastructure indicators	47. Area of green land per capita in urban areas	52. Per capita length of public transportation in operation in urban areas
		48. Coverage rate of urban population with access to tap water	53. Ratio of rural population benefiting from water improvement projects to total rural population
		49. Treatment rate of urban waste water	54. Ratio of green covered area to completed area in city
		50. Ratio of urban consumption wastes treated	
		51. Public transportation vehicles per 10000 urban population	
	Environmental management indicators	55. Area of afforestation per capita for this year	58. Removing rate of industrial NOx emissions
		56. Removing rate of industrial SO_2 emissions	59. Removing rate of ammonia nitrogen in industrial waste water
		57. Removing rate of COD in industrial waste water	60. Sudden accidents effecting environment

2. Ratio of output value of six high energy-bearing industrial sectors to gross industrial output value (the No. 17 indicator—Ratio of output value of high energy-bearing industrial products to gross industrial output value)
3. Proportion of employees of tertiary sector in total employees (the No. 21 indicator in 2010 report—work engagement of the tertiary industry),
4. Water resources per capita (the No. 22 indicator in 2010 report—Local water resources per capita).

As cross-border water is not included in total water resources, and the total amount of water resources is the numerator in the calculation, the word "local" is removed.

5. Proportion of total environment pollution investment in gross regional output (the No. 41 indicator in 2010 report—Proportion of environment pollution investment in gross regional output)
6. Per capita area of forestation for this year (the No. 51 indicator in 2010 report—Per capita area of forestation)
7. Removing rate of COD in industrial waste water (the No. 53 indicator in 2010 report—Removing rate of COD in industry)
8. Removing rate of ammonia nitrogen in industrial waste water (the No. 55 indicator in 2010 report—Removing rate of ammonia nitrogen in industry).

Fifth, the calculations of 5 indicators have been redefined. They are:

1. CO_2 emissions per unit of land area,
2. SO_2 emissions per unit of land area,
3. COD emissions per unit of land area,
4. Nitrogen oxides emissions per unit of land area,
5. Ammonia nitrogen emissions per unit of land area.

When calculating these five indicators, the area of deserts is subtracted from the national territory area to be the denominator, as suggested by experts.

Obviously, the changes are good for the consistence with the statistical yearbooks and the calculation. We highly appreciate the advice from the experts. The revision would not come through so effectively without their contribution.

1.3.2 The Establishment of China City Green Development Index System

In addition to last year's report, 34 more cities have been incorporated into our 2011 report to make weighted comparisons between provinces and cities and promote green development performance in China. The cities include 4 municipalities, 5 cities that has been specifically designated in the state plan, and 25 provincial cities (Lhasa and Urumqi were not included). The measurement system, revised by experts, includes three first-class indicators, nine second-class indicators and 43 third-class indicators. The following paragraphs will introduce its compilation principles, selection standards, indicator structures and evaluation standards.

1.3.2.1 The Compilation Principles

We have followed four principles to prepare the report. First, in evaluating city performance in green development, we have held the principle of striking balance between economic expansion and environmental protection. Second, we have introduced indicators that are unique in city comparison. For example, the air quality evaluation, which can not be possibly adopted in inter-provincial comparison, is of great importance in evaluating the carrying capacity potential of natural resources and environment in city development. Third, this year's measurement system almost stays the same as that of the last year by deleting some non-data supported indicators and adding some city-related ones. Fourth, the data we referenced are just, fair, open and authoritative.

1.3.2.2 Selection of Cities

Of the 113 initially selected cities, 79 were out of supported data. As a result, we have selected 34 cities this year to ensure accuracy.

1.3.2.3 Index Structure of the Evaluation System

The system, similar to last year's provincial measurement system, contains three first-class indicators, nine second-class indicators and 43 third-class indicators. The details about the indicators are shown in table (See Table 1.2).

It is to note that there are no official data for four third-class indicators, namely 'carbon dioxide emissions per unit of gross regional product', 'energy consumption per unit of industrial added value', 'carbon dioxide emissions per unit area' and 'per capita carbon dioxide emissions'.

1.3.2.4 Weight of Indicators

The indicators are not equally important for the evaluation of green development index, so the weight of each indicator is different. The weight allocation method in this research is similar to "Delphi method". The final weight of each indicator is set after soliciting suggestions from experts of different fields. It is worth mentioning that the indicator of Carrying Capacity Potential of Natural Resources and Environment is more important to the evaluation of province-based green development index than to that of city-based green development index, thus the weight of this indicator is lower in the evaluation of city green development index.

Table 1.2 City green development index system in China

First-class indicators	Second-class indicators	Third-class indicators	
Green degree of economic growth	Green growth efficiency indicators	1. GDP per capita	4 CO_2 emissions per unit of GRP
		2. Energy consumption per unit of GDP	5. SO_2 emissions per unit of GRP
		3. Per capita electricity consumption by urban households	6. COD emissions per unit of GRP
			7. NOx emissions per unit of GRP
			8. Ammonia nitrogen emissions per unit of GRP
	Primary industry indicators	9. Labor productivity of primary industry	
	Secondary industry indicators	10. Labour productivity of primary sector	12. Energy consumption per unit of industrial value added
		11. Water consumption per unit of industrial value added	13. Ratio of industrial solid wastes utilized
			14. Reuse rate of industrial water
	Tertiary industry indicators	15. Labour productivity of tertiary sector	17. Proportion of employees of tertiary sector in total employees
		16. Proportion of value added of tertiary sector in GRP	
Carrying capacity potential of natural resources and environment	Resource abundance and ecological conservation indicators	18. Water resources per capita	
	Environmental pressure and climate change indicators	19. CO_2 emissions per unit of land area	26. Nitrogen oxides emissions per capita
		20. CO_2 emissions per capita	27. Ammonia nitrogen emissions per unit of land area
		21. SO_2 emissions per unit of land area	28. Ammonia nitrogen emissions per capita
		22. SO_2 emissions per capita	29. Ratio of days with air quality at and above level 2 to the whole year
		23. COD emissions per unit of land area	30. Ratio of days with principal pollutants as respirable suspended particulate to the whole year
		24. COD emissions per capita	
		25. Nitrogen oxides emissions per unit of land area	

(continued)

Table 1.2 (continued)

First-class indicators	Second-class indicators	Third-class indicators	
Support degree of government policies	Green investment indicators	31. Ratio of environmental protection expenditures to government expenditures	33. Ratio of expenditures for science and technology, education, culture, and medical and health care to government expenditures
		32. Ratio of investment in the treatment of industrial environmental pollution to GRP	
	Infrastructure indicators	34. Area of green land per capita in urban areas	37. Treatment rate of urban waste water
		35. Ratio of green covered area to completed area in city	38. Ratio of urban consumption wastes treated
		36. Coverage rate of urban population with access to tap water	39. Public transportation vehicles per 10,000 urban population
	Environmental management indicators	40. Removing rate of industrial SO_2 emissions	42. Removing rate of industrial NOx emissions
		41. Removing rate of COD in industrial waste water	43. Removing rate of ammonia nitrogen in industrial waste water

Table 1.3 First-class and second-class indicators and their weights in china city green development index

First-class indicators	Weight (%)	Second-class indicators	Weight (% to first-class indicators) %
Green degree of economic growth	33	Green growth efficiency indicators	50
		Primary industry indicators	5
		Secondary industry indicators	30
		Tertiary industry indicators	10
Carrying capacity potential of natural resources and environment	34	Resource abundance and ecological conservation indicators	5
		Environmental pressure and climate change indicators	95
Support degree of government policies	33	Green investment indicators	25
		Infrastructure indicators	45
		Environmental management indicators	30

Table 1.3 lists the weights of First-Class and Second-Class Indicators, and Chaps. 6, 11 and 15 will elaborate on the Third-Class Indicators respectively.

1.4 Results and Analysis of China Green Development Index

The report evaluate the green development index of 30 provinces (except Tibet) and 34 cities in the year of 2009 according to the evaluation systems of province-based and city-based green development index.

1.4.1 Results of China Province Green Development Index

Table 1.4 shows the performance of 30 provinces in three First-Class Indicators.

This measurement uses dimensionless processing. Thus, "0" represents average provincial green development, and the province that scoring above "0" indicates its green development is better than the average, vice verse. The result shows, among 30 provinces, 12 provinces exceed the national average, and the index value from high to low is: Beijing, Shanghai, Qinghai, Hainan, Zhejiang, Yunnan, Fujian, Jiangsu, Guangdong, Inner Mongolia, Shandong. While other 18 provinces are below the average (see Fig. 1.1). The measuring indexes and methods of 2011 change a lot, but in general, the top 12 provinces are all listed top 12 in the measurement of 2010, yet their sequence has changed. This indicates that the measuring method of the green development index is stable to some degree.

From the highest to the lowest: Beijing, Shanghai, Qinghai, Tianjin, Hainan, Zhejiang, Yunnan, Fujian, Jiangsu, Guangdong, Inner Mongolia, Shandong,

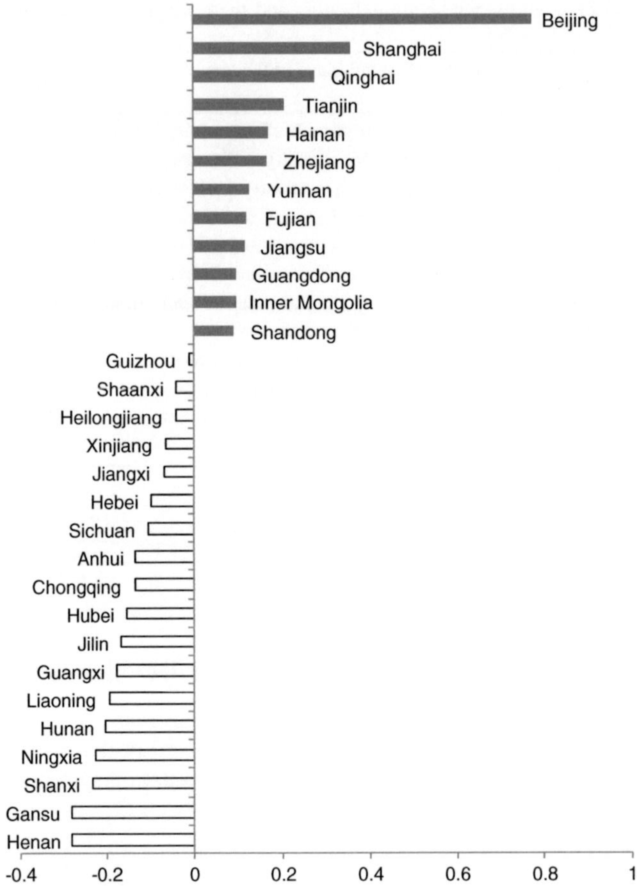

Fig. 1.1 Rankings comparison of green development index among 30 provinces (autonomous regions/ municipalities) in China, 2009

Guizhou, Shaanxi, Heilongjiang, Xinjiang Uyghur Autonomous Region, Jiangxi, Hebei, Sichuan, Anhui, Chongqing, Hubei, Jilin, Guangxi, Liaoning, Hunan, Ningxia, Shanxi, Gansu, Henan.

The above result indicates that the green development level of each province differs a lot. And this difference lies in two factors: one is in the efficiency of economic growth, resources per unit, and green degree of economic growth, as well as support degree of government policies; the other is carrying capacity potential of natural resources and environment. Among the top 12 provinces or districts, Beijing, Shanghai, Zhejiang, Fujian, Jiangsu, and Guangdong are mainly affected by the first factor, while Qinghai, Yunnan, Hainan, Inner Mongolia by the second.

1.4.2 Comparison of Provincial Green Development Index

As can be seen from Table 1.4, the top ten provinces and cities are Beijing, Shanghai, Qinghai, Tianjin, Hainan, Zhejiang, Yunnan, Fujian, Jiangsu and Guangdong respectively, eight of which are located in eastern China, except Qinghai and Yunnan.

Among the 10 provinces ranking from 11 to 20 on the list, five of them—Guizhou, Inner Mongolia, Shaanxi, Xinjiang and Sichuan are in western China where the per capita quantity of resources is relatively higher; two of them—Shandong (ranking 12th) and Hebei (ranking 18th) are located in eastern China; Heilongjiang (ranking 15th) is in the northeast; and the other two—Jiangxi (17th) and Anhui (20th) are in central China.

Among the 10 provinces ranking from 21 to 30, Jilin and Liaoning are located in northeastern China and the other eight are in central and western regions.

Generally, provinces in eastern China have an edge on the level of green development. The 10 provinces, except Hebei, enjoy a higher level of green development evidenced by eight in the top 10. After over 30 years of reform and opening up, provinces in eastern China have reached a higher level of economic development through the drive of export-oriented economy. However, these provinces also face the limitation of resources and environment, which justifies their poor performance in the rank of carrying capacity potential of natural resources and environment. Nowadays, provinces in eastern China are more concerned with balanced development. They have made progress in green development by implementing energy saving and emission reduction, optimizing industrial structure and adjusting industrial layout. At the same time, support degree of government policies in eastern China is relatively higher for their economic advantages. All in all, provinces in eastern China, though restricted by natural resources, have a higher level of green development.

Provinces in western China have a distinct advantage of natural resources and are abundant with natural resources, so their performance on the rank of green development index largely depends on the relatively higher score of carrying capacity potential of natural resources and environment. This shows that provinces in western China still have a long way to go on economic development. As western provinces are making greater efforts to tap natural resources to boost their economy, environmental problems may become increasingly acute. Recently, governments in western provinces have established more supportive policies for effective utilization of natural resources and environment protection, and they are in the front of green investment indicator.

Provinces in central China need to devote more efforts to enhance green development. The study results show that only two provinces (Jiangxi and Anhui) are at the middle level and other four provinces in central China are at a low level. The six provinces in central China have stroke a balanced development, which is shown in first-class indicators, but they are generally below the average level.

Table 1.4 Green development index and ranking in 30 provinces (autonomous regions, municipalities), 2009

Regions	Green development index		First-class indicators					
			Green degree of economic growth		Carrying capacity potential of natural resources and environment		Support degree of government policies	
	100 %		30 %		40 %		30 %	
	Index value	Ranking	Index value	Ranking	Index value	Ranking	Index value	Ranking
Beijing	0.770	1	0.556	1	−0.063	18	0.278	1
Shanghai	0.357	2	0.367	2	−0.117	24	0.107	3
Qinghai	0.273	3	−0.202	29	0.556	1	−0.082	27
Tianjin	0.206	4	0.320	3	−0.105	21	−0.009	17
Hainan	0.169	5	0.059	9	0.160	4	−0.050	20
Zhejiang	0.165	6	0.161	4	−0.092	19	0.096	4
Yunnan	0.127	7	−0.163	26	0.241	3	0.049	9
Fujian	0.121	8	0.107	8	−0.037	14	0.051	8
Jiangsu	0.116	9	0.153	5	−0.159	30	0.121	2
Guangdong	0.098	10	0.138	6	−0.117	23	0.077	7
Inner mongolia	0.095	11	−0.010	11	0.135	6	−0.030	19
Shandong	0.091	12	0.112	7	−0.108	22	0.087	5
Guizhou	−0.012	13	−0.204	30	0.269	2	−0.076	26
Shaanxi	−0.039	14	−0.046	14	−0.009	12	0.017	13
Heilongjiasng	−0.040	15	−0.048	16	0.156	5	−0.147	28
Xinjiang	−0.065	16	−0.058	17	0.056	9	−0.063	22
Jiangxi	−0.067	17	−0.096	21	0.019	10	0.010	14
Hebei	−0.095	18	−0.020	12	−0.118	25	0.043	10
Sichuan	−0.104	19	−0.112	23	0.081	7	−0.074	23
Anhui	−0.131	20	−0.081	19	−0.054	16	0.004	15
Chongqing	−0.133	21	−0.113	25	−0.055	17	0.035	12

(continued)

Table 1.4 (continued)

Regions	Green development index		First-class indicators					
			Green degree of economic growth		Carrying capacity potential of natural resources and environment		Support degree of government policies	
	100 %		30 %		40 %		30 %	
	Index value	Ranking	Index value	Ranking	Index value	Ranking	Index value	Ranking
Hubei	−0.151	22	−0.048	15	−0.104	20	0.000	16
Jilin	−0.166	23	−0.023	13	0.011	11	−0.154	29
Guangxi	−0.174	24	−0.112	24	−0.031	13	−0.030	18
Liaoning	−0.193	25	−0.009	10	−0.123	27	−0.061	21
Henan	−0.202	26	−0.081	18	−0.047	15	−0.075	24
Ningxia	−0.225	27	−0.170	27	−0.137	28	0.081	6
Shanxi	−0.232	28	−0.110	22	−0.158	29	0.036	11
Gansu	−0.279	29	−0.184	28	0.070	8	−0.165	30
Henan	−0.280	30	−0.083	20	−0.121	26	−0.076	25

Notes

1. This table is prepared according to the provincial green development index measurement system and the data of 2009
2. The provinces (districts and cities) are ranked according to their quantitative values of green development index from large to small

Sources China Statistical Yearbook 2010, China Statistical Yearbook on Environment 2010, China Statistical Annual Report on Environment 2009, China Industry Economy Statistical Yearbook 2010, China Statistical Yearbook For Regional Economy, China City Statistical Yearbook 2010, China Water Statistical Yearbook 2010, China agriculture Statistical Yearbook 2010, China City Construction Statistical Yearbook 2009, and China deserts and their management

Central provinces have achieved certain progress in economic development, but still lag behind eastern provinces and their weak economic strength in turn makes it harder to achieve balanced development.

The three provinces in northeastern China vary significantly in green development. The three provinces are home to many old industrial bases and resource-exhausted cities, which face greater challenges in green development. Study results show that green development index of the three provinces are below the average and they vary from each other. Heilongjiang province ranked 15 on the list, with a cutting edge on carrying capacity potential of natural resources and environment. Jilin and Liaoning ranked 23 and 25 respectively on the list, both below the average.

1.4.3 City Green Development Index

The measurement of city green development level has practical significance. Based on city green development index system and data of 2009, we measured 2009 green development index and ranking of 34 cities. The results are shown in Table 1.5.

Similar to the provincial measurement system, the city scoring above "0" indicates its green development level is better than the average, vice verse. Among the 34 cities, 15 cities exceed the national average, while other 19 cities are below the average. The top 15 are in turn: Shenzhen, Haikou, Kunming, Beijing, Hefei, Guangzhou, Dalian, Qingdao, Changsha, Fuzhou, Xiamen, Nanning, Ningbo, Shenyang, and Harbin.

For the convenience of the readers, based on the data in Table 1.5, below is rankings comparison of green development index among 34 cities in China, 2009 (Fig. 1.2).

From the highest to the lowest: Shenzhen, Haikou, Kunming, Beijing, Hefei, Guangzhou, Dalian, Qingdao, Changsha, Fuzhou, Xiamen, Nanning, Ningbo, Shenyang, Harbin, Shijiazhuang, Hangzhou, Nanjing, Shanghai, Jinan, Yinchuan, Nanchang, Hohhot, Zhengzhou, Guiyang, Taiyuan, Tianjin, Chongqing, Xi'an, Wuhan, Chengdu, Lanzhou, Xining.

1.4.4 Comparison of City Green Development Index

The new report has analyzed green development performance of 14 eastern cities (Beijing, Tianjin, Shijiazhuang, Shanghai, Nanjing, Hanghzou, Ningbo, Fuzhou, Xiamen, Ji'nan, Qingdao, Guangzhou and Shenzhen), six central cities (Taiyuan, Hefei, Nanchang, Zhengzhou, Wuhan and Shangsha), 10 western cities (Hohhot,

Table 1.5 Green development index and rankings among 34 cities in China, 2009

Cities	Green development index		First-class indicators					
			Green degree of economic growth		Carrying capacity potential of natural resources and environment		Support degree of government policies	
	100 %		33 %		34 %		33 %	
	Index value	Ranking	Index value	Ranking	Index value	Ranking	Index value	Ranking
Shenzhen	0.689	1	0.341	1	0.029	13	0.319	1
Haikou	0.555	2	0.140	3	0.368	2	0.047	12
Kunming	0.496	3	−0.098	29	0.546	1	0.047	11
Beijing	0.387	4	0.267	2	−0.101	24	0.222	2
Hefei	0.260	5	0.085	8	0.167	5	0.008	18
Guangzhou	0.200	6	0.127	5	−0.068	21	0.141	4
Dalian	0.184	7	0.090	7	0.029	11	0.065	9
Qingdao	0.166	8	0.072	10	0.011	14	0.083	8
Changsha	0.156	9	0.124	6	0.042	10	−0.011	19
Fuzhou	0.111	10	−0.024	17	0.087	6	0.048	10
Xiamen	0.090	11	0.028	12	−0.107	26	0.168	3
Nanning	0.086	12	−0.093	27	0.201	4	−0.021	20
Ningbo	0.075	13	−0.039	19	0.004	15	0.110	6
Shenyang	0.041	14	0.139	4	−0.064	19	−0.034	22
Harbin	0.028	15	−0.039	18	0.202	3	−0.135	31
Shijiazhuang	−0.029	16	−0.070	25	−0.091	23	0.133	5
Hangzhou	−0.029	17	0.000	16	−0.071	22	0.041	13
Nanjing	−0.032	18	0.021	15	−0.152	31	0.099	7
Shanghai	−0.053	19	0.083	9	−0.166	33	0.030	14
Changchun	−0.066	20	0.026	14	0.072	8	−0.163	32
Jinan	−0.075	21	0.026	13	−0.110	27	0.008	17
Yinchuan	−0.097	22	−0.101	30	−0.020	17	0.024	15
Nanchang	−0.103	23	−0.044	20	−0.014	16	−0.045	23
Hohhot	−0.121	24	−0.052	24	0.029	12	−0.098	27
Zhengzhou	−0.123	25	−0.049	22	−0.047	18	−0.027	21
Guiyang	−0.168	26	−0.184	32	0.073	7	−0.057	24
Taiyuan	−0.168	27	−0.075	26	−0.102	25	0.009	16
Tianjin	−0.207	28	0.029	11	−0.136	30	−0.099	28
Chongqing	−0.249	29	−0.188	33	0.060	9	−0.121	30
Xi'an	−0.274	30	−0.048	21	−0.131	28	−0.094	26
Wuhan	−0.292	31	−0.050	23	−0.183	34	−0.059	25
Chongdu	−0.331	32	−0.094	28	−0.135	29	−0.102	29
Lanzhou	−0.390	33	−0.138	31	−0.067	20	−0.184	33
Xining	−0.711	34	−0.206	34	−0.159	32	−0.346	34

Notes
1. This table is prepared according to the city green development index measurement system and the data of 2009
2. The cities are ranked according to their quantitative values of green development index from large to small

Sources China City Statistical Yearbook 2010, China Statistical Annual Report on Environment 2009, China City Construction Statistical Yearbook 2009, China Statistical Yearbook for Regional Economy 2010, etc

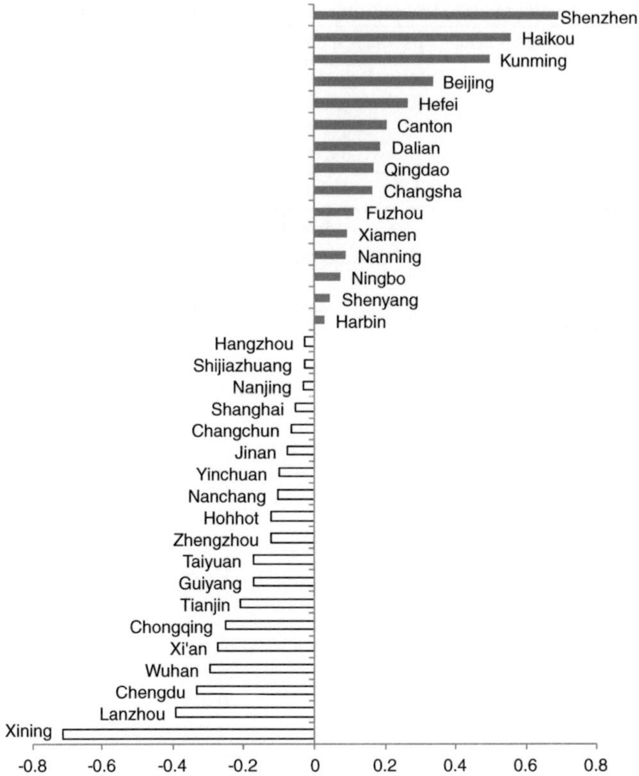

Fig. 1.2 Rankings comparison of green development index among 34 cities in China, 2009

Nanning, Chongqing, Chengdu, Guiyang, Kunming, Xi'an, Lanzhou, Xi'ning and Yinchuan) and four northeastern cities (Shenyang, Dalian, Changchun and Harbin).

Table 1.5 shows that the top 10 cities are Shenzhen, Haikou, Kunming, Beijing, Hefei, Guanghzou, Dalian, Qingdao, Changsha and Fuzhou. We can see that eastern China enjoys a higher level of green development, as six cities are ranked among the top 10 in the national ranking. However, western cities generally ranked low in the ranking, with 6 out of 10 among the bottom 10.

City green development index bears following regional disparities:

Eastern China enjoys a higher level of green development as a whole, with 13 out of 14 cities ranking above the average. The ranking is in accordance with China's regional disparities in economic development—the eastern coastal area enjoys a relative prosperity than central and western China. The three second-class indicators reflect that strong efforts has been put into lowering energy consumption per unit of gross regional product and pollutant emissions as well as booming the tertiary industry. The local governments have enhanced their green development by supporting green and strategic rising industries and improving cities' infrastructure. It is worth mentioning that only one city in eastern area, Tianjin, is below the average, ranking 28th as a whole with green degree of economic growth, carrying capacity potential of natural resources and environment and support degree of government policies ranking 11th, 30th and 28th respectively.

Central China showed an uneven situation, with two cities out of six ranking above the average and others well behind the 23rd. Clearly, central cities still have a long way to go for green development. For example, Wuhan, bottomed the ranking in carrying capacity potential of natural resources and environment, should put much efforts in energy saving and emissions reduction.

Western China experienced a low level of green development as eight out of 10 cities ranking well behind the 22nd and six bottomed the list. Surprisingly, provincial and city comparisons indicate distinct disparity in western area. In provincial comparison, most western provinces rank above the average, however, their cities are lagged behind. Obviously, cities in western part haven't made full use of their provinces' abundant resources. What's more, most western cities, with resource and central government's preference policies, still ranked low due to the relative poor financial situations in local governments. Many of them still followed an extensive economic development model.

Northeastern cities are balanced in green development, with its highest-placed city, Dalian, ranking 7th and other three cities remain average.

1.4.5 Findings of the Comparison of Green Development Indexes

By comparing green development indexes in different areas of China, we get the following inspirations which will lay solid foundations for further studies in this field.

First, the measuring results showed that green development is practical. When measuring the province and the city green development level, we found that green development in eastern China is backed for its 30 years of economic advantages. The green development in western and middle China has resource and environment endowment. The resource exploited northeastern China, however, has a long way to go for green development.

Second, disparities in province and city green development rankings showed that although some provinces enjoy high level of green development, some of their cities may not do the same way. For example, Qinghai province ranks third in green development list, but Xining, its capital, ranks at the end. Green development in Qinghai province has rich resource endowment, but it also has a fragile ecological environment. Xining city, as one of the important industrial district, is shouldered huge responsibilities in balancing economic development and environmental protection.

Third, governments at all level strongly support green development in different ways. Environmental resources management in different areas is almost the same, while infrastructures and city management vary largely from each other. Governments still need to invest more in environmental protection and green industries and improve their management in this field.

It is important to know that provinces and cities which are listed on the top of the rankings still have problems in green development and those listed at the bottom also have some advantages in this field. It is a long-term mission to learn from each other and realize a balanced development. Energy conservation, emission reduction and the transformation of economic development model are still the essential parts in the 12th Five-year Period plan, supporting economic green development.

1.5 Framework and Core of the Report

A Research report on China's Green Development Index 2010: Regional Comparison consists of five parts: the preface, experts' review, pandect, five sections (15 chapters, 31 columns and 11 specials) and appendixes. The report elaborates China's green development by enumerating the related achievements, problems, regional comparison, opportunities and challenges.

Here we want to thank Ma Jiantang, director of the National Bureau of Statistics of China, Zhong Binglin, president of the Beijing Normal University, and Zhao Dewu, president of the Southwestern University of Finance and Economics, for their strong support to our study.

We would also like to thank 30 experts from all works of life for their hard work in ensuring the scientificness and objectiveness of our report.

As the outline of this report, the pandect has introduced the social influences of the China Green Development Index 2010, analyzed the latest results in this field from home and abroad. It has also introduced the improvement of the inter-provincial indicators of China's green development index, the building of city indicator systems and the measurement rankings which aim to give inspirations for further study.

The report consists of five sections.

The first three sections discuss the three first-class indicators, the Green Degree of Economic Growth, the Carrying Capacity Potential of Natural Resources and Environment, and the Support Degree of Government Policies. Each has five chapters, with the first chapter as a lead-in and the following three chapters discussing the second-class indicators to reveal the characteristics of China's green development. The last chapter introduces details of the indicators' measurement and calculations. It also includes the green rankings of provinces and cities.

The first three sections cover 31 cases selected from 31 Chinese provinces/regions/municipalities as columns. Our columns are provided by local governments, schools and experimental zones. Of course, we will strengthen such co-operations.

The fourth section discusses eleven specials with further illustrations to China's green development index. The eleven specials involve current hot topics such as food safety, relationship between green human resource and China's green development, nuclear power development in green economy, Daoism and green development, etc. Green development is discussed from various perspectives by experts. The ideas are rich of academic and practical value.

The fifth section is about feedbacks, a new part in the 2011 report. It is very important to hear the voices of provinces, regions and municipalities when comparing their levels of green development. Communications help us improve our work and avoid misunderstandings. After the release of the 2010 report, our efforts have been rewarded and widely supported. This section also introduces the valuable comments and advice from Qinghai Environment Protection Agency. We are looking forward to more feedbacks and ready to improve our work.

In the last part of the book are three appendixes. Appendix one is an introduction to the China Green Development Index Indicators System selected from the pandect of 2010 report. This appendix informs the readers of the establishment of the measuring system and choice of indicators. Appendix two explains the definitions and sources of the 60 indicators in the inter-provincial comparison. Appendix three explicates the definitions and sources of the 43 indicators in city comparison.

This report scientifically reveals the green development of provinces, regions and municipalities in China, trying to find a new path of boosting economy through green development. At the very beginning of the 12th Five-year Period, we are clear of the difficulties ahead to create an energy saving and environment-friendly society. We still have a long way to go for a balanced development of China and the world.

Session I
Green Growth of the Economy

Chapter 2
Green Growth: Constructing a Resource-Saving and Environment-Friendly Production Pattern

Youjuan Wang, Weibin Lin and Qian Wan

In the 12th *Five-Year Plan of China*, "Changing developing mode and striving to create a scientific development pattern" has been specified as the theme of national economic and social development. Meanwhile, as the focus of accelerating the transformation of the economic development pattern, the building of resource-saving and environment-friendly society has also been emphasized. Obviously, the green economic growth is of extraordinary importance in forming a resource-saving and environment-friendly society, and is an inevitable choice in coping with increasingly serious resource and environmental constraints. In persuit of a green growth, endeavors on enhancing the sense of environmental crisis, accelerating energy saving and emission reduction, and constructing an energy-saving and environment-friendly production pattern are urgently needed. Besides, there is also a fundamental need for the green growth mode to perfect the price forming mechanism of resource products and the resource-environment tax system, and therefore establishing a rational pricing system.

2.1 Green Growth: The Key to Building a Resource-Saving and Environment-Friendly Society

The issues of resource shortage and environmental pollution are becoming the major constraints that obstruct China's sustainable development. In this context, the building of resource-saving and environment-friendly society plays an important role in accelerating the transformation of economic development mode. This new development mode pays more attention to resource saving and environmental protection both in producing process and households living. As China is still in its rapid industrialized phase, developing a resource-saving and environment-friendly production pattern and propelling a green economic growth are the major tasks in realizing a resource-saving and environment-friendly society.

X. Li and J. Pan (eds.), *China Green Development Index Report 2011*,
Current Chinese Economic Report Series, DOI: 10.1007/978-3-642-31597-8_2,

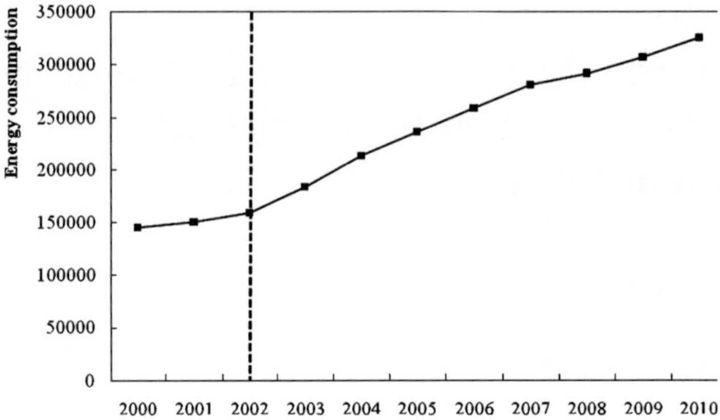

Fig. 2.1 Energy consumption of China from 2000 to 2010. *Source* Website of National Bureau of Statistics of China

2.1.1 Resource and Environmental Limitations in China

In recent years, China has experience a rapid development both economically and socially, which is reflected by fast economic growth, improvement of the standard of living and considerable progress on social undertakings. However, simultaneously, the deterioration of natural resources and environment is the price to pay for the socioeconomic development, including the huge consumption of natural resources such as water, energy and earth, the emission of environmental pollutants, and the severe destruction of our natural ecological environment. To make clear the extent of resources consumption and environmental pollution, two indicators, i.e., energy consumption and CO_2 emission are utilized. Figure 2.1 shows the trend of energy consumption in China in the last 10 years.

As demonstrated in Fig. 2.1, there is an uprising trend of energy consumption from 2000 to 2010. This trend is most notable from 2002, due to the increasing development of heavy chemical industry in China and energy consumption was thereby largely required, (The vertical line in Fig. 2.1 represents the turning point of the increasing rate of energy consumption). From 2002 to 2008, the energy consumption kept in a growth rate of >8 %, and the highest one even reach 16 %, which has surpass the growth rate of GDP. This trend was kept until the year of 2008, when the world was striked by the global financial crisis.

On the other side, huge amount of energy consumption and accompanied wastes and emissions have exerted great pressure on environment, threatening people's quality of life and obstructing the sustainable development process. Owing to the large number of energy consumption, CO_2 emissions in China present a stably increasing trend and roar to the largest CO_2 emitter worldwide. The large amount of CO_2 emissions in China not only make great contribution to global warming, but also have China accused and constrained by some

Table 2.1 The ratio of energy reserve to production (RP ratio) of China and the world

Energy type	Reserve	Production	RP ratio of China	RP ratio of the world
Raw coal	1.145×10^{11} ton	3.24×10^{9} ton	35	122
Crude oil	2.1×10^{9} ton	2.03×10^{8} ton	10	42
Natural gas	2.46×10^{12} m^3	9.676×10^{10} m^3	25	60

Source BP statistical review of world energy 2010

international organizations and other countries. Therefore, in the long run, CO_2 emission will remain a significant problem that needs to be addressed in no time.

Resource and environment carrying capacity is facing increasingly severe challenges due to irrational resource consumption and environmental pollution. As to the resource consumption, taking energy utility as example, according to the statistical data of BP (2010), China's economic recoverable reserves of coal, oil and natural gas are 1.145×10^{11}, 2.1×10^{9} ton and 2.46×10^{12} m^3 respectively in year 2010. However, energy production in 2010 sets a new record. The production of raw coal has reached 3.24×10^{9} ton with a growth rate of 9 %. With regard to crude oil, a breakthrough has been made with the production of 2.03×10^{8} ton and growth rate of 7.1 %, which is the highest growth rate in the recent years, and natural gas production is up to 9.676×10^{10} m^3 with a growth rate of 13.5 %. Thus, it can be calculated that the reserve and production ratio (RP ratio) of coal, oil and natural gas are 35:10:25, which is only 29, 24 and 42 % of the world's average, as shown in Table 2.1 and Fig. 2.2. Evidently, resource constraint creates a great challenge for the sustainable development of Chinese economy and society. Specially for crude oil, if the reserve keeps unchanged, it is only available for another decade or so even if we hold the current mining scale. Energy bottleneck is, thereby, around the corner.

In terms of the environmental degradation, through past 3 decades, rapid industrialization has also imposed enormous pressure on the environment. Specifically, air, water quality and soil are under severe degradation due to the emissions of toxic gas, sewage and wastes, thereby resulting to the ecological damage and abominable human living condition. The data published by the Ministry of Environmental Protection of China indicates that China was of poor air quality, especially around the cities of concentrated heavy industries and high population density like Beijing and Tangshan. On the other hand, the degradation of water quality directly exerted impact on water use security. Watersheds in China that reached the first grade of water quality standard only constitute 4.6 % of the total river length, and the watersheds of inferior-V grade approach 20 %. A quarter of people are kept from clean drinking water. One third urban residents have to bear with dirty air. It is clear that most of the population is exposed to a degraded environment with low quality of air, water and soil. In recent years, the severe environmental and ecological problems have inflicted great damage on economic development, more fatally, on human life and property.

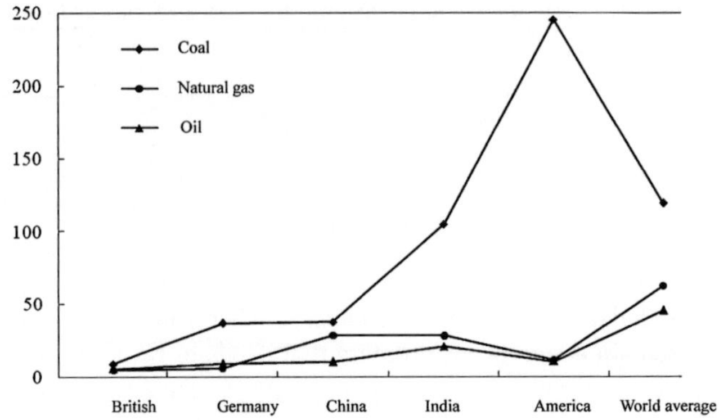

Fig. 2.2 The RP ratios of coal, oil and natural gas of main countries in 2009. *Source* BP statistical review of world energy 2010

2.1.2 Core Reasons of Resource and Environmental Problems

As discussed in Sect. 2.1.1, intensive resource consumption and environmental pollution tremendously threatened the carrying capacity of resource and environment. As China is still at its increasing industrialization stage, resource use and environmental emissions are mainly attached to the production process, i.e., large amount of resources are consumed for creating GDP at the cost of severe environmental pollution and ecological damage. This phenomenon reflects the extensive characteristic (intensive energy consumption, high environmental pollution and low utilization efficiency) of China's economic growth. Since the reform and open up policy in the 1970s, China's economic growth rate has reached 9.9 %. The continuous high-speed economic growth has attracted considerable attention globally, reckoning it as "the miracle of China". However, at the same time, the quality of this economic development has somewhat been doubted, with increasing attentions paid on resource consumption and environmental pollution in the process of such growth.

Here we first employ energy intensity as an indicator to reflect the energy efficiency of the economic development in China, i.e., energy consumed per GDP. Figure 2.3 indicates that the economic development in China is still characterized as the extensive and inefficient mode. Energy utilization efficiency is relatively low compared to that of developed countries. As shown in Fig. 2.3, energy intensity in 2009 stayed in a high level, which was around 5 times of German and Japan, and 3 times of the U.S. The high-energy intensity reflects huge resource cost behind economic development and indicates the unsustainability of current developing mode. These have brought huge negative externalities which needs to be further addressed.

Another concerned indicator is energy supply per GDP which is defined in the China Statistical Yearbook, it could represent the gap among China, the average of the world and major countries from the angle of energy supply. As shown in

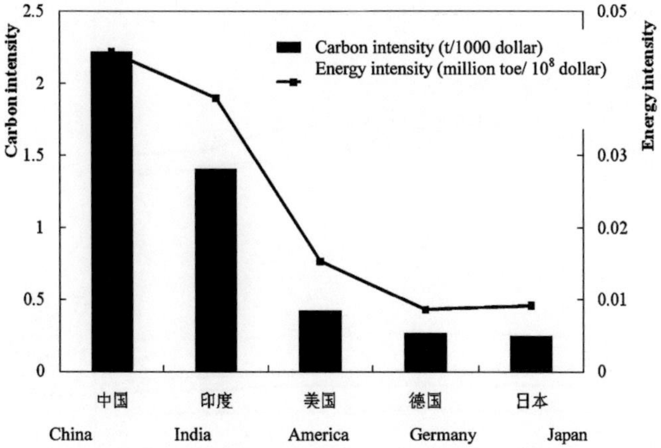

Fig. 2.3 Energy intensity and carbon intensity of major countries in 2009. *Sources* Energy intensity comes from WDI database of World Bank, which is calculated by PPP of 2005. Carbon intensity is obtained from EIA and is based on the price of U.S. dollar in 2005

Fig. 2.4, energy supply per GDP in China and India is higher than those of the average level of the world and other major developed countries. Specifically, the energy supply per GDP in China is 2.7 times of the world, 4 times of the U.S., 4.5 times of OECD countries, 4.5 times of French, 7 times of the U.K., and 8 times of Japan.

In addition, CO_2 emission per GDP could also be employed to represent the environmental cost of economic development. The changing trend of CO_2 intensity in Fig. 2.3 indicates that current economic growth mode has caused large amount of greenhouse gas emission as well as huge ecological pressure, which has attracted attentions from the whole society. As shown in Fig. 2.3, China's carbon intensity is one of the highest values in the world, not only higher than the developed counties, but also higher than some developing countries like India as well as the world average value. In 2009, the carbon intensity of China is about 5 times of the U.S., more than 8 times of German, and 9 times of Japan. This giant different depicted in Fig. 2.3 reveals the ignorance of environmental protection of China in pursuit of economic growth and also a great promotion potential in reducing carbon intensity.

2.1.3 Essential Requirements to Harmonize "Green Environment" and "Economic Growth"

In the face of intensified environmental and resource constraints, a sense of crisis should be enhanced, as well as the awareness of green and low carbon development. Moreover, the implementation of "energy saving and emission reduction"

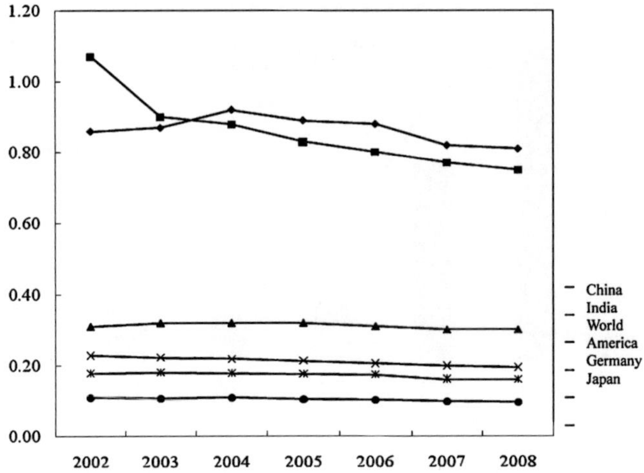

Fig. 2.4 Energy supply per unit GDP of some major countries toe/1,000 dollar, calculated by the 2000 price. *Source* China energy statistical yearbook 2010

should be further intensified. Meanwhile, we should consider that China is still a developing country whose GDP per capita is less than half of the world's average, one tenth of developed countries, and ranks 100 in the world (China's GDP was converted into 3,761 dollars which is less than half of the world in 2009 with the exchange rate of 6.8 between RMB and U.S. dollar).[1] This indicates that China should keep attaching importance to economic development. The problem is: how to cope with the contradiction between economic growth and resource saving and environmental protection? The only solution to this is transforming current economic growth mode to a resource-saving and environment-friendly one.

Countermeasures for resource saving are: (1) Restrain the rapid growth of energy intensive industries, and highlight the importance of energy saving in industry, construction, transportation and public institution. Management on energy saving should also be emphasized in high-energy-consumption enterprises. In addition, some regulations and incentives such as energy management contract, demand side management, energy efficiency label policy, energy-saving certificate. And the mandatory procurement of energy-efficient products should be perfected and promoted. Moreover, advanced energy-saving technology and products need further promotion. According to the 12th *Five-Year Plan*, a 16 % decrement in energy intensity in the end of 2015 is required compared to the level of year 2010. (2) Strengthen the efforts of water resource saving. We should carry

[1] Ministry of foreign Affairs of the People's Republic of China: Per capita GDP of China is less than half of the average level of the world. 2011.02.15.
http://finance.ifeng.com/hk/sckx/20110215/3409279.shtml.

out a rigorous water resource management, enhance volume control and quota management of water resource, speed up the establishment of water distribution program of river basins, reinforce the construction of water right system and water-saving society, improve industrial water utilizing efficiency, and finally, update water-saving technologies of key water-consuming industries and guiding the water-saving behaviors of local residents. (3) Make use of the land economically and intensively. For arable land, measures such as insisting on the most rigorous arable land protection policies, designating permanent basic farmland, establishing protection and compensation mechanism and strictly controlling the occupancy of arable land by construction activities, as well as keeping the balance between farmland occupation and complement according to the policy of "supplement first and occupation follows" are of great significance to make sure that the total quantity of arable land is not declined. In addition, policies on land use saving should be executed rigorously to control the whole construction scale. According to the 12th *Five-Year Plan*, the construction land per GDP should be decreased by 30 % in 2015 compared with that of 2010.

In order to implement the target responsibility system for energy conservation and emission reduction, and reinforce the control of pollutants, environmental protection is also emphasized in the 12th *Five-Year Plan*. Main pollutants were added up to four rather than two, i.e., COD, ammonia nitrogen, SO_2, NO_x, in which the emissions of COD and SO_2 are planned to decrease by 8 %, while the emission of ammonia nitrogen and NO_x are planned to decrease by 10 %. Before this, only COD and SO_2 are employed as binding indexes in environmental monitoring. However, with SO_2 controlled, NO_x has become the main air pollutant in recent years, the incorporation of ammonia nitrogen and NO_x is important for the enlargement of the environmental protection and the total-quantity control of pollutants discharge. Besides, a 17 % decent of CO_2 intensity is also requested in the 12th *Five-Year Plan* to cope with the global warming issue.

2.2 Constructing a Resource-Saving Production Pattern

The meaning of constructing a resource-saving production pattern is to prioritize resource saving in the production process, to improve resource consumption efficiency, to lower resource consumption intensity, and to obtain the largest economic benefit with least cost. The first step for this is to intensify the control of total resource consumption and avoid irrational exploitation. After that, it is also essential to rationalize the pricing system of resource-based products and use price mechanism to optimize the allocation of scarce resources. Circular production pattern is also regarded as a promising approach to improve the efficiency of resource use and reuse.

2.2.1 Intensifying the Control of Total Resource Consumption to Avoid Disordered Exploitation

The principle of "Use the resource in a fast speed" has long been followed in resource exploitation and utility, resulting in excessive and disordered resource exploitation. To cope with gradually serious resource constraints, efforts should be spent on intensifying resource consumption control, especially for the non-renewable and scarce resources. In the 6th session of Chap. 22 of the 12th *Five-Year Plan*, the requirements of "to prioritize resource-saving and resource consumption regulation…to promote the security level of resource utility" is highlighted.

With respect to energy saving, the key point is to control the total amount of nonrenewable fossil fuel supply. According to the total—amount—control target of the 12th *Five-Year Plan*, primary energy consumption should be controlled under 4×10^9 ton, in which coal consumption is assigned to 4×10^9 ton with net import of 2×10^8 ton, while oil consumption is assigned to 5×10^8 ton with net import of 3×10^8 ton. Therefore, during the period of the 12th *Five-Year Plan*, the control from resource and energy consumption on domestic oil and chemical industries will be obvious in China. In all, the rational management of total energy consumption is the key to developmental mode transformation.

As to water resource, volume control and quota management of water resource should be enhanced, and water distribution plan of river basins and water right system should be established as well. For land resource protection, the establishment of land use plan should be accelerated, so as the protection of arable land and the control of total construction land. In terms of minerals resource, the protection and exploitation management of dominant mineral resources should be attached more importance to. In addition, we should strive to perfect the compensation systems for use of mineral resources. Moreover, by executing the partition management regulation of mineral resource planning, rationally allocating mineral right, and optimizing the distribution of geological exploration and exploitation of mineral resources taken, we could keep the exploitation of mineral resources coordinate with the development of social and economic development and therefore improve the sustainable supply ability of mineral resource.

2.2.2 Taking Advantage of Pricing Mechanism for Optimal Resource Allocation by Streamlining the Pricing System of Resource Products

The extensive and inefficient economic growth mode in China is rooted in the fact that the twisted market price fails to deliver the exact signal. A low resource price may result in the abnormal demand and a waste of resources, leading to the overdevelopment of energy-intensive industries. Considering the pricing mechanism of resource products one step further, we could find that the twisted resource price originated from the blurred definition of resource property. At this stage, there is still no clear definition on resource property in China such as land, water, minerals and energy sources. It is difficult to adjust prices efficiently due to the lack of a mature market that reflects the actual demand and supply of these resources. As a result, the existing pricing mechanism fails to reflect the scarcity and cost of resources, which results in extensively relies on resource exploitation and input. A high input, high consumption, high pollution and low efficiency growth route has emerged. This growth mode makes the environmental and resource problems more and more severe and thus diminish the marginal benefit of economic development. In addition, the path dependence strengthens the inertia of the pricing system, and it lacks a motivation for the entities of market to rectify their behavior and promote the institutional changes. In conclusion, establishing a sophisticate resource pricing system and define the price as a reflection of scarcity and value of resources is not only an essential way to the building of resource-saving society, but also a fundamental approach in transforming economic growth mode and solving resource and environmental problems.

The general direction of resource pricing system reform in China is to clarify the resource property, to rationalize the relations of prices, to keep marketizing, and to make use of the fundamental role of market in resource allocation. Another is to make the price a reflection of resource scarcity, supply and demand balance of market, and the externality of resource and environment, thus fostering the transformation of economic growth mode and the adjustment of industrial structure in support of the built of resource-saving society.

Firstly, to clearly define the resource property. As soon as the property rights determined, the market-oriented reform will work. This has been emphasized in China's 12th *Five-Year Plan* as "Bringing in the market mechanism, establishing compensation and transaction systems for the mineral property, standardizing the development of transaction system of mineral exploit and mining right, and supplementing the laws and policy system to realize the well organization of resource and environmental property rights and achieve open, justice and fair transaction mode". To clarify the resource property, we should pay attention to the following issues: firstly, the affiliations of all the rights should be determined, including the distribution and partition of resource ownerships, utility right, current right, and so on. After that, different property right systems should be established in accordance to different kinds of resource. Regarding the resources whose property are difficult to determine, we should designate its public nature and open both the current and utilization right and management right of the first grade market to realize the compensative resource use. Also we should open the current, utility and the management rights of the second grade market to the resources with definite property right, cultivate the market system of resources, improve the marketization degree of resources and optimize the resource allocation mode. By optimizing the resource property right system, we can realize the capitalized management of the resource property, and separate the ownership and management right. In addition, rights and obligations should be determined in the format of legal contract to realize the compensative use of resources, and guarantee the right of a country as the property owner. Meanwhile, it is essential to optimize the resource exploitation and mining in persuit of sustainable socioeconomic development.

Secondly, to rationalize the price relations, especially the proportional price among energy sources. Attentions should be paid to adjust the price relations among different energy sources to maintain them to a reasonable level, and avoid extensive demand or contradictions of supply and demand brought by low price relations of one kind of energy sources, i.e., price delivers wrong signals. In China's 12th *Five-Year Plan*, the request has been proposed as "rationalizing the price relation between natural gas and the alternative energy sources". Moreover, converging with the international price is also required in rationalizing price relations, especially for the energy sources that are prone to be influenced by the international market, e.g., petroleum products. Price should be adjusted step by step towards the price level of the international market and finally realize the price convergence to the international level.

Finally, based on the previous two steps, we should insist on the market-oriented reform to make full use of market in optimizing resources allocation Specific practice is that we should break the resource monopoly management pattern by bringing in the competition mechanism and untying the governmental control. To ensure smooth operates of the market, we should establish and optimize the operating principles, create a fair and open atmosphere and form a unified, open, and well-organized resource market. The 12th *Five-Year Plan* has pointed out that "In order to propel the rational exploitation and resource use, we should increase the resource tax bearing properly, perfect the assessment and imposing mode, and transform the fixed quantity imposing way, which is based on the quantity to the fixed rate imposing way according to the linkage mechanism of price, tax, fee, and rent."

Once the resource pricing system is established, price will play a role in delivering signals about supply–demand balance and scarcity of the market, based on which the optimal resource allocation can be obtained. On this basis, in order to reduce production costs, each economic entity spontaneously should employ multiple approaches to improve the resource utilization efficiencies, thus providing intrinsic motivations for resource-saving efforts. Under such circumstance, each economic entity will rectify its behavior automatically and stick to principle of Reduce, Reuse and Recycle. Also it is important to prioritize the principle of reduction, to improve the resource yield efficiency, to promote the circular production pattern and establishing the resource recycling system. Only by adapting these measures, the whole society can be accelerated to the economic transform, and the strategic goals of the 12th *Five-Year Plan* and the transformation of economic growth mode can be realized.

2.2.3 *Striving to Develop the Circular Economy by Production Pattern Innovation to Improve the Comprehensive Resource Utilization Efficiency*

The 12th *Five-Year Plan* has pointed out that the circular production pattern will be promoted in the next 5 years. Industrial parks will be planned, constructed and rehabilitated according to the requirement of circular economy. It is also needed to obtain the intensive land use, waste exchange and reuse, cascade energy utility, waste water recycle and centralized pollutants handling. Finally, by propelling the circular combination of industries and constructing an inter-linked and cycling industrial system, the resource output ratio is expected to increase by 15 %.

In terms of agriculture, the transformation should be achieved by developing the circular agricultural production pattern and constructing an inter-linked circular economic system of high-efficiency. New agricultural modes such as ecological agriculture, stereoscopic agriculture, green agriculture, organic agriculture and recreational agriculture are highly suggested in this case. In the development process, based on the principle of "low exploitation, high utilization, low emission and reuse", we should make full use of materials and energy that entered the production-consumption system, improve the operation quality and efficiency of the whole system, and realize circular and high-efficiency energy and material utilization. With all these measures implemented, the coordinate development between agriculture and resource and environment is expected to realize, as well as the sustainable agricultural development.

As to industry production, propelled by the impetus from the regional planning, enterprises' self-discipline and government supervision together, we should accelerate the transformation of enterprises towards a circular production pattern. Such production pattern could be interpreted that in an industrial park, in which the wastes and by-products of an enterprise are used as the input and raw materials of another enterprise. By means of waste exchange, circular utility and cleaner production, the increase of resource use and energy conversion efficiency, and waste emissions reduction (even zero emission) can be obtained. To realize circular economy, measures on weeding out outdated production pattern should be taken, accompanied with adjusting industrial structure and realizing the innovation of production pattern. These measures are essential to transform the current low-efficiency and high-pollution production pattern to the circular economy mode of high-efficiency and low-emission.

Finally, a complete resource pricing system can foster the establishment of resource recycling and reusing system. The 12th *Five-Year Plan* proposed that in order to promote the large-scale renewable resource use, we should complete the recycling system of renewable resources, accelerate the construction of "Three-in-One" recycling network that covers the recycling sites, sorting centers and terminal markets of urban communities and rural areas. In addition, measures such as completing the recycling system of old components remanufacturing, propelling the development of remanufacturing industry, establishing the garbage sorting and recycling system, perfecting the sorting and recycling, airtight transportation, concentrated disposal systems, and promoting the recycle and harmless treatment should also be paid great attention to.

2.3 Constructing an Environment-Friendly Production Pattern

As long ago as 1992, the concept of "Environment-friendly" has been formally proposed in Agenda 21 in United Nations Conference on Environment and Development (UNCED) in Rio. In the mid- and late 1990s, the international society put forward a set of concepts such as environment-friendly land use, environment-friendly basin management, environment-friendly cities, environment-friendly agriculture and environment-friendly building. In 2004, the Japanese government was ambitious to build the environment-friendly society in the "White Paper on the Environment". As for China, President Hu Jintao appealed to build the environment-friendly society in the Symposium on Population, Resources and Environment in March, 2006. Afterwards, in the Fifth Plenary Session of the 16th CPC Central Committee, the building of resource-saving and environment-friendly society was officially designated as a strategic task of the mid-term and long-term planning on national socioeconomic development. In this year, the topic of the sixth chapter of the 12th *Five-Year Plan* is entitled as "Green development, building resource-saving and environment-friendly society". Therefore, the construction of an environment-friendly production pattern was put in an unprecedented high place.

Environmental problems in China are gradually deteriorating accompanied with the development of economy and society. Since the reform and open up policy in China, with the rapid economic development, industrialization and urbanization consumed a large amount of resources and exerted great pressure on the environment. The contradiction between economic development and resource and environment is thereby becoming increasing prominent. Environmental problems emerged periodically during hundreds of years in developed countries have concentrated within the last 30 years in China. These environmental issues can be characterized as structural, compound and compressed type. Now, environmental degradation is becoming the focus of the public, social consensus and international society, and environmental security is becoming an important component and guarantee of national security. Therefore, in the next decade, as new environmental problems will continue to emerge, and emergencies and hidden danger may keep increasing, social conflicts caused by environmental issues will correspondingly erupt and increase. The task of pollution control and ecological protection will be increasingly difficult. Thereby, how to coordinate the relation among economic growth, social development and environmental protection is an unprecedented challenge that China has to face. To construct an environment-friendly production pattern and harmonize human activities with the nature, we should comply with the requirement of *Scientific Outlook on Development*, transform economic growth mode fundamentally, and step on a new industrialization way.

The environment-friendly production pattern emphasizes environmental impact, aspect and highlights that human should control the production intensity within the environmental capacity. This production pattern is a new way that aims to form a key feedback mechanism for the entry of production and consumption activities by altering the eco-environmental factors in quality and state. Specifically, it supervises the whole process of production and consumption by means of analyzing the mechanism of metabolic waste production and emission. In addition, multiple measures are adopted in this production pattern to reduce the amount of pollution, and realize anti-harm pollution, and ultimately to reduce the passive impact of the socioeconomic system to the eco-environmental system.

Under current condition, in order to realize the transformation of economic growth mode and construct an environment-friendly production pattern, we should eliminate the structural and institutional factors that lead to environmental pollution and degradation, and rebuilt a society that could encourage the entity to protect the environment and reduce pollution spontaneously. By using economic, administrative and technological approaches, we are target to realize the construction and development of environment-friendly production pattern as well as a harmony relationship between economic growth and the environment in China.

2.3.1 Economic Approach: Environmental Tax and Trading System of Environmental Property Rights

Economic approach which plays a positive incentive role is regarded as a optimal way to solve problems. In order to build an environment-friendly society and transform the current economic growth mode, reforming on the environmental discharge fee system and environmental tax system, and establishing a mature environmental property rights trading system are of most effective.

The environmental discharge fee system in China has a history of 32 years. It is one of the earliest proposed and generally implemented environmental management strategies. The environmental discharge fee system requires the polluter to be responsible for the environmental damage by using economic approaches. This has been most contributive in aspects such as environmental emission reduction and the collection of environmental management capital. In the past, we used to employ environmental discharge fee system to protect the environment and use economic means to adjust the relationship between economic growth and the environment. Complying with the principle of "Who polluted, who solves", the costs of environmental damage was required to be paid by the polluter. Through the past 3 decades, the gradually perfected environmental discharge fee system has made great contribution to the environmental protection of China. It was officially established in 1979 in "The Environmental Protection Law of the People's Republic of China (for Trial Implementation)". The publication of "Interim Measures on the Collection of Pollution Discharge Fee" in 1982 by the State Council indicates the environmental discharge fee system was also established. According to this document, each province, municipality and autonomous region made their local measures and rules. Thus, a top–down law system on imposing pollution fee was formed in China. In order to adapt to the new environmental situation and new requirements on pollution abatement, "Management Regulation of Pollutants Discharge Fee Collection and Usage" was published by the State Council, corresponding rules and measures were also put forwarded, including "Management Rules of Discharge Fee Collection" jointly announced by the former State Development Planning Commission, the Ministry of Finance, the former State Environmental Protection Administration and the former State Economic and Trade Commission, the "Measures for the Collection and Administration of Pollution Discharge Fee" issued by the Ministry of Finance, the former State Development Planning Commission, and the former State Environmental Protection Administration, and "Notice on Reduction or Delay in Payment of Pollution Discharge Fee" published by the Ministry of Finance, and the former State Environmental Protection Administration. Computer management system for the pollution discharge fee has also been installed. Currently the pollution charges in China are imposed on items such as waste water, air pollutant, solid waste and noise, these pollutants are mainly embedded in industries like coal-based power generation, chemical industry, iron production, cement manufactory, papermaking, etc. In terms of regional distribution, pollution emissions of Jiangsu, Shanxi, Shandong, and Hebei province have all surpassed 1 billion RMB and make the top of the whole country.

Although the environmental discharge fee system has made great contribution to the environmental protection in China. However, with the development of economy and the promotion of environmental management, more and more problems start to impair the implementation. In this context, the environmental discharge fee system is no longer adaptable to current developing situation, and urgently needs to be reformed. The pollution charges in China were implemented by means of compensation on owing quantity. On account that it is in the primary stage of environmental tax, endeavor on adjusting and updating the tax system is essential. In addition, judging from the reform trend of environmental tax, the strategy of "substitution of taxes for fees" is regarded as the new direction of reform, which not only solves problems such as inequality, low efficiency and difficulties in operation, but also beneficial for environmental protection and economic transformation.

The 12th *Five-Year Plan* has clearly proposed that we should establish and complete the polluter paying system, increase the rate of pollution discharge fee, innovate the levy of waste disposal fees, appropriately raise the garbage disposal standard and the financial subsidy level, and perfect the levy of waste water treatment fees. However, the intrinsic limitation of the charging system inevitably leads to substitution of current pollution tax system. Thus, in order to take advantages of environmental tax system in the construction of environment-friendly production pattern, we should follow the principles of tax legislation, tax neutrality, equality and efficiency, earmarking a fund for its specified purpose only and unified collection and allocation. It is also essential to construct a reasonable environmental tax system, to supplement relevant independent environmental tax types, and to make green improvements on previous tax types. Measures mentioned in *the* 12th *Five-Year Plan* include promoting the reform of environmental tax, conducting the levy of environmental protecting tax by choosing environmental items with big prevention difficulty and mature technical standards, and progressively enlarging the levy boundary.

In addition, the establishment of environmental property rights trading system should be accelerated in China, including property definition system, property allocation system, property trading system, and property protection system. Firstly, we should make a clear definition of the environmental property rights, i.e., make clear definition and institutional arrangement on various rights in the property system, including ownership, share and the distribution and allocation of all these rights. Secondly, it is essential to optimize property rights allocation system, which should be focused on the placement, proportional and combination of property rights possessed by different entities in a certain scope (including the distribution of central and local beneficial rights). Thirdly, we should promote the development of property rights trading system which is defined as the owners acquire the benefits through property rights operation based on certain procedures. Finally, we should accelerate the establishment of property rights protection system, i.e., constructing a legal protection system which constitutes of procedures on rights acquisition, principles of implementation, methods and the range of protection. Nowadays, the deficiencies of environmental property rights system in property definition, allocation, trading and protection have led to the fact that the social entities are not enthusiastic to reduce emission and protect the environment. As stated in *the* 12th *Five-Year Plan*, measures should be taken to promote the well-organized flow and equal, open and justice trading of environmental property rights, including bringing in the market mechanism, establishing compensative using and trading systems for the pollutants emission rights, developing pollutants emission property trading market, normalization the performance of pollutants emission rights trading price, and completing the laws and regulation system. These measures will provide strategic guidance and policy guarantee for the establishment of environmental property system.

2.3.2 Administrative Approach: Green Political System

A trend of solely pursuing the growth of GDP has long been dominated in China, especially that some local governments hold the unscientific viewpoints of "high input, high growth", and "growth brings political achievements, political achievements bring leaders". These misleading concepts and behaviors are the root of environmental problems. Specifically, these behaviors lead to the absolute ignorance of the resource and environmental costs behind the economic growth such as the irrational resource consumption and environmental degradation. Obviously, the perceptional deviations and political faults directly result in the unsustainability of economic growth. Therefore, in order to construct an environment-friendly production pattern and realize the transformation of economic growth pattern, we should change our ideas and pursue the green growth mode.

Politically, a green administrative system should be implemented to guarantee the healthy economic development. However, if there lacks a solid institutional framework to guarantee this system, it will probably descend into a bunk in such a competitive and fast-developing society. Therefore, a green, scientific and sustainable administrative system is urgently needed. This system should compose of comprehensive, coordinated, sustainable and *Scientific Outlook on Development*, comprehensive view of political achievements and comprehensive decision making mechanism. Green cultural value, green and civilized administrative behaviors and regulations are also embodied in the green administrative system. Thus, it is regarded as the basic institutional guarantee of the building of environment-friendly society. Only with these basic institutions established and worked, the government is able to consider drawing up and implementing the environment-friendly management policies such as the green national economic accounting system, green political achievements examination system, green trading policies and green financial and tax policies. Via these innovations, the intrinsic incentives of all government officials will be totally changed, the concept of green administration will be commonly accepted and implemented, and the green growth and environment-friendly production pattern can be achieved at the governmental level.

In addition, except for the politicians and the government, the promotion of the green administrative system should also rely on the public. Therefore, democratization and scientific decision making should be in operation. In order to ensure that the public could really exercise the constitutional right and to protect the environment and the rights of the public, the core of green administrative system should be providing legal and administrative guarantee and support for the public. As a result, we should promote the democratization of environmental decision making, including propelling the disclosure system of environmental information and the legal system of environmental public interest litigation, and implementing significant environmental affairs public hearing system, etc.

Only by providing a green, scientific and sustainable administrative concept and the corresponding administrative system, the transformation of economic growth, the construction of environment-friendly production pattern as well as the harmony development of human and the nature can be administratively propelled.

2.3.3 Technological Approach: Green Technology

Science and technology is the primary productive forces and a powerful driving force for the development of the human society. Green technology is regarded as a kind of environment-friendly technology, which roots in the harmonious and co-existence of human and natural ecosystem and plays important role in the construction of environment-friendly production pattern. The concept of green technology is derived from the Agenda 21 issued in the United Nations Conference on Environment and Development (UNCED) in 1992. This concept was further developed by the United Nations Educational Scientific and Cultural Organization (UNESCO) who indicated that "no technology, no sustainable development". Now, green technology is defined as a new technical-economic mode which could combine environmental protection, ecological restore, efficiency improvement and economic development. As it plays an important role in fostering sustainable development, increasing attention has been paid to the green technologies. Now, a powerful trend concerning green technology and sustainable development has already emerged globally.

Three tiers constitute the green technology: Tier 1 indicates the pollution prevention and control technologies, including traditional end of pipe pollution control, such as the cleaner treatment technologies of waste water, gas and rubbish. Tier 2 is environment-friendly technologies, indicating technologies that can improve resource consumption efficiency and reduce pollutant emissions, including cleaner production, comprehensive utility of resources, and renewable technologies. Ties 3 are ecological protection technologies, which can improve environmental quality and maintain the ecological balance and enhance the eco service, including the technologies of ecological restoration, soil and water conservation, biodiversity protection and ecological landscape construction. Currently, the technologies of direct environmental protection are called "Deep Green Technology", which consists of the technologies that specifically used in the treatment of environmental pollution, waste water treatment technologies and dust removal and desulphurization equipment. Another, technologies that have multiple purposes and indirectly aim at environmental protection are called "Light Green Technology", including technologies designed for improving the quality of products, lowering the generation rate of wastes and reducing energy consumption.

Among these green technologies, clean coal technology deserves a close attention. Since China is a developing country that is rich in coal and poor in oil and natural gas, the clean coal technology is of significant meaning to the country. The clean coal technology consists of a set of specific technologies such as machining, combustion, transformation and emission control that could reduce pollutants emissions and improve utility efficiency in the whole process of coal exploitation and utility. This includes clean coal combustion technologies and technologies that transform coal to clean fuels. Now, the clean coal technologies are among the most advanced and mainstream technologies in the world considering that they could relieve the negative effects of coal exploitation on the environment. Many developed countries such as Japan and the U.S. have put forward a set of policies to promote the development and application of the clean coal technologies and have obtained remarkable achievements. Nonetheless, the characteristics of high investment and long return period never arouse enough motivation for the enterprises to spontaneously adapt the clean coal technologies. Under this situation, China should speed up the development and adaptation of clean coal technologies and effectively employ the international technology platform. Meanwhile, by introducing advanced technologies abroad, the wide application of the clean coal technologies should be further promoted.

Green technologies are helpful for the construction of a resource-saving, high-efficiency and emission-reducing production system, and could control the human activities within the range that the environment is resilient. With the green technologies applied, the sustainable and harmonious development for both human society and the resource-environment can be actually realized. Many countries have paid attentions to environmental technologies regarding the research, development, application and export. The R&D of environmental technologies has been enhanced as well. In China, the construction of an environment-friendly production pattern could be fueled by the innovation of environment-friendly technology. Specifically, by independent innovation and gradient shift of international technologies, we should be focused on producing and adapting large-scale green technologies, therefore accelerating the transformation of economic growth mode.

Of course, the construction of an environment-friendly production pattern can not be realized in the short term. Instead, persistent efforts should be spent. We believe that, by incorporating economic, administrative and technology approach, we will be capable of achieving the strategic goal of constructing an environment-friendly production pattern, transforming economic growth mode in the future, and finally, ensuring the sustainable development of our society.

Chapter 3
Green Growth of the Primary Industry

Biliang Hu, Ning Cai and Ligang Xu

The primary industry has been staying a foundation in China's national economic and social development, as well as the important safeguard of constructing socialist harmonious society and building a new socialist countryside. However, China's primary industry has been facing serious problems in the twenty first Century, such as the reduction of arable lands, polluted farmlands and deterioration of safety of agricultural products, which directly threatening sustainable economic and social development in China. In order to get rid of these problems, the Central Government of China has formulated policies and has adopted measures, making great efforts to realize green development of the primary industry.

Agriculture is focused in this chapter. Mining will be analyzed in Chaps. 7 and 8. From the perspective of green development, this chapter tries to analyze and appraise the efforts and attempts China has taken to solve the difficulties faced by the primary industry in "green growth" approach, and to put forward our preliminary ideas and related policy advises to further promote sustainable development of the primary industry in the period of the 12th *Five-Year Plan* based on the green growth experience and lessons from the last five-year period of China's primary industry.

3.1 Major Features and Problems of Green Growth of the Primary Industry in the 11th Five-Year Plan Period

During the period of the 11th *Five-Year Plan*, with the great attention paying from the Central Government of China, comprehensive coordination between local governments and active participation of the masses, China made great efforts to actively explore ways of green growth of the primary industry under the huge pressures from resources, population and environment, making abundant achievements. Meanwhile, owing to China's social situations and issues of agriculture, farmer's income and rural defenses, China still has problems in its

X. Li and J. Pan (eds.), *China Green Development Index Report 2011*,
Current Chinese Economic Report Series, DOI: 10.1007/978-3-642-31597-8_3,

green growth of the primary industry, which should be solved in the following development.

3.1.1 Major Features of Green Growth of the Primary Industry

The development of the primary industry always enjoys the priority in the Chinese government missions. From 2004 to 2011, the CPC Central Committee and the State Council issued No.1 Document focusing on agriculture, rural areas and farmers for successive 8 years, stressing on the growth of agriculture, development of rural areas and increasing incomes of farmers. With the background, China has achieved great progresses in ecosystem construction and has made prominent achievements in sci-tech research and innovation by gradually improving its resource utilization efficiency in the primary industry and reducing influences of environmental pollution year by year, supporting its green development of the primary industry in a sound manner.

3.1.1.1 Resource Utilization Efficiency has been Improved

Improving resource utilization efficiency is an important task of constructing a resource-saving society, as well as one of the goals of green growth of the primary industry. During the 11th *Five-Year Plan* period, with the constantly development of the agricultural science and technology, the increase of the grain production per unit area greatly improves utilization efficiency of the land resource in the primary industry. On one hand, the promotion and popularization of agricultural science and technology can help farmers increase agricultural production with scientific technologies; on the other hand, the improvement of crop varieties also has greatly raised the quality of agricultural products. Additionally, transformation of agricultural cultivation methods has greatly improved utilization efficiency of the land resource. Table 3.1 indicates the yield of major agricultural products per unit area in China.

According to Table 3.1, the yield per unit area of grain in 2005 was 5,225 kg per hectare, and was increased by 222 kg per hectare to reach 5,447 kg per hectare in 2009; the yield per unit area of cotton was increased from 1,129 kg per hectare in 2005 to 1,288 kg per hectare in 2008, up 14.08 %; in 2009, output of peanut, rapeseed and tobacco per hectare also increased by 285, 84 and 268 kg respectively, up 9.27, 4.68 and 13.70 % respectively compared with 2005, the last year of the 10th *Five*-Year Plan period.

Besides the improvement of utilization efficiency of land resource, the water utilization efficiency was greatly improved in the 11th *Five-Year Plan* period. In 2005, the area of water-saving irrigation in China was 21.338 million hectares; in 2009, the figure reached to 25.755 million hectares after several years' basic water conservancy construction of farmlands and spread of water-saving agriculture. During the 4 years, a total of 4.417 million hectares of farmlands of water-saving

Table 3.1 Yield of major agricultural products per unit area in China (Unit: kg/hectare)

Years	Grain	Cotton	Peanut	Rapeseed	Tobacco
2005	5,225	1,129	3,076	1,793	1,956
2006	5,310	1,295	3,254	1,833	2,072
2007	5,320	1,286	3,302	1,874	2,044
2008	5,548	1,302	3,365	1,835	2,133
2009	5,447	1,288	3,361	1,877	2,224

Source National Bureau of Statistics: *China Statistical Yearbook 2010*, Beijing, China Statistics Press, 2010

irrigation were newly added, increased 20.70 %.[1] Meanwhile, water-saving irritation rate also went up from 38.78 % in 2005 to 43.46 in 2009, up with 4.68 %.

3.1.1.2 Environmental Pollution has been Reduced Year by Year

Environmental pollution of the primary industry mainly refers to the agricultural chemical pollution caused by pesticides, fertilizers and plastic film in the process of agricultural production and the non-point source pollution caused by livestock and poultry industry. For the former one, the fertilizers and pesticides are overused so that they cannot be absorbed or degraded by plants and the soil, but even likely to reduce the soil fertility as the time passing by, leading to water and soil erosions. For the latter one, since the large amount of excrement and urine made by livestock and poultry cannot be timely dealt with, which can eutrophicate rivers and lakes, lead to organic pollution of the water, influence farmers' living environment negatively and cause various diseases. During the 11th *Five-Year Plan* period, China strengthened its efforts on rectifying and examining environmental pollution of the primary industry, achieving prominent results.

In February 2006, *Suggestions of the CPC Central Committee and the State Council on the Promotion of the Construction of New Socialist Countryside* (namely, No.1 Document of 2006) was put forward to protect and to improve rural environment and raise farmers' living quality and health level. In January 2007, Ministry of Environmental Protection, National Commission of Development and Reform and Ministry of Agriculture jointly issued *Suggestions on Strengthening the Work of Environmental Protection in Rural Areas*, clarifying the guiding philosophy, basic principles and main goals of environmental protection in rural areas in China during the 11th *Five-Year Plan* period. In December 2009, Ministry of Environmental Protection issued the Guiding Suggestions on Environmental Monitoring in Rural Areas in China to strengthen monitoring of the environment in rural areas.

Meanwhile, Ministry of Agriculture and Ministry of Environmental Protection issued several regulations such as *Regulations for the Implementations on*

[1] Ministry of Water Resources: China Water Statistical Yearbook 2010, Beijing, China WaterPower Press, 2010.

Agricultural Chemicals, Management Regulations of Restricting Pesticide Utilization, Suggestions of Ministry of Agriculture on Further Strengthening the Pesticide Administration, and *Technology Guidance on the Safety of Environment of Pesticide Utilization* from 2007 to 2009, in purpose of standardizing the utilization of pesticides and fertilizers and reducing the pollution of agricultural chemical products to the primary industry, trying to reduce the harm of pesticide to the environment in every aspect. In order to standardize China's livestock and poultry husbandry and control its non-point source pollution, Ministry of Agriculture issued *Notice of Ministry of Agriculture on Strengthening Management of Livestock and Poultry Husbandry and Suggestions of Ministry of Agriculture on Accelerating the Promotion of Standardized* and *Large-Scale Cultivation of Livestock and Poultry* to strengthen specialization management of livestock and poultry husbandry.

Thanks to a series of policies and measures, the influence of environmental pollution of China's primary industry was reduced gradually in the 11th *Five-Year Plan* period. The consumption of fertilizer for every ¥ 100 million of value added in the primary industry was 2,125.88 tons in 2005, and the figure was reduced to 1,534.04 tons in 2009, down with 27.83 % in 4 years. The consumption of pesticide for every 100 million of value added in the primary industry was reduced from 65.12 tons in 2005 to 48.51 tons in 2009, down with 16.61 tons; in addition to, the consumption of plastic film for every 100 million of value added in the primary industry was 42.79 tons in 2005, and was 23.02 tons in 2009, cutting by 25.18 %. The non-point resource pollution caused by the livestock and poultry husbandry was effectively controlled and its influence on the rural environment was reduced more or less.

3.1.1.3 Great Achievements has been made in Ecosystem Construction

During the 11th *Five-Year Plan* period, China achieved great progress in its eco-system construction of primary industry. After years of efforts, its eco-system construction shows the characters of focusing on forestry eco-system, stressing on grassland, wetland and fishery eco-systems and coordinating and promoting bio-diversity protection in an all-round way.

First, forestry ecosystem has been further strengthened. It is showed based on the seventh census of forest resources of China that China has 305.9041 million hectares of forestland, 195.45 million hectares of forests with forest coverage of 20.36 percent, 14.913 billion cubic meters of overall living wood growing stock, and 13.721 billion cubic meters of forest growing stock. Till 2009, the total forestation area nationwide reached to 6,262,330 hectares, up with 17 % than in 2008. Of this amount, the forestation area of key forestry projects hit 4,596,244 hectares, accounting for 73.40 % of the total forestation area.

Second, prominent achievements have been made in protection of grassland, wetland and fishery ecosystem. In 2008, the gross production of fresh grassland on natural grassland nationwide was 947.155 million tons, equivalent to 296.268 million tons of hay, with stock-carrying capacity of 231.78 million sheep units; the livestock overloading and overgrazing rate of key grasslands was 32 %, reducing 1 % year on year. China has established more than 550 wetland natural reserves, 36 important international wetlands, and 38 national wetland parks; a total of 49 % of wetlands are under effective protection. The national fishery environment monitoring system also supervises 105 important fishery water areas such as the Bohai Sea, East China Sea, South China Sea, the Yangtze River basin and the Yellow River basin, involving 31.98 million hectares of water areas.[2]

Third, prominent effects have been achieved in biodiversity protection. In 2008, China had 2,538 natural reserves, increasing with 189 annually compared with the 10th *Five*-Year Plan period. These reserves covered an area of 148.943 million hectares, accounting for 15.1 % of the jurisdiction area, up with 0.1 % point compared with 2005; more than 90 % of the land eco-system and biodiversity have been effectively protected.[3]

Column 3.1 Ya'an of Sichuan: Constructing the Ecological Barrier in the Upper Reach of the Yangtze River and Turning the Green Hill into the "Green Bank" for Forest Farmers

During the past 10 years, Ya'an witnessed rapid development in its forestry. It put equal stress on both ecological construction and industrial development, laying down solid basis for its road of developing the city through biological construction. From the program of returning farmland to forest to natural forest protection projects, and from wildlife protection to forest development, Ya'an's forest coverage ranked No.1 in Sichuan Province. More than 70 % of farmers of the city have been directly benefiting from the policy of returning farmland to forest and the former "tree cutters" have become "tree growers" and green hills have become the "green banks" of forest farmers. Wild animals represented by giant panda have been under the best protection. Ya'an is transforming from the traditional forestry to modern one.

In 2000, Ya'an launched the program of returning farmland to forest in an all-round way, unveiling its ecological construction project. During this project, Ya'an accelerated its ecological construction by mobilizing the vast majorities of people. Statistics showed that by the end of 2009, the city had invested 2.193 billion yuan of the national and municipal level funds, completing the task of returning 62,846.67 ha of farmland to forest and newly increasing about 186,666.68 ha of forestlands.

In the mean time, Ya'an officially launched natural forest protection project in 2000 and fully stopped commercialized deforestation of natural forests. During the 10 years of the project, Ya'an became the pioneer in formulating the "flow charts of the management of natural forest protection project" according to the

[2] Ministry of Agriculture: China Agricultural Yearbook 2010, Beijing, China Agriculture Press, 2010.

[3] National Bureau of Statistics: *China Statistical Yearbook 2006–2010*, Beijing, China Statistics Press, 2006–2010.

requirements of "strictly administering forests, being prudent in using funds and stressing on quality," conducted "process management" on the project, innovated the system of forest management and protection, completed various tasks of project construction, took the lead in realizing the goal of constructing an ecological barrier in the upper reaches of the Yangtze River and won high affirmation from the national inspection team. From 1998 to 2010, the state invested 628.516 million yuan as project fund; the forestland increased 94,666.67 ha and the overall forest growing stock increased by 19.43 million cubic meters.

During the past decade, ecological constructions such as the project of returning farmland to forest and natural forest protection project, have greatly improved Ya'an's ecological environment. The achievements are obvious such as that the city's forest coverage has increased from 43 to 61 %; its water and soil erosion was reduced by more than 32 %; the air quality has reached the state No.1 level, water quality reached the state secondary standard; and the power generation in low water season also increased by more than 10 %, etc. On the basis of complements of various construction tasks required by the state, Ya'an has achieved sound ecological, economic and social interests and primarily constructed the ecological barrier on the upper reach of the Yangtze River.

Ecological construction in Ya'an such as the program of returning farmland to forest and natural forest protection project was also highly praised by the Party and state leaders. In June 2001, while inspecting Ya'an, Zhu Rongji pointed out, "Ecological environment is your resource, your source of money and your lifeline." He also affirmed the projects such as returning farmland to forest, "Ya'an has done an excellent and solid work in the returning farmland to forest and grassland." On May 19, 2002, while inspecting Ya'an, Jiang Zemin praised, "Ya'an has an excellent biological environment and we must strengthen our efforts to propagate its experience."

With deepening implementation of the projects such as returning farmland to forest and natural forest protection, the farmers' tradition of extensive cultivation has been profoundly changed. A large number of farmers were liberated from the sole farming work and turned to industries such as breeding, process and labor export, promoting the huge changes of rural economic and social structures, and providing more channels for farmers to become rich and promoting constant income increase of farmers. At present, the city has many bamboo villages where farmers mainly rely on bamboo industry in income.

Ecological construction of Ya'an has been affirmed by forest farmers, leading enterprises, people from all walks of life and media. Standing in the forest park of Ya'an, visitors frankly note that their purposes of coming here are to enjoy the green hills and clean waters. Ecological construction has made outstanding contribution in the city's competition for the honors such as "China's Excellent Tourism City," "China's Top 10 Attractive Cities," and "China's Pioneering City of Low Carbon."

During the past decade, Ya'an gradually has been transformed its forest industry from traditional forestry to the modern one. The vibrant forest industry in Ya'an has increased farmers' income and enabled leading enterprises to grow stronger. Today, Ya'an is working hard to change from a big city of forest resources to a strong ecological and economic city.

Source: This article was provided by researchers of the Publicity Department Ya'an City, Sichuan Province.

3.1.1.4 Prominent Achievements has been made in Sci-tech Research

In order to energetically promoted green development of the primary industry and encouraged construction of technological system for modern agriculture, Ministry of Agriculture and Ministry of Science and Technology accelerated the pace of sci-tech research and development and innovation of the primary industry during the 11th *Five-Year Plan* period, achieving great progresses and fruits in areas such as high efficient in agriculture and forest production, construction of ecological environment and researches and promotions of advanced farming machineries, equipment and materials.

In terms of high efficiency in agriculture and forest production, efforts were made to explore innovation and utilization of crop seeds and to select a batch of high-quality and high-efficiency new crop, forest and fruit varieties with multi-resistance so as to greatly improve China's levels of crop seed production; the grain high-yield project was continued to establish a batch of high-yield technologies with regional characteristics; high- tech were combined with traditional technologies to cultivate several new livestock and poultry varieties with independent intellectual property rights; efforts were made to select several drought-resistant and salt-tolerance tree varieties and establish the technological systems such as rapid restoration of sand vegetation, recovery of grassland and sand prevention technologies for lakes and reservoirs.

In terms of ecological environment construction, efforts were made to conduct researches in areas such as improving farmland capacity and recovering deteriorated farmland according to different farmland types in different regions to improve the sustainability of farmland utilization. China energetically promoted advanced agricultural tillage technologies such as no-tillage and less-tillage technologies so as to pursue the social, ecological and economic interests of protective tillage. It innovated and designed the friendly-environmental style farming system, called for water- and fertilizer-saving growing modes, strengthened exploitation and utilization of clean agricultural energies, reduced pollution caused by traditional energies, researched and promoted pollution rectification technologies in rural areas to cope with rural environment problems.

In terms of research and promotion of advanced farming machineries and equipment, China developed and manufactured a batch of high-tech farming machineries and equipment, improving the labor production forces of the primary industry, and energetically promoted mechanized farming mode of crops to enhance the mechanized farming level, researched and manufactured several mechanisms to efficiently apply pesticides and fertilizers so as to reduce pollution of agricultural chemical products, solved the technical difficulties in manufacturing efficient pesticides and fertilizers and improved the utilization efficiency of agricultural resources in China.

After years of investment and construction, China's primary industry applied for 32,605 patents during the 11th *Five-Year Plan* period, accounting for 2.11 % of the total; a total of 16,478 patent applications were granted, accounting for 2.01 % of the total. 20,539 of them has yielded great sci-tech fruits, accounting for 14.42 % of China's total number of great sci-tech fruits; meanwhile, a total of 129 items of National Sci-Tech Progress Award and 17 items of National Invention Award were granted to primary industry. Meanwhile, from 2006 to 2009, China has approved 2,010 projects of funds for transferring agricultural sci-tech fruits, encouraging the state, local governments and enterprises to invest more than 6 billion yuan of fund for the transformation, which greatly promoted the transformation of agricultural sci-tech fruits.[4]

Column 3.2 Yangling of Shaanxi: Agricultural Sci-Tech Fruits Accomplishing China's "Green Silicon Valley"

Covering an area of 13,500 ha in the central part of Guanzhong Plain of Shaanxi Province, Yangling is the demonstration area of China's agricultural high-tech industry. According to the Book of History—Record of Zhou Dynasty, agriculturalist Hou Ji taught people to grow crops and trees in places around Yangling more than 4000 years ago, starting China's agricultural civilization. Since 1934, Yangling has attracted more than 10 agricultural science education institutions such as universities and colleges and nearly 5,000 professionals of agricultural education of more than 70 disciplines such as agriculture, forestry and water resources. Honored as the "City of Agriculture Science," it is considered as the China's "Green Silicon Valley" of agricultural high-tech industry.

"Green Silicon Valley" mainly refers to the aspects of research and development of agricultural science and technology, transformation of agricultural sci-tech fruits and demonstration and promotion of agricultural technology industry of Yangling. Its high-tech industry not only represents China's highest level and orientation of agricultural development in arid and semi-arid areas, but also becomes one of the important channels for Chinese agriculture to implement the green growth.

In terms of the achievements of agricultural sci-tech research, during the 11th *Five-Year Plan* period, Yangling applied for 1,500 patents and agricultural sci-tech fruits, among which 456 patents of invention were granted and 107 new plant and animal varieties were approved. Many of the fruits were in the leading position both natively and abroad. In 2008, Northwest A&F University had seven world-leading sci-tech fruits, two domestic leading fruits and three fruits that were

[4] National Bureau of Statistics and Ministry of Science and Technology: China Statistical Yearbook on Science and Technology 2006–2010, Beijing, China Statistics Press, 2006–2010.

awarded the second prize of National Sci-Tech Progress Award. In 2010, Yangling had 14 agricultural sci-tech fruits approved by the state, 28 approved new crop varieties and applied for 457 patents, including 259 patents of invention. Yangling's strong capacity in agricultural research and development has provided solid sci-tech supports to China's green agricultural growth.

Based on the strong qualification of scientific agricultural research, Yangling actively promotes transformation of agricultural technological fruits. In 2010, about 40 % of its agricultural technologies were transformed. During the 12th *Five-Year Plan* period, Yangling demonstration area will continue the investment to the transformation of agricultural sci-tech fruits and construction of transformation channels and will actively innovate transformation approaches. It is expected that the city will increase the transformation rate with 60 % in the next 5 years.

The mode of Yangling demonstration area of "scientific research, fruit transformation, industrial demonstration and agricultural technology promotion" promotes the industrialization of agricultural technologies and its demonstration and promotion, and enhances the proportion of agricultural science and technology in the agricultural production. In the 11th *Five-Year Plan* period, Yangling established a modern agricultural demonstration park of 10,000 ha, demonstrated more than 4,000 domestic and international new varieties in 17 categories, and formed a large number of agricultural high-tech industries and industrial chains such as good crop varieties, vegetables and economic forests and fruit trees. Yangling demonstration area has established 163 demonstration and promotion bases and 37 expert yards in 16 provinces and autonomous regions with total demonstration and promotion areas of 1.69 million hectares and annual profits of 6.95 billion yuan, effectively realizing the goal of increasing farmers' incomes and agricultural production.

In addition to, Yangling demonstration area also actively strengthens research in areas such organic fertilizer, industrial system of agricultural ecological cycle, comprehensive disposal of castoff of agriculture and animal husbandry, ecological film and new plastic film to reduce chemical pollution in agricultural production and ensure food safety of the primary industry, making greater contributions to China's agricultural green growth.

Yangling's development mode based on agricultural sci-tech innovation and demonstration and promotion of agricultural high-tech industry is an example of agricultural green development. The huge radiation effects of the demonstration area improves the agricultural production efficiency in China's arid and semi-arid regions, increases the contribution of science and technology to agricultural production and is conducive to realizing China's agricultural green development.
Sources:

1. Public information website of Yangling:
   http://www.ylagri.gov.cn/zjyl/introyl.htm?wbviewid=183.
2. Development Outline of the 12th *Five*-Year Plan of Yangling Agricultural High-Tech Industrial Demonstration Area:
   http://www.ylagri.gov.cn/_info/zwgk/zwgkcont_27440.htm.

3. Website of Northwest A&F University: http://www.nwsuaf.edu.cn/nwsuaf/export/sites/nwsuaf/column/kxyj/index.html.
4. Statistical Communiqué of Yangling Demonstration Area on National Economic and Social Development: http://www.sei.gov.cn/ShowArticle.asp?ArticleID=212062.

3.1.2 Problems in Green Growth of China's Primary Industry

The Chinese Government attaches great importance to the green development of the primary industry and has achieved eye-catching progress in theoretical researches and practice exploration for many years. But we still need to realize that there are many missins in this area to be improved so as to further promote green growth of China's primary industry.

3.1.2.1 Pollution is Still Serious

Though China formulated a series of policies and measures to reduce pollution from the primary industry to the environment in the 11th *Five-Year Plan* period, China still faces severer situation of pollution from the primary industry with greater negative influences on the environment compared with countries such as the Unites States, the United Kingdoms and Japan. According to the statistics from the World Development Indicator of the World Bank, in 2008, fertilizer consumption per hectare of farmland in China reached to 46.80 kg, 3.98 times that of the average consumption worldwide during that period of time, which was only 11.77 kg. During the same period of time, the consumption of fertilizer per hectare of farmland in the United States was 10.31 kg, which was only 22.03 % of China's level. Meanwhile, the figures of Canada and Argentina, the world major agricultural countries, in 2008 were only 5.69 kg and 3.88 kg, far below the world average level, and are only 12.16 and 8.29 % of China's level respectively. China should strengthen prevention of pollution from the primary industry so as to narrow down the gap with the world developed agricultural countries and further improve the environmental friendly level of the primary industry.

3.1.2.2 Productivity Level is Still Weak

China's productivity level of the primary industry is not high mainly in the two aspects include low agricultural production rate and low agricultural mechanization. Calculated on the U.S. dollars in 2000, China's agricultural value added per labor in the 11th *Five-Year Plan* period was $459 while the world average level was $959. China's agricultural production rate was less than a half of the world average level. From 2005 to 2009, the agricultural value added per labor in the United States, Canada, the United Kingdoms, Japan and the Republic of Korea were $45,015, $46,138, $28,065, $39,368 and $14,501 respectively, much higher than China's level in the same period of time. In terms of agricultural mechanization, China had 124.3 units of tractors on every 100 km^2 of farmland in the 11th *Five-Year Plan* period, lower than the world average level of 198.7 units at the same period of time. In the United States, France and the Republic of Korea, the figures were 258.9, 624.7 and 1458.2 sets respectively, much higher than China's average level of the same period.[5]

3.1.2.3 Food Quality is Still Need to Improve

In recent years, a series of food safety accidents such as melamine milk, lean meat powder, drainage oil and colored steamed buns took place in China, which seriously impacted the image of China's primary industry and harmed China's green growth of the primary industry. In September 2008, some dairy enterprises in China were discovered to add chemical materials such as melamine, which could harm people's health, into the milk, having more than 10,000 infants being treated in hospitals and four deaths. The whole dairy industry was extremely influenced and the product quality was suspected by people throughout the country. In March 2011, China's largest meat enterprise was found to add lean meat powder, which was clearly banned by the government, into its meat products. Eating meat with the powder may cause vertigo, malaise, Palpitation and arrhythmia, greatly harming people's health. These events went against the enterprises' concepts of "giving top priority to consumers' health and safety," and led to consumers' panic on meat products. Consumers also lost their confidence on food safety of the whole industry.

Column 3.3 Xinye of Henan: Green Agricultural Products in China's Bread Basket

Located in the center of Nanxiang Basin in the southwest part of Henan Province, Xinye County is honored "plain throughout the county." The county has a total area of 106,200 hectares with a population of 780,000. Xinye is a typical agricultural county on the plain, known as China's base of high-quality cotton production, demonstration county of production base of pollution-free vegetables, demonstration county of production base of green livestock products and excellent county of optimizing economic development environment.

[5] The World Bank, the World Development Indicator 2010, Beijing, China Financial and Economic Publishing House, 2010.

In recent years, Xinye County focuses on the basic agricultural tactics such as constructing the base of green agricultural products, establishing green ecological brand and expanding the green agricultural production chain, energetically promotes standardized agriculture production through the development mode of enterprises cooperating with farmers, and develops green agriculture and ecological agriculture, tries to develop itself into a powerful and green agricultural county featuring excellent eco-system and product safety.

On one hand, Xinye County actively constructs the industrial park of vegetables and strengthens supervision on the quality security of agricultural product, tries to establish the county into a "green city." In 2010, Xinye had permanent vegetable field of 29,333.33 ha, including 13,733.33 ha of pollution-free vegetable. The county has established a sci-tech park for vegetable industry covering an area of 5,333.33 ha. 18 vegetable varieties have got pollution-free certificates from Ministry of Agriculture. The green development strategy has greatly increased the proportion of pollution-free vegetables in Xinye, improved the quality of agricultural products from the fountainhead and promoted green development of agriculture. So far, vegetable growing and animal husbandry have become the pillar industries of Xinye, making the county the "National Demonstration County of Pollution-Free Vegetable Production" and "Henan Provincial Demonstration Base of Green Livestock Product." It is now direct supplier of meat and vegetable for Beijing Municipal Government.

On the other hand, based on the green and pollution-free vegetable and beef cattle industries, Xinye County actively establishes the green, ecologically friendly and safe agricultural product brands to increase their value added. In 2008, the county intensified its efforts in certification work of pollution-free and green agricultural (livestock) products and production area recognition work with stresses on the brands of Xinye vegetable and Pinan beef cattle. On the basis of strictly implementing industrial standards and the national norms, Xinye also launched "industrial brand strategy," and registered a large number of product trademarks in vegetables, peanuts and beefs. Xinye's construction of green agricultural brand greatly improved the market values of its agricultural products.

Additionally, while constructing its green industrial base and green agricultural brands, Xinye County actively develops green agriculture industrial chain. The county has fostered a batch of leading enterprises of deep processing in areas such as vegetable and livestock products and formed the industrial development structure of "leading enterprises, bases and farmers." In 2009, food processing enterprises above designated size of the county had sales revenue of more than 2 billion yuan; of these enterprises, Master Kong Jiayuan Food Company processed more than 2,666.67 ha vegetables with production value of more than 200 million yuan.

With green agricultural product as the starting point, Xinye County energetically develops green agriculture industry and actively launches the agricultural development mode of fostering green agricultural product brands, winning sound

reputation. The healthy, safe and ecological friendly quality of green agricultural products greatly improved the quality of the county's agricultural products; the green and environmental friendly agricultural production mode also increases farmers' awareness of green agriculture and promotes green growth of agriculture.
Sources:

1. Website of Xinye People's government: http://www.xinye.gov.cn/.
2. Xinye of Henan: Great Achievements Made in Green Agriculture: http://nc.mofcom.gov.cn/news/13290810.html.
3. Government Work Report of Xinye County (2006–2010): http://www.xinye.gov.cn.
4. Baidu Baike: http://baike.baidu.com/view/144364.htm.

3.1.2.4 Government Supervision is Still Weak

Pollution problem and food safety of the primary industry are related to the relatively weak government supervision on the primary industry. The influences of the primary industry on environment usually lag behind the normal development and are under guise and we need a long time to discover their harmfulness. Hence, in the current context, the government and the public fail to pay enough attention to related problems without sufficient supervision. At the same time, compared with the secondary industry and tertiary industry, the government supervision on the primary industry is insufficient since they scatter in rural areas. Only by realizing the seriousness of related problems, strengthening government legislation and supervision and further formulating related guiding and encouraging measures can we pragmatically solve various problems faced by the primary industry and realize its green growth.

3.2 Regional Comparison on Green Growth of the Primary Industry in the 11th Five-Year Plan Period

If we say the Central Government is the designer and promoter of green development of the primary industry, various local governments are responsible for implementing the green growth. During the 11th *Five-Year Plan* period, various provinces and municipalities energetically support green development of the primary industry to realize green growth of the industry, achieving relatively greater progress in areas such as level of labor productivity, efficiency of land production and utilization of water resources in agriculture. Next, we will describe the information of green development of the primary industry in east China, central China, west China and northeast China, compare their green growths and analyze their respective development characters and differences.

Table 3.2 Labor productivity of the primary sector in various regions of China in the 11th Five-Year Plan period (Unit: yuan/person)

	Labor productivity of the primary sector			
	2006	2007	2008	2009
Nationwide	8,118	9,582	11,636	12,272
East China	11,455	13,223	15,774	16,808
Central China	6,874	8,118	10,251	10,895
West China	5,996	7,276	8,790	8,981
Northeast China	11,754	14,090	16,644	17,853

Notes
1. Labor productivity of the primary sector = value added in the primary industry/number of employees of the primary industry
2. Owing to different statistic approaches, we use the number of employees in the primary industry in 2005 to replace that of 2006

Source Calculated according to *China Statistical Yearbook 2006–2010* published by National Bureau of Statistics

3.2.1 Labor Productivity of the Primary Sector

To achieve green growth of the primary industry, it is required not only to develop "green," but also to promote "growth". Thus, the level of labor productivity (the value added created by employees per unit) is an important indicator measuring green growth of the primary industry. Increasing labor productivity is the orientation of China's green development of the primary industry. The labor productivities of the primary sector in various regions in China from 2006 to 2009 are displayed in Table 3.2.

During the 11th *Five-Year Plan* period, China's labor productivity of primary sector increased rapidly. In 2006, the value added was created by an employee in the primary industry was 8,118 yuan. In 2009, the figure reached to 12,272 yuan, an increase of 4,154 yuan within 4 years, with a growth rate exceeding 50 % and an annual growth of 14.77 %.

In terms of specific regions, the labor productivity in the east of China, central China, the west of China and the northeast of China increases year by year with different characteristics. The northeast enjoys the highest labor productivity level and the value added created by every employee in the primary industry is more than 40 % higher than the country's average level. From 2006 to 2009, the labor productivity increased by 6,099 Yuan per person, up with 51 %, with an annual growth of 14.95 %. The East comes next only to the northeast in terms of labor productivity. Its level is still 35 % more than the country's average level; in the first 4 years of the 11th *Five-Year Plan* period, its labor productivity increased by 46.73 % with an annual growth exceeding 13 % points. The labor productivity in central China is comparatively low, accounting for only 85, 64 and 61 % of the average levels of the whole country, the east and the northeast respectively. However, in the 4 years, the labor productivity increased by 4,021 yuan per person with growth rate and annual growth rate of 58.50 and 16.59 % respectively, the highest among the four regions.

Table 3.3 Land production efficiencies in various regions of China (Unit: 10,000 yuan/1,000 hectares)

	Land production efficiency			
	2006	2007	2008	2009
Nationwide	1,371	1,607	1,795	1,930
East China	2,076	2,421	2,661	2,899
Central China	1,211	1,437	1,629	1,780
West China	1,085	1,315	1,482	1,577
Northeast China	1,092	1,193	1,340	1,372

Note Land production efficiency = gross agricultural output/crop growing areas
Source Calculated according to *China Statistical Yearbook 2006–2009* published by National Bureau of Statistics

The west has the lowest labor productivity, accounting for only 73 % of the country's average level. From 2006 to 2009, the labor productivity in the west increased by 49.78 % with an annual growth of 14.42 %.

The characteristics of labor productivities of primary sector in the four regions are in accordance with their respective development conditions. The northeast is China's traditional agricultural production area and the country's bread basket. Its unique natural conditions and high mechanization level make the region's labor productivity of the primary industry the highest nationwide. This is the largest characteristic of the region. Owing to the relatively backward natural conditions, poor economic and sci-tech development and low agricultural mechanization level, the labor productivity levels in central China and the west are comparatively low.

3.2.2 Land Production Efficiency

Land production efficiency reflects the production capacity of unit land; it is an important indicator measuring the situations of land utilization and crop production of a region. The higher the land production efficiency is, the higher the unit land yield and land utilization efficiency are, and the higher the green growth of the primary industry is. Low land production efficiency indicates extensive growth of the primary industry, poor technology and management in agricultural operation and insufficiency of "green" in the growth of the primary industry. The land production efficiencies of China and that of various regions in the 11th *Five-Year Plan* period are shown in Table 3.3.

National-widely, China's land production efficiency kept increasing in the 11th *Five-Year Plan* period from 13.71 million yuan/1,000 ha in 2006 to 19.30 million yuan/1,000 ha. Excluding price factor, China's land production efficiency from 2006 to 2009 grew at a rate of 12.07 % annually, equivalent to adding an extra of 1,863 yuan of production value to every hectare of land.

From the four regions, it is shown that the land production efficiency also increases year by year, which conforms to the growing trend of the national average level. However, the prominent gaps exist among various regions in terms of land production efficiency and growth rate in the 11th *Five-Year Plan* period. In 2006 and 2009, the east had the land production efficiencies of 20.76 million yuan/ 1,000 hectares and 18.99 million yuan/1,000 hectares respectively, higher than the average levels of the whole country and other three regions in the 11th *Five-Year Plan* period. The land production efficiencies in central China and the west are lower than those of the east and the country's average level. However, during the 11th *Five-Year Plan* period, the growth rate of land production efficiency in central China and the west was 1.5 % points higher than that of the east and China's average level, and the gaps among the east, central China and the west is narrowing down year by year. The land production efficiency of the northeast during the 11th *Five-Year Plan* period was lower than that of the country's average level, the east and central China, but higher than that of the west. In 2009, northeast China was the lowest in terms of land production efficiency among the four regions, only 13.72 million yuan/1,000 ha. The low annual growth rate of the region from 2006 to 2009 led to the increasing wider gap between the northeast and the rest three regions in terms of land production efficiency.

It is natural to see the situation of the east taking the lead, followed by central China and the west and the northeast in the bottom in terms of land production efficiency. Thanks to the sound natural conditions, relatively complete agricultural infrastructures and high economic development level, the east takes the lead in land production efficiency. Owing to the active promotion of the programs of rise of central China and great development of the west, China increased its efforts in promoting agricultural technologies, sound seed varieties and investment to other farming materials to the central China and the west, which greatly improve land production efficiencies of the two regions. Though the northeast boasts of fertile lands and is the base of China's commercial grain production, it has more land but less people, and high levels of extensive operation and mechanization. In addition to its special climate, which leads to a longer growth cycle, land production value in this area is lower than that of the other three regions where crops are intensively and meticulously cultivated with a shorter growth cycle.

3.2.3 Agricultural Water Resource Utilization

Water is the vital source of life, and green growth of agriculture also requires effective utilization of water resources. Saving water is the key of agricultural green growth. Effective irrigation of farmland can influence not only the fertility and the eco-system of agricultural production of lands, but also the output of lands to large extent. The proportion of areas with effective irrigation in the total area of farmland can be used to investigate utilization of China's agricultural water resources. For more information, please refer to Table 3.4.

The Table 3.4 shows that in the 11th *Five-Year Plan* period, the proportion of effective irrigation areas in the total areas of farmlands nationwide went up year by

Table 3.4 Proportion of land with effective irrigation in various regions of China in the 11th Five-Year Plan period (Unit: %)

	Proportion of land area with effective irrigation			
	2006	2007	2008	2009
Nationwide	42.87	46.43	48.04	48.69
East China	62.82	67.73	70.51	70.48
Central China	52.43	56.05	57.18	57.56
West China	25.26	36.37	37.97	38.75
Northeast China	27.00	28.34	29.23	30.77

Note Proportion of land area with effective irrigation = areas with effective irrigation/the total areas of farmland
Sources Calculated according to *China Statistical Yearbook 2006–2009* published by National Bureau of Statistics and China Statistical Yearbook in Rural Areas 2006–2010 published by Department of Rural Economic and Social Survey of National Bureau of Statistics

year from 42.87 % in 2006 to 48.69 % in 2009, up with 5.82 % points with annual growth of 1.94 % points. However, compared with the previous 3 years, the growth of the proportion of effective irrigation areas nationwide slowed down in 2009.

In perspective of various regions, during the 11th *Five-Year Plan* period, there were great differences in terms of the proportion of effective irrigation areas in the four major regions. The proportion of effective irrigation areas in the east went up year by year and was much higher than that of the rest three regions and the nation's average level. In 2006, the proportion of effective irrigation areas in east China was 62.82 %, 1.2, 2.49 and 2.33 times higher than that of central China, the west and the northeast respectively. In 2009, the proportion in the east increased to 70.48 %, with an annual growth of 2.55 % points within the four years, 0.61 % points higher than the country's average level. The proportions in central China, the west and the northeast went up to 57.56, 38.75 and 30.77 % respectively in 2009, increasing by 1.71, 4.50 and 1.26 % respectively within the 11th *Five-Year Plan* period. We can see that the gap of the proportions of effective irrigation areas among central and the northeast, and the east were widening while that between the west and east was narrowing down gradually. Compared with central China and the west, the proportion of effective irrigation area in the northeast grew slowly. In 2007, the proportion in the west exceeded that of the northeast. The No.1 Document of the CPC Central Committee of 2011 attaches great importance to the construction of water conservancy in China. We expect that the effective irrigation areas nationwide and in various regions will increase rapidly. The green growth of the primary industry is expected to be further accelerated.

3.2.4 Intensities of Fertilizer and Pesticide Utilization

In the process of comparing the green growth of the primary industry in various regions, evaluation of friendliness to the environment is one of the major contents

Table 3.5 Utilization of fertilizers and pesticides in various regions of China in the 11th Five-Year Plan period (Unit: ton/100 million yuan)

Indicator	Intensity of fertilizers utilization				Intensity of pesticides utilization			
Years	2006	2007	2008	2009	2006	2007	2008	2009
Nationwide	2002.12	1788.23	1552.51	1534.04	62.13	56.81	49.55	48.51
The East	1773.56	1603.66	1372.69	1304.14	64.73	57.91	51.41	49.45
Central China	2377.94	2142.28	1838.54	1825.25	83.00	75.54	65.29	63.39
The West	1982.15	1754.29	1546.63	1592.90	38.16	35.56	31.99	32.61
The Northeast	1908.90	1613.58	1431.00	1427.30	58.34	59.93	46.98	46.00

Note Intensity of fertilizer utilization = utilization amount of fertilizer/value added of the primary industry; Intensity of pesticide utilization = utilization amount of pesticide/value added of the primary industry

Sources Calculated according to *China Statistical Yearbook 2007–2010* published by National Bureau of Statistics, and *China Statistical Yearbook on Environment 2007–2010* published by National Bureau of Statistics and Ministry of Environmental Protection

for researches. Currently, the pollutions from the primary industry to the environment are mainly agricultural chemical pollution caused by the utilization of fertilizers and pesticides. The more the fertilizers and pesticides are used, the greater they may harm the environment and vice versa. Hence, the influences of the primary industry to the environment can be judged by the utilization of fertilizers and pesticides. The intensities of fertilizers and pesticides utilization in various regions of China in the 11th *Five-Year Plan* period can be seen in the Table 3.5.

From the Table 3.5, we can see that China's fertilizer utilization intensity and pesticide utilization intensity went down gradually from 2006 to 2009 with stable declining trend. In 2006, every 100 million yuan of value added in China's primary industry led to 2,002.12 tons of fertilizers consumption and 62.13 tons of pesticides; in 2009, the figures dropped to 1,534.04 and 48.51 tons, a drop of more than 20 % within the 4 years with an annual rate of 8 %.

Fertilizer and pesticide consumptions in various regions have been dropping year by year and fertilizer consumption is dropping faster than pesticide consumption. The east is the place with the lowest level of fertilizer consumption intensity, about 15 % lower than the country's average level. From 2006 to 2009, fertilizer utilization dropped by 26.47 % with an annual rate of 9.74 %; this region also occupies the third position in terms of pesticide utilization intensity among the four major regions, a little bit higher than the country's average level. In the 11th *Five-Year Plan* period, pesticide utilization intensity dropped by 23.61 % with an annual rate of 8.58 %. Fertilizer and pesticide consumption intensity in central China are the highest nationwide, 18 and 30 % higher than the country's average level respectively. Though its annual reduction rate in terms of fertilizer utilization and pesticide utilization are both 8 %, it has the greatest potential threats to the environment among the four regions. The west ranks No.3 in terms of fertilizer utilization intensity, almost in the same level as the country's average; but its pesticide consumption is the lowest nationwide, accounting only for about 60 % of the national average level. In 2009, the region only used 32.61 tons of pesticides for every 100 million yuan of value added in the primary industry, only half of that of the central China. The northeast takes the second position in terms of fertilizer and pesticide utilization intensities. From 2006 to 2009, fertilizer and pesticide utilization intensities dropped by 25.23 and 21.15 % with annual rates of 9.24 and 7.62 % respectively.

We can see from the above analyses that the west pays more attention to protecting environment during the process of the production of the primary industry and chemical products such as fertilizers and pesticides have the least potential threats to the environment. The east and northeast also pay attention to the issue of green growth in the process of developing the primary industry with effective control on non-point source pollution. Limited by natural conditions and agricultural production technologies, central China need to use more fertilizers and pesticides to improve the quality of agricultural products, and thus have greater potential pollution on environment. In general, compared with the 10th *Five*-Year Plan period, the fertilizer and pesticide utilizations of the four regions in the first 4 years of the 11th *Five-Year Plan* period were dropping gradually, and their environmental-friendly level were going up, effectively promoting the green growth of the primary industry.

3.2.5 Summary

By analyzing and comparing labor productivity levels, land production efficiency, utilizations of agricultural water resources and intensity of fertilizer and pesticide consumption, we can get the overall conditions of the green growth of the primary industry in China in the 11th *Five-Year Plan* period: With the strong advantageous in terms of economy and sci-tech level, the east takes the leading position nationwide; central China and the west developed slowly in the early stage of the 11th *Five-Year Plan* period. But in the previous one or two years, these two regions attach great importance to the green growth of the primary industry and are actively pursuing the east; in the previous years, the northeast was better than central China and the west in this regard, but it grows slowly in recent years and shows the trends of being caught up by central China and the west. In general, the green growths of the primary industry in various regions still have a large space of growth in the next *Five*-Year Plan period based on the sound development of the 11th *Five-Year Plan* period.

3.3 Prospects of Green Growth of the Primary Industry in the 12th Five-Year Plan Period

The 12th *Five*-Year Plan for National Economic and Social Development of the People's Republic of China (hereafter referred to as the 12th *Five-Year Plan*) clearly points out the basic concepts and orientation for China's agricultural and rural development in the coming 5 years and deliberately illustrates the policies and measures of expanding channels of increasing farmers' incomes. In the 12th *Five-Year Plan* period, China will enter a new historical stage in terms of agricultural and rural development, and the green growth of the primary industry will also face new historical opportunities and challenges.

3.3.1 Opportunities for Green Growth

The 12th *Five-Year Plan* promulgates the basic concept of agricultural and rural development: empowering and benefiting farmers and accelerating the construction of new socialist countryside. China has entered the new phase of simultaneously constructing and developing industrialization, urbanization and agricultural modernization and is facing the unprecedented development opportunities of green agriculture with the themes of protecting environment and sustainable development.

3.3.1.1 Agriculture

The Compendium on the 12th *Five-Year Plan* has comprehensive deployments in areas such as green development of agriculture, rectification of rural environment, key projects of new countryside construction, conservation of land and water resources, key projects of circular economy and food safety. The major tasks of green development of agriculture can be found in Table 3.6.

The information of Table 3.6 not only reflects the policy orientation of green development of China's agriculture in the 12th *Five-Year Plan* period, but also clarifies the overall goals of green agricultural construction of China in the 12th *Five-Year Plan* period. Its core is to promote energy saving, emission reducing and pollution cutting under the premier of ensuring food safety and increasing farmers' income through scientific and technological progress and management innovation so as to ensure ecological balance and sustainable agricultural development and realize high-yield, high-quality, ecologically friendly and safe agriculture. In the next 5 years or longer time, we will realize green production of agriculture from the following aspects in China:

Stable-yield and high-standard farmlands are constructed while low- and medium-yield farmlands are transformed on a large scale. Currently, about 70 % of China's farmlands are low- and medium-yield lands, only 48.6 % of which have effective irrigation. The proportion of stable-yield and high-standard farmlands is relatively low. In the 12th *Five-Year Plan* period, China will make effort to newly construct 2.67 million hectares of high-standard farmlands and upgrade and improve 0.67 million hectares of high-yield lands.[6]

---

[6] Han Changfu: Basic Thoughts on Developing Grain Production in the *12th Five-Year Plan* Period, Qiushi, 2011 (3).

Table 3.6 Opportunities of green growth of agriculture in China in the 12th Five-Year Plan

Farmland	Most strict land system, 12.12 million hectares of farmland in 2015
	Constructing high standard farmland projects with stable yield
	Land rectification projects in rural areas with compensative farmland of 1.33 million hectares
Water	Constructing farmland water conservation projects with effective agricultural water utilization coefficient reaching 0.53 in 2015
	Launching the project of ensuring drinking water safety in rural areas
Rectification of rural Environment	Rectifying non-point source pollution such as pesticides, fertilizers and farming films; promoting prevention of pollution from livestock husbandry
Energy	Implementing methane projects in rural areas and energetically developing crop straw, wind power and solar energy
Key projects of circular Economy	Supporting comprehensive utilization of straw, livestock and poultry excrement and discarded wood materials
Science and technology	Raising the comprehensive planting and harvesting mechanization level to 60 % and implementing modern seed fostering projects
Strategic emerging Industries	Biological industry focusing on bio-agriculture and developing new type of plastic film for agriculture
Agricultural structure and space layout	Agricultural strategic layout of seven regions and 23 stripes
Food safety	Establishing tracking system for food safety

Source National Development and Reform Commission: The Compendium on the Twelfth Five-Year Plan for National Economic and Social Development of the People's Republic of China, Beijing, People's Publishing House, 2011

Keep expanding the areas of protective farmland. It is proved by practices that implementation of protective farmland can reduce farming costs, increase organic matters in the soil, reduce water and soil erosion and improve the capacity of storage and preservation of soil moisture. In 2011, China will increase more than 0.1 million hectares of protective farmland.

The issue of accelerating water conservancy reform and launching basic water conservancy construction in farmlands are carried on in an all-round way. The No.1 Document of the CPC Central Committee of 2011 pointed out that China would basically complete the task of reforming supportive and water-saving facilities in large irrigation areas and key medium irrigation areas by 2020, energetically develop water-saving irrigation system through promoting the technologies such as seepage-proofing channels, pipeline water convey and spraying and dripping irrigation technologies, enhance the coverage of subsidizing water-saving and drought-resisting equipment, actively develop dry farming and adopt technologies such as plastic films, deep plough and protective plough. In the 12th *Five-Year Plan* period, China will newly add 0.27 ha of farmland with effective irrigation.

Keep increasing the contribution of science and technology to agricultural production. China will construct modern agriculture and promote agricultural technology integration, mechanization in the labor process and information-based production and operation. By 2015, the rate of comprehensive mechanization of plough, growing and harvesting in agricultural production will reach 60 % and mechanization level of rice growing and harvesting will account for 45 and 80 % respectively. In singe cropping rice areas in the northeast and the middle and lower reaches of the Yangtze River, the whole process of rice production has been mechanized, which hence further improves the level of agricultural production efficiency. In the mean time, Plan on Introducing International Advancing Agricultural Technologies (948 Plan) will focus on supporting China's technology introductions in areas such as agricultural and biological resources, agricultural scientific researches, biological-based seed cultivation, low-carbon agriculture, biological energy and new transgenic technologies, providing Sci-tech supports for China's green agricultural production.

Controlling and reducing non-point source pollution. In the 12th *Five-Year Plan* period, China will actively promote formula fertilization by soil testing and new type of organic fertilizers, energetically implement the program of returning straws back to farmlands and using farmyard manure, encourage utilization of bio-pesticides and low poisonous and low residual pesticides, and promote recycling of farming plastic films.

Keep expanding acreage of green products. By the end of the 12th *Five-Year Plan* period, the areas of pollution-free agricultural products will account for 60 %; the output of pollution-free products will account for 40 % of the total products; certified safe products with high quality will exceed 98 %. China will have 2 million hectares of green food production and testing bases, 7,500 such enterprises and 20,000 products, 100 demonstration parks of organic agriculture, 680 standard material bases, 2,000 enterprises of organic agricultural products and 11,000 such products; the rate of certified safe products with high quality will exceed 98 percent.[7]

3.3.1.2 Forestry

During the 12th *Five-Year Plan* period, China's will actively transform forestry development mode, accelerate forestry reform, promote construction of modern forestry in an all-round way and continue the mission of prospering forests, enriching farmers and protecting environment. The 12th *Five-Year Plan* requires continuing to implement the projects of protecting natural forest resources and returning farmlands to forests, meanwhile strengthens fire prevention and diseases and pests prevention of forests and outlines the major objectives and tasks of forestry development in the 12th *Five-Year Plan* period: Forest coverage will increase from 20.36 % in 2010 to 21.66 percent in 2015, up with 1.3 % within

[7] Ministry of Agriculture: China to Improve Six Systems in Three Aspects ion Green Food and Organic Food Processing in the *12th Five-Year Plan*, http://www.moa.gov.cn/zwllm/zwdt/201103/t20110321_1951404.htm,2011-03-21.

Table 3.7 Opportunities of Green Development of Forestry in the 12th Five-Year Plan Period

The Second phase of natural forest resource protection project	Supervising the 107 million hectares of forests in the region of Natural Forest Resource Protection Project
Returning farmlands to forests	Continuing the project of returning farmlands to forests in important ecological regions or vulnerable ecological regions with focus on the sloping farmland more than 25°
Construction of shelter forest system	Continuing the shelter forest projects in the northwest, north and northeast, along the coast, the Yangtze River area and the Pearl River area
Qinghai-Tibet plateau, loess plateau and sanjiangyuan protection reserve	Energetically protecting forests and vegetation, preventing and controlling sand, preserving water and soil and maintaining water resources

Source National Development and Reform Commission: The Compendium of the Twelfth Five-Year Plan for National Economic and Social Development of the People's Republic of China, Beijing, People's Publishing House, 2011

5 years; wood growing stock will increase from 13.7 billion cubic meters in 2010 to 14.3 billion cubic meters in 2015, up with 4.38 % points within 5 years; China will have 309 million hectares of forest lands which gross carbon storage of forest and vegetation will reach to 8.4 billion tons, and its total output value of forest will hit 3.5 trillion yuan. The 12th *Five-Year Plan* period will provide great strategic opportunities for the green development of forestry in China (see Table 3.7).

Source: National Development and Reform Commission: The Compendium on the 12th *Five*-Year Plan for National Economic and Social Development of the People's Republic of China, Beijing, People's Publishing House, 2011.

Column 3.4 Sand Industry of Qira County of Xinjiang: Green Industry in the Desert

Located in the southern part of Xinjiang Uygur Autonomous Region with a total area of 0.3 million hectares, Qira County has a typical continental desert climate. As China's traditional agricultural economic county, it has a population of 150,000, including an agricultural population of 130,000. Owing to the vast deserts, rampant sandstorms and severe desertification, the county suffered greatly from deserts.

For many years, Qira County moved south to the direction of Kunlun Mountain three times because of the sand flow. However, ever after migrations, the vulnerable ecological system and backward natural conditions still restrained the economic and social development of Qira County.

In order to change the situation of "sand aggressing while human receding" and fundamentally improve the ecological environment, in recent years, Qira County started from establishing sand consolidation nets, planting short shrubs, constructing shrub forests, developing forests within oases and improving economic forest coverage and energetically developed sand industry by taking full advantages of local natural conditions so as to turn the disadvantages into advantages.

After years of theoretical researches and practical exploration, Qira gradually established its sand industry including growing forest and medical plants on sands. Forestry mainly includes poplar, willow, oleaster, diversifolious poplar and fruits trees like juglans, pomegranate, apricot, peach, pear and apple. Medical plants mainly include liquorices, herba cistanches and Chinese ephedra. These projects have made prominent achievements.

In 2008, Qira County planted 28,454 ha of qualified forests with fruit forestry, livestock husbandry and efficient corps developing rapidly, and constructed 11,513.33 ha of economic forest bases mainly of juglans, red date, pomegranate and apricot. A total of 20,404 tons of fruits were harvested that year, up with 35.93 % compared with that in the previous year. In 2009, the county had 83.6 ha of root and tuber crops, up with 242 % over the previous year. In 2010, the county had 5.67 ha medical plants and 760.6 ha of melon, up with 26.9 and 121.1 % year on year. More importantly, in 2004–2008, a total amount of 1390 ha of Cistanche deserticola Ma were planted in Qira County while at present 1066.67 ha of them are producing profits with annual output value of 16 million yuan. Growing Cistanche deserticola Ma has become an important and win–win industry for Qira County to protect eco-environment and increase farmers' incomes.

In addition to the prosperous development of sand industry, Qira County also made great achievements in sand prevention and control, greatly weakening the threats of quicksand on human beings. In 2009, Qira County totally constructed 17 km of windbreak forests, consolidated 11,333.33 ha of quicksand and drove sand dunes 7.8 km back. In 2010, Qira County afforested 2864 ha, up with 88 % over the previous year. In 2011, the county established the comprehensive system of oasis shelter forests in different strips, pieces and networks on the outskirts of the existing oases, combining arbors, bushes, and grasses. In addition, a total of 1691.54 ha of farmlands that were destroyed by quicksand have been recovered.

The sand industry of Qira County have not only mitigated the damage of desertification to the eco-environment and improved the environment and climate of the county, but also increased the economic efficiency of the primary industry. Meanwhile, sand industry also solved the contradiction between environmental rectification and increase of farmers' income to some extent and realized coordinative development of improving the environment and economic increase. It is a new way of realize green growth of the primary industry in areas with backward natural conditions such as deserts and Gobi.

Sources:

1. Public Information Net of Qira People's Government: http://www.clx.gov.cn/Article/ShowArticle.aspx?ArticleID=68794.
2. Phoenix Finance: http://finance.ifeng.com/roll/20081226/283798.shtml.
3. Statistical Communiqué of Qira on the National Economic and Social Development (2008–2010): http://www.clx.gov.cn/Article/ShowArticle.aspx?ArticleID=68953.

Table 3.8 Opportunities of green growth of livestock husbandry

Livestock and poultry raising	Constructing resource farms and improved variety farms of livestock, poultry and aquatic varieties, large-scale livestock and poultry breeding farms and demonstration farms of aquatic product healthy breeding, and increasing the proportion of livestock husbandry in the primary industry
Protection and construction of grassland ecological environment	Improving 2 million hectares of grasslands and planting 1 million hectares of grasslands
Animal protection	Establishing a six-level animal epidemic prevention and control system

Source National Development and Reform Commission: The Compendium of the Twelfth Five-Year Plan for National Economic and Social Development of the People's Republic of China, Beijing, People's Publishing House, 2011

4. Hetian Information website: http://www.xjht.gov.cn/Article/ShowArticle.aspx?ArticleID=34695.
5. Baidu Baike: http://baike.baidu.com/view/972497.htm.

3.3.1.3 Livestock Husbandry

According to the general requirements on the 12th *Five-Year Plan*, the livestock husbandry will transform its development mode in that period, trying to promote standardized production of livestock products, develop modern forage seed industry for livestock, modern feedstuff industry and modern services for livestock husbandry, safeguard livestock security and protect ecological environment of grasslands. Efforts will be made to strengthen safe production of big livestock such as pig, ox and sheep to ensure quality and safety of livestock products such as meat, eggs and dairy products.[8] The 12th *Five-Year Plan* also puts forward to implementing the projects such as returning pasturing areas to grasslands, developing and utilizing grasslands in the south and disaster prevention and alleviation on grasslands and construct fences for grasslands so as to promote green development of livestock husbandry. The green growth of livestock husbandry in China in the 12th *Five-Year Plan* period can be seen in Table 3.8.

The Table 3.8 shows that variety improvement, epidemic prevention and control, large-scale cultivation, feedstuff improvement, disposal and utilization of animal excrement and safety of meat products will be the major orientation of China's green development of livestock husbandry in the future. The increase of the protection of grassland ecological environment will help maintain sustainable development of livestock husbandry and improve the quality of meat products and dairy products to certain extents. The modern agricultural development plan and

[8] Editorial Department Chinese Journal of Animal Science: The *12th Five-Year Plan* for Livestock Husbandry: Changing at the Beginning, Chinese Journal of Animal Science, 2011 (2).

the key projects of new countryside construction of the 12th *Five-Year Plan* have provided new opportunities for the green development of livestock husbandry.

3.3.1.4 Fishery

The compendium of the 12th *Five*-Year Plan points out clearly that China will promote Infrastructure construction of fishery to provide more complete hardware supports for fishery development. Transforming fishery development modes and accelerating development of modern fishery will be the main topics for China's fishery development in the 12th *Five-Year Plan* period. China will transform its resource-consumption fishery to resource-saving, resource-maintenance and environmental-friendly fishery, strictly implement the system of summer fishing moratorium and fishing controlling system, and promote healthy cultivation mode to guide pond aquiculture to water-saving and healthy breeding direction, industrial aquiculture to the mode of water recycling, inshore marine aquiculture to the mode of multistoried aquiculture and large lake and reservoir aquiculture to clean-water aquiculture.

On March 18, 2011, Ministry of Agriculture issued the *Notice on Strengthening Fishing Boat Management to Control Marine Fishing Intensity in the* 12th *Five-Year Plan Period*, clearly pointing out that China will continue to implement the system of controlling the number and overall power of boats of marine fishing during the 12th *Five*-Year Plan period. The notice puts forward that in principle, the number and the overall power of boats of marine fishing cannot exceed those of 2010. This will ease the pressure of inshore fishery on the natural environment and ecological environment of Chinese oceans, and can help sustainable development of inshore fishery. Meanwhile, China will reasonably develop distant sea fishery; the adjustment of fishery structure will promote green development of China's fishery industry.

3.3.2 Green Growth Facing Challenges

Though the primary industry has new development opportunities in the new period, there are still traditional factors that restrain the green development of the primary industry. New uncertainties that will influence green growth of the primary industry will emerge and thus, China's green growth of the primary industry in the 12th *Five-Year Plan* period will still face huge pressures and challenges.

3.3.2.1 Lack of Green Education

Viewing from the development course of agriculture, forestry, livestock husbandry and fishery production, it can be seen that farmers of agriculture, forestry, livestock husbandry and fishery make their decisions of production and operation activities according to their analyses of investment and output and profit maximization. They seldom consider environmental protection and ecological balance, thus leading to the phenomena such as over cultivation, random deforestation, overgrazing, over fishing and pollution of the primary industry. In addition to, there is no enough environmental protection education to farmers of agriculture, forestry, livestock husbandry and fishery and one cannot find environmental protection education and publicity contents for farmers on media such as books and TV programs. The weakness of green education has made neglect the importance and necessity of green growth of the primary industry in their work. In this case, the Chinese Government and the society should strengthen their efforts on educating farmers of agriculture, forestry, livestock husbandry and fishery on knowledge of environmental protection, and establishing their consciousness of developing low-carbon, green and ecological friendly agriculture, forestry, livestock husbandry and fishery.

3.3.2.2 Fragmentation of Land Utilization

In average, a rural household in China operates 0.5 ha of lands in 5.7 pieces. Every piece of land is only 0.09 ha in average. The serious Fragmentation of Land Utilization has greatly blocked the green growth and modernization process of China's agriculture. On the one hand, fragmented lands lead to land waste; on the other hand, intensive and meticulous farming on small pieces of lands fundamentally restrains the implementation of agricultural technologies and popularization of mechanization, blocks the promotion of improved varieties and reasonable utilization of farming plastic films and pesticides and reduces irrigation efficiency and fertilizing effects, which cannot help increase production efficiency and green growth of agriculture.[9] In the 12th *Five-Year Plan* period, China will still face the negative influences brought by land fragmentation operation; land use system is influencing green growth of agriculture and development of modern agriculture.

3.3.2.3 Contradiction Between Increasing the Farmer Incomes and Environmental Protection

In the 12th *Five-Year Plan* period, the growth of per-capita net income of rural residents in China is expected to exceed 7 %, higher than the expected growth of GDP for the first time. This will increase the pressures on the green growth of China's primary industry, making the efforts of scientifically and reasonably

[9] Liu Qiang: Brief Analysis on the Fragmentation Problem in Land Contraction in Rural Areas, Management and Administration on Rural Cooperative, 2010 (3).

coordinating the relations between economic development and environmental protection the most important work in the future for the primary industry. Taking every measure to increase incomes of grain farmers, fruit farmers, forest farmers, herdsmen and fishers and realizing ecological friendly, sustainable and low-carbon development of agriculture, forestry, livestock husbandry and fishery are two core contents of China's primary industry in the 12th *Five-Year Plan* period. It is the necessary way of increasing farmers' incomes and protecting environment to transform the extensive, high energy-consumption and high-pollution production modes of agriculture, forestry, livestock husbandry and fishery, enhance investment for innovation in technology and management and develop recyclable, ecological friendly and green primary industry.

3.3.2.4 Lack of Promotion of Green Growth

Ecological green development of agriculture, forestry, fishery and livestock husbandry have clear externality. Though a series of laws such as Forest Law of the People's Republic of China, Land Administration Law of the People's Republic of China, Grassland Law of the People's Republic of China and Fisheries Law of the People's Republic of China have regulated reasonable use of forest, land and water, environment problems such as destroying forest, land reduction, land desertification and water pollution are still serious. The reason is that system of public resource administration is incomplete or unreasonable. We should fully consider the positive significance of public resources such as land, forest, grassland and water on the sustainable development, establish institutions for the management and utilization of public resources and support and encourage green growth of the primary industry so as to provide more reasonable institutional guarantees for the green growth of the primary industry.

3.3.3 Suggestions of Green Growth of the Primary Industry in the 12th Five-Year Plan Period

The opportunities and challenges of the green growth of the primary industry in the 12th *Five-Year Plan* period put forward requirements for the next step work for China. On this basis, in order to realize green development of the primary industry, we put forward related countermeasures and ideas for reference.

3.3.3.1 Promoting Green Development

It is necessary to continue to carry out the most strict land institutions to ensure 12.12 million hectares of farmland, focus on the implementation of forest protection works such as natural forest resource protection, returning farmlands to grasslands and constructing shelter forest system, optimize forest ecosystem in China, improve forest environment, strengthen protection of eco-system of wetland and fishery so as to maintain its sustainable development and support the primary industry, prevent chemical pollution from pesticides, fertilizers and farming plastic films for the primary industry, maintain fertility of farmland, reduce land pollution, standardize livestock and poultry raising in rural areas, control the harm from non-point source pollution caused by excrement of livestock and poultry, comprehensively utilize straws, excrement of livestock and poultry and discarded wood materials, and promote bio-energy projects to provide clean agricultural energy.

3.3.3.2 Improving Efficiency of Resource Utilization

It enjoys a significance to improve China's productivity levels of the primary industry such as agriculture, forestry and livestock husbandry by using modern scientific agricultural technologies to narrow down the gap with the developed countries, optimize land use approaches, improve land output efficiency, reform medium- and low-yield farmlands, upgrade and improve the quality of the existing high-yield farmland, implement high-standard and stable-yield farmland projects, strengthen construction of farmland water conservancy infrastructure, save water resources, increase the efficiency of fertilizers and pesticides and improve the utilization efficiency of agricultural chemical means of production.

3.3.3.3 Promoting Sci-tech Innovation and Popularization

It is urgent to increase investment in scientific researches of the primary industry to ensure smooth operation of agricultural technical researches, strengthen efforts in solving technical difficulties in the primary industry so as to pool all strength to solve the technical bottleneck in the primary industry, attach importance to the fostering of scientific researchers and technicians of the primary industry by cooperating with universities, enterprises and agricultural demonstration bases, intensify efforts of constructing agricultural experimental bases and agricultural demonstration bases, ensure application of agricultural technologies, innovate technology transformation approaches and transformation channels of the primary industry, promote technology industrialization of the primary industry, strengthen promotion of technologies of the primary industry and popularizing agricultural technologies and knowledge in the form of teaching agricultural technologies in rural areas.

3.3.3.4 Emphasize Food Security

It takes a high point to strengthen education on the safety of agricultural products to improve people's awareness of safety of agricultural products, focus on the consciousness of safety of agricultural products' producers, stress on the importance of safety of agricultural products, improve supervision mechanism of agricultural products' safety, severely punish various activities that threaten agricultural products' safety, energetically develop green foods, improve criteria of green foods and prosper China's market of green agricultural products.

3.3.3.5 Improving Policies of Green Development

It is significant to strengthen legislation for green growth of the primary industry so as to protect green development of the primary industry from the angle of the national law, increase subsidies of green production of the primary industry, provide supports in finance, encourage green production activities of farmers and enterprises, meanwhile, give benefits to green production of the primary industry on taxation, strengthen publicity of green growth of the primary industry in order to improve people's awareness of green development, improve green supervision on the primary industry and avoid non-green growth events, and optimize and reform the operation and management systems of farmlands and forest lands so as to conduct green production in a sustainable manner.

Chapter 4
Green Growth of the Secondary Industry

Jiangxue Zhang, Hongli Ma and Xiwei Wang

The secondary industry has already scored notable achievements since the reform and opening-up in China. However, surging energy prices and worsening environmental quality has brought new challenges to the green growth of the secondary industry. This chapter, based on the features of the secondary industry's green growth and its existing problems during the 11*th Five-Year Plan* period, respectively analyzes the differences among provinces and cities, makes predictions on the secondary industry's green growth during the 12*th Five-Year Plan* period and probes into its development opportunities and challenges.

4.1 Features and Existing Problems of the Green Growth of the Secondary Industry in China During the 11th Five-Year Plan Period

During the 11*th Five-Year Plan* period, the secondary industry maintained rapid development, with an average annual growth rate of 12.1 % in output value from 2006 to 2010. The relative proportion of the secondary industry in the whole industry slightly dropped from 47.4 to 46.8 %. Meanwhile major changes occurred in industrial development. The transformation of the secondary industry into the green growth model reflected the practice of the *Scientific Outlook on Development*, put forward at the 17th Party Congress, from the perspective of industries. "China Green Industries Forum", held in June, 2010, clearly gave out the development path for traditional industries to transform into the "green industry".

X. Li and J. Pan (eds.), *China Green Development Index Report 2011*,
Current Chinese Economic Report Series, DOI: 10.1007/978-3-642-31597-8_4,

Table 4.1 The growth speed of six energy-intensive industries' industrial value-added during the 11th Five-Year Plan Period (Unit: %)

Sector	2006	2007	2008	2009	2010
Processing of petroleum, coking, processing of nuclear fuel	5.4	13.4	4.3	5.2	9.7
Manufacture of raw chemical materials and chemical products	20.0	21.0	10.0	14.6	16.4
Manufacture of non-metallic mineral products	21.0	24.7	16.9	14.7	20.1
Smelting and pressing of ferrous metals	19.3	21.4	8.2	9.9	13.0
Smelting and pressing of nonferrous metals	23.8	17.8	12.3	12.8	14.4
Production and distribution of electric power and heat power	13.2	13.8	8.6	6.0	12.1

Source Calculate according to the relevant data on the National Bureau of Statistics Website and Monthly Statistics Database

4.1.1 Features of the Green Growth of the Secondary Industry

The key to green growth is to improve the input–output efficiency. Only when the resource utilization efficiency is improved, can the organic integration of "green" and "growth" be achieved and coordinated development of economy and environment be reached.

The structure of the secondary industry was gradually improved. During the 11*th Five-Year Plan* period, the elimination of backward production capacity made great progress, with half of the industrial backward production capacity eliminated. Among them were 110 million tons in iron-smelting, over 68 million tons in steel-making, 330 million tons in cement-producing, 100 million tons in coke-producing, 10.3 million tons in paper-making and 38 million weight cases in glass-producing.[1] Besides, the growth of the six energy-intensive industries was curbed. The energy-intensive industries were prevented from reckless expansion, through differentiating electricity bills, limiting credit scales and implementing proper fiscal and taxation policies during the 11*th Five-Year Plan* period. By 2010, the growth speed of six energy-intensive industries as petrochemical industry had shown a declining tendency compared with that of 2006 (see Table 4.1).

The resource utilization mode has been intensified. For one thing, the energy consumption of major industrial products has fallen dramatically. The comprehensive energy consumption of smelting a unit of copper dropped by 35.9 %, that of caustic soda dropped by 34.8 %, that of cement 28.6 %, that of crude 28.4 %, that of steel 12.1 %, that of electrolytic aluminum 12.0 %, that of polystyrene 11.5 % and the standard coal consumption for thermal power generation declined by 16.1 %.[2] For another thing, the industrial water utilization efficiency has

[1] Chinese Statistics Web Links: Industrial Economy has played a splendid role in maintaining growth and adjusting structure, http://www.stats.gov.cn/tjfx/ztfx/sywcj/t20110304_402707882.htm,2011-03-04.

[2] Beijing Economic Information Net: *China has made great progress on economic restructuring during the* 11*th Five-Year Plan period*, http://www.beinet.net.cn/fxyj/yjbg/201103/t769725.htm,2011-03-11.

tremendously improved. The water consumption per unit of industrial value-added had reached 116.4 cubic meters per 10,000 yuan as of 2009, which had dropped by 30.05 %, compared with that of 2005. The target of the 11*th Five-Year Plan*—30 % had been over fulfilled ahead of time.[3] Moreover, energy consumption per unit of industrial value-added had decreased from 0.33 tons of standard coal equivalent (SCE) per 10,000 yuan of 2005 to 0.25 tons of SCE per 10,000 yuan as of 2008.

The environmentally friendliness has improved and the control of "three industrial wastes" (waste gas, wastewater and solid wastes) has gained great achievements. As of 2009, the wastewater discharges per unit of industrial value-added had reached 17.46 tons per 10,000 yuan, which had dropped by 44.54 % compared with that of 2005. During the 11*th Five-Year Plan* period, the waste gas emissions per unit of industrial value-added, decreasing from 3.88 cubic meters per yuan—the peak in 2007 to 3.64 cubic meters per yuan in 2009, was basically under control. The industrial solid wastes discharged descended substantially year by year reaching 7.1045 million tons in 2009, which was only 42.93 % of that in 2005.

4.1.2 Existing Problems of the Green Growth of the Secondary Industry

Industrial restructuring is very slow. Although the share of six energy-intensive industries' output value has declined since 2007, its average proportion in the gross industrial output value reached 33.74 % from 2006 to 2009, over 3 % higher than the average annual rate of 30.63 % during the "10th Five-Year Plan" period. By the first quarter in 2011, the value-added of six energy-intensive industries had a 12.2 % increase, 2.6 % more compared with that of the forth quarter last year. It bounced back much more quickly than the whole industry did.[4] Apart from that, the industrial structure's transformation from "heavy-duty" to "high processing degree" is a relatively recent phenomenon. The value-added of the equipment manufacturing made up 29.6 % of that produced by industrial enterprises above designated size in 2010, which only enjoyed a 0.9 % rise compared with 2005.[5]

---

[3] The Task Group of the State Development Research Center. 12 development issues during the 12*th Five-Year Plan* period, Beijing: China Development Press, 2010.

[4] Mechanical Equipment Net: *Industrial Economy's Performance in the First Quarter of* 2011, http://www.jx108.cn/news/zixun/jyzdao/20110423111704.html,2011-04-23.

[5] Beijing Economic Information Net: *China has made great progress on economic restructuring during the* 11*th Five-Year Plan period*, http://www.beinet.net.cn/fxyj/yjbg/201103/t769725.htm,2011-03-11.

Table 4.2 International comparison on energy consumption of energy-intensive production china

Index	Chinese level	International advanced level
Electricity consumption of coal production (KWH/ton)	21.5	17.0
Comprehensive energy consumption of crude (Kg of SCE/ton)	106	73
Coal consumption of thermal power generation (Kg of SCE/ KWH)	320	299
Coal consumption of thermal power supply (Kg of SCE/KWH)	340	312
Comparable energy consumption of steel production of large and medium-sized enterprises (Kg of SCE/ton)	697	610
Comprehensive energy consumption of copper smelting (Kg of SCE/ton)	509	500
Comprehensive energy consumption of cement production (Kg of SCE/ton)	139	118
Comprehensive energy consumption of sheet glass Production (Kg of SCE/weight cases)	16.5	15
Comprehensive energy consumption of polystyrene production (Kg of SCE/ton)	976	629
Comprehensive energy consumption of caustic soda (Kg of SCE/ton)	1040	–

Note "International advanced level" refers to the average of energy consumption indexes of countries in world's advanced level
Source Qingyi, Wang. *Evaluation of China's Energy Efficiency*. Energy Conservation and Environmental Protection, 2011(1)

The industrial energy utilization efficiency is very low. Although Energy consumption per unit of GDP accumulatively dropped by 19.06 %[6] during the 11*th Five-Year Plan* period, the overall target of the 11*th Five-Year Plan* was not yet accomplished. This was, to a large extent, due to inefficiency in industrial energy utilization. By now, there still exists 50 % of the backward production capacity ready to be eliminated. Industrial enterprises' technical level of production is of varied qualities. The energy consumption level of major industrial products still lags behind international advanced standard (see Table 4.2).

The consumption of construction resources rises rapidly. According to the recent statistics, energy consumption per unit of construction's value-added began to decline during the 11*th Five-Year Plan* period. However, the construction materials and the resources consumed during the construction process only account for 20 % of the energy consumption volume during the construction life span.[7] By 2008, energy consumption for domestic use, a significant part of

[6] Beijing Economic Information Net: *China has made great progress on economic restructuring during the* 11*th Five-Year Plan period*, http://www.beinet.net.cn/fxyj/yjbg/201103/t769725.htm,2011-03-11.

[7] Tsinghua University Construction Energy Conservation Research Center. *The Annual Research Report of Development of Construction Energy Conservation of 2008*. Beijing: China Architecture & Building Press, 2008.

construction energy consumption, made up 10.94 % of China's energy consumption volume and increased by 0.2 % compared with 2005 whereas electricity consumption for building operation, including that consumed by the tertiary industry and residents, even accounted for 22.87 % of the China's electricity consumption volume, which had a 1.18 % increase compared with that of 2005. Apart from the rise in the energy consumption, the apparent consumption volume of steel in China occupied 46.4 %[8] of that of world in 2009. Compared with that of 2008, cement consumption went up 4 % while the world's consumption of the corresponding period decreased by 1.7 %.[9]

Environmental pollution control is still serious. Firstly, the environmental governance and the economic development do not well coordinate. Although the regional ecological environment improved a lot, there still existed lots of differences in environmental management level among different regions during the 11*th Five-Year Plan* period. Therefore, the organic combination of the environmental governance and regional development are expected. Secondly, efficiency of governing various kinds of pollutants differs. The pace of governing waste gas is relatively low, especially in the removal of sulfur dioxides. The surveillance mechanism for monitoring greenhouse gases is not yet established, which is quite inconsistent with the international pressure on emission reduction we are faced with. Thirdly, environmental pollution governance of the construction lags behind. By the end of the year 2010, 217 demonstration projects of green constructions had been implemented, with the total construction area exceeding 40 million square meters, however the percentage of which had not even reached 0.2 % of the already-existing construction areas in cities and towns. The percentage of the energy-efficient construction areas in cities and towns only reached 23.1 % of already-existing construction areas.[10]

4.2 Regional Comparison on the Green Growth of China's Secondary Industry During the 11th Five-Year Plan Period

On the basis of the former analysis on the overall situation of the secondary industry's green growth and the index system in this report, we will conduct

---

[8] International Iron and Steel Institute. 2011 *will witness an increase in global steel consumption*, http://futures.stockstar.com/IG.

[9] *Global Cement Consumption Decreased by* 1.7 % *in* 2009, http://www.ccement.com/news/Content/29652.html,2009-07-13.

[10] Reorganize according to the report of MOHURD (Ministry of Housing and Urban–Rural Development of People's Republic of China) on the results of special supervision on energy conservation and emission reduction in the field of national housing and urban–rural development [Division of Establishment and Operation, (2011) 25], http://www.mohurd.gov.cn/zcfg/jswj/jskj/201104/t20110421_203196.htm,2011-04-21.

Table 4.3 The secondary industry's labor productivity by region

	The secondary industry's labor productivity (10,000 yuan per person)			
	Eastern region	Central region	Western region	Northeastern region
2006	6.98	4.33	5.65	8.25
2009	9.20	5.98	6.96	13.64
Average annual change rate (%)	10.60	9.53	5.80	16.33

Notes
(1) The value-added of the secondary industry is calculated according to 2005 price
(2) Tibet is included in the calculation of this table
Sources National Bureau of Statistics. *China Statistical Yearbook of* 2007. Beijing: China Statistics Press, 2007; National Bureau of Statistics. China Statistical Yearbook of 2010. Beijing: China Statistics Press, 2010

regional comparative analysis on the secondary industry's green growth during the 11*th Five-Year Plan* period, on two levels—the provincial level and the level of major cities, from three different perspectives—industrial structure and efficiency, resource utilization and pollutants governance.

4.2.1 Comparative Analysis on the Green Growth of the Secondary Industry on Provincial Level

For comparative analysis, this part will adopt quartering method and divide our country into four regions, that is, northeastern region, eastern region, central region and western region.[11]

Labor productivity of the secondary industry rises rapidly in northeastern, eastern and middle parts of China (see Table 4.3). From 2006 to 2009, the secondary industry's labor productivity in three northeastern provinces of China rose from 82,500 yuan per person to 136,400 yuan per person, with an average annual increase of 16.33 %, making it the fastest-growing area in labor productivity. Eastern region with an average annual increase of 10.60 % and central region with that of 9.53 % came in the second place. Western region, whose labor productivity only increased by 5.8 % annually, ranked bottom.

[11] Eastern region includes 10 provinces namely Beijing, Tianjin, Hebei, Shanghai, Jiangsu, Zhejiang, Fujian, Shandong, Guangdong and Hainan; central region includes 6 provinces namely Shanxi, Anhui, Jiangxi, Henan, Hubei and Hunan; western region includes Inner Mongolia, Guangxi, Chongqing, Sichuan, Guizhou, Yunnan, Tibet, Shaaxi, Gansu, Qinghai, Ningxia and Xinjiang namely 12 provinces; northeastern region refers to Liaoning, Jilin and Heilongjiang, those three provinces. Due to the lack of data in Tibet, Tibet is not included when calculation is conducted. Therefore, the provinces included in comparison add up to 30.

Table 4.4 Output value of energy-intensive industries as percentage of the gross industrial output value by region (Unit: %)

	Eastern region	Central region	Western region	Northeastern region
2006	29.15	43.79	45.39	40.70
2009	29.10	39.64	40.92	35.01
Average annual change rate	−0.06	−3.16	−3.28	−4.66

Sources National Bureau of Statistics. *China Statistical Yearbook on Industrial Economy* 2007. Beijing: China Statistics Press, 2007; National Bureau of Statistics. *China Statistical Yearbook on Industrial Economy* 2010. Beijing: China Statistics Press, 2010

Industrial restructuring has had great performance in northeastern, central and western regions (see Table 4.4). From 2006 to 2009, output value of the energy-intensive industries in eastern region dropped slightly to 29.10 % from the previous 29.15 %. Central and western regions' results were relatively notable, declining respectively by 3.16 and 3.28 %. The average annual declining magnitude in northeastern region, reaching 4.66 %, was the most notable among other regions.

Due to the constraint of the target "to reduce energy consumption per unit of GDP by 20 %", energy consumption per unit of industrial value-added by enterprises above designated size in all provinces(autonomous regions, municipalities) has improved except in Xinjiang Province. Guangdong was the province whose energy consumption level was the lowest against industrial value-added whereas that of Ningxia was the highest. By 2009, the energy consumption level of Guangdong was 7.04 times higher than that of Ningxia, compared with 7.35 times in 2006, the difference of which had, to some extent, narrowed. Shaanxi Province, whose energy consumption level dropped by an average annual rate of 14.81 % during these three years, was the fastest one among other provinces (autonomous regions, municipalities). Apart from Shaanxi Province, Hunan, Jilin, Jiangxi, Herlongjiang, Inner Mongolia, Beijing and Tianjin all enjoyed an average annual decrease of over 10 % on energy consumption level. By contrast, the industrial energy utilization efficiency improved a bit more faster in central region (see Table 4.5).

The comprehensive utilization rate of industrial solid wastes has increased in most provinces (autonomous regions, municipalities) but this indicator in Beijing, Xinjiang, Zhejiang and Tianjin began to decline. In 2009, Tianjin was the city with highest comprehensive utilization rate of industrial solid wastes being 98.30 % whereas that of Gansu was the lowest being 33.40 %. Judged from different regions, central region's comprehensive utilization rate of industrial solid wastes climbed the fastest by an average annual rate of 5.09 % during the 11*th Five-Year Plan* period while that of eastern region increased the slowest with merely a rate of 1.37 % (see Table 4.6).

Industrial water consumption differs a lot among provinces (autonomous regions, municipalities) all over China. Water-deficient areas tend to utilize water

Table 4.5 The secondary industry's energy consumption per unit of industrial value-added of enterprises above designated size by region (Units: tons of SCE/10,000 yuan, %)

Region	Provinces (autonomous regions, municipalities)	2006	2009	Average annual change rate
Eastern region	Beijing	1.33	0.91	−10.55
	Tianjin	1.33	0.91	−10.50
	Hebei	4.19	3.00	−9.47
	Shanghai	1.20	0.96	−6.75
	Jiangsu	1.57	1.11	−9.83
	Zhejiang	1.43	1.12	-7.16
	Fujian	1.37	1.15	−5.35
	Shandong	2.02	1.54	−7.87
	Guangdong	1.04	0.81	−7.40
	Hainnan	3.15	2.61	−5.68
Central region	Shanxi	5.89	4.55	−7.58
	Anhui	2.86	2.10	−8.86
	Jiangxi	2.72	1.67	−12.82
	Henan	3.78	2.71	−9.45
	Hubei	3.33	2.35	−9.81
	Hunan	2.74	1.57	−14.23
Western region	Inner Mongolia	5.37	3.56	−11.25
	Guangxi	2.88	2.24	−7.47
	Chongqing	2.63	1.85	−9.84
	Sichuan	2.82	2.25	-6.75
	Guizhou	5.21	4.32	−5.69
	Yunnan	3.40	2.74	-6.48
	Shaanxi	2.46	1.37	−14.81
	Gansu	4.59	3.53	−7.70
	Qinghai	3.64	2.94	−6.45
	Ningxia	8.68	6.51	−8.34
	Xinjiang	2.91	3.10	2.12
Northeastern region	Liaoning	2.92	2.26	−7.57
	Jilin	2.80	1.62	−14.04
	Heilongjiang	2.23	1.38	−12.68

Sources National Bureau of Statistics. China Statistical Yearbook on Industrial Economy 2007. Beijing: China Statistics Press, 2007; National Bureau of Statistics. China Statistical Yearbook on Industrial Economy 2010. Beijing: China Statistics Press, 2010

resources more efficiently whereas areas with sufficient water supply do not perform well in water conservation. Guizhou with 30.08 million cubic meters per 10,000 yuan ranked top in water consumption per unit of industrial value-added, which was nearly 25 times higher than 120,700 cubic meters per 10,000 yuan in Tianjin, the city with the lowest water consumption per unit of industrial value-added in 2009. During the 11*th Five-Year Plan* period, water consumption per unit of industrial value-added had somewhat declined for each province (autonomous region, municipality), among which Qinghai had the largest declining magnitude. Seen from different regions, water consumption per unit of industrial value-added

Table 4.6 Comprehensive utilization rate of industrial solid wastes by region (Unit: %)

	Eastern region	Central region	Western region	Northeastern region
2006	80.90	57.79	47.08	48.35
2009	85.04	66.62	53.36	54.59
Average annual change rate	1.37	5.09	4.45	4.30

Sources National Bureau of Statistics of China, Ministry of Environmental Protection. *China Statistical Yearbook on Environment* 2007. Beijing: China Statistics Press, 2007; National Bureau of Statistics of China, Ministry of Environmental Protection. *China Statistical Yearbook on Environment* 2010. Beijing: China Statistics Press, 2010; Ministry of Environmental Protection. *The Annual Statistical Report of China's Environment of* 2006. Beijing: China Environmental Science Press, 2006

Table 4.7 Industrial water consumption and reuse ratio of industrial water by region

	Eastern region	Central region	Western region	Northeastern region
Water consumption per unit of industrial value-added (m^3/10,000 yuan)				
2006	105.34	206.59	180.22	113.88
2009	72.21	148.35	127.81	75.96
Average annual change rate (%)	−10.48	−9.40	−9.69	−11.10
Reuse ratio of industrial water (%)				
2006	81.69	88.29	75.63	82.13
2009	87.56	89.99	81.97	80.80
Average annual change rate (%)	2.40	0.64	2.79	−0.54

Note
1. Industrial value-added is calculated according to 2005 price and the industrial value-added index of 2009 is replaced by the industrial value-added index of the secondary industry
2. The data of Tibet has been included in the calculation of this table
Sources National Bureau of Statistics of China, Ministry of Environmental Protection. *China Statistical Yearbook on Environment* 2007. Beijing: China Statistics Press, 2007; National Bureau of Statistics of China, Ministry of Environmental Protection. *China Statistical Yearbook on Environment* 2010. Beijing: China Statistics Press, 2010; Ministry of Environmental Protection. *The Annual Statistical Report of China's Environment of* 2006. Beijing: China Environmental Science Press, 2006. Ministry of Environmental Protection. *The Annual Statistical Report of China's Environment of* 2009. Beijing: China Environmental Science Press, 2009

in eastern and northeastern regions not only declined faster than in central and western regions but was also much lower in the very beginning. Besides, the reuse ratio of industrial water in most of the provinces(autonomous regions, municipalities) in China has ascended except for Xinjiang, Chongqing, Heilongjiang, Sichuan, Inner Mongolia, Ningxia, Guangdong, Jiangxi, Yunnan, Fujian, Hainan and Jilin, those 12 provinces(autonomous regions, municipalities). During the *11th Five-Year Plan* period, reuse ratio of industrial water improved in central, eastern

Table 4.8 Governance of industrial waste gases by region

	Eastern region	Central region	Western region	Northeastern region
Industrial sulfur dioxide removal rate (%)				
2006	40.77	45.03	35.25	35.52
2009	62.89	63.74	59.64	42.77
Average annual change rate (%)	18.09	13.85	23.07	6.80
Industrial nitrogen oxide removal rate (%)				
2006	11.56	4.14	5.13	4.50
2009	5.41	7.72	3.24	1.96
Average annual change rate (%)	−17.73	28.82	−12.28	−18.81

Note Tibet and Qinghai Province are not included in the calculation of the indicator of industrial nitrogen oxide removal rate

Sources National Bureau of Statistics of China, Ministry of Environmental Protection. *China Statistical Yearbook on Environment* 2007. Beijing: China Statistics Press, 2007; National Bureau of Statistics of China, Ministry of Environmental Protection. *China Statistical Yearbook on Environment* 2010. Beijing: China Statistics Press, 2010; Ministry of Environmental Protection. *The Annual Statistical Report of China's Environment of* 2006. Beijing: China Environmental Science Press, 2006. Ministry of Environmental Protection. *The Annual Statistical Report of China's Environment of* 2009. Beijing: China Environmental Science Press, 2009

and western regions whereas that in northeastern region showed a tendency of decline (see Table 4.7).

Each region had different performance in curbing waste gases and pollutants produced in the secondary industry (see Table 4.8). In 2009, industrial sulfur dioxide removal rate in Gansu Province reached 82.68 %, which topped the list among other provinces (autonomous regions, municipalities). From 2006 to 2009, Ningxia was the province whose industrial sulfur dioxide removal rate surged the fastest from 8.81 to 51.27 %. From the perspective of different regions, central region was the one with the highest industrial sulfur dioxide removal rate reaching 63.74 %. Eastern region came right after central region, with a rate of 62.89 %. The average annual increasing speed for industrial sulfur dioxide removal rate in each region could be ranked as follows from the highest to the lowest: western region, eastern region, central region and northeastern region. During the *11th Five-Year Plan* period, governance of nitrogen oxides had not been attached to adequate importance by some of the provinces. As a result, industrial nitrogen oxide removal rate in Tianjin, Hebei, Liaoning, Jilin, Jiangsu, Shanghai, Fujian, Jiangxi, Shandong, Henan, Guangxi, Hainan, Sichuan, Yunnan, Shanxi, and Ningxia begun to drop to various degrees. Although Beijing had performed the best in industrial nitrogen oxide governance among all other provinces (autonomous regions, municipalities) in 2009, its industrial nitrogen oxide removal rate was still very low, merely reaching 36.27 %. In the view point of different regions, industrial nitrogen oxide removal rate only showed increase in central region from 4.14 to 7.72 % whereas that in other regions all presented a declining tendency.

Table 4.9 Governance of industrial wastewater by region

	Eastern region	Central region	Western region	Northeastern region
Industrial wastewater's COD removal rate (%)				
2006	75.99	64.18	55.90	61.29
2009	84.60	69.40	63.58	62.46
Average annual change rate (%)	3.78	2.71	4.58	0.64
Industrial wastewater's ammonia nitrogen removal rate (%)				
2006	63.68	51.68	51.86	41.47
2009	81.03	61.73	60.19	76.78
Average annual change rate (%)	6.58	6.48	5.35	28.38

Note Tibet is included in the calculation of industrial wastewater's COD removal rate
Sources National Bureau of Statistics of China, Ministry of Environmental Protection. *China Statistical Yearbook on Environment* 2007. Beijing: China Statistics Press, 2007; National Bureau of Statistics of China, Ministry of Environmental Protection. *China Statistical Yearbook on Environment* 2010. Beijing: China Statistics Press, 2010; Ministry of Environmental Protection. *The Annual Statistical Report of China's Environment of* 2006. Beijing: China Environmental Science Press, 2006. Ministry of Environmental Protection. *The Annual Statistical Report of China's Environment of* 2009. Beijing: China Environmental Science Press, 2009

The removal rate of the industrial wastewater's COD (chemical oxygen demand) in most provinces (autonomous regions, municipalities) has increased but began to drop in Jilin, Tianjin, Sichuan and Hainan. That rate increased much faster in eastern and western regions than in central and northeastern regions. Xinjiang, Shaanxi, Sichuan, Ningxia, Qinghai and Inner Mongolia, six provinces (autonomous regions, municipalities) in western region, all decreased in industrial wastewater's ammonia nitrogen removal rate. Industrial wastewater's ammonia nitrogen removal rate reaching 92.84 % in Beijing was higher than in any other provinces (autonomous regions, municipalities) all over China. Industrial wastewater's ammonia nitrogen removal rate achieved the fastest growth in Heilongjiang Province rising from 24.49 to 80.07 %. From the angle of regions, the highest industrial wastewater's ammonia nitrogen removal rate being 81.03 % appeared in northeastern region by 2009. That indicator in northeastern region reached 76.78 %, which made it the fastest-growing region in industrial wastewater's ammonia nitrogen removal rate, faster than any other regions, during the *11th Five-Year Plan* period (Table 4.9).

4.2.2 Comparative Analysis on the Green Growth of the Secondary Industry on the Level of Major Cities

This part will conduct a comparative analysis on the situation of the secondary industry's green growth in China cities during the *11th Five-Year Plan* period,

Table 4.10 The secondary industry's labor productivity by city (Unit: 10,000 yuan/person)

	2006	2009	Average annual change rate (%)
Municipalities and separate planning cities	17.58	21.67	7.75
Provincial capitals	7.77	10.15	10.22

Note Industrial value-added of the secondary industry is calculated according to 2005 price
Sources National Bureau of Statistic. *China Statistical Yearbook on Regional Economy* 2007–2010. Beijing: China Statistics Press, 2007–2010

including cities of three types: 4 municipalities,[12] 5 separate planning cities[13] and 25 provincial capitals.[14] Having taken city size and policy factors into consideration, we regards municipalities and separate planning cities as one category and provincial capitals as another when conducting analysis.

On the part of labor productivity, the secondary industry's labor productivity in provincial capitals grew far more rapidly than in municipalities and separate planning cities. During 2006 and 2009, the secondary industry's labor productivity increased by an average annual rate of 7.75 % from 175,800 yuan per person to 216,700 yuan per person in municipalities and separate planning cities whereas that in provincial capitals rose much faster with an average annual rate of 10.22 % from 77,700 yuan per person to 101,500 yuan per person. It is thus clear that China's major cities maintained fast growth in the secondary industry's labor productivity, the gap of which between provincial capitals and municipalities and separate planning cities had continually narrowed down (see Table 4.10).

Those two categories of cities had their own strengths and weaknesses in resource utilization indicators. On the whole, the secondary industry in provincial capitals did a better job on improving resource utilization efficiency than in municipalities and separate planning cities (see Table 4.11). However, industrial water utilization in those two categories of cities begged to differ. Water consumption per unit of industrial value-added in municipalities and separate planning cities dropped from 260.58 tons per 10,000 yuan in 2006 to 185.75 tons per 10,000 yuan in 2009. That indicator decreased from 307.23 tons per 10,000 yuan to 243.75 tons per 10,000 yuan in provincial capitals during the corresponding period. Water consumption per unit of industrial value-added was evidently smaller in amount in municipalities and separate planning cities compared with in provincial capitals. Besides, its improving speed for municipalities and separate planning cities was much faster than for provincial capitals during the 11*th Five-Year Plan* period. However, the reuse ratio of industrial water was much higher in provincial capitals but increased faster in municipalities and separate planning cities. Cities with the reuse ratio of industrial water exceeding 90 % were as

[12] Municipalities refer to Beijing, Tianjin, Shanghai and Chongqing.

[13] Separate planning cities refer to Shenzhen, Qingdao, Dalian, Ningbo and Xiamen.

[14] Provincial capitals include Shijiazhuang, Taiyuan, Hohhot, Shenyang, Changchun, Harbin, Nanjing, Hangzhou, Hefei, Fuzhou, Nanchang, Jinan, Zhengzhou, Wuhan, Changsha, Guangzhou, Nanning, Haikou, Chengdu, Guiyang, Kunming, Xi'an, Yinchuan, Lanzhou and Xining.

Table 4.11 Resource utilization of the secondary industry in major cities

		2006	2009	Average annual change rate (%)
Water consumption per unit of industrial value-added (ton/ 10,000 yuan)	Municipalities and separate planning cities	260.58	185.75	−9.57
	Provincial capitals	307.23	243.75	−6.89
Industrial water reuse ration (%)	Municipalities and separate planning cities	69.84	76.26	3.06
	Provincial capitals	80.51	84.37	1.60
Comprehensive utilization rate of industrial solid wastes (%)	Municipalities and separate planning cities	85.17	85.66	0.19
	Provincial capitals	68.10	72.02	1.92

Note Industrial value-added data is calculated according to 2005 price
Sources National Bureau of Statistic. *China Statistical Yearbook on Regional Economy* 2007. Beijing: China Statistics Press, 2007; National Bureau of Statistic. *China Statistical Yearbook on Regional Economy* 2010. Beijing: China Statistics Press, 2010

follows: Beijing, Tianjin, Xiamen, Shijiazhuang, Taiyuan, Hohhot, Harbin, Hefei, Jinan, Zhengzhou and Kunming. Disappointingly, there still existed such cities as Ningbo, Guangzhou and Changsha whose industrial water reuse ratio had not reached 50 %. Thus, there were huge distances on this indicator among those cities. From 2006 to 2009, the comprehensive utilization rate of industrial solid wastes in municipalities and separate planning cities increased in a low speed from 85.17 to 85.66 %, mainly due to the decrease in that indicator in Beijing, Ningbo and Xiamen, while that indicator showed a relatively high surge in provincial capitals, changing from 68.10 to 72.02 %. Among municipalities and separate planning cities, Dalian was the one with fastest growth in the comprehensive utilization rate of industrial solid wastes from 78.7 to 94.9 %. By 2009, the comprehensive utilization rate of industrial solid wastes had not reached 50 % in Hohhot, Taiyuan, Guiyang and Kunming, those four provincial capitals, making them fall considerably behind other cities.

Each city differs a lot in pollutants governance. The investment system for industrial environment pollution governance needs to be perfected. In 2006, investment of municipalities and separate planning cities in industrial environment pollution governance made up 0.17 % of GDP while that of provincial capitals accounted for GDP's 0.14 %. That indicator for those two categories of cities declined respectively to 0.07 and 0.09 % in 2009. During the *11th Five-Year Plan* period, each city has made great achievements on governing sulfur dioxides, COD, ammonia nitrogen and nitrogen oxides, four pollutants produced in industrial output. The removal rate of industrial sulfur dioxides for municipalities and separate planning cities and provincial capitals rose respectively from 46.19 and 44.04 to 65.77 and 67.67 %. However, there still existed 11 cities, namely Shenyang, Changchun, Harbin, Hangzhou, Hefei, Nanchang, Zhengzhou, Wuhan, Changsha,

Table 4.12 Pollutants governance of the secondary industry in major cities

		2006	2009	Average annual change rate (%)
Investment in industrial environment pollution governance as percentage of GDP (%)	Municipalities and separate planning cities	0.17	0.07	−20.53
	Provincial capitals	0.14	0.09	−12.38
Industrial sulfur dioxide removal rate (%)	Municipalities and separate planning cities	46.19	65.77	14.13
	Provincial capitals	44.04	67.67	17.89
Industrial wastewater's COD removal rate (%)	Municipalities and separate planning cities	75.06	79.37	1.92
	Provincial capitals	69.87	80.46	5.05
Industrial wastewater's ammonia nitrogen removal rate (%)	Municipalities and separate planning cities	69.39	86.07	8.02
	Provincial capitals	69.52	74.86	2.56
Industrial nitrogen oxide removal rate (%)	Municipalities and separate planning cities	11.07	10.50	−1.73
	Provincial capitals	5.48	5.48	0.02

Note Due to data limitation, Haikou and Xining are not included in the calculation of industrial sulfur dioxide removal rate of provincial capitals
Sources Ministry of Environmental Protection. *The Annual Statistical Report of China's Environment of* 2006. Beijing: China Environmental Science Press, 2006. Ministry of Environmental Protection. *The Annual Statistical Report of China's Environment of* 2009. Beijing: China Environmental Science Press, 2009

Nanning and Xi'an whose sulfur dioxide removal rate had not exceeded 50 % by 2009, which indicates there is plenty of room for that indicator to improve in the green growth of the secondary industry. Provincial capitals performed better than municipalities and separate planning cities in governing COD discharged in the industrial wastewater. The provincial capitals' removal rate of industrial wastewater's COD exceeded 80 % in 2009, exceeding 79.37 % of municipalities and separate planning cities by an average annual rate of 5.05 %. That indicator for most cities had reached over 70 %. However, Xining, Changchun and Chongqing had a relatively low rate in removing COD with respectively 18.86, 47.65 and 52.48 %. Apart from that, that indicator showed a declining tendency in Tianjin, Shenyang, Changchun, Qingdao, Changsha and Guangzhou. On the part of governing the amount of ammonia nitrogen discharged in the industrial wastewater, municipalities and separate planning cities did a better job. The removal rate of the amount of ammonia nitrogen discharged in the industrial wastewater in municipalities and separate planning cities and provincial capitals reached respectively 69.39 and 69.52 % in 2006, with little difference. By 2009, that indicator in municipalities and separate planning cities had amounted to 86.07 % while that of provincial capitals was only 74.86 %. That indicator for each city varied greatly. Among the municipalities and separate planning cities, Ningbo, Dalian and Beijing had reached over 90 % in the removal rate of ammonia nitrogen but that of Tianjin was only 29.87 %. Harbin, Nanjing, Jinan and Guangzhou, among provincial capitals, had the ammonia nitrogen removal rate of over 90 % while that of Xining and Shenyang had merely reached 22.58 and 27.18 %. Compared with the governance over previous categories of pollutants, the removal rate of industrial nitrogen oxides depends on the enhancement of relevant measures from China for further improvement. By 2009, that indicator for municipalities and separate planing cities barely reached 10.50 %, a 0.57 % drop compared to 11.07 % in 2006. According to the data available, only Xiamen exceeded 50 % in the removal rate of industrial nitrogen oxides (Table 4.12).

4.3 Outlook on the Green Growth of China's Secondary Industry During the 12th Five-Year Plan Period

"The Twelfth Five-Year Plan for National Economy and Social Development of the People's Republic of China"(hereinafter referred to as 12*th Five-Year Plan*) clearly states that we should stick to the road of new industrialization with Chinese characteristics, adapt to changes in market demand, give full play to China's industrial comparative advantages in line with the new trend of technology progress, and develop a modern industrial system of structural optimization, advanced technology, cleanliness and safety, high value-added and huge employment potential. During the period, the green growth of the secondary industry will face new opportunities and challenges.

4.3.1 Opportunities for the Green Growth of the Secondary Industry During the 12th Five-Year Plan Period

The 12*th Five-Year Plan* provides opportunities for the green growth of the secondary industry, which can be classified as follows.

4.3.1.1 Emphasis on Technical Upgrading Sets a Direction for the Development of Traditional Industries

Technology is the fundamental driving force to upgrade the structure of the secondary industry. The research and development, introduction, digestion and assimilation of resource-saving and environment-friendly technologies can succeed in giving impetus to the upgrading and transformation of the traditional industries. By now, China has consistently ranked Number 1 in the output of some over 200 industrial products like steel, coal, cement, cotton, etc. all over the world. However, the manufacture has long been confronted with the problem of "being big yet not being strong". Especially, the competitiveness of traditional industries requires to be improved urgently. The 12*th Five-Year Plan* has put forward the goal to transform and upgrade the manufacture and put emphasis on the adoption of new technology, new material, new process as well as new equipment and the integration of industrialization and informationization in depth, which points out directions for developing traditional industries.

In order to encourage enterprises to reform technologies, various kinds of policies in favor of enterprises will be implemented during the 12*th Five-Year Plan* period to help them improve equipment, optimize production process, phase out the outdated techniques and equipment and improve the comprehensive utilization level of energies and resources. Our country is expected to motivate enterprises to enhance their ability to develop new products, raise the technological content and value-added in the products and promote the upgrade and update of the products. The imformationization of research and design, production and distribution and business management should be promoted and advanced quality management should be implemented to push forward the innovation of business management.

Column 4.1 Emerging Industries Take the Lead in the "Green Growth" of Zhangshu City in Jiangxi Province

In order to be actively integrated in the construction of Poyang Lake Ecological Economic Zone, Zhangshu City highlights the concept of green growth, adopts strict project access rules, optimizes industrial layout and guides enterprises towards transformation and upgrade through the technical reform. A series of new industrial projects characterized by low energy consumption, low pollution and low emissions, begins to sprout, represented by biopharmaceuticals, tourism, food and new energy. Industries in the city presents a satisfying trend of green uprising. In 2010, the industrial enterprises above designated size in Zhuangshu City achieved a 25.9 % increase in industrial value-added and comprehensive energy consumption against 10,000 yuan GDP presented a year-on-year decrease of 6 %.

Zhuangshu City brings those scattered industrial spots into integration, carries out "retreating from urban areas, shifting to industrial parks" strategy and cultivates environmentally-friendly economic industry belts of distinctive features. This city, based on the industrial parks, focuses itself on industrial butt joint, strengthens the construction of industry bases and releases a series of preferential policies, ranging from land use to industrial park plan, to introduce production factors as capitals, talents and technologies into the park. Three provincial industrial parks: Chenbei Economic and Technological Development Zone with food, clothing, equipment manufacturing, etc. as its leading industries, Fucheng Pharmaceutical Park with biological pharmacy and pharmaceutical circulation as leading industries together with Xinjishan Salt Chemistry Base with salt chemistry as its leading industry are established. In order to realize the intensive use of resources and centralized control over pollution, the city cultivates a new system for economic development in which the leading industries are taken as guide; corporations' advantages are complemented with each other; rational allocation of resources are achieved and layout structure are optimized. 19 major projects of over 50 million yuan was introduced by the city, 17 of which intended to locate itself in the industrial parks in the sole year of 2010. By now, about 90 % of enterprises above designated size had clustered in the industrial parks.

Besides, through such measures as financial support, this city guides enterprises to eliminate backward production capacity, promote transformation and upgrade of technologies, build the production process for material circulation, namely "resources–products–renewable resources–renewable products" and minimize wastes in amount and maximize the reuse ration of wastes. According to statistics, the input of enterprises above designated size in technological transformation has been no less than 10 % of the sales revenue earned that year for three consecutive years in the city. Over 90 % of enterprises have conducted technological transformation once or twice. The money derived from the reuse of wastewater, waste residues and waste gases can reach approximately a billion yuan. By now, the industries in the city has step into a development track of "increasing output without increasing pollution, improving efficiency without increasing consumption". The whole industrial economy has presented a trend of "low input, high output and low emissions".

Source:

China Jiangxi Net,

http://www.jxcn.cn/34/2011-4-10/30093@886628.htm,2011-04-10.

4.3.1.2 The Integration of New Technologies and Industries Provides Opportunity for Fostering Strategic Emerging Industries

Technological innovation and the upgrade of industries bring the world economy into a new development phase. UK focuses on developing low-carbon economy and the digital economy in order to construct its future. EU proclaims to inject hundreds of billions of euro into the development of green economy. Russia values nanotechnology and nuclear technology. Emerging industries especially leading industries and pillar industries are crucial to the core competitiveness and development potential of China's industries.

During the 12*th Five-Year Plan* period, we are supposed to promote the integration of emerging technologies and emerging industries on the basis of major technological breakthroughs and major development needs. While continuing to make high-tech industries bigger and stronger, we take the responsibility to cultivate those strategic emerging industries into leading industries and pillar industries and to achieve the goal of increasing the proportion of value-added of strategic emerging industries to 8 % of GDP. According to the 12*th Five-Year*

Plan, our focus will be on energy conservation and environmental protection, a new generation of information technology, biology, high-end equipment manufacturing, new energy, new material and new energy vehicles, which means China will put emphasis on fostering emerging industries in the coming period.

Although emerging industries have a bright prospect, their risks shall not be neglected. During the 12*th Five-Year Plan* period, government will set up special funds for the development of strategic emerging industries and industry investment funds, enlarge investment scales on starting new industries, make full use of financing in multi-layer capital markets and directs those social funds to the innovative enterprises still at early or middle stage. Besides, we should make comprehensive use of fiscal incentive policies like risk compensation, encourage financial institutions to increase credit support and improve taxation policies that foster innovation and guide investment and consumption.

Column 4.2 Energy Conservation and Emission Reduction in Huaining County, Anhui Province Promote the Development of Green Economy

How to increase industrial output while reducing energy consumption at the same time? That is a tough question. However, Huaining County in Anhui Province has done a great job. Huaining County regards energy conservation and emission reduction as the keystone to restructuring and mode transformation, persists in promoting the transformation of development mode towards "low input, low consumption, high output" direction and fosters the growth of green economy.

Good combination of backing up the good and phasing out the bad

Huaining County takes industrial energy conservation, construction energy conservation, energy conservation of municipal engineering and energy conservation of government bodies as four important domains of energy saving. Energy consumption against 10,000 yuan GDP has presented a declining trend year by year. Energy consumption against 10.000 yuan GDP dropped by 4.56 % in 2009 compared to last year. Ten major energy-consuming enterprises, listed under the province, city and county assessment, has all over fulfilled the energy conservation quota. It was predicted that year 2010 would witness a year-on-year decrease of over five percent in energy consumption against 10,000 yuan GDP and 9 enterprises under assessment would accomplish their energy conservation tasks. Public institutions' energy consumption dropped by 4.3 % in 2009 compared to that in 2008 and that in 2010 would have a further decline. Relevant authorities releases many incentives to eliminate backward production capacity and are active in helping enterprises declare national and provincial subsidy programs for shutting down small businesses. During the 11*th Five-Year Plan* period, 56 cement firms, with 71 production lines and a total production capacity of 1.4 million tons, were shut down. 14 clay firms were closed in 2009 and another 2 in 2010. Over 75 % of the 38 remaining clay firms are scheduled to be eliminated in 2011 and 2012. 5 fireworks enterprises were also closed.

To make breakthroughs on energy conservation and emission reduction

Huaiyang County emphasizes the establishment of sewage treatment plants, which are also the major projects of reducing emissions, and invests a total amount of 60.57 million yuan. The county sewage treatment plant, with daily disposal capacity of 50,000 tons, is under construction with an input of 30 million yuan. The first phase, with daily disposal capacity of 15,000 tons, has already been put into use since January in 2010. By now, sewer networks of 97.4 km have been built in the county and with 16.3 km still under construction. Those sewer networks enable sewage treatment facilities to run properly. A sewage treatment plant introduced by Shipai Town, with daily disposal capacity of 2,000 tons, has been built and put into operation, which is responsible for centralized treatment of the sewage produced by enterprises in the industrial clustering areas in this town. The county should prompt the Yueshan Copper Mine's bankruptcy according to law, promote the industrial transformation of Haitun Cement from burning cement to grinding cement and shut down a series of small enterprises according to law. Through the transformation of economic structure, closure and bankruptcy of a series of enterprises, the emissions of polluters has been substantially reduced and the emission reduction target during the 11*th Five-Year Plan* period can be fully achieved. The funds used for reducing emissions during the 11*th Five-Year Plan* period added up to 60 million yuan.

Source :

Social News Center, http://www.soci.com.cn/resource/37021.html, 2010-12-07.

4.3.1.3 The Reform on Energy Production and Utilization Ways Provides a Powerful Guarantee for Exploring and Exploiting New Energy

Due to the stimulus of oil crisis in the 1970s, some countries begin to explore new energies. In recent years, the upsurge in oil price has prompted each country to pay more attention on energy-saving technologies and products. The exploitation of new energies like solar energy, bio-energy, wind power, geothermal energy and oceanic energy can reduce consumption of high-carbon fossil energies. Developing green energies to transform energy structure into one with energies containing less carbon dioxide is the important component and key link of guaranteeing the green growth of the secondary industry.

It was put forward in the 12*th Five-Year Plan* to keep taking conservation as a priority, being based on domestic, developing in diversified ways, protecting environment, to enhance mutual beneficial cooperation internationally, to adjust and optimize energy structure, and to establish a modern energy industrial system which is safe, stable, economic and clear. This has provided a specific direction for energy development during the 12*th Five-Year Plan* period. Specifically, the priority is to promote the diversified clear development of energies and alter the energy-consuming structure mainly based on coals. Secondly, the energy exploitation layout should be optimized with eastern coastal areas and several areas from the mid being the priorities in developing nuclear power. Thirdly, construction of energy delivery channels should be strengthened to improve efficiency of energy delivery and utilization in the East and West.

Column 4.3 Baoding National High-tech Zones in Hebei Province has been Shaped into Domestic First-class New Energy Industrial Park

Baoding National High-tech Zone, since established in 1992, by virtue of its favorable geographical position and industrial distinguishing features, has chosen a development path based on new energy industries and borne countless fruits. Since it began to nurture new energy industries in 2000 and introduced the concept of "China Electric Valley" into Hong Kong in 2006, Baoding National High-tech Zone had adhered to the path based on new energy industries and created lots of "China firsts" as China's first solar photovoltaic power station integrated with a five-star hotel, China's first silicon ingot solar cell of 240 kg, China's first high-power wind power blade, China's first research and development center on wind power blades, China's first detecting platform for wind power transmission, China's sole State Key Laboratory of Wind Power Technology and Equipment and China's sole State Key Laboratory of Photovoltaic Materials and Technology. It has turned into a new energy industrial zone with the highest degree of industrial centralization, the most intact industrial chain, the most outstanding creativity and the clearest industrial positioning.

Six industries including wind power, photovoltaic, panel point, power storage, power transmission and transformation and power automation have thrived in "China Electric Valley". Baoding has housed over 200 new energy enterprises and industrial zones of wind power, photovoltaic, power storage, panel point and power automation has respectively come into being. During the 12*th Five-Year Plan* period, Baoding High-tech Zone will take "China Intelligent Electric Valley, Northern New Low-carbon City" as development vision of the new period, highlight low-carbon concept as guide, featured industries as support, technological innovation as motive and shaping new city image as target and make efforts to transform itself into the new energy equipment and technology center with international influence, the industrial pilot area and innovation center of national intelligent grid and the core growth pole and technological new district of Baoding. By the end of the 12*th Five-Year Plan* period, there will have been a leap in the aggregate GDP's growth in the zone, with major economic indicators increasing by an average annual rate of nearly 30 %. Baoding High-tech Zone, with total income to reach 266 billion yuan and gross industrial output value to amount to 170 billion yuan by 2015, will become a champion of "Billion Zone".

Source:

ChinaTaiwan.org, http://www.chinataiwan.org/local/hebei/dongtaixinwen/201104/t20110426_1835202.htm.

4.3.1.4 Striving to Develop Circular Economy Provides a Practical way to Improving Resource Output Efficiency

Circular economy takes highly efficient use of resources and recycling of resources as core and "decrement, recycling, resource recovery" as target in order to achieve the green growth of "low energy consumption, low emissions and high efficiency". "Circular Economy Promotion Law of the People's Republic of China" passed in 2008 has not only stipulated circular economy planning system but also strengthened governance over energy-intensive and water-inefficient enterprises and specified requirements on recycling and resource recovery. Demonstration and pilot work of circular economy was conducted in October, 2005 and December, 2006 respectively to explore how to develop circular economy in cities.

The 12*th Five-Year Plan* has been put forward to encourage the development of circular economy, adhering to the principle of "decrement, recycling and resource recovery" and taking improvement of resource output efficiency as target, promote the development of circular economy in every link of production, circulation and consumption and accelerate the pace of establishing a resource recycling system which covers the whole society. To be more specific, the first thing to do is to adopt circular production mode so that resource consumption can be reduced while resource utilization level can be improved. Second is to establish a sound resource recycling system. Third is to promote green consumption mode. Forth is to enhance support from policies and technologies and promote the construction of circular economy pilot sites in Gansu Province and Qiaodam Circular Economy Pilot Site in Qinghai Province and the construction of comprehensive reform experimental zones in accordance with the transformation of resources-based economy in Shanxi Province.

Column 4.4 The Construction of Qiaodam Circular Economy Pilot Site has made Great Progress

Qiaodam Circular Economy Pilot Site (pilot site) in Qinghai Province, approved by the State Council in 2005 and permitted by six commissions as National Development and Reform Commission, State Environmental Protection Administration, etc., is one of the first 13 circular economy pilot sites of China and has become the China's biggest regional industrial pilot park of circular economy. The pilot site, with a total area of 256,000 km^2, has a total population of about 560,000. "*The Overall Plan for Qiandam Circular Economy Pilot Site of Qinghai Province*", approved in March, 2010 by the State Council, has upgraded Qiaodam's development of circular economy to a national strategy from a regional strategy. Thanks to great attention and full support of Qinghai Provincial Government, the pilot site has done plenty of highly effective work on creating a development mode of circular economy for the highlands. The total output value of the pilot site reached 28.5 billion yuan in 2009, with an average annual increase of 16.7 %. Industrial value-added amounted to 22.3 billion yuan, which grew by an average of 18.6 % yearly.

Over the years, the pilot site has mainly dealt with the following work.

First is to make scientific plans on the pilot site's development. Planning work has made breakthroughs. The pilot site has formulated "the Overall Plan" with a high starting point, high standards and high quality, which has achieved national approval. "The Overall Plan" was put forward to help establish Qiaodam District's new development model of circular economy, that is "recycling at the very start, producing in a clean way" and form general development ideas on exploring new approaches to sustainable development in resources-based yet ecologically fragile areas according to the requirements of "highlighting the feature of resources, strengthening technological innovation, cultivating competitive industries and achieving cyclic development", meanwhile adhering to the five principles of "overall planning and rational layout, ecological protection and harmonious development, scientific development and first action and trial, comprehensive utilization and cyclic development, technology as support and market as guide". 26 major development plans including "the Development Plan for Characteristic Competitive Industries of Qiaodam Circular Economy", "the Plan for Haixi National Sustainable Development Pilot Site" and "the Plan for the Construction of Scientific and Technological Innovation System in Qiaodam Circular Economy Pilot Site" have been successively compiled, which has contributed to basically forming a comprehensive planning system.

Second, the pilot site has spared no effort to promote project construction and the industrial framework led by circular economy has initially taken shape. Over 30 major industrialization projects have been in place over the years. 17 of them are solar photovoltaic power stations in the scope of development of new energy, which have been permitted by Provincial Development and Research Commission to conduct preliminary work and 6 of which has started construction.

Third, the pilot site has enlarged its investment in infrastructures and its development environment has achieved notable improvements. The pilot site has raised and implemented 9.1 billion yuan through multiple channels and carried out a series of major infrastructure projects on transportation, electricity, etc. Taking the grid as an example, 330 kV power transmission line of 744 km, 110 kV of 990 km and 35 kV of 352 km have been newly build, with the big electrical network's total coverage area amounting to 250,000 km^2.

Forth, the pilot site has strengthened scientific research and its technological support ability has obviously improved. Since 2005, the pilot site had carried out various kinds of scientific projects up to 386, invested funds of 95.83 million yuan and achieved 136 scientific and technological awards on exploration, exploitation and processing of oil and gas, extraction of potassium, lithium, magnesium and boron from salt lakes, mining and dressing of nonferrous metals and utilization of waste ores. The pilot site has been defined as "the Characteristic Competitive Industries Base for West Development" by State Council West Development Office, approved by the National Ministry of Industry and Information as "Qiaodam National Industry Demonstration Base of New Industrialization for Salt

Lake Chemical Industry and New Metal Materials" and permitted by the Ministry of Science and Technology as "Sustainable Development Pilot Site" and "National High-tech Industrialization Base for Salt Lake Distinctive Materials".

Fifth, omni-directional work on investment attraction has been conducted in the pilot site and its foreign cooperation has achieved significant results. 188 projects have been signed in ways of door-to-door investment promotion and investment promotion through business, enterprises as well as meetings, with contract funds of 96.5 billion yuan. 116 of those projects have been put into operation and 28.1 billion yuan put in place.

Sixth, the pilot site has paid close attention to energy conservation and emission reduction and ecological construction and has created a much more optimized sustainable development environment. 320 million yuan has been accumulatively invested into ecological construction to launch several ecological environment projects as reforestation, shelterbelts, national key public welfare forests, comprehensive ecological environment governance over Tianjun sector in Qinghai Lake Basin and surrounding areas along Qinghai Lake Basin, 315 National Road. greening works along Qinghai-Tibet railway in Haixi District. 13,033 hectares of desertification lands have been governed and 577,107 hectares of vegetation in sand area protected.

Sources:

1. The State Council. The Overall Plan for Qiandam Circular Economy Pilot Site of Qinghai Province, 2010-04.
2. Work report of government of Haixi autonomous prefecture of Mongolians and Tibetans on the 28th conference of the Twelfth Standing Committee of NPC of Haixi autonomous prefecture—report on the construction work of Qiaodam Circular Economy Pilot Site, 2010-01.
3. This column is offered by Tian Junliang, director of Qinghai Ecological Environment Protection Research and Guidance Center.

4.3.2 Challenges Facing the Green Growth of the Secondary Industry During the 12th Five-Year Plan Period

During the 12*th Five-Year Plan* period, the secondary industry in China is endowed with unprecedented opportunities. Meantime, those factors containing green growth of the secondary industry still exist, which poses great pressure as well as challenges on the secondary industry's green growth.

4.3.2.1 The High Percentage of Energy-Intensive Industries make Energy Conservation and Emission Reduction a Tough Task

We are now in the midterm of industrialization when industries with high energy consumption and high emissions occupy a large amount. Those small steel plants, coke plants, cement plants and coal plants with high pollution are pervasive across China. The high percentage of energy-intensive industries as petrochemicals, steel and cement in the whole industry and high dependence of industrial growth on energies lead to the inefficiency in energy and resources utilization so that carbon dioxide and sulfur dioxide emissions of our country are among the world's highest. We will face more pressure on emission reduction during the 12*th Five-Year Plan* period. Not only will the hard constraints of resources and environment on domestic economic development be enhanced, but the developed countries will require China to shoulder the responsibility of reducing emissions, which sets higher requirements on China's transformation of economic development mode. Therefore, energy conservation and emission reduction is a long-term and arduous task, with a lot of challenges to take.

4.3.2.2 The Lack of Industrial Technology Restrain the Sustainable Development of the Secondary Industry

Although China has become a global manufacturing power and "the world factory", it mainly focuses on the manufacturing of non-core components with low value-added and the labor-intensive assembly link. By comparison, scientific force in leading high-tech fields is relatively weak and key technology is controlled by others. Absence of independent intellectual property rights and domestic brands make it difficult to develop a powerful pillar for the sustainable and healthy development of economy and society. In addition, China is now located in the low-end of the global value chain and earns slim margins while the upstream and downstream of value chain links with high value-add as research & design, core technologies, distribution channels, brand operation are under the firm control of transnational corporations overseas. Therefore, how to change the industrial actuality, seek technological innovation and achieve the development of new technological industries becomes the key challenge to the sustainable development of the secondary industry.

4.3.2.3 The Mechanism for the Green Growth of the Secondary Industry Still Needs to be Improved

The 12*th Five-Year Plan* period is the key five years for China's comprehensive transformation. Promotion of the second transformation and reform, focusing on the industrial restructuring, to a large extent, depends on the institutional arrangement and innovation of the secondary industry's green transformation.

For that reason, it is of great necessity to perfect the incentive and restraint mechanism for energy conservation and emission reduction, optimize the energy structure, bring the gross energy consumption under reasonable control and improve the pricing mechanism for resource products as well as the taxation system for resources and environment. Besides, it is essential to establish sound laws and regulations for energy conservation and emission reduction, strengthen the target responsibility assessment of energy conservation and emission reduction, bring resource conservation and environmental protection into every field and link of production, construction, circulation and consumption, and enhance the sustainable development capacity.

Chapter 5
Green Growth of the Tertiary Industry

Jinshi Liu, Jiang Xin and Jin Jingjing

The *12th Five-Year Plan* for the National Economic and Social Development of the People's Republic of China clearly stated that it's by the green growth that economic development mode will be transformed and resource-conserving & environment-friendly society will be constructed. Concretely speaking, resource conservation and management should be strengthened, recycling economy should be promoted and environmental protection should be reinforced. The green transformation of economic development mode not only offers new opportunities to economic development, but also brings challenges to sustainable economic growth.

Characteristics of tertiary industry are low intensity of resource consuming, light environmental pollution and high labor productivity, and these contribute to the green feature of tertiary industry. Therefore it corresponds with the green transformation of economic development mode to develop the tertiary industry, particularly strive to develop modern service sectors with higher technological level and higher value-added. During the *11th Five-Year Plan* period, Chinese tertiary industry has rapidly developed, and raised the importance in the national economy. Simultaneously, the green degree of tertiary industry growth has been enhanced, the resource utilization efficiency has been steadily increased, the environmental pollution has been mitigated and the labor productivity has been improved. On the other hand, there are still pending issues to be tackled for the green growth of tertiary industry. This chapter emphasizes on characteristics and issues of the green growth of tertiary industry, regional comparison of green growth of tertiary industry during the *11th Five-Year Plan* period and the outlook of the green growth of tertiary industry during the *12th Five-Year Plan* period.

X. Li and J. Pan (eds.), *China Green Development Index Report 2011*,
Current Chinese Economic Report Series, DOI: 10.1007/978-3-642-31597-8_5,

5.1 Characteristics and Issues of the Green Growth of Chinese Tertiary Industry During the 11th Five-Year Plan Period

The growth of Chinese tertiary industry has exhibited evident green feature during the *11th Five-Year Plan* period, under the attention and supports of whole nation on the green development of economy. Proportion of value-added of tertiary industry in GDP increases rapidly, energy consumption per unit of value-added falls consistently, and labor productivity rises steadily. Simultaneously, issues also appear in the green growth of tertiary industry, such as the insufficient absorption of labor, the high proportion of traditional industry, and the disparity of regional development. This session will summarize and analyze characteristics and issues of the green growth of Chinese tertiary industry during the *11th Five-Year Plan* period.

5.1.1 Characteristics of the Green Growth of Tertiary Industry During the 11th Five-Year Plan Period

Characteristics of the green growth of tertiary industry during the *11th Five-Year Plan* period have been exhibited in the following three aspects: rapid increase of proportion of value-added of tertiary industry in GDP, continuous fall of energy consumption per unit of value-added and steady rise of labor productivity.

5.1.1.1 Rapid Increase of Proportion of Value-Added of Tertiary Industry in GDP

During the *11th Five-Year Plan* period, the proportion of value-added of tertiary industry in GDP has increased consistently, and the importance of tertiary industry in national economy has enhanced continuously. As shown in the Table 5.1, the proportion of primary industry and secondary industry in GDP falls persistently, and the proportion of tertiary industry rises continuously during the *11th Five-Year Plan* period. From 2006 to 2009, the proportion of value-added of tertiary industry in GDP has increased from 40.9 to 43.4 %, during the same time, the proportion of primary industry has fallen from 11.1 to 10.3 %, and the proportion of secondary industry has dropped from 47.9 to 46,3 %. During these four years, the average growth rate of tertiary industry has reached 18.6 %, far beyond that of primary and secondary industry.

Table 5.1 Changes on the value-added of Chinese three industries during the 11th Five-Year plan period

Years		2006	2007	2008	2009
Value-added (100 million RMB)	GDP	2,16,314	2,65,810	3,14,045	3,40,507
	Primary industry	2,4,040	28,627	33,702	35,226
	Secondary industry	1,03,720	1,25,831	1,49,003	1,57,639
	Tertiary industry	88,555	1,11,352	1,31,340	1,47,642
Proportion (%)	GDP	100	100	100	100
	Primary industry	11.1	10.8	10.7	10.3
	Secondary industry	47.9	47.3	47.4	46.3
	Tertiary industry	40.9	41.9	41.8	43.4
Average growth rate (%)	GDP	16.3			
	Primary industry	13.6			
	Secondary industry	15.0			
	Tertiary industry	18.6			

Note Proportion and growth rate is calculated by the current price
Source China Statistical Yearbook 2010, National Bureau of Statistics of China, Beijing, China Statistical Publishing House, 2010

5.1.1.2 Continuous Fall of Energy Consumption per Unit of Value-Added of Tertiary Industry

In three industries, the green degree of tertiary industry is relatively high, and the energy consumption of each sector within tertiary industry, compared to the secondary industry, is relatively low. From the perspective of the aggregate energy consumption, the energy consumption of tertiary industry is low, and only takes a small portion in the aggregate energy consumption. As shown in the Table 5.2, the proportion of energy consumption of tertiary industry in three industries is stable at 14 %, far below that of secondary industry, and the proportion sustains at low level during the *11th Five-Year Plan* period. In 2009, the proportion of energy consumption of tertiary industry is less than 1/5 of that of secondary industry.

From the perspective of the energy consumption per unit of value-added, the energy consumption per unit of value-added of tertiary industry is relatively low, and drops consistently. As shown in the Table 5.3, the energy consumption per unit of value-added of tertiary industry sustains at the low level during the *11th Five-Year Plan* period. In 2009, the energy consumption per unit of value-added of tertiary industry is only 32 % of that of GDP. From 2006 to 2009, the energy consumption per unit of value-added of tertiary industry has consistently fallen, from 0.405 t standard coal per 10,000RMB in 2006 to 0.290 in 2009, down 28.5 %.

Table 5.2 Structure of the energy consumption of Chinese three industries during the 11th Five-Year Plan period

Indicator	Years	Aggregate energy	Primary industry	Secondary industry	Tertiary industry
Energy consumption (10,000 t standard coal)	2006	2,58,676	6,331	1,88,706	35,874
	2007	2,80,508	6,228	2,04,659	38,807
	2008	2,91,448	6,013	2,13,115	40,422
	2009	3,06,647	6,251	2,23,759	42,794
Ratio of energy consumption (%)	2006	100	2.4	73.0	13.9
	2007	100	2.2	73.0	13.8
	2008	100	2.1	73.1	13.9
	2009	100	2.0	73.0	14.0

Note Industry category follows the classification of the National Bureau of Statistics of China, Ratio of energy consumption = energy consumption of each industry/aggregate energy consumption

Source China Statistical Yearbook 2010, National Bureau of Statistics of China, Beijing, China Statistical Publishing House, 2010

Table 5.3 The energy consumption per unit of value-added of Chinese three industries during the 11th Five-Year Plan period

Energy consumption per unit of value-added (ton standard coal per 10,000RMB)	GDP	Primary industry	Secondary industry	Tertiary industry
2006	1.196	0.263	1.819	0.405
2007	1.055	0.218	1.626	0.349
2008	0.928	0.178	1.430	0.308
2009	0.901	0.177	1.419	0.290

Note 1. this table is calculated at current price; 2. Energy consumption per unit of value-added = energy consumption/value-added

Data source China Statistical Yearbook 2010, National Bureau of Statistics of China, Beijing, China Statistical Publishing House, 2010

5.1.1.3 Steady Rise of Labor Productivity of Tertiary Industry

During the *11th Five-Year Plan* period, the employee number of Chinese tertiary industry has risen persistently, and the proportion of employees of tertiary industry in total employees has increased steadily. As shown in the Table 5.4, during the *11th Five-Year Plan* period, the employee number has risen from 246.14 million in 2006 to 266.03 million in 2009, up 8 %. It's the highest growth among three industries. In the meanwhile, the proportion of employees of tertiary industry in total employees has increased from 32.2 % in 2006 to 34.1 % in 2009, up 1.9 %.

With the absolute growth of employee number and the relative growth of proportion of tertiary industry in total employees, the labor productivity of tertiary industry rises steadily as well, and thus, the green growth of tertiary growth is further enhanced. As shown in the Table 5.5, the labor productivity of tertiary industry exceeds the average level significantly. During the *11th Five-Year Plan*

Table 5.4 The employment of three industries in China during the 11th Five-Year Plan period

Years		2006	2007	2008	2009
Employees (10,000 persons)	Total	76,400	76,990	77,480	77,995
	Primary industry	3,2561	31,444	30,654	29,708
	Secondary industry	19,225	20,629	21,109	21,684
	Tertiary industry	24,614	24,917	25,717	26,603
Proportion in total employees (%)	Total	100	100	100	100
	Primary industry	42.6	40.8	39.6	38.1
	Secondary industry	25.2	26.8	27.2	27.8
	Tertiary industry	32.2	32.4	33.2	34.1

Source China Statistical Yearbook 2010, National Bureau of Statistics of China, Beijing, China Statistical Publishing House, 2010

Table 5.5 The labor productivity of three industries in China during 11th Five-Year plan

Years		2006	2007	2008	2009
Value-added (100 million RMB)	Total	2,16,314	2,65,810	3,14,045	3,40,507
	Primary industry	24,040	28,627	33,702	35,226
	Secondary industry	1,03,720	1,25,831	1,49,003	1,57,639
	Tertiary industry	88,555	1,11,352	1,31,340	1,47,642
Employees (10,000 persons)	Total	76,400	76,990	77,480	77,995
	Primary industry	32,561	31,444	30,654	29,708
	Secondary industry	19,225	20,629	21,109	21,684
	Tertiary industry	24,614	24,917	25,717	26,603
Labor productivity (RMB100 million/10,000 persons)	Total	2.83	3.45	4.05	4.37
	Primary industry	0.74	0.91	1.10	1.19
	Secondary industry	5.40	6.10	7.06	7.27
	Tertiary industry	3.60	4.47	5.11	5.55

Note Labor productivity = Value-added/number of employees
Source China Statistical Yearbook 2010, National Bureau of Statistics of China, Beijing, China Statistical Publishing House, 2010

period, the labor productivity of tertiary industry has increased from 360 million RMB/10,000 persons to 555 million RMB/10,000 persons, with 54 % growth.

Column 5.1 Chongqing: Transformation from Brown Economy Toward Green Economy

Traditionally, Chongqing has been always the heavy industrial and manufacturing base in Western China. Chongqing formed the "brown economy" pattern with characteristics of the unreasonable industrial distribution and the large proportion of high energy consuming & heavy polluted sectors. As one of the old industrial cities of the country, Chongqing has almost reached its limit on the development dynamic and the carrying capacity of environment under the original development pattern. Chongqing has realized the opportunity to transform economic development pattern during the *12th Five-Year Plan* period, and actively explored the development of green economy. Now Chongqing has already made significant progress on the realization of economic, social and environmental win–win, through the application of green development mentality. Following experiences can be summarized from the transformation practices from "Brown Economy" to "Green Economy" by Chongqing.

Structural adjustment and the economic development pattern transformation are the fundamental path toward green economy. Chongqing persists the new road of industrialization, strives to the transformation of development pattern, enhances the capacity of independent innovation, boosts the strategic industries such as modern information sector, enlarges the vigor of traditional industries reform and upgrading, accelerates the development of modern services, promotes the steady development of modern agriculture, constructs the modern industrial system of the collaboration development of three industries, and consistently improves the whole industrial competitiveness. It can be inferred that Chongqing intends to boost the economic growth and transformation, promote the rational industrial distribution, enhance the economic competitiveness, optimize the urban and rural structure, realize the green growth of economy gradually, and expand the green employment, through the adjustment of economic and industrial structure, particularly the transformation of traditional manufacturing sectors and the cultivation of modern Hi-Tech sectors.

Strict environmental management is the booster of green transformation of economy. Chongqing rests upon the compound decisions in regional development involving with national support and macro-policy, builds the compound decision system of environment and development, and promotes the win–win of environmental protection and development by the long-acting mechanism. On one hand, Chongqing eliminates the backward production capacity, enhances energy-conserving and emission-reduction, and forces the adjustment of industrial structure from the bottom. On the other hand, Chongqing accelerates the exit and move of polluting companies from the main city, boosts industrial upgrade of companies. More importantly, Chongqing launches the strict environmental standard, and raises admittance threshold of enterprises, and pushes the industrial structure adjustment from the up-streaming. It's the strict environmental management that contributes significantly to the green transformation of Chongqing.

The green investment and green technological innovation of companies is the fundamental guarantee of the green transformation. The green investment is the necessary condition of greening traditional companies and fostering emerging industries. Regardless of whether greening traditional polluting industries or fostering emerging industries, the green investment is the prerequisite and important guarantee. The green technological innovation is the powerful support to ensure companies realize the green growth, and the fundamental guarantee of the green transformation of companies. Possessing the advanced and core green technologies and insisting the improvement of technological strengths are the crucial factor for companies to realize the long-term green development.

Laws and regulations, and the innovation on standards are the system guarantee of the green development of economy. Laws, regulations and standards, the rules to guide economic operation, are vital to develop the green economy. Chongqing publicizes the Resolution on Reinforcing Environment Protection and Tentative Scheme on the Emission Trading of Main Pollutants of Chongqing, and revises Regulations of Environmental Protection of Chongqing. Through these regulations and policies, Chongqing adjusts the industrial structure and distribution, pushes the clean production, carries out the environment admittance system, strictly controls the emission of pollutants, launches the bulletin system of emission of pollutants, comprehensively implements the system of pollution discharge permit etc., and thus boosts the development of green economy from policy perspective.
Sources:

1. Chongqing: Exploration on the Transformation from 'Brown Economy' to Green Economy of Chongqing.
http://www.022net.com/2010/9-24/433658343095524-4.html.
2. The Effect of Chongqing Model from 'Brown Economy' to Green Economy http://www.chinanews.com/ny/2010/09-25/2552658.shtml.

5.1.2 Problems of the Green Growth of Tertiary Industry During the 11th Five-Year Plan Period

Under the circumstances that the green growth of tertiary industry has made progress, issues still exists in the green growth of tertiary industry during the *11th Five-Year Plan* period. Compared to the developed countries, the proportion of Chinese tertiary industry in total employees is still low. It means the potential of tertiary industry to absorb labor force needs to be further tapped. Within the tertiary industry, traditional services still dominate, and modern services are lagging. In addition, the developmental disequilibrium of tertiary industry is evident among different regions. All these affect the quality and speed of the green growth of tertiary industry.

Table 5.6 International comparison of the employment of three industries during the 11th Five-Year Plan period

Nation	Primary industry	Secondary industry	tertiary industry	Employee ratio of tertiary industry to secondary industry	Employee ratio of tertiary industry to primary industry
US	1.4	20.6	78.0	3.8	55.7
UK	1.4	22.3	76.0	3.4	54.3
Canada	2.5	21.6	75.9	3.5	30.4
France	3.4	23.2	73.1	3.2	21.5
Germany	2.2	29.8	67.9	2.3	30.9
Japan	4.2	27.9	66.7	2.4	15.9
Korea	7.4	25.9	66.6	2.6	9.0
Thailand	41.7	20.7	37.4	1.8	0.9
China	40.8	26.8	32.4	1.2	0.8

Note 1. Data in the table is the data of 2007; 2. Employee ratio of tertiary industry to secondary industry = employee proportion of tertiary industry/employee proportion of secondary industry; employee ratio of tertiary industry to primary industry = employee proportion of tertiary industry/employee proportion of primary industry

Source China Statistical Yearbook 2010, National Bureau of Statistics of China, Beijing, China Statistical Publishing House, 2010

5.1.2.1 The Potential of Tertiary Industry to Absorb Labor Force Needs to be Further Tapped

During the *11th Five-Year Plan* period, the employee number of Chinese tertiary industry has increased rapidly, the proportion in total employees has risen quickly, and the tertiary industry has gradually become the main force to absorb labors. However, the proportion of employees of Chinese tertiary industry in total employees is still low, compared to some developed countries, the potential of tertiary industry to absorb labor force needs to be further tapped.

As shown in the Table 5.6, in 2007, the proportions of employees of tertiary industry in U.S., U.K., and Canada are all above 75 %. It means more than 3/4 of employed population works in the tertiary industry. In China, the proportion of employees of tertiary industry in total employees is only 32.4 %, far below the developed countries, even lower than Thailand. The proportion of employees of tertiary industry in numerous countries and regions exceeds that of secondary industry, and the proportions of employees of tertiary industry in US, UK, Canada and France are 3 times of these of secondary industry, whereas the employee ratio of tertiary industry to secondary industry in China is only 1.2.

Comparing the tertiary industry with the primary industry, it can be inferred that the proportion of employee of Chinese primary industry in total employees is much higher than that of the developed countries, the proportion of employees of primary industry in China is 29 times of US, and the proportion of employees of tertiary industry in China is only 30 % of US. The employee ratio of tertiary industry to primary industry in China is only 0.8, far below that of developed countries. Although the employee number and the proportion of employee of

Chinese tertiary industry has increased significantly during the *11th Five-Year Plan* period, compared to other nations, the potential of tertiary industry to absorb labor force in China needs to be further tapped.

5.1.2.2 Internal Structure of Tertiary Industry Needs Updating

During the *11th Five-Year Plan* period, Chinese tertiary industry has made significant progress. However, from the perspective of the internal structure of tertiary industry, traditional services still dominate, and modern services are lagging, and thus the green growth of tertiary industry is affected. Modern services rest upon high technologies such as electronic information and etc. and advanced management, management philosophy and organizational form. Thus modern services have charactertics of low resource consumption, light pollution and high output value-added. The development of modern services can improve the green degree of tertiary industry growth. Contrarily, traditional services have higher green degree than the secondary industry, but the green degree is far less than modern services. Therefore, in order to adjust the internal structure of tertiary industry, modern services need to be developed, so as to boost the green growth of tertiary industry.

As shown in the Table 5.7, during the *11th Five-Year Plan* period, traditional services, such as wholesale and retail trade, transport, storage, post and tele-communication etc., takes high weight in the tertiary industry. It's the wholesale and retail trade within traditional services that possesses the highest proportion in the tertiary industry. From 2006 to 2009, the proportion of wholesale and retail trade in the tertiary industry has increased from 18.8 to 19.6 %. In 2009, the value-added of wholesale and retail trade, transport, storage, post and tele-communication and hotel and catering of traditional services reaches 37 %. Simultaneously, modern services have grown slowly, and only possess a low proportion in the tertiary industry. Take the data transmission, computer service, and software as example, the proportion of its value-added takes 6.4 % of the tertiary industry in 2006, whereas drops to 6.0 % in 2009. The proportion of banking and public administration of modern services rises, but only in small degree.

5.1.2.3 Evident Regional Disparities in the Tertiary Industry Development

During the *11th Five-Year Plan* period, when the tertiary industry rapidly grows, regional disparities can be found, especially huge regional gaps in proportion of tertiary industry appear. As shown in the Table 5.8, in 2009 the proportion of tertiary industry in GDP is 43.4 % in the whole country. Among all regions, the proportion of tertiary industry in Beijing is the highest, reaching 43.4 % of GDP, far beyond the national average; whereas Henan is the lowest, only 29.3 %, far less than the national average.

Table 5.7 The value-added by sectors within tertiary industry during the 11th Five-Year Plan period (Unit: 100 million RMB)

Industry	2006	2007	2008	2009
Tertiary industry	88,555	1,11,352	1,31,340	1,47,642
Transport, storage, post and tele-communication	12,183	14,601	16,363	17,058
Data transmission, computer service, and software	5,683	6,706	7,860	NA
Wholesale and retail trade	16,531	20,938	26,182	28,985
Hotel and catering	4,793	5,548	6,616	7,118
Banking	8,099	12,338	14,863	17,728
Real estate	10,370	13,810	14,739	18,655
Leasing and business services	3,791	4,695	5,608	NA
Scientific research, technical services, and geological prospecting	2,685	3,441	3,993	NA
Water conservancy, environment and public utility management	946	1,111	1,266	NA
Personal and other services	3,542	3,996	4,628	NA
Education	6,407	7,693	8,887	NA
Public health, social securities and social welfare	3,326	4,014	4,629	NA
Culture, sports and entertainment	1,363	1,631	1,922	NA
Public administration and social organization	8,837	10,830	13,784	NA

Note Data of the value-added by sectors of tertiary industry in 2009 is only available for some sectors

Source China Statistical Yearbook 2010, National Bureau of Statistics of China, Beijing, China Statistical Publishing House, 2010

5.2 Regional Comparison of the Green Growth of Chinese Tertiary Industry During the 11th Five-Year Plan Period

During the *11th Five-Year Plan* period, Chinese tertiary industry has grown rapidly and steadily, the importance of tertiary industry in national economy has been more and more evident. Either from the perspective of the environmental influence of resource consumption, or from the angle of the proportion of tertiary industry in national economy, the tertiary industry has made evident progress. It exhibits that the green growth of Chinese tertiary industry has made great achievements during the *11th Five-Year Plan* period. However, due to the regional differences on physical geographic environment, social environment and historical background, the regional development gaps exit in China. Studying this unevenness will help us to understand the causes of this disparity, help us to propose the advice and measurement on reducing the gaps, and help to the balanced economic development in China.

The green growth of tertiary industry in Eastern China, Central China, Western China and Northeast will be compared below, and the green growth of tertiary industry in major cities will be compared as well, in order to analyze the regional disparities in the green growth of Chinese tertiary industry.

Table 5.8 Composition of the regional GDP by three industries

Region	Composition (Regional GDP = 100)			Region	Composition (Regional GDP = 100)		
	Primary industry	Secondary industry	Tertiary industry		Primary industry	Secondary industry	Tertiary industry
Nation	10.3	46.3	43.4	Henan	14.2	56.5	29.3
Beijing	1.0	23.5	75.5	Hubei	13.9	46.6	39.6
Tianjin	1.7	53.0	45.3	Hunan	15.1	43.5	41.4
Hebei	12.8	52.0	35.2	Guangdong	5.1	49.2	45.7
Shanxi	6.5	54.3	39.2	Guangxi	18.8	43.6	37.6
Inter Mongolia	9.5	52.5	38.0	Hainan	27.9	26.8	45.3
Liaoning	9.3	52.0	38.7	Chongqing	9.3	52.8	37.9
Jilin	13.5	48.7	37.9	Sichuan	15.8	47.4	36.7
Heilongjiang	13.4	47.3	39.3	Guizhou	14.1	37.7	48.2
Shanghai	0.8	39.9	59.4	Yunan	17.3	41.9	40.8
Jiangsu	6.6	53.9	39.6	Tibet	14.5	31.0	54.6
Zhejiang	5.1	51.8	43.1	Shaanxi	9.7	51.9	38.5
Anhui	14.9	48.7	36.4	Gansu	14.7	45.1	40.2
Fujian	9.7	49.1	41.3	Qinghai	9.9	53.2	36.9
Jiangxi	14.4	51.2	34.4	Ningxia	9.4	48.9	41.7
Shandong	9.5	55.8	34.7	Xinjiang	17.8	45.1	37.1

Source China Statistical Yearbook 2010, National Bureau of Statistics of China, Beijing, China Statistical Publishing House, 2010

5.2.1 Regional Comparison of the Green Growth of Chinese Tertiary Industry During the 11th Five-Year Plan period

Comparing the green growth of tertiary industry among various regions, huge disparities can be found on the tertiary industry development among Eastern China, Central China, Western China and Northeast. The proportion of value-added of tertiary industry varies greatly among regions. It illustrates the developmental disparities among different regions in the tertiary industry. In addition, gaps also exist in the environmental pollution and the effect on labor productivity of the tertiary industry; this exhibits the regional gaps in the developmental quality of tertiary industry.

5.2.1.1 Regional Comparison of the Proportion of Value-Added of Tertiary Industry in GDP

As the industry with higher green degree, the tertiary industry consumes relatively low energy during the development process. Since the energy consumption per unit of value-added is relatively low, the environmental influence of tertiary industry is small, and thus the rise of proportion of tertiary industry in national economy implies the green growth of national economy. During the *11th Five-Year Plan* period, the proportion of value-added and employees of tertiary industry has substantially increased, and the importance of tertiary industry in national economy has continuously risen. From 2006 to 2009, the proportion of value-added of the tertiary industry in GDP has increased from 40.9 to 43.4 %, the proportion of the primary industry has fallen from 11.1 to 10.3 %, and the proportion of the secondary industry has dropped from 47.9 to 46.3 %.

In the process of integral lift of the proportion of value-added of the tertiary industry, regional disparities appear in China in the proportion of the tertiary industry. As shown in the Table 5.9, in 2009, the proportions of value-added of the tertiary industry in Eastern China, Central China, Western China, and Northeast are 44.1, 36.0, 38.8, and 38.7 % respectively. The highest proportion of value-added of the tertiary industry is in the Eastern China, then the Western China and Northeast follow, and the lowest is in the Central China. Regional gaps of the employee proportion of tertiary industry are less than that of value-added.

5.2.1.2 Regional Comparison of the Environmental Pollution of Tertiary Industry

In the *11th Five-Year Plan* period, economic development emphasizes more on the reduction of negative environmental effects, and the gradual approach toward low-carbon economy. In the transformation process toward low-carbon economy, the tertiary industry development also adopts the measure to mitigate the

Table 5.9 Industrial structure by region in China in 2009

Indicator	Industry	Eastern China	Central China	Western China	Northeast
Proportion of Value-added (%)	Primary industry	6.5	13.6	13.7	11.4
	Secondary industry	49.3	50.4	47.5	49.9
	Tertiary industry	44.1	36.0	38.8	38.7
Proportion of Employment (%)	Primary Industry	26.6	42.8	49.0	39.3
	Secondary industry	36.1	26.5	18.9	22.6
	Tertiary industry	37.2	30.8	32.1	38.1

Note 1. Proportion of value-added by industry = value-added by industry/GDP; Proportion of employees by industry = employee number by industry/total employed people. 2. Regional classification in the table follows the standard of the State Council and National Bureau of Statistics: the Eastern China includes Beijing, Tianjin, Hebei, Shandong, Shanghai, Jiangsu, Zhejiang, Fujian, Guangdong and Hainan; the Central China includes Shaanxi, Shanxi, Henan, Inter-Mongolia, Hubei, Hunan, Jiangxi and Anhui; the Western China includes Yunnan, Guizhou, Sichuan, Chongqing, Guangxi, Gansu, Qinghai, Ningxia, Tibet and Xinjiang; the Northeast covers Heilongjiang, Jilin and Liaoning

Source China Statistical Yearbook 2010, National Bureau of Statistics of China, Beijing, China Statistical Publishing House, 2010

environmental effects. Within the tertiary industry, transport, storage, post and tele-communication has the most direct and important effect on environment, because the development of this sector relies on the use of motor vehicles, and it has significant environmental effects. During the 11th five-year, transport, storage, post and tele-communication has grown rapidly. It results in the huge jump of energy consumption of transport, storage, post and tele-communication, and has more and more severe environmental influences. From 2006 to 2009, the average discharge of nitric oxide of highway communication has risen from 21.7 ton/10 thousand people to 23.8 ton/10 thousand people, up 9.8 %.

There are huge regional gaps of the pollution intensity of Chinese tertiary industry. As shown in the Table 5.10, in the *11th Five-Year Plan* period, the discharge of nitric oxide of highway communication per capita in Eastern China is far beyond the national average; the Central China and Western China is relatively lower than the national average; while the Northeast is far less than the national average. From 2006 to 2009, the growth rates of discharge of nitric oxide of highway communication per capita in China, Eastern China, Central China, Western China and Northeast are 9.8, 0.8, 37.0, 8.4, and −12.1 % respectively. It can be inferred that the absolute number of discharge of nitric oxide of highway communication per capita in Eastern China is large, but the growth rate is very low. The growth rate of discharge of nitric oxide of highway communication per capita in Central China is remarkably high; it implies that the environmental

Table 5.10 The discharge of Nitric Oxide of highway communication per capita by region in China during the 11th Five-Year plan period

Region	2006	2007	2008	2009
Country	21.7	21.0	21.3	23.8
Eastern China	32.1	32.0	31.9	32.3
Central China	16.6	16.3	17.0	22.8
Western China	16.7	15.8	15.9	18.1
Northeast	15.3	10.6	11.2	13.4

Note The discharge of nitric oxide of highway communication per capita by region = the discharge of nitric oxide of highway communication by region/population by region. The regional discharge of nitric oxide of highway communication and population is the sum of the provincial data

Sources China Statistical Yearbook 2006–2010, National Bureau of Statistics of China, Beijing, China Statistical Publishing House, 2006–2010

China Statistical Yearbook on Environment 2005–2010, Ministry of Environmental Protection of the People's Republic of China, China Environmental Science Press, 2005–2010

protection is neglected in the development process of tertiary industry. The discharge of nitric oxide of highway communication per capita in Western China is lower than the national average, and the growth rate is close to the national average. The discharge of nitric oxide of highway communication per capita in Northeast is far less than the national average, with the negative growth.

To sum up, it's the Northeast that requires most attention to reduce the environmental effect of green growth of tertiary industry. Meanwhile, it's necessary for the Central China to further control the environmental effects. Only if all regions develop harmoniously, and reduce the environmental effect of tertiary industry development, the quality of green growth of tertiary industry can be guaranteed.

5.2.1.3 Regional Comparison of the Labor Productivity of Tertiary Industry

During the *11th Five-Year Plan* period, the labor productivity of Chinese tertiary industry has grown rapidly. From 2006 to 2009, the value-added of tertiary industry has risen from RMB8, 855.5 billion to RMB14, 764.2 billion, up 66.7 %; the employee number of tertiary industry has increased from 246.14 to 266.03 million, up 8.1 %; and the labor productivity has improved from 360 million/10 thousand people to 555 million/10 thousand people, up 54 %. However, significant regional disparities in the labor productivities appear in China.

Generally, in the *11th Five-Year Plan* period, the labor productivity of tertiary industry in Eastern China is higher than the national average, and the labor productivity in Northeast is relatively higher than the national average, while the labor productivity in Central China and Western China is far less than the national average. As shown in the Table 5.11, the labor productivity of Eastern China has risen from 610 million RMB/10 thousand people to 840 million RMB/10 thousand

Table 5.11 The labor productivity of tertiary industry by region in China (Unit: 100million RMB/10 thousand people)

Region	2006	2007	2008	2009
Whole country	3.6	4.4	5.1	5.5
Eastern China	6.1	6.9	7.6	8.4
Central China	2.8	3.3	3.7	4.1
Western China	2.6	3.0	3.3	4.0
Northeast	4.4	5.0	5.6	6.4

Note The labor productivity of tertiary industry by region = the value-added of tertiary industry by region/the employee number of tertiary industry by region. The regional value-added and employee number of tertiary industry is the sum of the provincial data

Source China Statistical Yearbook 2010, National Bureau of Statistics of China, Beijing, China Statistical Publishing House, 2010

people; the labor productivity of Central China has increased from 280 million RMB/10 thousand people to 410 million RMB/10 thousand people; the labor productivity of Western China has increased from 260 million RMB/10 thousand people to 400 million RMB/10 thousand people; and the labor productivity of Northeast has risen from 440 million RMB/10 thousand people to 640 million RMB/10 thousand people.

From the perspective of the change on labor productivity of tertiary industry, from 2006 to 2009, the labor productivity of tertiary industry in the whole country, Eastern China, Central China, Western China and Northeast rises 53.3, 38.9, 45.8, 54.2 and 45.1 % respectively. The growth rate of labor productivity of Western China is relatively higher than the national average, the Central China and Northeast is relatively below the national average, and the Eastern China is far below the national average. We can draw the conclusion that there are evident regional gaps of labor productivity of tertiary industry in China, but the gaps are converging continuously. The labor productivity of tertiary industry is toward regional convergence.

Column 5.2 Rizhao: Blue Economy and Green Pursuit

Rizhao is a coastal city with 99.6 km natural coastline, among which 64 km is beach. Natural resources on tourism are richly endowed in Rizhao. Rizhao shows special preference on the blue economy, the development of marine economy becomes priority of the city. Meanwhile, Rizhao is pursuing the green economy, and pays high attention on the protection of ecological environment when developing marine economy. In 1989, Shandong province set up Rizhao city from Linyi county and partial Weifang county. In 1992, Rizhao made the city planning and set the city orientation as: the modern garden-style coastal city focused upon port, industry, trade and tourism. Since 1989, 10 mayors have been switched in Rizhao. Remarkably, except for a few minor amendments on the planning in 2006, all these mayors insist one blueprint, and follow the same planning to construct the city. In 2003, Rizhao hired the international planning & designing experts to make study, and finally decided to build water paradise and aquatic sport base around the Wanpingkou lagoon, in order to build Rizhao toward the capital of aquatic sports. This idea is attributed to the master plan of 1992. This plan pre-reserved the area around Wanpingkou lagoon to build large-scale seashore park. Now, in Wanpingkou area, the 9.2 km Olympic water park has been completed. In the park, the world sailing game base and the water sports base are the best playgrounds in the world, and have hosted two China Water Sports Games and numerous world sailing matches.

The advantage of transformation and upgrade of Rizhao is more prominent, with the accelerating construction of development platforms, such as Shandong peninsula blue economic zone, harbor industrial zone of Southern Shandong, Rizhao port, bonded logistics center, international marine city, and major industrial parks etc. Within the industrial park, Rizhao strengthens the transformation of all the industrial zones in city, realizes the recycling use of resources among companies and zero waste emission within the industrial park. As one of the four environmental experiment projects under the Sino-European cooperation scheme of environmental management, Rizhao Economic and Technological Development Zone entirely plans and manages the companies in park to cyclically utilize and share in the areas of energy, materials, infrastructure and information etc. The park targets to maximize the resource utilization and minimize the waste emission, and build the complete industrial chain of recycling economy.

As an emerging port city, Rizhao has few projects of the traditional industry, and thus the base number of energy consumption and pollutant discharge is low. However, economic growth is rapid in recent years, the scale of new projects is large, and so the increment is relatively high calculated by the detailed rules of energy-conserving and emission-cutting. The reduction potential of the existing projects and industries is limited; therefore the situation of energy-conserving and emission-cutting is sharply tense. Wealth or clean beach is a difficult choice faced by Rizhao government. The local government has no hesitation on the insistence of energy-conserving and emission-cutting. "We need to develop 'Tri-economy'; the first is blue economy, the key tone of local economy; the second is green economy, environmental protection is the priority; and the third is characteristic economy, the new advantage to be built in Rizhao." "Since the *11th Five-Year Plan*, the city has eliminated 16 thousand backward production equipments from more than 250 companies, completed and been operating 61 energy-conserving and emission-cutting project. All districts and counties have constructed the sewage treatment plants. In March 2010, the highest-standard conference on energy-conserving and emission-cutting was held in the city, all districts & counties and the major companies signed the responsibility letter. 24 strict measures on emission-cutting have been carried out. RMB 1 billion has been invested on the establishment, remodeling and expansion of 10 sewage treatment plants. 19 shaft kiln cement production lines have been shut down.

Rizhao won the UN- Habitat Scroll of Honor Award 2009 due to the remarkable planning on habitat and ecological environment. The harmony of blue and green is the real portrayal of the development of Rizhao.
Sources:

1. Rizhao: the Blue Economy and Green Aspiration, http://news.163.com/10/1203/04/6MUVV20B00014AED.html.
2. Blue economy and green aspiration, one planning of ten mayors in Rizhao, Shandong. http://house.focus.cn/news/2011-02-10/1187823.html.

5.2.2 Inter-City Comparison of the Green Growth of Tertiary Industry in China During the 11th Five-Year Plan period

A few major cities are chosen as the representatives to analyze the regional gaps on green growth of Chinese tertiary industry from city level. The selected cities include: four municipalities directly under the jurisdiction of the central government (Beijing, Tianjin, Shanghai and Chongqing), four typical provincial capital cities (Nanjing, Wuhan, Chengdu and Changchun, located at the Eastern China, Central China, Western China and Northeast respectively) and five specifically designated cities in the state plan (Dalian, Qingdao, Ningbo, Xiamen and Shenzhen).

5.2.2.1 Inter-City Comparison of Proportion of Value-Added of Tertiary Industry in GDP

During the *11th Five-Year Plan* period, the proportion of value-added of tertiary industry in GDP has increased from 40.9 % in 2006 to 43.4 % in 2009, but gaps among cities on the proportion of value-added of tertiary industry is huge. As shown in the Table 5.12, significant disparities can be found among the key cities in the proportion of value-added of tertiary industry. Within the selected cities, the largest proportion of value-added of tertiary industry is 75.5 % in Beijing; and the lowest proportion is 37.9 % in Chongqing. The proportion of Changchun, Ningbo and Chongqing is below the national average, i.e. 41.5, 41.2 and 37.9 % respectively.

Within the selected cities, the second highest proportion of value-added of tertiary industry is 59.4 % in Shanghai in 2009. Shenzhen, Xiamen, Nanjing, Wuhan with the proportion above 50 % follow, i.e. 53.3, 51.6, 51.3, and 50.4 % respectively. Obviously, the proportion of Beijing is much higher than other cities, indicating the huge gaps in the green growth of tertiary industry among cities.

Table 5.12 The proportion of value-added of tertiary industry of Chinese major cities in 2009

City categories	City	Proportion of value-added of tertiary industry in GDP (%)
Municipalities directly under the jurisdiction of the central government	Beijing	75.5
	Tianjin	45.3
	Shanghai	59.4
	Chongqing	37.9
Typical provincial capital	Changchun	41.5
	Nanjing	51.3
	Wuhan	50.4
	Chengdu	49.6
Specifically designated cities in the state plan	Dalian	43.9
	Ningbo	41.2
	Xiamen	51.6
	Qingdao	45.4
	Shenzhen	53.3

Note The proportion of value-added of tertiary industry in GDP = the value-added of tertiary industry/GDP; The proportion of employees of tertiary industry in total employees = the employee of tertiary industry/total employees

Source China Statistical Yearbook on Regional Economy 2010, National Bureau of Statistics of China, Beijing, China Statistical Publishing House, 2010

5.2.2.2 Inter-City Comparison of Proportion of Employees of Tertiary Industry in Total Employees

With the fast development of tertiary industry, the employees absorbed in tertiary industry are increasing gradually, and the proportion of employees of tertiary industry in total employees is higher and higher. During the *11th Five-Year Plan* period, the employee number of tertiary industry has risen 8 %, and the labor force grown 2 % in general. The growth of employee number of tertiary industry is much higher than the labor force growth. Simultaneously, the proportion of employees of tertiary industry in total employees has increased from 32.2 % in 2006 to 34.1 % in 2009. The proportion of employees of tertiary industry in total employees has risen continuously. Gaps can be found among cities in the proportion of employees of tertiary industry in total employees. As shown in the Table 5.13, the highest proportion of employees of tertiary industry in total employees is 73.8 % in Beijing, and the lowest proportion is 34.7 % in Chongqing. The proportion of Beijing is 39.1 % higher than the proportion in Chongqing.

Similar to the proportion of value-added of tertiary industry in GDP, Beijing has the much higher proportion of employees of tertiary industry in total employees than any other cities. Shanghai has the second highest proportion, i.e. 55.7 %, and has relatively large gap with Beijing. The following Xiamen, Dalian, Wuhan and Nanjing have the proportion of 52.4, 50.6, 49.0 and 47.4 % respectively. Meanwhile, the proportion of Ningbo is only 30.5 %, below the national

Table 5.13 The proportion of employees of tertiary industry in total employees of Chinese major cities in 2009

City categories	City	Proportion of employment of tertiary industry (%)
Municipalities directly under the jurisdiction of the central government	Beijing	73.8
	Tianjin	47.3
	Shanghai	55.7
	Chongqing	34.7
Typical provincial capital	Changchun	36.5
	Nanjing	47.4
	Wuhan	49.0
	Chengdu	44.3
Specifically designated cities in the state plan	Dalian	50.6
	Ningbo	30.5
	Xiamen	52.4
	Qingdao	35.6
	Shenzhen	46.0

Note The proportion of employees of tertiary industry in total employees = Employee number of tertiary industry/total employees
Source China Statistical Yearbook on Regional Economy 2010, National Bureau of Statistics of China, Beijing, China Statistical Publishing House, 2010

level (34.1 %). Generally, the development level of tertiary industry in major cities is higher than the national level.

5.2.2.3 Inter-City Comparison of the Labor Productivity of Tertiary Industry

During the *11th Five-Year Plan* period, the labor productivity of tertiary industry has risen sharply in China. In 2009, the labor productivity of Chinese tertiary industry is RMB56 thousand/person, the labor productivity of tertiary industry of the Eastern China, Central China, Western China and Northeast is RMB84 thousand/person, RMB41 thousand/person, RMB40 thousand/person and RMB64 thousand/person respectively. Comparing the data of major cities in the Table 5.14 with the national average, all major cities listed, except Chongqing, have the higher labor productivity of tertiary industry than the national average. However, the proportion of labor productivity of tertiary industry of Chongqing exceeds the proportion of Western China.

The rank of labor productivity of Chinese tertiary industry is: Shanghai, Shenzhen, Ningbo, Qingdao, Beijing, Nanjing, Tianjin, Wuhan, Dalian, Changchun, Xiamen, Chengdu and Chongqing. Shanghai has the highest labor productivity of tertiary industry, and the labor productivity is RMB152 thousand/person in 2009, not only higher than the national average (RMB56 thousand/person), but also higher than the average of Eastern China (RMB84 thousand/person).

Table 5.14 The labor productivity of tertiary industry of Chinese major cities in 2009

City categories	City	Labor productivity of tertiary industry (RMB10 thousand/person)
Municipalities directly under the jurisdiction of the central government	Beijing	12.7
	Tianjin	11.0
	Shanghai	15.2
	Chongqing	4.3
Typical provincial capital	Changchun	9.5
	Nanjing	11.6
	Wuhan	10.4
	Chengdu	6.9
Specifically designated cities in the state plan	Dalian	10.1
	Ningbo	12.8
	Xiamen	7.9
	Qingdao	12.7
	Shenzhen	14.0

Note Labor productivity = Value-added/employee number
Source China Statistical Yearbook on Regional Economy 2010, National Bureau of Statistics of China, Beijing, China Statistical Publishing House, 2010

5.3 Green Growth Outlook of Chinese Tertiary Industry During the 12th Five-Year Plan Period

The outline of the *12th Five-Year Plan* clearly states that it's necessary to make the environment to push the fast development of service sectors. Specifically, the promotion of service sectors is the strategic focus of optimizing and upgrading industrial structure. The favorite environment of policy and organization should be built for the development of service sectors. The service sectors are encouraged to extend to new fields, develop new business models and foster new hot spots. The management of service sectors is promoted toward scale, branding and networking. The weight and level of service sector should be enhanced continuously. During the *12th Five-Year Plan* period, the green growth of tertiary industry is going to face the crucial development opportunities: the strategic adjustment of economic structure offers important opportunities for the green growth of tertiary industry; technological advance and innovation provides new supports for the green growth of tertiary industry; and the construction of Resource-conserving and environment-friendly society makes the favorite environment for tertiary industry.

Simultaneously, it's worth noticing that the negative factors are bringing significant challenges to the green growth of tertiary industry in China, such as the increasing gaps of resident income, impact of global financial crisis and the lag of system reform of tertiary industry.

5.3.1 *Opportunities of Green Growth of Chinese Tertiary Industry*

In the outlook of *12th Five-Year Plan*, the accelerating transformation of economic development mode is still the priority of scientific development of China. In order to ensure the new remarkable progress of scientific development and the substantial progress of transformation of economic development mode, China needs to insist that the strategic adjustment of economic structure is the main direction of transformation of economic development mode, the technological advance and innovation is the important support of accelerating transformation of economic development mode, and the construction of Resource-conserving and environment-friendly society is the pivot of transformation of economic development mode. All these insistences offer important development opportunities for the green growth of tertiary industry.

5.3.1.1 Strategic Adjustment of Economic Structure Offers Important Opportunities for the Green Growth of Tertiary Industry

During the *12th Five-Year Plan* period, the strategic adjustment of economic structure will be the main direction of the accelerating transformation of economic development mode in China. The crucial point in the strategic adjustment of economic structure is the accelerating development of tertiary industry, and the promotion of economic growth toward the dependence on the driving synergy of the primary, secondary and tertiary industry. The tertiary industry has developed fast during the *11th Five-Year Plan* period, and the proportion of tertiary industry in national economy has risen, but Chinese tertiary industry is still lagging compared with the developed countries. According to the data published by the National Bureau of Statistics, in 2007, the proportion of primary, secondary and tertiary industry of US is 1.5, 20.6 and 78.0 % respectively, and the proportion of China is 40.8, 26.8 and 32.4 %. In order to push the economic growth toward the driving synergy of the primary, secondary and tertiary industry, the outline of the *12th Five-Year Plan* points out that it's necessary to speed up the development of producer services, strive the development of household services, make the favorite environment for the development of service sectors. The strategic adjustment of economic structure offers important opportunities for green growth of tertiary industry.

The accelerating development of producer services is to deepen the division of labor based on specialization, speed up the innovation on the service product and mode, boost the amalgamation of producer services and advanced manufacturing, and push the accelerating development of producer services. Specifically speaking, China needs to expand financial services, strive the development of modern logistics, foster high-tech services, and standardize and promote commercial services. In details, China needs to optimize the development of commerce and trade services, actively develop tourism, encourage the development of household services, and comprehensively support the development of sports cause and sports industry. The construction of favorite environment for service industry is to use openness to promote reform, use competition to push development, drive the system innovation of services, improve the policy system of services, and optimize the development environment of services. Specifically, China needs to accelerate the reform on service sectors, and improve policy of services. The series of policies will offer crucial opportunities for the rapid development of tertiary industry during the *12th Five-Year Plan* period, further increase the proportion of tertiary industry in national economy, and push the green development of Chinese economy.

5.3.1.2 Technological Advance and Innovation Provides New Supports for the Green Growth of Tertiary Industry

The technological advance and innovation is the important support of the accelerating transformation of economic development mode. During the *12th Five-Year Plan*, China will thoroughly carry out the strategy of revitalizing China through science and education and the strategy of reinvigorating China through human resource development, fully realize the science and technology as the primary productive force and talented person as the No. 1 resource, improve the level of educational modernization, strengthen the independent innovation capacity, increase the innovation talents, push the development toward the dependence on technological advance, quality improvement of labor force and management innovation, and speed up the construction of innovation-oriented country. The technological advance and innovation provides new development room for the green growth of tertiary industry.

Technological advance can boost the green growth of tertiary industry from the following aspects. The first, technological advance and innovation can improve the resource utilization ratio, and reduce the energy consumption and environmental pollution of tertiary industry. On one hand, technological advance and innovation can directly enhance the utilization ratio of natural resources through optimizing the usage mode of natural resource. On the other hand, it can reduce the natural resource consumption by the better management mode. The second, technological advance and innovation can further increase labor productivity. Technological advance, particularly the information and internet technology are widely used on each field of tertiary industry, such as logistics, transportation, financial services etc., and technology significantly improve the labor productivity of these fields. The last, technological advance and innovation can push the structural upgrade of tertiary industry. Technological advance can create some new sectors & new business model, and continuously penetrate into the traditional industries and push the upgrade of these traditional industries.

5.3.1.3 Construction of Resource-Conserving and Environment-Friendly Society Makes Favorite Environment for the Development of Tertiary Industry

During the *12th Five-Year Plan* period, the environmental and resource restriction on economic growth will be tightened in China, and thus China announces that the construction of Resource-conserving and environment-friendly society will be the important pivot of transformation of economic development mode. The basic national policy of energy-conserving and environmental protection should be thoroughly implemented. It's encouraged to save energy, reduce the discharge intensity of greenhouse gas, develop the recycling economy, promote the low-carbon technologies, actively respond to the global climate change, boost the harmony between socio-economic development and population, resource and environment, and insist the road of sustainable development. The construction of energy-conserving and environment-friendly society makes the favorite environment for the development of tertiary industry.

As the green industry, the tertiary industry corresponds to the notion of the construction of energy-conserving and environment-friendly society, the development of tertiary industry is relatively little affected by the resource and environmental restriction, and thus the transformation of economic development mode has little limit on the development of tertiary. Contrarily, its effect on the secondary industry is relatively large. Meanwhile, the construction of energy-conserving and environment-friendly society has positive effects on the tertiary industry. In order to build the energy-conserving and environment-friendly society, the outline of *12th Five-Year Plan* points out that China shall actively respond to the global climate change, strengthen the resource saving and management, strive the development of recycling economy, intensify the force on environmental protection, boost the ecological protection and restoration, and reinforce the construction of water conservancy and the natural disasters prevention and reduction system. It will provide new development opportunities for the sectors in tertiary industry, such as scientific research, technical service, water conservancy, and environment and public facilities management.

Column 5.3 Green Economy: The Construction of Ecological Home of Didaohe Village, Jixi City, Heilongjiang Province

Since a long time ago, tourists from cities have preferred the natural views of villages, and even delighted to stay a few days in village house. It's the so-called "ecological tourism". However, few tourists enjoy oneselves so much as to forget to return home, as the living facilities of village, with poor sanitation condition, are simple, crude, and backward. There is a fantasy to transform village into city in one night, but it's not only unrealistic, but also hard to reach the happy ending. Problems of urban are far more than village, and generally cannot be reversed. When facing this dilemma, the construction of ecological home of Didaohe

Village, Jixi City, Heilongjiang Province might be the most realistic and sustainable solution.

Jixi City is the well-known coal production region, citizens of Jixi certainly don't lack coal, but coal will produce severe pollution on environment. In village, straw is traditional and popular energy source, but problems also exist on the use of straw. After harvest, a large number of straws need to be piled and stored in village. This not only pollutes the environment, but also increases the fire risk. Furthermore, the straw is inconvenient to use, and the thermal efficiency is low. Aiming at these issues, local government of Didaohe village has opened eyes and learned experiences from others, and built the new energy project. In 2006, Didaohe invested RMB1.56 million on the vaporization station of straw in Wangjiacun, under the support of higher-level governments. This project not only facilitates the living of villagers, but also directly saves the coal more than 1,000 t per annum, each family saves almost RMB1000 on energy.

After the success of this trial project in Wangjiacun, Didaohe village organized the observation and study on other regions, and decided to build the ecological home with more demonstration meaning. In 2007, Didaohe invested RMB1.2 million on 10 comprehensive ecological homes in Wangjiacun. The operation model of ecological home is to use solar energy to build the ecological system with virtuous cycle of energy, through the marsh gas and vaporizing straw technologies, and relied on the sunlight greenhouse.

In 2008, Didaohe invested RMB 2 million on the bio-mass solidification plant. This plant crushes and solidifies the processing by-products of agro-product, such as straw, rice hull and corncob, and leaves, sticks and core woods. The granular fuel produced can be used for stove and low-capacity boilers after the transformation. The quantity of combustion of raw material increases significantly after solidification, 1 ton solidified fuel from straw is equal to 0.8 t standard coal. The average sale price of the fuel is RMB350/ton, only 60 % of the local coal price. This project produces 5000 t granular fuel per annum, and 2000 peasant households are benefited, accounting to 40 % of total households in the village.

As its name, ecological home realizes the characteristics and function of environmental protection, green, high-efficiency and energy-saving, and minimizes the energy consumption and pollution of pesticide and fertilizer. Now ecological home of Didaohe has already built the green industrial chain from catering, planting, fishing to tourism.

Sources:

1. Blue Flame, Green Economy, Red Artery—Construction of Ecological Home of Didaohe Village, Jixi City, Heilongjiang Province, Chinese & Foreign Entrepreneurs, 2009(7).

2. New Energy Construction Cheer Up the Lives of Villagers of Didaohe Village, Jixi City, Heilongjiang Province http://nc.mofcom.gov.cn/news/6736305.html.

5.3.2 Challenges Facing Green Growth of Tertiary Industry During the 12th Five-Year Plan Period

When we are excited at the green growth opportunities of tertiary industry during the *12th Five-Year Plan* period, we should also notice the challenges of green growth of tertiary industry. This section will analyze the major green growth challenges of tertiary industry from the angle of income disparities, influence of global financial crisis, and institutional factors.

5.3.2.1 Increasing Income Gap Restricts the Green Growth of Tertiary Industry

The outline of *12th Five-Year Plan* stipulates that China shall strive to develop household services, cater to the life of urban and rural residents, increase the service types, enlarge the service supply, enhance the service quality and satisfy the diversified demands. Household services covers the services for the life of urban and rural residents, such as commerce and trade services, tourism, house services, sports cause and sports industry etc. The household services are characterized as low energy consumption and light environmental pollution, and thus have higher green degree. The development of household services can promote the green growth of tertiary industry.

The increasing income gaps restrict the development of household services, and further restrict the room for green growth of tertiary industry. The development of household services relies on the increase of resident income. Since the reform and opening, income of Chinese residents has risen rapidly, but the income gaps have been continuously increasing. It has affected the green growth of household services. According to the fundamental principles of economics, income determines consumption, and marginal propensity to consume (MPC) is diminishing with the increase of income. The increasing income gaps mean that the low proportion of increased income of high-income group will be spent in consumption; and low-income group is incapable to consume due to low income. Under the limit of fixed aggregate amount of national income, the increasing income gaps will result in the inadequate consumption of whole society. This leads to the scant demand of households services, and thus restricts the development room. In sum, there is the trend of increasing income gaps during the *12th Five-Year Plan* period, and this trend will be the constraint for the green growth of tertiary industry.

5.3.2.2 Global Financial Crisis Affects the Green Growth Vitality of Tertiary Industry

In the 11th five year plan period, under the circumstances that global financial crisis has profound influences on Chinese economic development, Chinese tertiary industry has still grown steadily and rapidly. However, during the 12th five year plan, the effect of global financial crisis cannot be neglected. The *12th Five-Year Plan* clearly points out that global financial crisis has profound influences; global economic growth will slow down. Under this circumstance, Chinese economic development will face more complicated environment. The green growth of Chinese tertiary industry will face more challenges.

Global financial crisis has significant effects on both producer services and household services. From producer services perspective, the most direct effect is on financial services. The global financial crisis has profound influences on financial services, from the regulation to the operation & products of financial services. In addition, global financial crisis has negative effects on logistics, Hi-tech, commercial services. From the angle of household services, the diminishing disposable income of Chinese residents has the relatively long-term effect on the household consumption during the financial crisis and even in a certain time after the crisis. The demand of household services is depressed to some degree. Statistical Communiqué of the People's Republic of China on the 2010 National Economic and Social Development states that the value-added of Chinese tertiary industry in 2010 is RMB17.1005 trillion, and it takes 43 % of GDP, down 0.4 % yoy. It indicates that the effect of stimulus package to tackle global financial crisis isn't strong on tertiary industry, and thus the development of tertiary industry is affected. The tertiary industry undergoes the reverse development after the financial crisis, and its transformation task is difficult during the *11th Five-Year Plan* period.

5.3.2.3 Lagged Reform of Economic System Limits the Efficiency of Green Growth of Tertiary Industry

During the *11th Five-Year Plan* period, state and local governments launch a series of policies to stimulate the service development. These policies boost the system reform to some extent, and stimulate the fast development of tertiary industry. However, the system reform of tertiary industry is still lagging than the development of industry, and it affects the vitality and efficiency of tertiary industry development. The system reform lags at two aspects: the first is the lagging in policy launching; the second is the execution of policy. These two factors delay the system reform of service sectors and affect the vitality and efficiency of service development. Currently, the issues of lagged reform of economic system lie in the followings:

The first is the lag of commercialization process of tertiary industry. Because of the natural monopoly, competition of some sectors in Chinese tertiary industry is incomplete, some sectors are even monopolistic. The lag of commercialization process restricts the green development of tertiary industry. The second is the lack of supporting fiscal and tax policies for service sectors. Although Chinese government makes favorite fiscal and tax policies to support the service development, the lack of detailed supporting policies makes the execution of these favorite policies difficult. The last, the service sectors also face the lack of financial supporting policies. National government encourages financial institutes to develop financial products catering to the demand of service sectors, and launches some financial policies. However, the supporting verification system on company capital and company credit isn't established on time, and thus the financial policies cannot play an effective role.

Chapter 6
Measurement and Analysis of Green Degree of Economic Growth

Faqi Shi, Yajing Cong and Liang Zhang

2011 marks the successful implementation of the 11th *Five-Year Plan* and the beginning of the 12th Five-Year Plan. During the past five years, China has withstood the global financial crisis and a series of unexpected events, achieved a steady and rapid economic growth in general, and become the world's second largest economy. In 2009, given an intricate and volatile economic situation at home and abroad, the central government introduced a series of timely and important measures to cope with the crisis, which was also an opportunity for accelerating the transformation of the economic development pattern and putting more efforts into energy conservation and emission reduction. As an important aspect of the pattern transformation, green economic development received unprecedented attention in 2009. Green growth blessed everywhere across China.

As a facet of Green Development Index (GDI), Green Degree of Economic Growth (GDEG) is a comprehensive evaluation of the green degree of the regional economic development. According to the measurement criterion of GDEG embedded in the "China Province Green Development Index System" (hereinafter referred to as "Province Measurement System") and "China City Green Development Index System" (hereinafter referred to as "City Measurement System"), we used the annual data for 2009 in this chapter to measure the GDEG of 30 provinces (autonomous regions and municipalities directly under the jurisdiction of the central government)[1] and 34 large and medium sized cities (including four municipalities directly under the jurisdiction of the central government, 25 provincial capitals and 5 cities specifically designated in the state plan, except Lhasa and Urumqi) from four perspectives such as Green Growth Efficiency and three industries.

[1] Tibet, Hong Kong, Macao and Taiwan were unmeasured due to lack of basic data.

X. Li and J. Pan (eds.), *China Green Development Index Report 2011*,
Current Chinese Economic Report Series, DOI: 10.1007/978-3-642-31597-8_6,

6.1 Calculated Results of Green Degree of Economic Growth

The development level may vary among provinces and even among cities within the same province. In order to fully uncover the characteristics of provinces and cities, the Third-Level Indicators and their weights of GDEG were not all the same in the Province Measurement System and City Measurement System. Therefore, we introduced the calculation results of provinces and cities respectively.

6.1.1 Calculated Results of Inter-Provincial GDEG

According to the measurement system and weight standard in the China Province Measurement System, the calculation results of GDEG of 30 provinces (autonomous regions and municipalities directly under the jurisdiction of the central government) are shown in Table 6.1.

As shown in Table 6.1, the top 10 eastern provinces[2] with the highest ranking of GDEG were Beijing, Shanghai, Tianjin, Zhejiang, Jiangsu, Guangdong, Shandong, Fujian, Hainan and Liaoning. Respectively, the top 10 provinces with the highest ranking of Green Growth Efficiency were Beijing, Shanghai, Tianjin, Zhejiang, Jiangsu, Hainan, Guangdong, Fujian, Shandong and Jiangxi; the top 10 provinces with the highest ranking of Primary Industry Indicators (PII) were Beijing, Shanghai, Zhejiang, Tianjin, Fujian, Jiangsu, Xinjiang, Hainan, Shandong and Hebei respectively; the top 10 provinces with the highest ranking of Secondary Industry Indicators (SII) were Tianjin, Shandong, Shanghai, Beijing, Guangdong, Jiangsu, Zhejiang, Fujian, Jilin and Shaanxi; the top 10 provinces with the highest ranking of Tertiary Industry Indicators (TII) were Beijing, Shanghai, Tianjin, Guangdong, Zhejiang, Liaoning, Inner Mongolia, Jiangsu, Hainan and Fujian.

According to Table 6.1 and Fig. 6.1, we get the general characteristics of the GDEG of China's 30 provinces (autonomous regions and municipalities directly under the jurisdiction of the central government) as follows:

There was a significant regional variation of GDEG in terms of the ranking, the eastern region was higher, the northeastern and central regions were in the middle and the western region was lower.

From the perspective of geographical distribution of GDEG, the eastern region was higher, the northeastern and central regions were in the middle and the western region was lower.

---

[2] East China includes Beijing, Tianjin, Hebei, Shanghai, Jiangsu, Zhejiang, Fujian, Shandong, Guangdong and Hainan; Central China includes Shanxi, Anhui, Jiangxi, Henan, Hubei and Hunan; West China includes Inner Mongolia, Guangxi, Chongqing, Sichuan, Guizhou, Yunnan, Tibet, Shaanxi, Gansu, Qinghai, Ningxia and Xinjiang; Northeast China includes Liaoning, Jilin and Heilongjiang.

Table 6.1 Indexes of GDEG and rankings of 30 provinces in China in 2009

Indicators	Green degree of economic growth		Second-class indicators							
			Green growth efficiency indicators		Primary industry indicators		Secondary industry indicators		Tertiary industry indicators	
Weight	100 %		45 %		15 %		25 %		15 %	
Province	Score	Ranking	Score	Ranking	Score	Ranking	Score	Ranking	Score	Ranking
Beijing	0.556	1	0.261	1	0.090	1	0.065	4	0.140	1
Shanghai	0.367	2	0.118	2	0.066	2	0.068	3	0.116	2
Tianjin	0.320	3	0.095	3	0.040	4	0.126	1	0.059	3
Zhejiang	0.161	4	0.070	4	0.045	3	0.037	7	0.008	5
Jiangsu	0.153	5	0.068	5	0.040	6	0.038	6	0.007	8
Guangdong	0.138	6	0.052	7	0.004	11	0.060	5	0.022	4
Shandong	0.112	7	0.034	9	0.019	9	0.074	2	−0.016	20
Fujian	0.107	8	0.036	8	0.040	5	0.030	8	0.001	10
Hainan	0.059	9	0.058	6	0.022	8	−0.023	20	0.002	9
Liaoning	−0.009	10	−0.027	18	0.000	12	0.010	12	0.008	6
Inner Mongolia	−0.010	11	−0.026	17	−0.002	13	0.011	11	0.007	7
Hebei	−0.020	12	−0.008	11	0.017	10	−0.007	16	−0.022	24
Jilin	−0.023	13	−0.020	14	−0.022	22	0.024	9	−0.004	12
Shaanxi	−0.046	14	−0.035	22	−0.015	17	0.021	10	−0.017	21
Hubei	−0.048	15	−0.027	19	−0.008	14	−0.008	17	−0.004	13
Heilongjiang	−0.048	16	−0.020	13	−0.016	18	−0.004	14	−0.009	15
Xinjiang	−0.058	17	−0.035	23	0.037	7	−0.047	26	−0.013	17
Hunan	−0.081	18	−0.025	16	−0.015	16	−0.025	21	−0.015	18
Anhui	−0.081	19	−0.022	15	−0.025	24	−0.002	13	−0.032	28
Henan	−0.083	20	−0.017	12	−0.013	15	−0.005	15	−0.049	30
Jiangxi	−0.096	21	−0.001	10	−0.021	20	−0.045	24	−0.029	27

(continued)

Table 6.1 (continued)

Indicators	Green degree of economic growth		Second-class indicators							
			Green growth efficiency indicators		Primary industry indicators		Secondary industry indicators		Tertiary industry indicators	
Weight	100 %		45 %		15 %		25 %		15 %	
Province	Score	Ranking	Score	Ranking	Score	Ranking	Score	Ranking	Score	Ranking
Shanxi	−0.110	22	−0.066	26	−0.025	25	−0.008	18	−0.010	16
Sichuan	−0.112	23	−0.034	21	−0.016	19	−0.037	23	−0.025	25
Guangxi	−0.112	24	−0.041	25	−0.021	21	−0.017	19	−0.033	29
Chongqing	−0.113	25	−0.028	20	−0.040	28	−0.029	22	−0.015	19
Yunnan	−0.163	26	−0.037	24	−0.045	29	−0.052	27	−0.028	26
Ningxia	−0.170	27	−0.097	30	−0.023	23	−0.045	25	−0.005	14
Gansu	−0.184	28	−0.067	27	−0.028	26	−0.067	28	−0.021	23
Qinghai	−0.202	29	−0.080	29	−0.032	27	−0.071	30	−0.019	22
Guizhou	−0.204	30	−0.078	28	−0.052	30	−0.070	29	−0.004	11

Notes
1. Figures in this table are calculated based on data of each indicator for 2009 in accordance with the indicator system of GDEG embedded in the Province Measurement System
2. Index of each province in this table is ranked in descending order

Sources China Statistical Yearbook 2010, China Environmental Statistical Yearbook 2010, Annual Statistical Report on Environment in China 2009, China Industrial Economic Statistical Yearbook 2010, China Statistical Yearbook for Regional Economy 2010, China City Statistical Yearbook 2010, China Water Conservancy Statistical Yearbook 2010, China Agricultural Statistical Yearbook 2009, China Urban Construction Statistical Yearbook 2009, and Desert and Its Treatment in China

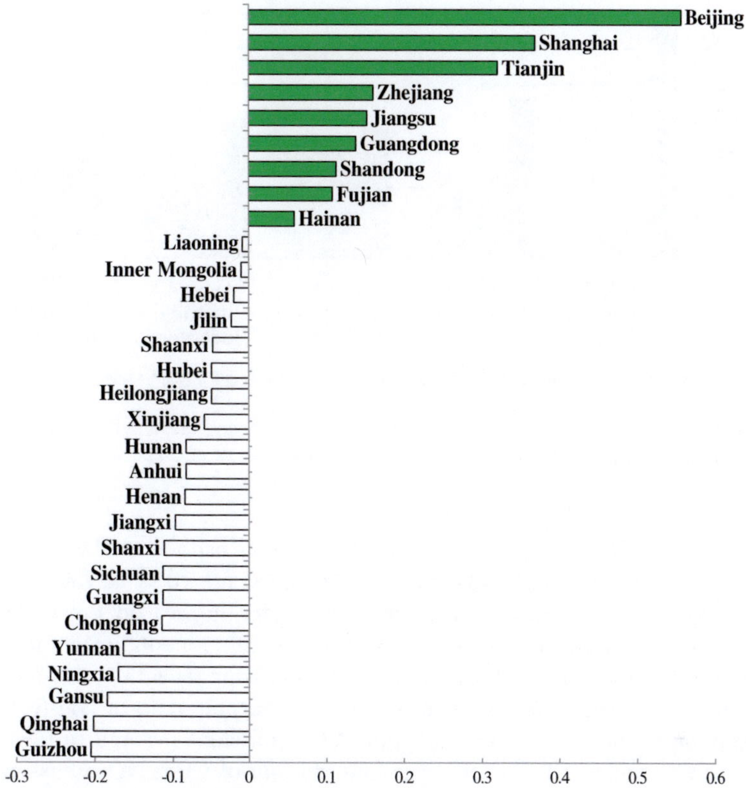

Fig. 6.1 Inter-provincial comparison of GDEG rankings

The eastern region consists of 10 provinces, 9 of which were in the top 10, except that Hebei ranked 12th. Beijing ranked first with a high score of 0.556, from the perspective of the Second-Level Indicators, except for the Second Industry Indicators (SII), the score of Beijing in the Green Growth Efficiency Indicators (GGEI), the Primary Industry Indicators (PII) and the Tertiary Industry Indicators (TII) all ranked first in the country. The three provinces in Northeast China respectively ranked 10, 13 and 16th, which were in the middle and upper level. The central six provinces, of which Hubei ranked 15th and the remaining five provinces ranked 18–22 respectively, were in the middle and low level. Among the 11 eligible provinces in West China, Inner Mongolia ranked 11th, Shaanxi 14th and Xinjiang 17th, while the other eight provinces, which were Sichuan, Guangxi, Chongqing, Yunnan, Ningxia, Gansu, Qinghai and Guizhou ranked 23–30 respectively. The main reason for the eight provinces' low rankings was that the green growth efficiency indicators and the SII pulled down the final rankings. But it is noteworthy that the TII in Guizhou ranked 11th, in the mid to upper level.

Comparing the sub-index of GDEG, there was the biggest difference between the eastern region and the central, western and northeastern regions in the GGEI,

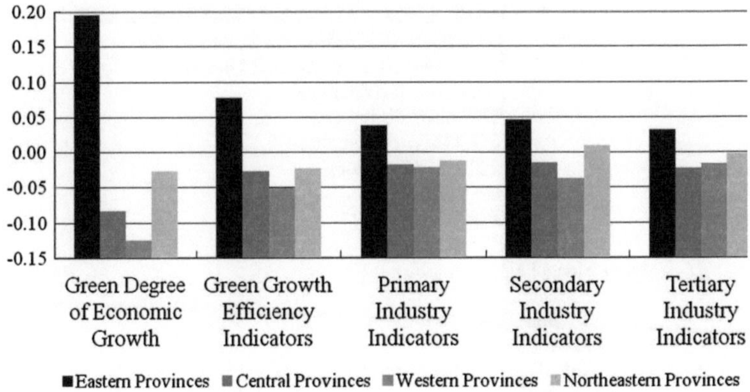

Fig. 6.2 Comparison of GDEG among four major areas in China. *Notes* Figures in the figure are the arithmetic mean of provinces within each of four major areas in China

followed by SII. Relatively speaking, the four regions had the smallest gap in the TII (Fig. 6.2). As the GGEI and the SII accounted for 70 % of the weight of GDEG indicators, these two indicators had greater impact on the ranking of GDEG. There were seven indicators out of the GGEI's nine indicators that were related to the work of energy conservation and emission reduction, and the seven indicators were energy consumption per unit of GRP, the ratio of non-fossil fuel consumption to total energy consumption, CO_2 emissions per unit of GRP, SO_2 emissions per unit of GRP, COD emissions per unit of GRP, NO_x emissions per unit of GRP, ammonia nitrogen emissions per unit of GRP, the weight of which accounted for 10.4 % of the total weight and approached 35 %[3] of the weight of the GDEG. During the 11th *Five-Year Plan* period, thanks to the transformation of the economic development pattern, the work of energy conservation and emission reduction received unprecedented attention, while for many provinces in the central and western regions which were still in the initial stage of development, they still relied on high input, highly energy-consumed and highly polluted industries to stimulate their economic growth, and this was bound to bring great pressure on their green economic development. Nevertheless, the eastern region had a more developed economy, a more mature industrial system, and a less pressure in energy conservation and emission reduction than that facing the central and western regions, hence the higher efficiency of its green growth.

[3] The proportion of non-fossil fuel consumption against total energy consumption and emission of carbon dioxide per unit of GRP were not calculated due to unavailability of data.

Table 6.2 Inter-provincial comparison of ranking gap between PGDI and GDEG

Province	Ranking of GDI (1)	Ranking of GDEG (2)	Ranking Gap (1)–(2)	Province	Ranking of GDI (1)	Ranking of GDEG (2)	Ranking Gap (1)–(2)
Beijing	1	1	0	Xinjiang	16	17	−1
Shanghai	2	2	0	Jiangxi	17	21	−4
Qinghai	3	29	−26	Hebei	18	12	6
Tianjin	4	3	1	Sichuan	19	23	−4
Hainan	5	9	−4	Anhui	20	19	1
Zhejiang	6	4	2	Chongqing	21	25	−4
Yunnan	7	26	−19	Hubei	22	15	7
Fujian	8	8	0	Jilin	23	13	10
Jiangsu	9	5	4	Guangxi	24	24	0
Guangdong	10	6	4	Liaoning	25	10	15
Inner mongolia	11	11	0	Hunan	26	18	8
Shandong	12	7	5	Ningxia	27	27	0
Guizhou	13	30	−17	Shanxi	28	22	6
Shaanxi	14	14	0	Gansu	29	28	1
Heilongjiang	15	16	−1	Henan	30	20	10

Note This table is derived from Tables 1.4 and 6.1

6.1.1.1 The Ranking of GDEG and the Level of Economic Development Moved in the Same Direction

The ranking of GDEG and the level of provinces' economic development moved in the same direction. That is to say, the higher the level of economic development was, the higher the ranking of GDEG was, which meant that economic growth and green development were compatible.

From calculation results, we can see that nine provinces from the eastern region ranked top 10 in the GGEI and the PII and eight eastern provinces ranked top 10 in the SII and the TII, while most provinces in the last 10 of the four indicators were from the western region. China's economic development featured a very prominent regional disparity. During the 11th *Five-Year Plan* period, though the gap of consumption, investment and foreign trade in the four major regions was reducing, the overall difference remained significant. Since the reform and opening up, the eastern region has achieved rapid economic and social development, and industrialization and urbanization has been more universalized. As a result, the industrial structure was constantly optimized and the quality of economic operation was enhanced. The three industrial indicators, whose weight accounted for 55 % of the GDEG, played a big role in the ranking of the GDEG. That was why the eastern region where economy was more developed had a higher GDEG.

6.1.1.2 Increased GDEG Helps Promote Green Development

Comparing the rankings of the GDEG and the GGEI, we can find that 20 provinces, which accounted for 2/3 of all the eligible provinces, experienced a ranking gap of five places between their GDEG rankings and GGEI rankings, while six provinces, accounting for 1/5 of all the eligible provinces, experienced a gap of 10 places, and they were Qinghai, Yunnan, Guizhou, Liaoning, Jilin and Henan respectively.

Half of the six provinces with the largest gap were from the western region. According to Table 6.2, the three western provinces all had low rankings of the GDEG and high rankings of the GDI. It was not accidental. GDI was made up of GDEG, Load Potential of Natural Resources and Environment and Support Degree of Government Policies. Qinghai with abundant resources in the western region ranked first in the ranking of Load Potential of Natural Resources and Environment, much loosely followed by Yunnan and Guizhou. To some western provinces, the resource advantage greatly benefited the ranking of GDI. But economic development of those provinces was relatively backward, as can be seen in the ranking of the four indicators of the GDEG.

6.1.2 The Results of City GDEG

In the process of economic and social development, cities, which consumed 75 % of the world's resources and produced 60–80 % of greenhouse gases, assumed an increasingly important role in today's economic and ecological environment.[4] After a lot of detours, China have come to realize the importance of green rural development and creating a green home. Measuring the inter-city GDEG has a positive significance of guidance nowadays, when urban and rural overall planning steadily improves and the government strongly promotes urbanization.

According to the measurement system and weight standard in the China City Measurement System, the calculation results of GDEG of 34 eligible cities were shown in the following Table 6.3.[5]

As shown in the Table 6.3, the top 10 cities in the ranking of GDEG were Shenzhen, Beijing, Haikou, Shenyang, Guangzhou, Changsha, Dalian, Hefei, Shanghai and Qingdao. Respectively, the top 10 cities in the ranking of GGEI were Shenzhen, Beijing, Guangzhou, Hefei, Changsha, Haikou, Qingdao, Dalian, Ningbo and Hangzhou; the top 10 cities in the ranking of PII were Shenzhen, Dalian, Nanjing, Ningbo, Harbin, Shenyang, Fuzhou, Jinan, Hangzhou and Shanghai; the top 10 cities in the ranking of SII were Shenyang, Shenzhen,

---

[4] Refer to the speech "focus on green city economy" delivered by Achim Steiner, Deputy UN Secretary General. http://finance.sina.com.cn/roll/20100703/11308228198.shtml.

[5] Lhasa and Urumqi.were unmeasured due to lack of basic data.

Table 6.3 Indexes of GDEG and rankings of 34 Cities in China in 2009

Indicators	Green degree of economic growth		Second-class indicators							
			Green growth efficiency indicators		Primary industry indicators		Secondary industry indicators		Tertiary industry indicators	
Weight	100 %		50 %		5 %		30 %		15 %	
City	Score	Ranking	Score	Ranking	Score	Ranking	Score	Ranking	Score	Ranking
Shenzhen	0.341	1	0.153	1	0.083	1	0.072	2	0.033	6
Beijng	0.267	2	0.134	2	−0.002	16	0.009	14	0.126	1
Haikou	0.140	3	0.099	6	−0.003	19	−0.004	20	0.048	4
Shenyang	0.139	4	0.023	11	0.005	6	0.096	1	0.016	9
Guangzhou	0.127	5	0.113	3	0.001	12	−0.050	30	0.063	3
Changsha	0.124	6	0.101	5	−0.005	24	0.042	4	−0.014	22
Dalian	0.090	7	0.051	8	0.023	2	0.012	11	0.003	14
Hefei	0.085	8	0.111	4	−0.005	23	0.013	10	−0.034	30
Shanghai	0.083	9	−0.002	14	0.002	10	0.012	12	0.071	2
Qingdao	0.072	10	0.060	7	0.002	11	0.020	8	−0.010	19
Tianjin	0.029	11	−0.011	17	−0.004	20	0.039	5	0.004	13
Xiamen	0.028	12	0.015	12	−0.001	15	0.003	18	0.011	11
Jinan	0.026	13	−0.009	16	0.003	8	0.021	7	0.010	12
Changchun	0.026	14	0.008	13	−0.003	18	0.052	3	−0.032	28
Nanjing	0.021	15	−0.012	18	0.007	3	0.006	16	0.020	8
Hangzhou	0.000	16	0.024	10	0.003	9	−0.024	27	−0.003	16
Fuzhou	−0.024	17	−0.022	21	0.005	7	0.009	13	−0.016	24
Harbin	−0.039	18	−0.027	23	0.005	5	−0.010	22	−0.008	17
Ningbo	−0.039	19	0.049	9	0.006	4	−0.067	34	−0.028	26
Nanchang	−0.044	20	−0.032	25	−0.005	22	0.026	6	−0.032	29
Xi'an	−0.048	21	−0.024	22	−0.011	29	−0.012	23	−0.001	15

(continued)

Table 6.3 (continued)

Indicators	Green degree of economic growth		Second-class indicators							
			Green growth efficiency indicators		Primary industry indicators		Secondary industry indicators		Tertiary industry indicators	
Weight	100 %		50 %		5 %		30 %		15 %	
City	Score	Ranking	Score	Ranking	Score	Ranking	Score	Ranking	Score	Ranking
Zhengzhou	−0.049	22	−0.016	19	−0.010	28	0.006	17	−0.029	27
Wuhan	−0.050	23	−0.071	30	0.000	13	0.007	15	0.015	10
Hohhot	−0.052	24	−0.039	26	−0.003	17	−0.052	31	0.042	5
Shijiazhuang	−0.070	25	−0.018	20	0.000	14	−0.002	19	−0.051	32
Taiyuan	−0.075	26	−0.069	29	−0.009	26	−0.024	28	0.027	7
Nanning	−0.093	27	−0.030	24	−0.010	27	−0.018	26	−0.036	31
Chengdu	−0.094	28	−0.067	28	−0.005	21	−0.009	21	−0.013	21
Kunming	−0.098	29	−0.008	15	−0.012	30	−0.053	32	−0.026	25
Yinchuan	−0.101	30	−0.103	33	−0.006	25	0.017	9	−0.009	18
Lanzhou	−0.138	31	−0.101	32	−0.013	32	−0.014	24	−0.011	20
Guiyang	−0.184	32	−0.096	31	−0.013	33	−0.062	33	−0.014	23
Chongqing	−0.188	33	−0.061	27	−0.012	31	−0.046	29	−0.069	34
Xining	−0.206	34	−0.123	34	−0.015	34	−0.016	25	−0.053	33

Notes

1. Figures in this table are calculated based on data of each indicator for 2009 in accordance with the indicator system of GDEG embedded in the City Measurement System
2. Index of each province in this table is ranked in descending order

Sources China Statistical Yearbook 2010, China Environmental Statistical Yearbook 2010, Annual Statistical Report on Environment in China 2009, China Statistical Yearbook for Regional Economy 2010, China City Statistical Yearbook 2010, China Urban Life and Price Statistical Yearbook 2010, China Urban Construction Statistical Yearbook 2009

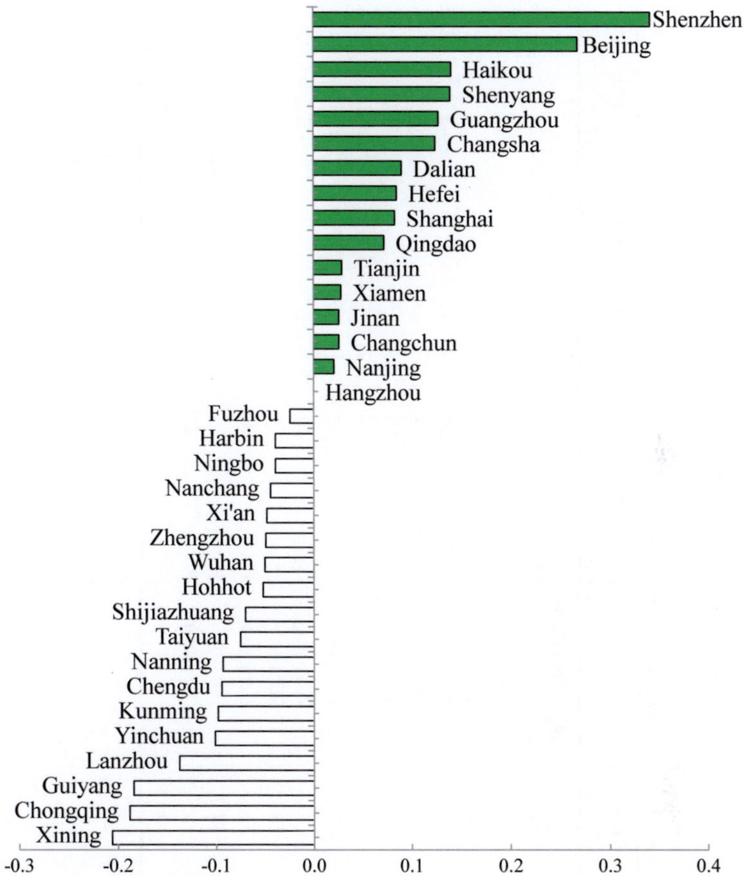

Fig. 6.3 Inter-city comparison of GDEG rankings

Changchun, Changsha, Tianjin, Nanchang, Jinan, Qingdao, Yinchuan and Hefei; the top 10 cities in the ranking of TII were Beijing, Shanghai, Guangzhou, Haikou, Hohhot, Shenzhen, Taiyuan, Nanjing, Shenyang and Wuhan (Fig. 6.3).

According to the calculation results, we get the general characteristics of the GDEG of China's 34 eligible cities as follows:

6.1.2.1 The City GDEG was Closely Correlated with its Corresponding Provincial GDEG

Comparing the calculation results of the GDEG of different provinces and cities, we can find that although the measurement systems were slightly different, there was little difference in the ranking between the cities and their corresponding provinces except the five cities specifically designated in the state plan. Comparing

Table 6.4 Comparison of ranking gap of GDEG between cities and their corresponding provinces

City	Ranking of City (1)	Ranking of its corresponding Province (2)	Ranking gap (2)–(1)	City	Ranking of City (1)	Ranking of its corresponding Province (2)	Ranking gap (2)–(1)
Beijng	2	1	−1	Zhengzhou	22	20	−2
Tianjin	11	3	−8	Wuhan	23	15	−8
Shijiazhuang	25	12	−13	Changsha	6	18	12
Taiyuan	26	22	−4	Guangzhou	5	6	1
Hohhot	24	11	−13	Nanning	27	24	−3
Shenyang	4	10	6	Haikou	3	9	6
Changchun	14	13	−1	Chongqing	33	25	−8
Harbin	18	16	−2	Chengdu	28	23	−5
Shanghai	9	2	−7	Guiyang	32	30	−2
Nanjing	15	5	−10	Kunming	29	26	−3
Hangzhou	16	4	−12	Xi'an	21	14	−7
Hefei	8	19	11	Lanzhou	31	28	−3
Fuzhou	17	8	−9	Xining	34	29	−5
Nanchang	20	21	1	Yinchuan	30	27	−3
Jinan	13	7	−6				

Note This table is derived from Tables 6.1 and 6.3

the two rankings, the ranking gaps between most capital cities and their corresponding provinces were within 10 places, and there were 14 cities whose ranking gaps were within five. That is to say that capital cities as the economic and financial centers of their corresponding provinces picked up the pace of the corresponding provinces in the green economic growth. It is noteworthy that compared with the ranking of the provinces' group the rankings of the four municipalities under the jurisdiction of central government in the cities' group more or less dropped. More notably, besides Beijing's ranking down by one, the other three ones ranked down by 7–8 respectively. That is to say, the four municipalities under the jurisdiction of central government have more advantage in the ranking in the province group (Table 6.4).

6.1.2.2 Significant Inter-city Disparity of the GDEG in Terms of the Ranking, the Eastern Region was Higher, the Northeast and Central Regions Were Scattered and the Western Region was Lower

According to the calculation results of GDEG, the characteristics of regional distribution of the cities' group were obvious. Among 14 eastern cities, Shenzhen, Beijing, Haikou, Guangzhou, Shanghai and Qingdao ranked the top 10 throughout the country, of which Shenzhen and Beijing ranked the highest with 0.341 and 0.267, far beyond the other cities. Both the GGEI and PII of Shenzhen ranked first, meanwhile, its SII ranked second and TII ranked sixth, among the top. Beijing's TII (ranking first) and GGEI (ranking second) effectively contributed to the ranking of its GDEG. Except that Shijiazhuang ranked 25th, the ranking of the remaining seven eastern cities were among 11–19th, which were more concentrated. In general, GDEG of the eastern cities had obvious advantages, while the rankings of central cities demonstrated a polarization trend. Among them, Changsha and Hefei ranked the top 10. Nanchang, Zhengzhou, Wuhan and Taiyuan respectively ranked 20th, 22nd, 23rd and 26th, which were in the middle position. Similarly, Shenyang, Dalian, Changchun and Harbin in the northeast respectively ranked 4, 7, 14 and 18th. Except that Xi'an and Hohhot ranked 21st and 24th, the other eight cities in the western region ranked among the last eight, having significant disadvantages.[6]

[6] Eastern cities include Beijing, Tianjin, Shijiazhuang, Shanghai, Nanjing, Hangzhou, Ningbo, Fuzhou, Xiamen, Jinan, Qingdao, Guangzhou, Shenzhen and Haikou; central cities include Taiyuan, Hefei, Nanchang, Zhengzhou, Wuhan and Changsha; western cities include Hohhot, Nanning, Chongqing, Chengdu, Guiyang, Kunming, Xi'an, Lanzhou, Xining and Yinchuan; northeastern cities include Shenyang, Dalian, Changchun and Harbin.

Table 6.5 Inter-city comparison of ranking gap between GRP per capita and GDEG

City	Ranking of GRP per capita (1)	Ranking of GDEG (2)	Ranking gap (1)–(2)	City	Ranking of GRP per capita (1)	Ranking of GDEG (2)	Ranking gap (1)–(2)
Shenzhen	1	1	0	Zhengzhou	18	22	−4
Guangzhou	2	5	−3	Hefei	19	8	11
Shanghai	3	9	−6	Nanchang	20	20	0
Dalian	4	7	−3	Fuzhou	21	17	4
Beijing	5	2	3	Changchun	22	14	8
Xiamen	6	12	−6	Chengdu	23	28	−5
Nanjing	7	15	−8	Yinchuan	24	30	−6
Hangzhou	8	16	−8	Xi'an	25	21	4
Tianjin	9	11	−2	Harbin	26	18	8
Hohhot	10	24	−14	Shijiazhuang	27	25	2
Ningbo	11	19	−8	Lanzhou	28	31	−3
Qingdao	12	10	2	Haikou	29	3	26
Changsha	13	6	7	Kunming	30	29	1
Shenyang	14	4	10	Guiyang	31	32	−1
Wuhan	15	23	−8	Chongqing	32	33	−1
Jinan	16	13	3	Xining	33	34	−1
Taiyuan	17	26	−9	Nanning	34	27	7

Source China Statistical Yearbook 2010, China Environmental Statistical Yearbook 2010, Annual Statistical Report on Environment in China 2009, China Statistical Yearbook for Regional Economy 2010, China City Statistical Yearbook 2010, China Urban Life and Price Statistical Yearbook 2010, and China Urban Construction Statistical Yearbook 2009

6.1.2.3 There was a Significant Correlation Between the Ranking of GDEG and the Economic Development Level

Viewed from the ranking, GDEG and the economic development level had a certain level of positive correlation. If we simply take GRP per capita as an alternative indicator measuring the level of urban economic development, and compare the ranking of GDEG and of GRP per capita in each city, we'll find that 18 cities with ranking gaps less than five, accounted for nearly 60 % of the 34 cities (see Table 6.5). The basic trend was that the higher level of economic development, the higher ranking of GDEG.

Among the 34 cities, the Haikou had the biggest ranking gap. Its GRP per capita ranked 29th, but GDEG ranked third, the only one with more than 20 ranking gaps. Although the GRP per capita of Haikou City was only 26,366 yuan in 2009, but Haikou seized the opportunity to construct itself into an international tourist island and actively developed the tertiary industry by relying on the current favorable economic policy environment. The proportion of added value of Haikou's tertiary industry and the proportion of tertiary industrial employment ranked second. Meanwhile, as the "National Excellent City of Urban Environment Comprehensive Improvement", Haikou's sulfur dioxide emissions per unit of GRP ranked

Table 6.6 Inter-city comparison of ranking gap between GDI and GDEG

City	Ranking of GDI (1)	Ranking of GDEG (2)	Ranking Gap (1)–(2)	City	Ranking of GDI (1)	Ranking of GDEG (2)	Ranking Gap (1)–(2)
Shenzhen	1	1	0	Nanjing	18	15	3
Haikou	2	3	−1	Shanghai	19	9	10
Kunming	3	29	−26	Changchun	20	14	6
Beijing	4	2	2	Jinan	21	13	8
Hefei	5	8	−3	Yinchuan	22	30	−8
Guangzhou	6	5	1	Nanchang	23	20	3
Dalian	7	7	0	Hohhot	24	24	0
Qingdao	8	10	−2	Zhengzhou	25	22	3
Changsha	9	6	3	Guiyang	26	32	−6
Fuzhou	10	17	−7	Taiyuan	27	26	1
Xiamen	11	12	−1	Tianjin	28	11	17
Nanning	12	27	−15	Chongqing	29	33	−4
Ningbo	13	19	−6	Xi'an	30	21	9
Shenyang	14	4	10	Wuhan	31	23	8
Harbin	15	18	−3	Chengdu	32	28	4
Shijiazhuang	16	25	−9	Lanzhou	33	31	2
Hangzhou	17	16	1	Xining	34	34	0

Notes This table is derived from Tables 1.5 and 6.3

34th, water consumption per unit of industrial added value 32nd, and the comprehensive utilization of industrial solid waste third. On the whole, it is not surprising that Haikou had lower GRP per capita but greener economic development.

6.1.2.4 There was a Significant Regional Disparity of the Contribution of Inter-City GDEG to GDI

Seen from the final measured results, Urban GDEG had a great impact on GDI, but the regional disparity was significant. After comparing Urban GDEG with GDI, we found the number of the provinces which had less than five ranking gaps, accounted for 20 of the total provinces, nearly 2/3. Among them, Kunming had the biggest ranking gap of 26 places (see Table 6.6). Relying on a machinery, metallurgy, tobacco processing-based industrial system, Kunming's pollutant emission per unit of GRP was relatively large, holding back its ranking of Urban GDEG.

The regional disparity of the contribution of Urban GDEG to GDI was that, comparing with the central and western cities, Urban GDEG of the eastern cities contributed the most to GDI, while central and western cities enjoyed more support of government policies and larger bearing potential of resource environment. Viewed from the three sub-indices of green development index, eastern cities were in majority among the cities whose GDEG was above the average level. However, eastern cities were in the middle position in the potential index of resources and

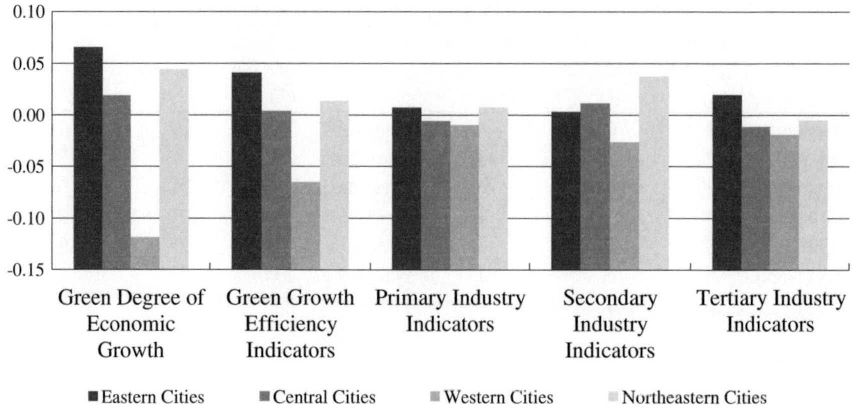

Fig. 6.4 Inter-city comparison of GDEG in four major areas in China. *Note* Figures in the figure are the arithmetic mean of provinces within each of four major areas in China

environment. Shanghai ranked only 33rd, just higher than Wuhan. The contribution of GDEG to GDI was driven by GGEI and TII. These two indicators in eastern cities were significantly higher than the other regions, especially GDI (see Fig. 6.4). But speaking of the inter-provincial rankings, eastern cities enjoyed less stronger absolute advantage. Northeastern cities performed slightly better in the Primary Industry and Secondary Industry Indices.

6.2 Analysis of Inter-Provincial Comparison of Green Degree of Economic Growth

Accounting for 30 % of the total weight of GDI, Inter-provincial GDEG comprised 22 Third-Level Indicators with 12 positively correlated indicators versus 10 negatively correlated ones. This essay measured 20 of the indicators. The weights of the positively correlated indicators and the negatively correlated ones were 1.40 and 1.70 % respectively.

6.2.1 Analysis on the Measurement of Provincial Green Growth Efficiency Indicators

Green Growth Efficiency Indicators (GGEI) accounted for 45 % of Inter-Provincial GDEG and 13.5 % of GDI in terms of weight, making a larger contribution to GDEG than the other three indicators. GGEI were mainly a weighted combination of the nine indicators in Table 6.7.

Table 6.7 Third-class indicators, their weights and attributes of inter-provincial green growth efficiency

Sequence number	Indicator	Weight (%)	Attribute
1	Gross regional product(GRP) per capita	1.70	Positively correlated
2	Energy consumption per unit of GRP	1.70	Negatively correlated
3	Ratio of non-fossil energy consumption to total energy consumption	1.70	Positively correlated
4	CO_2 emissions per unit of GRP	1.40	Negatively correlated
5	SO_2 emissions per unit of GRP	1.40	Negatively correlated
6	COD emissions per unit of GRP	1.40	Negatively correlated
7	NOx emissions per unit of GRP	1.40	Negatively correlated
8	Ammonia nitrogen emissions per unit of GRP	1.40	Negatively correlated
9	Electricity consumption per capita in urban areas	1.40	Negatively correlated

Note The contents in this table are determined by the task force after having conducted a lot of expert workshops

Among the nine indicators of GGEI, indicator 1 "GRP per capita", indicator 2 "Energy consumption per unit of GRP", indicator 3 "Ratio of non-fossil energy consumption to total energy consumption" accounted for the same weight of 1.70 %, higher than the weight of the other six indicators (1.40 %). The reason is that, indicator 1 was the sole economic indicator selected in the measurement in assessing the level of regional economic development. During the 11th *Five-Year Plan* period, various regions vowed to achieve lower energy consumption per unit of GRP, an important goal in regional economic development. In 2009, a pivotal year for the implementation of the Plan, despite the adverse effects of the financial crisis and the pressure of stimulating growth and promoting development, no region deviated from the crucial task of restructuring economy and transforming economic development pattern through energy conservation and emission reduction. That accounted for a heavier weight of indicator 2 than that of the other six. As an important criterion assessing the transformation of regional energy structure, indicator 3 with improved weight helped identify the orientation of green development. Unfortunately, indicator 3 was excluded in the measurement as some provincial data was unavailable. The first three indicators were supplemented by the other six with an equivalent weight of 1.40 %. The nine indicators were arranged in an unbalanced but complementary manner in order to achieve a comprehensive measurement and assessment of the regional green growth efficiency.

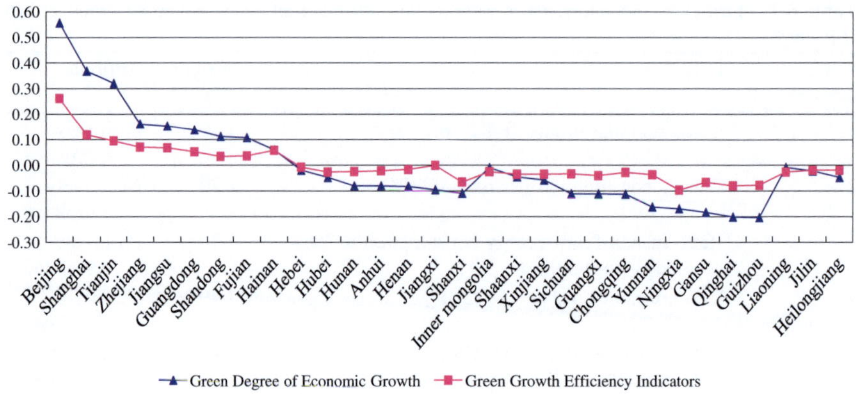

Fig. 6.5 Inter-provincial comparison of green growth efficiency index and GDEG index. *Note* From the area division perspective of eastern, central, western and northeastern areas, this figure arranges these regions from *left* to *right* in terms of indexes of GDEG in descending order

From Table 6.1 and Fig. 6.5, the eastern provinces (except Hebei Province) as a whole had higher GGEI, above the national average, while a large internal gap existed, with Beijing, at 0.261, topping all the provinces. Slightly below the national average, the GGEI of six central provinces and three northeastern provinces did not varied significantly. GGEI of western provinces as a whole were low.

Of the top 10 cities with the highest GGEI, the first nine were from eastern China, and almost coincided with the top nine GDEG cities. The rankings of cities were not identical, but Beijing, Shanghai, Tianjin, Zhejiang and Jiangsu were among the top five in both GGEI and GDEG. According to the statics in 2009, Shanghai, Beijing, Tianjin's GRP per capita ranked the top three in the country, amounting to 78,989, 70,452 and 62,574 yuan[7] respectively. As Beijing and Shanghai had a higher proportion of tertiary industry, they emitted less pollutants per unit of GRP and hence had higher GGEI. However, higher concentration of population in Beijing, Shanghai and Tianjin made them the largest consumers of urban electricity per capita.

The top 11 ~ 20 GGEI provinces were Hebei, Henan, Heilongjiang, Jilin, Anhui, Hunan, Inner Mongolia, Liaoning, Hubei and Chongqing respectively. Of them were one eastern province, four central provinces, two western provinces, and three northeastern provinces. The eastern province Hebei, despite a low per capita urban electricity consumption (ranking 27th), ranked from 7 to 20 in terms of the other indicators. The three GGEI of Henan, Anhui, Hunan and Hubei were mainly in the middle. The top one GGEI province from western China, Inner Mongolia owed its high GGEI to GRP per capita and disposable income of urban residents that ranked 7 and 9th. However, the sulfur dioxide and nitrogen oxide emissions per unit of GRP ranked third and second. Therefore, as a major province

[7] Resource: *China Statistical Yearbook 2010*, China Statistics Press, 2010.

Table 6.8 Third-class indicators, their weights and attributes of inter-provincial primary industry

Sequence number	Indicator	Weight (%)	Attribute
10	Labor productivity of primary sector	1.13	Positively correlated
11	Output ratio of land	1.13	Positively correlated
12	Proportion of water-saving irrigated area in effectively irrigated area	1.13	Positively correlated
13	Proportion of effectively irrigated area in area of cultivated land	1.13	Positively correlated

Note The contents in this table are determined by the task force after having conducted a lot of expert workshops

relying on mineral resource exploitation, Inner Mongolia still faced an arduous task in environmental protection.

The last ten GGEI provinces included only one central province, Shanxi and nine western provinces. It is noteworthy that Shanxi ranked 26th, with the lowest GGEI in central provinces. Shanxi's energy consumption per 10,000 yuan GRP ranked second, reaching 2.36 tons of standard coal, much higher than the simple average of all provinces, 1.37 tons of standard coal. The province also ranked high in the emissions of sulfur dioxide, nitrogen oxides and industrial solid waste per unit of GRP, showing that the extensive development pattern in Shanxi Province remain unchanged. The energy-and-metallurgy-based industrial system held back the industrial structure optimization and burdened energy conservation and emission reduction. These obstacles contributed to the low GGEI ranking of Shanxi Province. Against this backdrop, in December 2010 the State Council approved in Shanxi Province the establishment of the comprehensive reform pilot area for transforming national resources-based economy. This pilot was designed to accelerate the industrial structure upgrading and strategic economic restructuring so that Shanxi was to be built into a resource-saving and environmental friendly province. If the pilot was run properly, the green growth efficiency of Shanxi would inevitably be increased.

It is noteworthy that, from Fig. 6.5, GGEI contributed significantly to GDEG by accounting for a weight of 45 % in Inter-Provincial GDEG.

6.2.2 Analysis on the Measurement of Primary Industry Indicators on the Provincial Level

In the measurement system of GDEG, Primary Industry Indicators (PII), comprised of four Third-Level Indicators (see Table 6.8), the weight of each was 1.13 %, held a weight of 15 %, the lowest among all the third-level indicators.

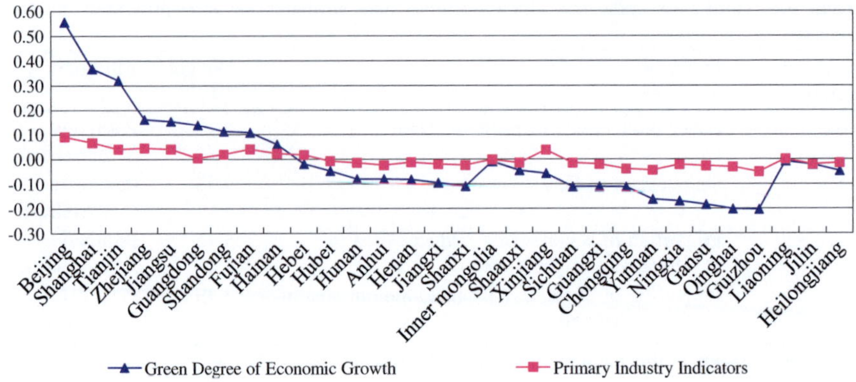

Fig. 6.6 Inter-provincial comparisons of primary industry index and GDEG index. *Note* From the area division perspective of eastern, central, western and northeastern areas, this figure arranges these regions from *left* to *right* in terms of indexes of GDEG in descending order

From Table 6.1 and Fig. 6.6, there was a slight difference in PII of the four major regions. The top ten provinces included nine eastern provinces and only one western province, Xinjiang. Beijing had a prominent PII of 0.09, followed by Shanghai. With regard to the three indicators, Beijing ranked top in its land productivity rate, irrigation saving rate and the proportion of effective irrigation area of arable land, with its productivity of primary industry in the eighth place. Shanghai's productivity of primary industry and land productivity rate rank second, and its irrigation saving rate and the proportion of effective irrigation area of arable land ranked third. The year 2009 witnessed a drop of 0.11 and 0.06 percentage points[8] over 2008 in the added value of primary industry of Beijing and Shanghai, at 0.97 and 0.76 %. The low proportion of primary industry, far below the national average, demonstrated its "small but smart" role in national economy. Xinjiang stood out in western provinces with the seventh largest PII, ranking second in irrigation saving rate and the proportion of effective irrigation area, seventh in productivity of primary industry and 17th in land productivity.

The 11 ~ 20 ranking provinces were Guangxi, Jilin, Ningxia, Anhui, Shanxi, Gansu, Qinghai, Chongqing, Yunnan and Guizhou, all from western China except Jilin, Anhui and Shanxi. Though falling far behind the other provinces in the productivity of primary industry, western provinces made rapid progress during the 11th *Five-Year Plan* period. The rural per capita net income had been greatly improved, from 2588.37 yuan in 2006 to 3816.47 yuan in 2009, at the largest annual growth rate[9] in 2009, while in 2006 the western region grew least rapid among the four major regions.

[8] Measurement of data from *China Statistical Yearbook 2010*, and *China Compendium of Statistics*.

[9] Resources: *China Statistical Yearbook* 2007 and *China Statistical Yearbook* 2010.

Table 6.9 Third-class indicators, their weights and attributes of inter-provincial secondary industry

Sequence number	Indicator	Weight (%)	Attribute
14	Labour productivity of secondary sector	1.25	Positively correlated
15	Water consumption per unit of industrial value added	1.25	Negatively correlated
16	Energy consumption per unit of industrial value added at a cut-off level	1.25	Negatively correlated
17	Ratio of industrial solid wastes utilized	1.25	Positively correlated
18	Reuse rate of industrial water	1.25	Positively correlated
19	Ratio of output value of six high energy-bearing industrial sectors to gross industrial output value	1.25	Negatively correlated

Note The contents in this table are determined by the task force after having conducted a lot of expert workshops

Among 13 major grain-producing provinces,[10] Jiangsu, Shandong and Hebei, three eastern provinces, ranked among the top ten in PII. Liaoning, Inner Mongolia, Hubei, Henan, Hunan, Heilongjiang, Sichuan and Jiangxi provinces ranked among 11 ~ 20. Jilin ranked 22nd and Anhui 24th. The national average of crop output value per hectare was 21,700 yuan, above which were Hebei, Liaoning, Jiangsu and Shandong from eastern China. In terms of irrigation and water conservancy, only Hebei, Inner Mongolia and Heilongjiang exceeded the national average in irrigation saving rate, and the rate of Jiangsu and Shandong was close to the national average. As major suppliers of grain in China, some major grain-producing provinces needed to upgrade the agricultural mechanization and improve agricultural growth efficiency.

6.2.3 Analysis on the Measurement of Secondary Industry Indicators on the Provincial Level

In the measurement system of GDEG, Secondary Industry Indicators (SII), comprised of six indicators: 14 ~ 19 (see Table 6.9), the weight of each was 1.25 %, held a weight of 25 %.

[10] 13 major grain-producing provinces (regions) were: Hebei, Inner Mongolia, Liaoning, Jilin, Heilongjiang, Jiangsu, Anhui, Jiangxi, Shandong, Henan, Hubei, Hunan and Sichuan.

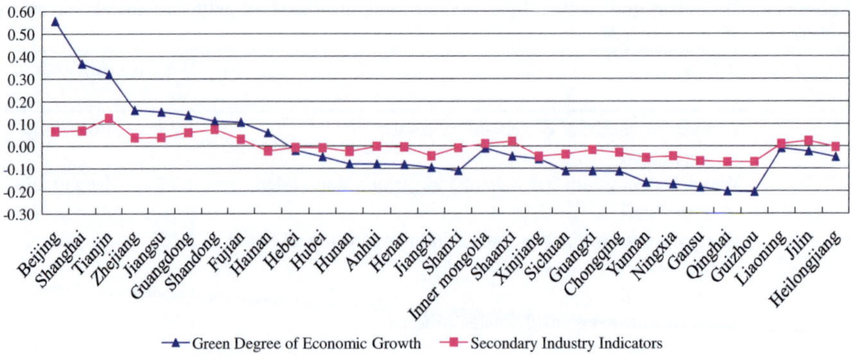

Fig. 6.7 Inter-provincial comparison of secondary industry index and GDEG index. *Note* From the area division perspective of eastern, central, western and northeastern areas, this figure arranges these regions from *left* to *right* in terms of indexes of GDEG in descending order

As an important spectrum of green development, green industry urges rapid development, which will help accelerate the transformation of economic growth pattern. To develop green industry, various regions need to work out rational layout and optimize regional economic structure in order to recycle resources and energy and free the hazard of waste.[11] Indicator 14 represented labor efficiency of secondary industry. The other five indicators represented energy consumption efficiency and resource cyclic utilization degree in secondary industry. Such an arrangement embodied the ideal of reducing material consumption and industrial solid waste while reclaiming used resources and removing hazards. These reflected the "green degree" of industry from different perspectives.

From Table 6.1 and Fig. 6.7, the top eight ranking provinces were eastern provinces: Tianjin, Shandong, Shanghai, Beijing, Guangdong, Jiangsu, Zhejiang and Fujian. The other two eastern provinces Hebei and Hainan ranked 16 and 20th. The SII of six central provinces were below average, with Anhui ranking first among central provinces and 13th nationwide, and Jiangxi becoming the last. There was a huge gap among western provinces, with Shanxi ranking 10th and Inner Mongolia 11th, above the national average, and the other nine ranked behind, six of them being western provinces. The three northeastern provinces had a pattern rate ranking with average SII.

In regard to SII, Tianjin ranked first with absolute advantage. In response to the financial crisis, Tianjin adopted a patternrately loose monetary policy. As a result, 2008 saw rapid industrial development. Rapid progress in metallurgy, equipment manufacturing, automotive industry and other industries pushed up the added value of secondary industry by 2.8 % points compared with 2007. Since early 2009, due to the weakened adverse effects of the crisis, the figure dropped significantly by

[11] See the keynote speech of Tang Yuan, Secretary of Research Office of the State Council, in "2010 China Green Industry Forum". *Source* Sina Finance (http://finance.sina.com.cn/hy/20100627/09208187224.shtml).

Table 6.10 Comparisons of secondary industry indicators between Tianjin and the average of 30 provinces

Indicator	Unit	Average	Tianjin
Labour productivity of secondary sector	10,000 yuan/person	10.17	19.30
Water consumption per unit of industrial value added	cu.m/10,000 yuan	103.51	12.01
Energy consumption per unit of industrial value added at a cut-off level	Tons of SCE/10,000 yuan	2.29	0.91
Ratio of industrial solid wastes utilized	%	68.79	98.30
Reuse rate of industrial water	%	70.51	96.40
Ratio of output value of six high energy-bearing industrial sectors to gross industrial output value	%	39.87	34.84

Note Here the average refers to the arithmetic mean of indicators of 30 provinces
Sources China Statistical Yearbook 2010, Annual Statistical Report on Environment in China 2009, and China Industrial Economic Statistical Yearbook 2010

7.1 % points compared with 2008, a symbolization of optimized industrial structure.[12] Under this momentum, Tianjin boasted its water consumption per 10,000 yuan of industrial added value at only 12.01 cubic meters, far below the average of 30 provinces; the comprehensive utilization of industrial solid waste and reuse rate of industrial water at 98.3 and 96.4 %, top among all regions; energy consumption per 10,000 yuan of industrial added value at 0.91 tons of standard coal, ranking 28th (see Table 6.10). Overall, despite the adverse effects of the financial crisis, Tianjin had made significant progress in its green industry.

6.2.4 Analysis on the Measurement of Tertiary Industry Indicators on the Provincial Level

In the measurement system of GDEG, Tertiary Industry Indicators (TII), comprised of three third-level indicators: 20 ~ 22, the weight of each was 1.50 % based on the equal weight method, held a weight of 15 %. The weights and properties of these indicators were shown in the following table:

The proportion of TII in national economy was an important criterion measuring the green degree of economic development. The development level of tertiary industry and the overall level of economic development were closely related to each other. Thus the regional gap of economic development will lead directly to regional disparities of development of tertiary industry. In 2010, the added value of tertiary industry in the eastern region accounted for 58.6 % of the national total, up by 0.9 % points compared with 2009. That figure was higher than the proportion of GRP of eastern provinces in the country's total (57.7 %). The higher added value of

[12] Measurement of data from series of *China Statistical Yearbook.*

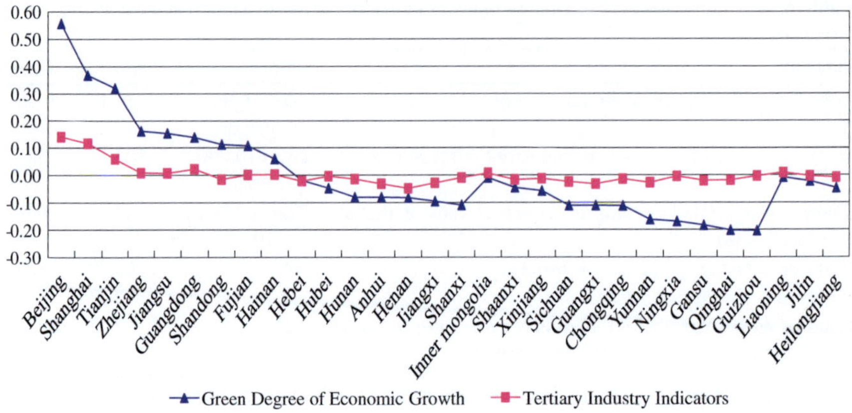

Fig. 6.8 Inter-provincial comparison of tertiary industry index and GDEG index. *Note* From the area division perspective of eastern, central, western and northeastern areas, this figure arranges these regions from *left* to *right* in terms of indexes of GDEG in descending order

tertiary industry of the eastern region and lower figure of the other three regions reflected a large regional disparity[13] in the development of tertiary industry.

From a regional perspective, the high TII gave eastern provinces a comparative advantage, while that of the other three regions, though lower, did not varied greatly (see Fig. 6.8). Eight of ten eastern provinces without Shandong (20th) and Hebei (24th) had higher TII than the national average. They were Beijing, Shanghai, Tianjin, Guangdong, Zhejiang, Jiangsu, Hainan and Fujian. The top three were Beijing, Shanghai and Tianjin, at 0.140, 0.116 and 0.059. The other five did not differ much in TII. From a TII perspective, Beijing and Shanghai ranked among the top three in terms of the three Third-Level Indicators. The proportion of added value of tertiary industry in Tianjin ranked fifth, and the other two indicators ranked among the top three. Shandong had a robust industry but ranked behind because of a smaller proportion of added value of tertiary industry.

The central provinces featured a second-half ranking except Hubei at the 13th place. Jiangxi, 27th, Anhui, 28th and Henan, 30th had the lowest ranking. Ranking varied greatly among western provinces, the ahead ones being Inner Mongolia, seventh and Guizhou, eleventh and six behind ones ranking from 21 to 29. Guizhou's TII ranked 11th, close to the national average. It is noteworthy that the proportion of added value of tertiary industry and the proportion of employment in tertiary industry in Guizhou ranked third and eighth, but its productivity of tertiary industry ranked 30th. Among northeastern provinces, Liaoning ranked sixth, Jilin 12th and Heilongjiang 15th, a pattern rate ranking.

[13] Measurement of data from *China Monthly Economic Indicators*, 2011, Issue No. 2, the China Economic Monitoring and Analysis Center affiliated to National Bureau of Statistics.

Table 6.11 Third-class indicators, their weights and attributes of inter-provincial tertiary industry

Sequence number	Indicator	Weight %	Attribute
20	Labor productivity of tertiary sector	1.50	Positively correlated
21	Proportion of value added of tertiary sector in GRP	1.50	Positively correlated
22	Proportion of employees of tertiary sector in total employees	1.50	Positively correlated

Note The contents in this table are determined by the task force after having conducted a lot of expert workshops

The TII ranged from −0.049 to 0.140, a difference of 0.189, well below that of GGEI, 0.358, slightly lower than that of SII, 0.197, but slightly higher than that of PII, 0.142. This disparity reflected a smaller contribution of TII to GDEG than that of GGEI and SII, and a larger contribution than that of PII.

6.3 Comparative Analysis of Green Degree of Economic Growth on the City Level

In order to fully compare the differences of GDEG between various cities, to understand China's GDEG, taking into account data availability, the cities' typicality, regional differences, and various factors, we selected 34 typical city as objects of measurement from GGEI, PII, SII, to TII. We adopted a similar measurement system of GDEG on the city level to that of Inter-provincial GDEG. Allowing for the characteristics of cities in developing green economy, we made some adjustment of measurement system and weight criterion. Comprised of 17 Third-Level Indicators, 8 positive and 9 reverse, GDEG on the city level held 33 % of the total weight of GDI. The measurement involved 15 indicators, and the weight of Third-Level Indicators ranged between 1.65 and 2.36 %.

6.3.1 Analysis on the Measurement of Green Growth Efficiency Indicators on the City Level

In the measurement system of GDEG, comprised of eight Third-Level Indicators, GGEI held a weight of 50 % in GDEG and 16.5 % in the total weight of GDI, higher than the other three second-level indicators. The indicators and their properties were shown in the following table:

Table 6.11, showed that, similar to GGEI in the measurement system of Inter-Provincial GDEG, the weight of energy consumption per unit of GRP was only 2.36 %, higher than that of the other seven Third-Level Indicators (2.02 %) (Table 6.12).

Table 6.12 Third-class indicators, their weights and attributes of inter-city green growth efficiency

Sequence number	Indicator	Weight (%)	Attribute
1	Gross regional product(GRP) per capita	2.02	Positively correlated
2	Energy consumption per unit of GRP	2.36	Negatively correlated
3	Per capita electricity consumption by urban households	2.02	Negatively correlated
4	CO_2 emissions per unit of GRP	2.02	Negatively correlated
5	SO_2 emissions per unit of GRP	2.02	Negatively correlated
6	COD emissions per unit of GRP	2.02	Negatively correlated
7	NO_x emissions per unit of GRP	2.02	Negatively correlated
8	Ammonia nitrogen emissions per unit of GRP	2.02	Negatively correlated

Note The contents in this table are determined by the task force after having conducted a lot of expert workshops

Figure 6.9 showed that the GGEI rankings demonstrated a general trend where eastern coastal cities ranked forward, western cities ranked behind, northeastern cities ranked upper middle, and central cities had sharp disparity within.

Of 14 eastern cities, Shenzhen, Beijing and Guangzhou ranked the top three in terms of GGEI. Haikou, Qingdao, Ningbo and Hangzhou ranked among the top ten. Xiamen, Shanghai, Jinan, Tianjin and Nanjing ranked between 12 and 18th. Shijiazhuang and Fuzhou ranked relatively rearward, 20th and 21st. From the third-level indicators, GRP per capita of Shijiazhuang amounted to 30,428 yuan, ranking 27th. The other six indicators had pattern rate rankings, with the indicator of pollutant emission ranked among the top eight, while per capita urban electricity consumption of Shijiazhuang rests in the last place. The third-level indicators of Fuzhou ranked in the middle.

The rankings of the six central cities varied sharply. The top ranking cities were Hefei, 4th, and Changsha, 5th, far ahead of Zhengzhou, 19th in the middle position. Nanchang, 25th, Taiyuan, 29th, Wuhan, 30th, ranked among the last 10 cities. From the third-level indicators, Changsha and Hefei ranked ahead in energy consumption (31st and 34th) per unit of GRP, emissions of sulfur dioxide per unit of GRP (30 and 28th) and emissions of nitrogen oxide per unit of GRP (34th and 33rd), while other indicators ranked pattern rately. The GRP per capita of Wuhan totaled 51,144 yuan, ranking 15th, while the other reverse indicators ranked in the middle and slightly above average, lowering the ranking of GGEI in Wuhan.

Of the 10 western cities, Kunming ranked 15th, while the other nine ranked behind 22nd. The last four ranking cities were Guiyang, Lanzhou, Yinchuan and

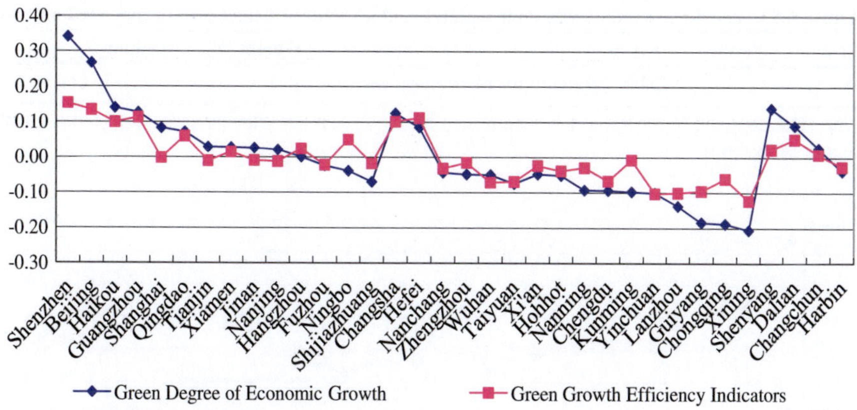

Fig. 6.9 Inter-city comparison of green growth efficiency index and GDEG index. *Note* From the area division perspective of eastern, central, western and northeastern areas, this figure arranges these regions from *left* to *right* in terms of indexes of GDEG in descending order

Xining. Thus the western cities were unfavorably ranked. From the Third-Level Indicators, the western cities had low GRP per capita but high emissions of pollutants GRP per capita. The extensive economic development pattern needed to be changed through accelerated transformation.

Among the northeastern cities, Dalian ranked eighth, Shenyang 11th, Changchun 13th and Harbin 23rd. From the third-level indicators, Harbin ranked 26th with its GRP per capita reaching 32,053 yuan, while its energy consumption per unit of GRP ranked eighth, lowering its GGEI. As China's old industrial base, Harbin relied heavily on secondary industry. Harbin should upgrade the industrial structure and reduce energy consumption in its future development.

6.3.2 Analysis on the Measurement of Primary Industry Indicators on the City Level

In the measurement system of GDEG, comprised of only one indicator: primary industry productivity, Primary Industry Indicators hold a weight of 5 % in GDEG and 16.5 % in the total weight of GDI (see Table 6.13 for details).

Since the 34 measured cities were rich cities among prefectural-level and higher level ones, their primary industry held less weight in the industrial structure. In 2009, there were only three cities with the weight of added value of primary industry over 10 %. They were Nanning, Harbin and Shijiazhuang, respectively 13.93, 12.57 and 10.27 %.[14] Therefore, in order to measure the GDEG on the city

[14] Measurement of data from *2010 China Statistical Yearbook for Regional Economy*.

Table 6.13 Third-class indicators, their weights and attributes of inter-city primary industry

Sequence number	Indicator	Weight %	Attribute
9	Labor productivity of primary sector	1.65	Positively correlated

Note The contents in this table are determined by the task force after having conducted a lot of expert workshops

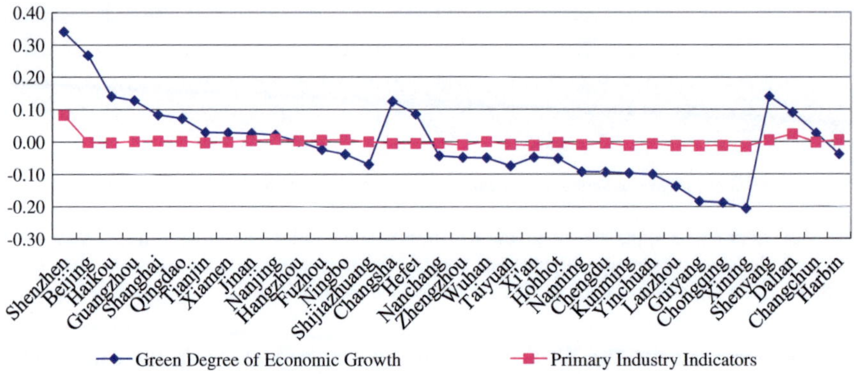

Fig. 6.10 Inter-city comparison of primary industry index and GDEG index. *Note* From the area division perspective of eastern, central, western and northeastern areas, this figure arranges these regions from *left* to *right* in terms of indexes of GDEG in descending order

level, the "primary industry productivity" was chosen as the prime criterion to measure the green degree of primary industry.

From Table 6.3 and Fig. 6.10, the difference between the both ends of PII, −0.015 and −0.083, on the city level, was only 0.098. This reflected a small variation of PII and a small gap in the development of primary industry among the 34 cities.

Among the 14 eastern cities, seven ranked among the top 10. They were Shenzhen, Nanjing, Ningbo, Fuzhou, Jinan, Hangzhou and Shanghai. The other seven ranked from 11th to 20th. Among the four northeastern cities, Dalian ranked second, Harbin fifth, and Shenyang sixth, while Changchun ranked 18th. Shenzhen's primary industry productivity was 107,900 yuan per capita, much higher than the national average of 20,700 yuan per capita, topping the list. At the second place was Dalian, the figure being 45,300 yuan per capita, though lower than half of Shenzhen, much higher than that of Nanjing (28,200 yuan per capita). These figures reflected the large gap among the top three cities. But from the regional perspective, eastern and northeastern cities had comparative advantage in PII over other regions. Among the six central cities, Wuhan ranked 13th and the other five ranked in the middle or below average. Wuhan was the only central city with its primary industry productivity above the national average. The last-place central city was Zhengzhou (28th), with a low productivity of only 10,300 yuan per capita. Among the ten western cities, eight ranked among the ten lowest PII cities, except Hohhot (17th) and Chengdu (21st), so overall a low ranking for the western region. The average primary industry productivity of this region was only 10,300 yuan per capita, far below the national average and the average of other regions.

6.3.3 Analysis on the Measurement of Secondary Industry Indicators on the City Level

In the measurement system of GDEG on the city level, comprised of five third-level indicators: 10–14, the weight of each being 1.98 %, Secondary Industry Indicators held a weight of 30 % in GDEG and 9.9 % in the total weight of GDI (see Table 6.14 for details).

Table 6.14 Third-class indicators, their weights and attributes of inter-city secondary industry

Sequence number	Indicator	Weight (%)	Attribute
10	Labor productivity of secondary sector	1.98	Positively correlated
11	Water consumption per unit of industrial value added	1.98	Negatively correlated
12	Energy consumption per unit of industrial value added	1.98	Negatively correlated
13	Ratio of industrial solid wastes utilized	1.98	Positively correlated
14	Reuse rate of industrial water	1.98	Positively correlated

Note The contents in this table are determined by the task force after having conducted a lot of expert workshops

In 2009, the proportion of added value of secondary industry against GRP in 34 measured cities ranged from 40 to 55 %. Owing to the industrial structure, however, that figure of Beijing and Haikou was much lower, only 23.5 and 23.1 %. For most cities, secondary industry constitutes the main driving force of economic development, yet it, the industry in particular, brought a huge challenge in energy conservation and emission reduction. The Third-Level Indicators, a measurement of green development of secondary industry on the city level, highlighted economic use of resources and environmental protection (Fig. 6.11).

By comparison, northeastern and central cities were leading cities in SII. Northeast China ranked first with absolute advantage, the first time to exceed the eastern region, which ranked top in all indicators of GDEG before 2009. Of the four northeastern cities, Shenyang took the first place, Changchun the second, Dalian the 11th and Harbin 22nd, slightly rearward. In terms of Third-Level Indicators, as China's old industrial cities, Shenyang, Dalian and Changchun were advantageous in the labor productivity of secondary industry and water consumption per unit of industrial added value. SII of six central cities varied greatly: among the top ten were Changsha (fourth), Nanchang (ninth) and Hefei (tenth), and Wuhan ranked

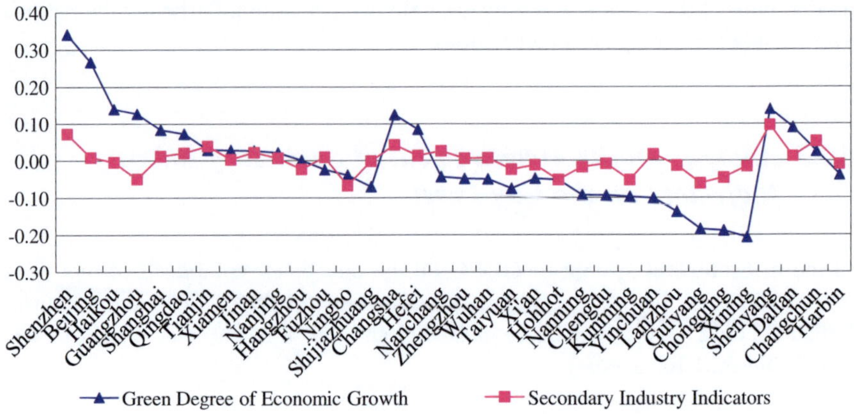

Fig. 6.11 Inter-city comparison of secondary industry index and GDEG index. *Note* From the area division perspective of eastern, central, western and northeastern areas, this figure arranges these regions from *left* to *right* in terms of indexes of GDEG in descending order

15th and Zhengzhou 17th, while Taiyuan ranked 28th. The reuse rate of industrial water Taiyuan reached up to 96.9 %, more than any other city; its labor productivity of secondary industry was 125,000 yuan per capita, ranking 12th; the other two indicators showed poor performance, especially the water consumption per unit of industrial added value (791.7 tons per 10,000 yuan), second only to Hefei, was much higher than the city average (278.3 tons per 10,000 yuan).

Compared with the other two indicators, the average SII did not win much advantage for the eastern region, higher than the western region, but lower than the northeastern and central regions. SII of 14 eastern cities varied greatly. Among the top ten were Shenzhen (2nd), Tianjin (5th), Jinan (7th) and Qingdao (8th), while Hangzhou (27th), Guangzhou (30th) and Ningbo (34th) had low rankings. The last place was Ningbo, with a score of −0.067, despite its pattern rate ranking in labor productivity of secondary industry (19th), the rankings of the other three indicators were low, the industrial water recycling rate (47.24 %) in particular, ranking 33rd, far behind the average level of measured cities (83.93 %).

The SII of western cities was undesirable. Yinchuan was the only city (No. 9) ranking forward, while the other nine cities ranked behind 20th. From Third-Level Indicators, labor productivity of secondary industry of Yinchuan and Hohhot amounted to 125,800 and 124,800 yuan per capita, ranking 10 and 11th respectively; Chengdu had a low water consumption per unit of industrial added value, 113.56 tons per 10,000 yuan, lower than any other western city; the comprehensive utilization rate of industrial solid waste of Chengdu and Xi'an reached 98.9 and 97.8 %, ranking second and seventh; Guiyang and Kunming ranked seventh and eighth in terms of the reuse rate of industrial water, slightly above 94 %. The other indicators, however, were low ranked. The western region needed to improve the efficiency of industrial water use and productivity.

6.3.4 Analysis on the Measurement of Tertiary Industry Indicators on the City Level

In the measurement system of GDEG on the city level, comprised of three third-level indicators: 16.17, the weight of each being 1.68 %, Tertiary Industry Indicators held a weight of 15 % in GDEG. See the following table for details: (Table 6.15).

Table 6.15 Third-class indicators, their weights and attributes of inter-city tertiary industry

Sequence number	Indicator	Weight (%)	Attribute
15	Labor productivity of tertiary sector	1.65	Positively correlated
16	Proportion of value added of tertiary sector in GRP	1.65	Positively correlated
17	Proportion of employees of tertiary sector in total employees	1.65	Positively correlated

Note The contents in this table are determined by the task force after having conducted a lot of expert workshops

The tertiary industry was an important gauge of regional economic development. Generally speaking, a more developed economy reflects a heavier weight of tertiary industry in national economy. Owing to low carbon and low pollution, pattern service is touted as "the green industry". Higher level of urbanization brought up the proportion of tertiary industry in urban economy, in terms of both the weights of tertiary added value and its employment. It is clear that indicators 16.17 were highly representative of tertiary industry development, a mirror of the status of tertiary industry development and the development of pattern service in urban economy.

From Table 6.3 and Fig. 6.12, among the ten eastern cities, Beijing ranked first with absolute advantage, followed by Shanghai and Guangzhou with a slight gap, and Haikou. Shenzhen, at 0.033, ranking sixth. By comparing the third-level indicators of the five cities, their weight of added value of tertiary industry were around 60 %, while that of Beijing was up to over 75 %; the employment share of tertiary industry was around 50 %, while that of Beijing was up to 73.78 %. Tertiary industry was highly developed in these cities. The other eastern cities ranked pattern rately or slightly higher, except Shijiazhuang (32nd), Ningbo (26th), Fuzhou (24th) and Qingdao (19th). Among them, the labor productivity of tertiary industry, the employment share of tertiary industry and the proportion of added value of tertiary industry of Shijiazhuang respectively ranked 24th, 31st and 32nd, contributing to its low ranking in TII.

Rankings of central cities varied greatly. Taiyuan ranked first in central China and seventh nationwide, compared with Hefei at the 30th place; Wuhan, Changsha, Zhengzhou and Nanchang respectively ranked 10th, 22nd, 27th and 29th. The shares of employment and added value of tertiary industry of Taiyuan ranked

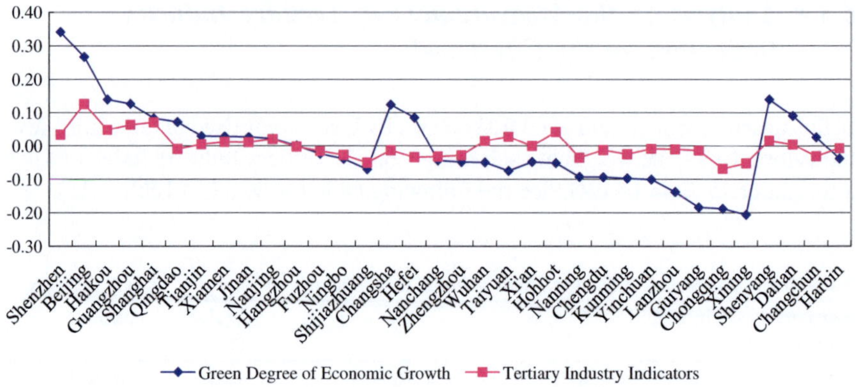

Fig. 6.12 Inter-city comparison of tertiary industry index and GDEG index. *Note* From the area division perspective of eastern, central, western and northeastern areas, this figure arranges these regions from *left* to *right* in terms of indexes of GDEG in descending order

fourth and sixth, while its labor productivity of tertiary industry ranked 17th, relatively rearward.

Except Xi'an and Hohhot, western cities were in the second half of the ranking list, including Yinchuan, Lanzhou, Chengdu, Guiyang, Kunming, Nanning, Xining and Chongqing. Hohhot ranked fifth, the highest ranking city outside the eastern region. The city's labor productivity of tertiary industry reached 145,300 yuan per capita, slightly lower than that of Guangzhou (156,900 yuan per capita) and much higher than the city average, 92,500 yuan per capita. Chongqing stayed at the bottom of the list, with its labor productivity, shares of employment and added value of tertiary industry reaching 43,400 yuan per capita, 34.72 and 37.89 %, far below the national average, ranking 33rd, 32nd, and 34th. Chongqing was not as developed in tertiary industry as other cities.

As to the four northeastern cities, Shenyang ranked ninth, Dalian 14th, Harbin 17th and Changchun 28th. Changchun, the lowest ranking city in northeastern China, lagged far behind the other three cities. In terms of third-level indicators, the labor productivity of tertiary industry of Changchun ranked 16th, a pattern rate ranking, while the shares of employment and added value of tertiary industry were only 36.53 and 41.49 %, ranking 28 and 30th, a drawback of the overall ranking.

6.4 Conclusion

From the measurement of Inter-Provincial GDEG, we can make three conclusions. First, GDEG varied greatly among regions, with eastern provinces leading others. Eastern China's success testified that economic development was compatible with green growth. Less developed regions should make use of their advantage as late comers and coordinate economic development with green growth. Second, energy conservation and economic use of resources remained a major issue concerning economic development. The measurement showed that provinces that had taken a large step forward in energy conservation and emission reduction had relatively higher GDEG ranking. It is note-worthy that despite a relatively lower energy consumption per unit of GRP owing to a sound industrial structure (high weight of service vis-à-vis low energy consumption per unit of output in service), the eastern region was the largest consumer of energies and resources. Third, while the three industries all contributed to green development, the eastern region set a good example for other regions with a higher green growth efficiency, or less energy consumption per unit of output.

From the measurement of GDEG on the city level, we also had three conclusions. First, compared with the gap in terms of inter-governmental GDEG, the gap between eastern cities and their middle and western counterparts was smaller. It was because the provincial level, or the provincial average, was mostly much lower than the level of provincial capital, which had an apparent advantage in green development over non-capital cities. Second, compared with the provincial structure, cities had a more sound industrial structure: low weight of primary industry, pattern rate weight of secondary industry and high weight of tertiary industry. Such an industrial structure constituted a major engine driving green economic growth. The green and low-carbon tertiary industry played an important role in breaking the bottleneck of resource shortage and improving resource utilization efficiency and was a critical contributor to higher GDEG. Third, cities were becoming more aware of green growth. Due to a more concentrated effort in optimizing industrial structure and recycling industrial wastes, cities were rewarded more in GDEG than provinces.

Session II
Resource and Environment Carrying Capacity

Chapter 7
Resource Bottlenecks and Environment Constraints in Green Development

Jinshe Liang, Hongrui Wang and Tao Song

Resource and environmental problems have been existed for a long time. Since the industrial revolution, social productive forces have been rapidly developed, as well as the ability of the human being to transform nature. Meanwhile, the population has been rapidly expanding and resource consumption per capita continues to increase. Human being deprives more and more resources from nature, but returns to it with increasing waste and pollution. In China, although the total amount of resources is relatively abundant, natural resources per capita are far below the world's average level due to its large population. China is still at the stage of rapid development of heavy industry and is facing the challenge of high resource consumption and low output efficiency in its economic growth. The high demand for resources and heavy environmental pollution become a serious concern, which greatly restricts China's economic and social development.

The concept of green development in this chapter emphasizes natural resource scarcity and pollutant emission limit. In order to ensure the sustainable development of our society, the two natures require reducing resource consumption and pollutant emissions. In the first part of this chapter, we have calculated the relative scarcity index of natural resources and measured the permeability of natural resources in each industrial sector. The results of these two indicators reflect the resource bottlenecks and their impacts on the development of the national economy. Following that, we have measured the resource consumption in three dimensions—scale expansion, structural change and technological progress, to conduct path analysis for reducing the effects of resource shortage. In the second part, we have analysed the current situation of environmental pollution, pollutant emission and pollution control to evaluate the quality of the environment in China. At the end of the chapter, we have analysed the impact of China's economic expansion, structural change and technological progress on pollutant emissions for a better understanding of the potential constraints of environmental pressure on economic development.

X. Li and J. Pan (eds.), *China Green Development Index Report 2011*,
Current Chinese Economic Report Series, DOI: 10.1007/978-3-642-31597-8_7,

7.1 Resource Bottlenecks in China's Green Economic Development

Before discussing the impact of resource bottlenecks on China's economic growth, we will explain the meaning of "bottleneck" in the context. In general, all resources are scarce under the definition of economics. Scarcity is defined relative to the use of resources, namely, the limited resource supply is always less than the unlimited demand at a specific spatial and temporal scale. In reality, many production activities involve two or more kinds of resources in the same place at the same time. A resource is scarce in relation to other scarce resources. For example, both arable land and water resources in China are scarce resources. From an agricultural perspective, arable land resources in many areas of China have low production as a result of the shortage in water resources. In this case, water resources tend to be relatively scarcer than arable land resources and water resources have greater bottleneck effects than arable land resources for agricultural productivity. In this part, we first measure the scarcity of China's natural resources by comparing the resource use (consumption) and resource stock (reserves). Comparing the scarcity of each resource, we can identify which ones are relatively scarcer and have stronger bottleneck effects.

One aspect of resource bottleneck effect on economy is their different levels of participation in each economic sector. For example, water resources are involved in almost all kinds of production processes, while the iron ore is not. As a supplement analysis to relative scarcity for bottleneck effect, we use the concept of "induction" in input–output analysis to measure the participation level of natural resources in each economic sector, which is known as the permeability of resources on economy. Permeability analysis can be used to evaluate the importance of each resource to the economy.

Two factors cannot be ignored when evaluating the scarcity of natural resources. First, all resources have their "locations", that is, each resource is located at a particular location on the earth with its property rights. A relative abundant resource in the world does not mean it is also rich in a certain country or region. For example, water resources are relatively affluent in most parts of south China as the precipitation is abundant throughout the year and there are a large number of rivers and lakes in the region. By contrast, water resource is relatively scarce in northwest China as the region has a small amount of annual precipitation and few rivers and lakes. Second, resource consumption is highly related to technology and economic conditions. For example, oil consumption is related to oil extraction and processing technologies and oil price. As oil extraction technology improves, international oil supply will increase and oil prices will drop, which induces oil consumption growth. Considering all the factors above, this chapter mainly focuses on evaluating the bottleneck effects of China's natural resources by using relative scarcity and permeability analysis instead of the total amount.

7.1.1 Relative Scarcity Analysis of China's Natural Resources

Based on the study conducted by Zhuang et al. (2011),[1] we first explain how to calculate the relative scarcity index by taking land and energy resources as examples. The relative scarcity index of China's energy resources is the ratio of China's share of the world's energy reserves to China's share of the world's energy consumption, calculated as formula (7.1) below:

$$\text{Relative scarcity index of China's energy resource} \\ = \frac{\text{China's energy reserves } \div \text{ World's energy reserves}}{\text{China's energy consumption} \div \text{World'senergy consumption}} \tag{7.1}$$

The denominator of formula (7.1) refers to China's relative energy consumption and the numerator refers to China's relative energy reserves. The ratio of the two values takes into account consumption and possession, as well as geographical location. Meanwhile, world's energy consumption indirectly considers the economic and technological factors.

When calculating the relative scarcity index of China's arable land, we need to take into account the special nature of land resources, which refers to the differences in their inputs and outputs in each country or region as a result of the differences in demands for agricultural products. In addition, arable land is a renewable resource, and the reserves will not be decreased under rational use. If we simply apply formula (7.1), it is not able to provide a reasonable comparison under a unified standard, and the results cannot accurately indicate the scarcity of arable land resources. Instead, we change the denominator of formula (7.1) from consumption to population, and then formula (7.1) changes to:

$$\text{Relative scarcity index of China's arable land resources} \\ = \frac{\text{China's arable land reserves } \div \text{ World'sarable land reserves}}{\text{China's population} \div \text{World's energy population}} \tag{7.2}$$

Formula (7.2) takes into account the total amount and locations of land resources, as well as demands and the renewable nature of arable land resources. For example, China currently feed approximately 1/5 of the world's population by 1/12 of the world's arable land. Based on formula (7.2), the relative scarcity index of arable land is approximately 2/5, which implies that arable land is a scarce resource in China. Formula (7.2) can be further adjusted as:

---

[1] Zhuang et al. (2011).

$$\text{Relative scarcity index of China's arable land resources} = \frac{\text{China's arable land reserves} \div \text{China's population}}{\text{World's arable land reserves} \div \text{World's energy population}} \quad (7.3)$$

Formula (7.3) refers to the proportion of China's per capita share of arable land to the world's per capita share, which explains the relative scarcity of China's arable land resources from the perspective of per capita share. If China's per capita share of arable land is lower than the world's average level, the relative scarcity index is then less than one; and if China's per capita share of arable land meets or exceeds the world's average level, the relative scarcity index is then equal to or greater than one.

By using formula (7.1 and 7.2), we can calculate the relative scarcity index of each resource to evaluate its bottleneck effect. The smaller the relative scarcity index is, the scarcer the resource is. If the relative scarcity index of a resource is less than one, the resource is relatively scarce in China as relative to the world's average; and if the index is greater than one, the resource is relatively abundant in China.

As Fig. 7.1 shows, the relative scarcity of China's natural resources is not positive. Most natural resources are relatively scarce in China except for a few are relatively well-off. From the year 2003–2007, the relative scarcity index of vast majority of natural resources showed significant downward trends, and the index of some resources fell even more than 50 %. Many originally abundant resources were increasingly consumed, resulting some resources ran short. Particularly in 2007, in addition to silver, uranium, phosphorus, vanadium, tungsten and molybdenum, other natural resources of China were in shortage relative to the world's average level, and certain resources were extremely scarce. Taking oil as an example, it should be classified as a seriously scarce resource as its relative scarcity index was less than 0.2.

Table 7.1 shows the net imports of oil, natural gas and their relative products in China from the year 2000–2009. The data further confirmed the above results. From the year 2000–2009, some processed products of crude oil, such as gasoline and diesel, had net exports in certain years, which were shown as negative in the table. China's net imports of crude oil, the raw material of petroleum products, showed an uptrend except a marginal decline in 2001. The net import of crude oil in 2005 kept at the same level as in 2004 because of the impact of the serious downturn in the stock market on China's economy. In other years, the net imports, however, showed a rapid growth. Especially in the 4 years after 2005, the average increase of net imports nearly reached 20 million tons per year. The net imports of crude oil increased from 59.96 million tons in 2000 to 198.58 million tons in 2009, rising more than 200 %. China has excessive dependence on imported oil that is restricted by the international oil price. The high dependence on imports will be a serious threat to China's economic security.

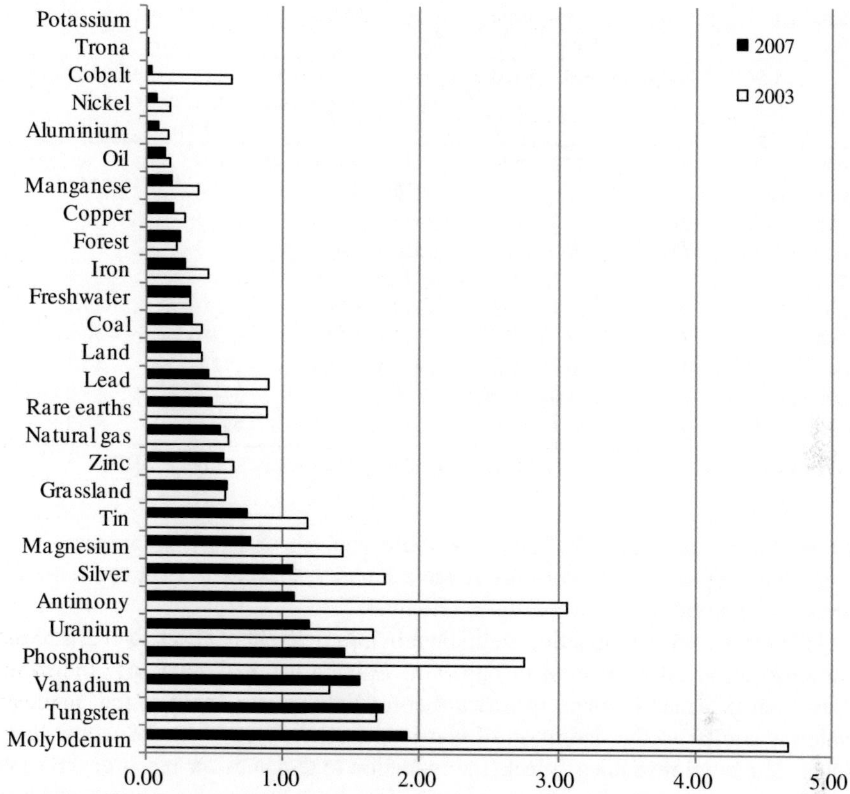

Fig. 7.1 The relative scarcity of China's natural resources in 2003 and 2007. *Source* Zhuang et al. (2011)

Furthermore, water resources and arable land resources have strong regional disparities and low mobility. In order to comprehensively measure the shortage of water and arable land resources in China, we should pay more attention to their regional disparities in addition to their total amount when studying the impacts of their shortage and bottleneck effects on economic development. In a word, the mobility of resources should be considered other than the relative scarcity.

7.1.2 Permeability Analysis of China's Natural Resources

The permeability of natural resources refers to their levels of participation in national economy. As mentioned earlier, each natural resource has different participation level in the economy. Resources with the same scarcity may have different bottleneck effects on China's economic development as a result of their

Table 7.1 The net imports of oil, natural gas and relative products in China from the Year 2000–2009 (Unit: 10,000 ton)

	Crude oil	Gasoline	Diesel	Kerosene	Fuel oil	Liquefied petroleum gas	Other petroleum products	Natural gas (100 million m^3)
2000	5,996	−455.2	−29.5	56.6	1446.6	480.1	−119	0
2001	5,271	−572.5	1.9	19.7	1779.5	486.8	−124.2	0
2002	6,175	−612	−76.3	44.5	1595.7	620.6	138.3	0
2003	8,288.7	−754.2	−139.2	8.6	2319.3	634.3	170.3	0
2004	11,722.8	−540.7	211.3	77	2877.5	637.8	23.5	−24.4
2005	11,875	−560	−94.4	59.6	2378.6	614.3	−29.6	−29.7
2006	13,883.3	−344.9	−7.1	189.8	2541.3	520.5	−29.6	−19.5
2007	15,928	−441.6	96.1	76.3	2037.4	371.6	272.8	14.2
2008	17,464.8	−4.7	561.4	111.4	1454.4	191.4	247.3	13.6
2009	19,858	−487.5	−267.1	17.7	1544.8	323.1	847.8	44.3

Sources China Statistical Yearbook 2010, China Energy Statistical Yearbook 2010

different participation levels. The permeability analysis of natural resources will be used as a supplement measure for relative scarcity analysis to evaluate the constraints of natural resources on the economy.

This study uses the induction coefficient in input–output analysis to evaluate the participation levels of natural resources in China's national economy. Induction coefficient is a measurement of forward correlations in the input–output method, which is similar to the definition of elasticity in economics. The induction coefficient of natural resources reflects the induction to demands for resources in each economic sector when increasing one unit of final product in all economic sectors. The value of induction coefficient reflects the level of demand for certain resources in each sector of the national economy.The greater value implies a higher demand and stronger permeability of the resource in national economy. As each economic sector consumes products produced by the natural resource extraction sector, the induction coefficient of the natural resource extraction sector can be used to measure the permeability of natural resources in economic sectors. For example, the induction coefficient of agriculture can be used to measure the permeability of arable land resources and the induction coefficient of coal mining industry can be used to measure the permeability of coal resources.

It should be noted that water production and supply can only reflect a small part of water consumption in the input–output table, which cannot accurately show the induction coefficient of water resources. The study conducted by Zhuang et al. (2011) does not calculate the induction coefficient of water resources. There is no doubt that water consumption exists in the production activities of all economic sectors. The manufacturing process and operations of any sector need to consume water resources. Although we did not calculate the specific values, it is certain that water resources have a relatively high induction coefficient and quite strong permeability in the national economy.

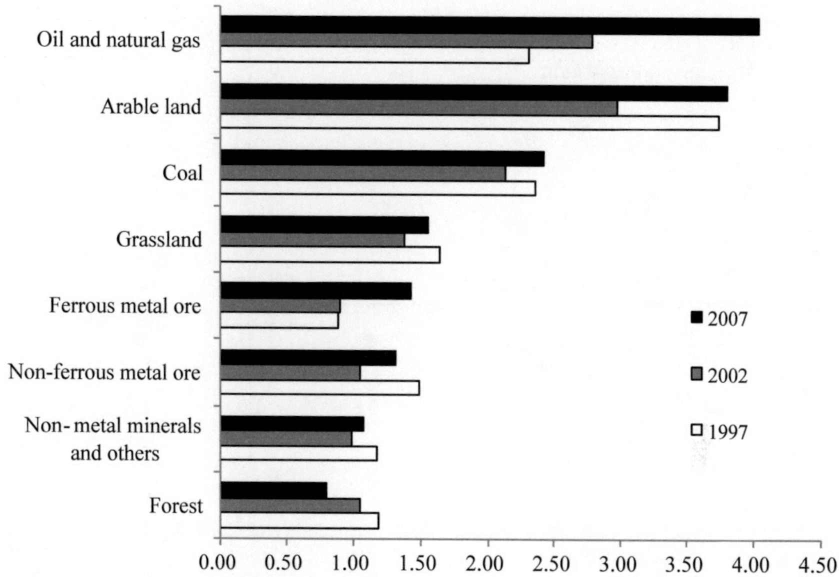

Fig. 7.2 The induction coefficient of China's natural resources in the Year 1997, 2002, and 2007. *Source* Zhuang et al. (2011)

Figure 7.2 shows that the permeability of energy resources and arable land resources are very strong in the national economy. Regarding the changing trend, their induction coefficient had different increasing extent from the year 1997–2007, which indicates that the permeability of these resources has been increasing in the economy and affected the economic development. With regard to the value of the induction coefficient, the increase in oil, gas and ferrous metal ores was obvious in recent years, and especially the induction coefficient of oil and gas had a substantial increase from the original high level. The permeability of oil and gas is the strongest in natural resource sector as the value of induction coefficient increased from 2.3 in 1997 to 4.04 in 2007. Among the fundamental resources in economic development, arable land resources have the highest value of induction coefficient. Arable land resources have played an important role in China's economic and social development, and its permeability ranking has been in the forefront of all resources. Overall, China's coal, oil, gas, and arable land resources have significant impacts on the national economy and their impacts are going to be more and more important as the basis for the development of national economy.

7.1.3 The Combination of Relative Scarcity and Permeability Analysis

Based on the results in the first two parts, we then combine the relative scarcity and permeability analysis of China's natural resources (Fig. 7.3). In Fig. 7.3, the

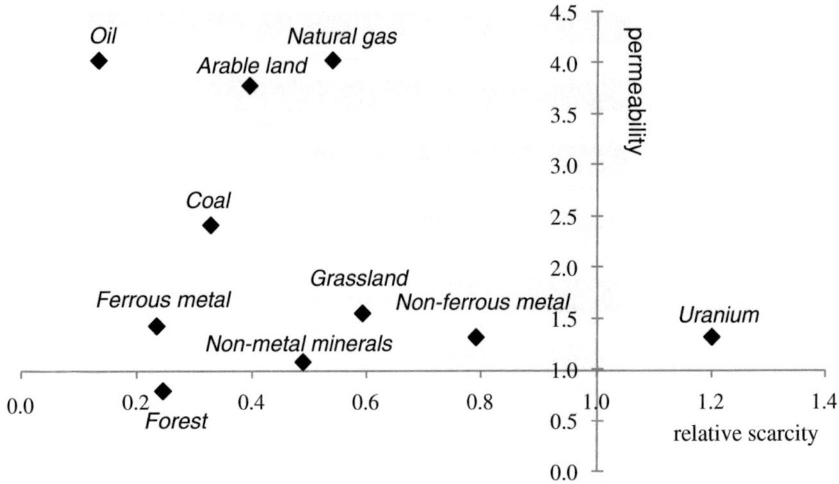

Fig. 7.3 The scatter plot of relative scarcity versus permeability of China's natural resources in 2007. *Source* Zhuang et al. (2011)

horizontal axis refers to the relative scarcity index and the vertical axis refers to the induction coefficient of natural resources. Water resources are not included in the scatter plot as we are not able to accurately calculate the induction coefficient of water resources in the national economy. In accordance with the above analysis, water resources have a strong permeability and low relative scarcity index. We then infer about water resources should be located near the upper left corner of the second quadrant at where is close to oil and arable land resources.

Only a few mineral resources are located in the first quadrant at where are close to the horizontal axis, indicating that the participation level of these relatively abundant mineral resources is not high in China's national economy. Most of China's natural resources are located in the second quadrant at where are close to the horizontal axis. The distribution pattern indicates that these resources are relatively scarce, but their permeability in all sectors is not high. They are not the resources with most serious bottleneck effect to restrict the economic development. Energy, arable land, and water resources are located in the second quadrant at where are away from the origin. Their locations in the scatter plot show two features at least: first, the relative scarcity index of these resources is far less than one which means they are relatively scarce. The relative scarcity index of coal, arable land, water and oil resources in particular are less than 0.5. Second, the permeability of these resources is very strong in the national economy and they affect all aspects of economic development. Therefore, energy, arable land, and water resources are the bottleneck resources that constrain China's economic development. To break through the bottleneck effect of the current resource constraints, we need to release the pressure from meeting the challenge of the acute shortage in these resources.

7.1.4 The Impact of Economic Development on the Bottleneck Effect of Resources

In the three parts above, we have used relative scarcity index, induction coefficient, and the combination of the two to evaluate the bottleneck effect of natural resources in China. The results show that energy, water and arable land resources are the three resources with the strongest bottleneck effect that constrains the China's economic development. In fact, the national economy can adjust its scale, structure and use of technology to reduce the resource consumption. The bottleneck effect of resources is related to the resource reserves and consumption. The path choice of economic development has impacts on the resource consumption, and it thereby will affect the bottleneck effect of resources. Given the limited space available, this section only takes energy consumption as an example. We first evaluate the economic effects on energy consumption from three aspects, and then summarize the solutions to alleviate and improve the tense relation between China's economic development and resource consumption.

In this part, we use the decomposition method of energy consumption growth developed by Liang et al. (2009) to analyse the effects of scale expansion, structural change and technological progress on energy consumption in economic development.[2] The formula is[3]:

---

[2] Liang et al. (2009).

[3] Let Y express the GDP. A denotes the column vector, in which each component represents one industrial sector's output share in GDP. The sum of these components is one. B denotes the column vector with components representing each industrial sector's energy consumption for output per unit. We use subscript to represent the time point t and 0.

There are two ways for the decomposition.

The first decomposition is: energy consumption increase (ΔE_{tot}) is

$$\Delta E_{tot} = A_t^T B_t Y_t - A_0^T B_0 Y_0 = (A_t^T - A_0^T) B_t Y_t + A_0^T (B_t - B_0) Y_t + A_0^T B_0 (Y_t - Y_0) \quad \text{(A1)}$$

In the equation, the first part on the right side of the second equal sigh refers to the changes in energy consumption caused by the share change in each sector, which is known as structural effect on energy consumption and calculated by energy consumption per unit at the time t. The second part refers to the changes in energy consumption caused by the change in energy consumption per unit in each sector, which is known as the technological effect on energy consumption and calculated based on the industrial structure at the time 0. The third part simply refers to the change in energy consumption caused by the changes in GDP, which is known as the scale effect on energy consumption. The second decomposition is:

$$\Delta E_{tot} = A_t^T B_t Y_t - A_0^T B_0 Y_0 = (A_t^T - A_0^T) B_0 Y_t + A_t^T (B_t - B_0) Y_t + A_0^T B_0 (Y_t - Y_0) \quad \text{(A2)}$$

The meaning of formula (A2) is similar to formula (A1). They are different because of the difference in the referenced energy consumption per unit and industrial structures when calculating structural and technological effects. The two formulas can be combined:

$$\begin{aligned} \Delta E_{tot} &= A_t^T B_t Y_t - A_0^T B_0 Y_0 \\ &= 0.5(A_t^T - A_0^T)(B_0 + B_t) Y_t + 0.5(A_t^T + A_0^T)(B_t - B_0) Y_t + A_0^T B_0 (Y_t - Y_0) \end{aligned} \quad \text{(A3)}$$

$$\Delta E_{tot} = \Delta E_{str} + \Delta E_{tec} + \Delta E_{sca} \tag{7.4}$$

In formula (7.4), $\Delta E_{tot} = A_t^T B_t Y_t - A_0^T B_0 Y_0$ refers to the energy consumption growth between the two time points (0 and t); $\Delta E_{str} = 0.5\left(A_t^T - A_0^T\right)(B_0 + B_t)Y_t$ refers to the energy consumption growth due to changes in industrial structure during this time period, known as structural effect; $\Delta E_{tec} = 0.5\left(A_t^T + A_0^T\right)(B_t - B_0)Y_t$, refers to the changes resulted from energy consumption per unit output in each industry during the time period, or the energy consumption growth due to the change in production technology, known as the technological effect; $\Delta E_{sca} = A_0 B_0 \left(A_t^T - A_0^T\right)$ refers to the energy consumption growth due to economic scale expansion without considering the structural effect and technological effect, known as scale effect. It should be noted that structural and technological effects on energy consumption trends are uncertain, which may cause the energy consumption growth either positive or negative. The scale expansion, however, will only increase the demand for energy consumption, so the increment is always positive.

Based on the above decomposition method, structural effect, technological effect, and scale effect in the 20 years from 1985 to 2007 are presented in Table 7.2 except the missing data in 1993 and 1995. The table shows the overall trend of China's energy consumption growth was positive except the negative growth in 1998 and 1999, and the growth was rapid especially after the year 2002. The major factor of the energy consumption growth was the effect of economic scale expansion. Technological progress was a key factor in reducing energy consumption, and the structural effect did not play a significant role. Except for a few years, the industrial restructuring had positive effects on energy consumption growth. From 1986–1991, China's energy consumption growth was small and relatively stabilized. The energy consumption growth caused by the economic scale expansion was not very significant. One reason is that the total economic output was small, as well as the economic growth and energy consumption, although the average economic growth rate reached 8.52 % during this period. Another feature of the energy consumption growth during this period is the big changes in structural effect. China was in the stage of industrial restructuring during this period, and the overall trend in energy consumption is increasing.

From 1992 to 2002, the annual energy consumption growth was in large differences because of the instability of technological effect and structural effect. Overall, the technological effect made some progresses during this period compared to the 1980s and it played a crucial role in restraining energy consumption growth. Structural effect continued on a large variation during this period, but the overall

---

(Footnote 3 continued)

Let: $\Delta E_{tot} = A_t^T B_t Y_t - A_0^T B_0 Y_0$, $\Delta E_{str} = 0.5(A_t^T - A_0^T)(B_0 + B_t)Y_t$, $\Delta E_{tec} = 0.5(A_t^T + A_0^T)(B_t - B_0)Y_t$, $\Delta E_{sca} = A_0^T B_0 (Y_t - Y_0)$.

The change in energy consumption between any two time points can be decomposed into structural effect, technological effect and scale effect, that is, formula (7.4).

Table 7.2 The decomposition of China's energy consumption growth (1985–2007) (Unit: 10^4 t standard coal)

	Energy consumption growth	Structural effect	Technological effect	Scale effect
1985 ~ 1986	4,058	938	−1,573	4,691
1986 ~ 1987	4,567	670	−6,224	10,121
1987 ~ 1988	4,662	5,257	−6,538	5,942
1988 ~ 1989	3,606	−2,333	4,866	1,073
1989 ~ 1990	1,145	4,594	−6,901	3,452
1990 ~ 1991	4,381	513	−1,514	5,382
1991 ~ 1992	5,688	6,427	−11,305	10,566
1992 ~ 1994	20,690	11,520	−21,481	30,651
1994 ~ 1996	13,635	3,416	−9,888	20,107
1996 ~ 1997	472	−661	−10,341	11,474
1997 ~ 1998	−3,600	−3,016	−6,559	5,976
1998 ~ 1999	−2,264	2,207	−15,043	10,571
1999 ~ 2000	7,021	1,519	−7,521	13,023
2000 ~ 2001	4,016	2,533	−11,673	13,157
2001 ~ 2002	7,665	781	−9,446	16,330
2002 ~ 2003	20,893	3,788	−5,046	22,152
2003 ~ 2004	26,783	4,911	9,305	12,566
2004 ~ 2005	17,980	1,168	−37,775	54,586
2005 ~ 2006	20,956	1,458	−17,983	37,481
2006 ~ 2007	17,911	2,199	−33,934	49,646

Sources Statistics from "China Statistical Yearbook" and "Industrial Economic Statistical Yearbook". Energy unit is 10^4 t stardard coal, energy consumption for output per unit is expressed in 10^4 t Standard coal/10^8 yuan output value. In order to eliminate the impact of inflation on the results, constant prices are used for the added value of each sector. To unify the data and make results comparable, the energy consumption in each sector is conversed to 10^4 t standard coal by using conversion coefficient based on energy statistics

trend was increasing with negative growth only in certain years. Industrial restructuring is still in progress and has a long way to go. Scale effect was relatively stable and fluctuated between 10,000 × 10^4t ~ 15,000 × 10^4t standard coal.

The energy consumption growth from 1992 to1996 was significantly higher than the 1980s. Despite technological effect was greatly improved over the previous period, it still cannot keep up with the pace of scale effect and structural effect. During the three-year period of 1997–1999, energy consumption was improved, and it even showed negative growth in the year 1998 and 1999. There are mainly two reasons for the negative growth in energy consumption during the 2 years: first, the buyer's market appeared in China's economy in 1997 for the first time. Overproduction led to the market supply exceeded demand, and enterprises could only reduce their production in order to keep down the inventory. Meanwhile, the Asian financial crisis caused weak domestic demand and reduction in energy consumption. Second, during the industrial restructuring in recent years, there were quite a number of the "five small" enterprises[4] with high energy consumption, low efficiency and heavy pollution being shut down, which effectively reduced the energy production and consumption.

Compared to the 1980s, the scale effect of China's energy consumption in the 1990s significantly increased as a result of the average GDP growth at an alarming rate up to 10.25 %. During this period, China's foreign trade exports also increased significantly. However, in the 1990s, most of the China's export business was "Processing & Assembly", which refers to the processing materials supplied by customer, processing according to customer's samples, assembling parts supplied by customer, and compensation trade. These trades have high energy consumption and relatively low economic efficiency. On the other hand, the technological effect of this period performed well. To a certain extent, it offset the growth in energy consumption caused by the scale effect. The progresses in production and energy-saving technologies enabled China to support a rapid economic growth with relatively low energy cost.

From the year 2003–2007, China's energy consumption increased rapidly, of which scale effect contributed most, technological effect was fluctuating, and structural effect remained unchanged. Although China's economic growth had an average rate around 10.5 % in these years, the average growth rate of energy consumption was 10.9 % and the elasticity coefficient of energy consumption to economic growth was higher than one. There are three main reasons for the more rapid growth in energy consumption than economic growth during this period: first, the scale effect increased too fast with an average increment over $30{,}000 \times 10^4$ t standard coal and the peak of $54{,}586 \times 10^4$ t standard coal in 2005. Second, the effect of technological progress was poor. During these years, a large amount of energy and raw materials were consumed in order to achieve rapid economic growth. More attention was paid to high investment than technological progress, particularly in high energy-consuming sectors such as oil and chemical industries. In these high energy-consuming sectors, the energy consumption of output per unit was increasing instead of dropping down, despite the technological effect made significant progress in 2005. The overall energy consumption was not dropping down as a result of the offset of the growth in scale effect. Third, the structural effect was prominent. In 2004, the proportion of secondary industry and tertiary industry accounted for 52.9 and 31.9 % respectively, which were increased by 2.5 % and decreased by 2.4 % compared to the year 2002. The proportion of industry accounted for 45.9 %, which was increased by 2.2 % compared to the year 2002. As the energy consumption of output per unit in secondary industry is much larger than the tertiary industry, the increase in the proportion of the secondary industry and decline in the proportion of the tertiary industry accelerated the increase in energy consumption. In addition, high energy-consuming industries such as steel, cement, electrolytic aluminum in the secondary industry were rapidly expanding, which also accelerated the energy consumption growth.

[4] The "five small" enterprises refer to small iron smelting company, small coke-making company, small chemical fertilizer company, small paper-making company, and small coal mine.

So far, we took the energy consumption as an example to show how to analyse the impact of economic development on resource bottleneck effect from three dimensions: technological effect, structural effect and scale effect. It should be noted that, standard coal used in the analysis of energy consumption cannot apply to other natural resources. When analysing the consumption of certain resources, units need to be unified by using the statistical standard conversion coefficient in order to make the data comparable among the three dimensions.

Based on the analysis above, the increasingly tension between economic development and resource consumption can be eased from three aspects: first, strictly control on the expansion of the economic scale. The goal for economic development should be made according to the carrying capacity of natural resources instead of excessive emphasis upon the economic scale and GDP growth. Second, making technological progress and developing innovation to reduce resource consumption, and increasing product value with less resource consumption. Third, speeding up the industrial restructuring and shifting the economic growth from the secondary industry with high resource consumption to the tertiary industry with relatively low resource consumption. Moreover, adjustment in the foreign trade mode and improvement in independent innovation capacity will also help change China's role as a world's factory in the global industrial chain.

7.2 The Environmental Constraints on China's Economic Development

Population growth and the increased consumption levels jointly promote production expansion. While consuming a large amount of resources, a variety of waste is emitted to the environment. If the emission of hazardous substances exceeds the limited self-purification ability or environment capacity, the environmental quality will deteriorate, which will endanger human health and cause environmental destruction and significant obstruction to human survival and development. All the effects are the environmental constraints on economic development.

This part examines China's environmental problems from three aspects: the current situation of environmental pollution, the growth trends of pollutant emissions, and the progress in pollution control. Obviously, if the pollution situation is serious and pollutant emissions exceed the pollution control, the environmental quality will be inevitably led to further deterioration and the green development of China's economy would be impossible. Based on the current environmental situation, we will use the same decomposition method of energy consumption growth to divide the pollutant emission growth in China's economic development into three parts: scale effect, structural effect and technological effect. We will also analyse the potential constraints of environmental pressure on China's economic development taking into account the inputs of pollution control.

7.2.1 China's Environmental Problems

7.2.1.1 Heavy Environmental Pollution and Grim Environmental Situation

After nearly 30 years' rapid development, China has been one of the world's largest economies, but also a country with heavy environmental pollution. The surface water has been heavily polluted. The water quality of the seven major river systems has been labelled as light pollution, the eutrophication problem of the "three lakes and one reservoir"[5] is serious, and "bloom" phenomenon frequently occurs. The underground water quality of 202 cities in China is ranked as good to poor (among the 5 classes), and shows an overall trend of deterioration. The improving trend only shows in a dispersed distribution pattern. Overall quality of coastal waters is in light pollution. The degree of air pollution is as bad as water pollution. Photochemical smog, atmospheric haze and acid deposition pollution frequently occur. In 2010, with regard to the ambient air quality in 113 key environmental protection cities, 26.5 % of the cities cannot meet the second class of the national ambient air quality standards.[6] The area suffered acid rain (annual average pH value is less than 5.6) was approximately 12.6 % of the China's land area. In addition, the total amount of domestic garbage, industrial solid waste, and hazardous waste continues to grow. Gas, liquid penetration, and leaching water generated by stockpiling of hazardous waste have become a major source of pollution. Solid waste pollution becomes more and more serious and the influenced area is expanding.

7.2.1.2 The "Three Wastes" Emissions and Environmental Pressures Continue to Increase

At the same time as the rapid social and economic development in China, a variety of waste emitted to the environment has continued to increase. Wastewater, waste gas and solid waste emissions have been increased year by year. China's wastewater emissions increased from 43.29 billion tons in 2000 up to 58.92 billion tons in 2009, of which the proportion of domestic sewage discharge was higher than industrial wastewater. China's industrial waste gas emissions also increased rapidly from 13.81 trillion standard cubic meters in 2000 up to 43.61 trillion standard cubic meters in 2009. In addition, China's industrial solid waste discharge increased from 816.08 million tons in 2000 to 2.04 billion tons in 2009. The transferred domestic garbage showed a slowly rising trend after the declined in 2006, and then reached 157.34 million tons in 2009. Within the rapid

---

[5] The "three lakes and one reservoir" refers to Tai Lake, Chao Lake, Dian Lake, and Three Gorges Reservoir.

[6] Ministry of Environmental Protection of China. (2011). *The Notice of Water Quality in Key Drainage Valleys and Ambient Air Quality Status in Key Environmental Protection Cities in 2010*, 2011, No. 8. (In Chinese).

industrialization and urbanization process, the amount of pollutants continues to increase, which will increase the environmental pressures that China faces.

7.2.1.3 Initial Success of Environmental Governance and Continuing Effort for Environmental Protection

As the environmental pollution is getting worse, environmental protection receives more and more attention, and environmental governance has gained initial success. China's total investment in environmental pollution control increased year by year, as well as the proportion of investment in environmental pollution control to GDP. China's pollutant emission compliance rate and processing rate also show uptrends. Both the industrial wastewater processing rate and urban sewage processing capacity have increased rapidly, which reached 94.2 and 75.3 % in 2009 respectively. The emission compliance rates of industrial SO_2, smoke and dust have been increasing steadily and reached 91, 90.3 and 89.9 % in 2009 respectively. Environment-friendly disposal rate of domestic garbage increased rapidly after 2006 and the comprehensive utilization rate of industrial solid waste has steadily increased, which reached 71.4 and 67 % in 2009 respectively.

In general, China's environmental situation can be summarized as: certain aspects have been improved, overall situation is still grim, and the pressure continues to increase.[7] China's natural environment is fragile, population is large, economic growth pattern is extensive, and the environmental regulations are lagging behind. All these features, coupled with the rapid economic and social development, make problems that happened at various stages of a century's industrialization process in developed countries collectively happened in China within several decades. The conflicts between environment and development are increasingly prominent. Environmental degradation caused by pollution, destruction of ecological balance, and public health hazards, become constraints on sustainable economic growth and social harmony. Environmental problems have become one of the most serious challenges that China currently faces, as well as in the future.

7.2.2 The Environmental Constraints on Economic Development

The constraint of environment on economic development is that the limited environmental capacity cannot eliminate the growing pollutant emissions. The environmental pressure of economic growth depends on pollutant emissions and

[7] Chinese Academy of Engineering, and Ministry of Environmental Protection of China (Chinese 2011).

pollution control to some extent. Increased emissions have a negative impact on the environment, while environmental governance has a positive impact. However, to strengthen environmental governance requires more inputs including human, material and financial resources and technologies. Existing research shows that economic growth affects pollutant emissions through three ways: scale effect, structural effect and technological effect.[8] Scale effect refers to the emission expansion resulting from the growth of economic scale when other conditions are the same. If economic scale and the emission intensity of each sector are the same, the changes in pollutant emissions resulting from changes in economic structure are known as structural effect. In the process of economic growth, if the economic structure changes to the direction of reducing pollution, the pollutant emissions are likely to remain stable or even decline; and if the economic structure changes to the direction of increasing pollution, the pollutant emissions may rapidly increase and accelerate environmental degradation. Technological effect refers to the change in pollution emission resulting from the differences in technologies under the same economic scale and structure. Technological effect is effective in two aspects: one aspect is that technological progress is able to increase productivity, improve the efficiency of resource utilization, cut down inputs for output per unit, and reduce the negative effects of production on environment when other aspects are stable; and the other aspect is that the development of clean technology, replacement of the old technologies, and effective recycling of resources will help reduce the emissions of output per unit.[9] Obviously, pollutant emissions should be limited and environmental governance should be strengthened in order to reduce the high pressure of the serious environmental problems in China. Economic development, however, has scale, structural and technological effects on pollutant emission. Therefore, pollutant emission limits have constraints on the economic growth rate and scale, industrial structure and technology. To support this argument, this part uses the same decomposition method of energy consumption growth to analyse China's the pollutant emission effects resulting from the economic expansion in three aspects—scale expansion, structural change and technological progress in the past decade. The results will help us understand the constrains of environmental emission limits on economic scale and growth rate, industrial structure and technologies, and also shed light on potential solutions for reducing environmental pollution and improving environmental quality.

Taking into account the availability of data, this study only decomposes the emission growth in the industrial sector. The data on added value and pollutant emissions of each industrial sector are used to calculate the contribution of structural, technological and scale changes to China's industrial wastewater, waste gas and solid waste emission growth from the year 1991–2007. The results further explain the environmental constraints on economic development.

---

[8] Grossman and Krueger (1991).

[9] Zhong and Zhang (2010).

Table 7.3 The decomposition of the emission growth of China's industrial "three wastes" in 1991–2007

	Industrial wastewater discharge (Unit: 10,000 tons)				Industrial waste gas emission (Unit: 10,000 tons)				Industrial solid waste discharge (Unit: 10,000 tons)			
	Increment	Structural effect	Technological effect	Scale effect	Increment	Structural effect	Technological effect	Scale effect	Increment	Structural effect	Technological effect	Scale effect
1991 ~ 1992	−20,154	26,095	−545,557	499,308	4,979	10,608	−23,550	17,920	−789	414	−1,918	714
1992 ~ 1993	−143,613	−198,924	−414,502	469,812	3,790	−7,566	−6,651	18,007	−434	−217	−736	520
1993 ~ 1994	−39,810	30,279	−485,176	415,088	4,042	1,201	−14,827	17,667	−221	54	−682	407
1994 ~ 1995	63,830	97,839	−336,631	302,622	10,012	10,542	−14,216	13,686	312	255	−215	271
1995 ~ 1996	−279,758	30,315	−587,564	277,491	−1,138	−5,578	−9,000	13,441	−661	−82	−859	280
1996 ~ 1997	133,890	−56,387	−29,248	219,525	17,366	721	4,607	12,038	1,236	−25	1,082	179
1997 ~ 1998	−67,082	−38,003	−213,581	184,502	−2,513	1,570	−15,093	11,010	4,309	−303	4,361	251
1998 ~ 1999	−85,047	7,321	−263,206	170,839	5,576	2,822	−7,567	10,321	−3,256	−4	−3,859	607
1999 ~ 2000	−27,390	−17,582	−197,790	187,983	11,356	−4,338	3,289	12,405	−599	403	−1,381	379
2000 ~ 2001	−36,355	704	−201,235	164,176	25,124	1,191	11,958	11,976	−1261	−232	−1,313	284
2001 ~ 2002	−17,266	−7,010	−195,428	185,171	11,988	−2,188	−2,100	16,277	−177	−115	−262	201
2002 ~ 2003	59,351	−20,343	−154,912	234,606	23,839	−585	2,080	22,344	−115	19	−368	234
2003 ~ 2004	79,095	−4,061	−135,451	218,607	38,105	11,278	3,913	22,914	−149	197	−544	198
2004 ~ 2005	181,401	−8,651	−38,976	229,029	30,872	−12,398	15,813	27,457	−77	39	−297	182
2005 ~ 2006	−79,340	−45,602	−311,833	278,095	62,936	−1,995	30,417	34,515	−293	6	−491	192
2006 ~ 2007	127,128	4,958	−187,958	310,128	57,151	5,248	2,563	49,340	−124	−38	−265	179

Notes

1. Data presented in the table are calculated by the industrial added value, industrial wastewater discharge, industrial waste gas emission, and industrial solid waste discharge of each sector in the calendar year
2. To eliminate the impact of inflation on the results, the data on industrial added value of each sector are calculated in constant price
3. In order to unify the calculation and make data comparable, the industries are merged into19 industrial sectors (mining; food, tobacco and beverage manufacturing; textiles; leather, fur, feather products industry; paper and paper products industry; printing, reproduction of recorded media; petroleum processing and coking industry; chemicals and chemical products manufacturing; pharmaceutical manufacturing; chemical fiber manufacturing; rubber products; plastic products; nonmetallic mineral manufacturing industry; ferrous metal smelting and rolling industry; non-ferrous metal smelting and rolling industry; fabricated metal products; machinery, electrical appliances and electronic equipment manufacturing; electricity, gas and water production and supply industry; and other industries)

Sources China Environment Statistical Yearbook; China Industrial Economy Statistical Yearbook. (1992–2008)

The results of the three types of decomposed emission growth are presented in Table 7.3. From the year 1991–2007, China's industrial wastewater discharge was in a negative growth in most of the years, but it showed a deteriorated trend in recent years. During the five years from 2003 to 2007, the wastewater discharge in 4 years had positive growth. The industrial waste gas emissions were in an increasing trend except the year 1996 and 1998, and the emissions increased significantly especially after 2000. The situation of industrial solid waste discharge was better. In addition to a positive growth in individual years, the solid waste discharge was generally in a negative growth trend.

Similar to the results in the decomposition analysis of energy consumption growth, the results of the decomposition analysis on emission growth show that the scale effect of "three wastes" was positive in all years and scale effect is the main factor of emission growth. The technological effects of industrial wastewater and solid waste emissions were generally negative, which is a key factor of emission reduction. The technological effect of industrial waste gas emission growth, however, was positive after the year 2000. The structural effect of "three wastes" emissions were either positive or negative, namely, the industrial restructuring tended to reduce pollution emissions in certain years or increase emissions in the other years.

One important implication from the results in Table 7.3 is that China's industrial restructuring and technological innovation did not play an active role in comprehensively and sustainably reducing the pollutant emissions. If the situation does not improve in the future economic development, the environmental factors will inevitably slow down the speed of economic growth.

Although the contribution of structural effect is not in a large proportion to the environmental pollutant emission growth, the decomposition analysis of structural effect clearly shows its fluctuating pattern in recent years. The results indicate that China is in the process of industrial restructuring which does not keep insisting on reducing pollution emissions in industrial sectors. In the future development, if China plans to maintain a high economic growth rate and the structural effect cannot play a more significant role in pollutant emission reduction, technological effect will definitely face much more pressure in reducing pollutant emissions.

Technological effect is the main driving force to reduce pollutant emission. It, however, has poor performance in recent years and the emissions even increased in certain years. On one hand, the technologies adopted in China's industrial production are lagging behind and China's economic growth is in an extensive pattern; On the other hand, there is still space for reducing China's pollutant emissions through technological progress. Governments should restrict the use of traditional production techniques with high pollutant emission intensity in order to stop enterprises making profits from the cost of environmental pollution. In addition, promoting an eco-friendly and civilized consumption behaviour and reducing the pollutant emissions in final consumptions will also help in pollutant emission reduction.

References

Chinese Academy of Engineering, and Ministry of Environmental Protection of China (2011) Macro strategy for China's environment, China Environmental Science Press, Beijing, (In Chinese)

Grossman G, Krueger A (1991) Environmental impact of a North American free trade agreement. NBER Working Paper, 3914

Liang J-S, Hong L-X, Cai J-M (2009) The decomposition of energy consumption growth—during the process of China's urbanization:1985–2006. J Nat Resour 24(1):20–29 (In Chinese)

Zhong M-C, Zhang X-G (2010) Summary about the critique of environmental kuznets curve. China Popul, Resour Environ 20(2):62–67 (In Chinese)

Zhuang L, Liu Y, Liang J-S (2011) Research on natural resources scarcity and penetration in China. Geogr Res 30(8):1351–1360 (In Chinese)

Chapter 8
Green Development and Resource Support

Xuemin Liu, Shengling Zhang and Lina Fan

During the period of the 11th *Five-Year Plan*, China's natural resources played an important role in the economic growth. To support economic development consistently and narrow down the gap of development among regions, a cross-regional circulating has been achieved with regard to energy resources and water resources, while forest and land resources have also made a contribution. In the coming 12th *Five-Year Plan*, green development and low-carbon economy is highlighted. We need to make use of energy resources reasonably, improve the energy efficiency, and protect water resources and arable land resources, so that we can realize the progressive transition to the resource-saving and environment-friendly society.

8.1 Resource Support to China's Green Economic Development in the 11th Five-Year Plan Period

Green development is inseparable from the support of natural resources. During the period of 11th *Five-Year Plan*, China's green economy has taken a solid step forward.

8.1.1 Utilization of Energy Resources

Energy resources are important materials for human survival and development. There is not economic development and social progress without energy resources support. During the period of 11th *Five-Year Plan*, energy constitute which bases on oil, natural gas, coal, biomass and some other types of energy has provided

X. Li and J. Pan (eds.), *China Green Development Index Report 2011*,
Current Chinese Economic Report Series, DOI: 10.1007/978-3-642-31597-8_8,

Table 8.1 The amount of china's oil production, imports and exports

Years	Output (100 Million tons)	Year-on-year growth rate (%)	Consumption (100 million tons)	Volume of exports (10 thousand tons)	Volume of imports (10 thousand tons)
2005	1.81	3.1	3.25	806.7	12,681.7
2006	1.84	1.66	3.49	633.7	14,517.0
2007	1.87	1.63	3.67	389.0	16,317.0
2008	1.90	1.60	3.73	423.8	17,888.5
2009	1.89	−0.53	3.84	507.3	20,365.3

Sources "China Statistical Yearbook (2010)" and "China Energy Development Report 2010"

important support for economic development. China's energy industry has achieved substantial increase in production capacity, scale of developing and utilizing natural gas and new energy has been increased, and energy conservation has gained remarkable results.

8.1.1.1 Exploration and Utilization of Oil and Gas Resources

During the period of 11th *Five-Year Plan*, China's oil production remained stable. China became the world's fourth largest oil producer with the production of 189 million tons in 2009. A group of bases rose up whose annual production are around tens of millions tons of refining oil and 1 million tons of ethylene. Processing capacity of crude oil has leapt to the second in the world. Oil and gas pipeline construction created a miracle in construction history with features of large-scale and high speed. (See Table 8.1).

As can be seen from Table 8.1, China's oil consumption has been increasing; however, oil production over the same period did not show simultaneous increase. There is a huge gap between consumption and production. With the increasing domestic demand, the amount of imported oil has been increasing. In 2005–2009, crude oil production remained at the level of 180–190 million tons, and nearly reached its peak. Nevertheless, the amount of imported crude oil was leaps and bounds. Since 2005, it has been increasing by nearly 20 million tons annually. By 2009, the amount of imported oil reached more than 200 million tons. The external oil dependence became more than 50 %. So the contradiction between supply and demand increased.

With development of economy, China's energy consumption showed a trend to a rapid growth. Developing natural gas has become one of the options to the improve environment and promote sustainable economic development. In 2009, China's natural gas production reached 85.3 billion cubic meters with a year-on-year growth of 6.22 %, and the consumption reached 89.5 billion cubic meters, with a year-on-year growth of 10.01 %. Natural gas consumption is more than production, and increasing rate of consumption is also higher than that of

Table 8.2 Production and consumption of natural gas (Unit: 100 million cubic meters)

Years	2005	2006	2007	2008	2009
Output	493	586	692	803	853
Consumption	468	561	705	813	895

Source National Bureau of Statistics: *China Energy Statistical Yearbook 2010*, Beijing, China Statistics Press, 2011

production (see Table 8.2). However, compared with the world level, the natural gas consumption per capita is very low which is just 1/16 of world's level, accounts for about 3.6 % of primary energy which is much lower than the world average. Currently China's natural gas production is still relatively low. It is only available for places near the gas producing base and industrial areas to benefit from taking advantage of natural gas. Therefore, China's current utilization of natural gas is determined by the producing area. According to industry analysts, in the next few years, demand of natural gas will grow faster than that of coal and oil, and markets of natural gas will be developed across the country.

8.1.1.2 Exploration and Utilization of Coal Resources

According to *The Program of Coal Industry Development in the* 11th *Five-Year Plan*, it is essential to develop the recycling economy vigorously, expedite development and utilization of coal-bed methane, and propel coal conversion demonstration projects orderly during the period of 11th *Five-Year Plan*, the idea of China's coal industry development has changed greatly, and market-oriented reforms were further accelerated.

First, coal production continued to increase. During the period of 11th *Five-Year Plan*, with the help from market and national policy, the pace of development was accelerated, construction of coal was optimized, and the national coal production increased steadily. In 2009, the amount of raw coal production reached 297 million tons (see Table 8.3), with a year on year growth of 12.7 %. Regions of abundant Coal-producing are Shanxi Province and Inner Mongolia Autonomous Region, followed by the Shaanxi Province, Henan Province, Shandong Province.

Second, coal consumption has stabilized. Since 2005, China's macroeconomic entered into orbit of the fast development. Rapid growth in coal consumption was driven by economic growth, and transited to a stable stage of development gradually. The reason is that the main indicators for coal-consumption industries after the national macro-control policy was implemented and whole consumption of coal began to fall since in 2003.

As China is in the stage of developing heavy chemical industry, coal consumption is closely related to increasing speed of economic growth. However, with the national macro-control efforts, increasing speed of economic and coal consumption will level off gradually. The dependence on coal consumption will be mitigated.

Table 8.3 Conditions of China's coal production (Unit: 100 million tons)

Years	2005	2006	2007	2008	2009
Output	23.5	25.3	26.9	28.0	29.7
Consumption	23.2	25.5	27.3	28.1	29.6

Source National Bureau of Statistics: *China Energy Statistical Yearbook 2010*, Beijing, China Statistics Press, 2011

Table 8.4 Conditions of coal import and export (Unit: 10,000 tons)

Years	2005	2006	2007	2008	2009
Volume of imports	2617.1	3810.5	5101.6	4034.1	12584.0
Volume of exports	7172.4	6327.3	5318.7	4543.4	2239.6
Net volume of exports	4555.3	2516.8	217.1	509.3	−10344.4

Source National Bureau of Statistics: *China Energy Statistical Yearbook 2010*, Beijing, China Statistics Press, 2011

Third, the import and export of coal decline rapidly. After 2003, coal exports began to decline, while imports increased rapidly, and the amount of exports begun to show an obvious downward trend. In order to ease the situation of tight supply, the nation has adjusted coal import and export tax policies for several times since 2004 aiming to control the export of coal and encourage import of coal.

With the country's macro-control, the rapid growth of coal exports has been effectively curbed, and coal imports have been substantially increased. By 2009, imports have been much higher than exports, with coal net imports reaching 103.44 million tons and setting the highest record in recent years (see Table 8.4).

Fourth, a breakthrough in reform of the coal market was made. During the period of 11th *Five-Year Plan*, the nation pays more attention to market-oriented reforms as strengthening the development of strategic research. Market-oriented pricing mechanism was basically set up, and self-ordering mechanism of supply and demand for both sides was gradually established. In the 5 years, the nation has obtained substantial progress in reform of the coal market. Regional coal trading center and national coal reserve system began to be built, financing system was reformed, the dominant position of enterprises was enhanced, and independent innovation capability was enhanced. These are the specific features. In 2005, re-ordering of national coal production which had lasted for several years was changed. In 2006, the coal ordering system which had been led by the government was abolished. In 2007, the transportation of coal production was converged to teleconference. In 2009, government announced that any fairs linking up coal production, transportation and consumption would be held no longer. During the period of *11th Five-Year Plan*, the total fixed asset investment of coal industry reached ¥1.15784 trillion, with 31 companies listing and financing more than ¥180 billion.

8.1.1.3 Exploration and Utilization of Biomass Energy

Biomass energy ranks fourth following the fossil energy (coal, oil, gas). The consumption of biomass energy accounts for about 14 % of the global energy consumption. To a certain extent, it can meet the needs of human beings for energy. China, a traditional agricultural country, is rich in biomass energy, the crop straw resources every year is equal to 150 million tons of the standard coal, and annual forest residue resource is about 200 million tons tons of the standard coal. With the addition of other resources, current biomass energy resources are equivalent to about 500 million tons of the standard coal, of which only 250 million tons of the standard coal is actually used, and just in low efficiency.

In China, biomass energy is the third largest energy in total energy following coal and oil. Its consumption accounts for about 15 % of total energy consumption. It is the exclusive energy which is transportable, storable and renewable. It can be used as fuel, as well as to generate electricity. In recent years, China's energy contradiction between supply and demand has increased. Developing biomass resources and forming new energy industry is an important way to solve the energy problem. Therefore, during the 11th *Five-Year Plan* period, the national support scheme, high-tech development plan and other high-tech industry development projects have increased their investment in research and development of biological energy. Government attaches great importance to it. Both *National Outline for Medium and Long-term Scientific and Technological Development* in 2006 and *Bio-industry development plan* in 2007 list bio-energy research and development as a priority. In September 2007, the government issued *Medium long-term renewable energy development plan*. The plan list bio-energy as an important component of renewable energy, and set specific development goals of bio-energy from then to 2020. In recent years, relevant departments and local governments develop and implement a series of regulations and policies, which promote the development of biomass energy vigorously. National Development and Reform Commission put forward to "the development of the bio-fuels industry three-step" plan: to industrialize technology during the period of 11th *Five-Year Plan*, to achieve industrial scale during the period of 12th *Five-Year Plan*, to realize large amount of production after 2015, with consumption of bio-fuels in 2020 accounted for more or less 15 % of all transport fuel.

8.1.1.4 Exploration and Utilization of New Energy

Traditional energy facing the situation of limited storage, pollution and price fluctuation. New energy sources become an alternative form of adequate and secure energy, enjoying magnificent function in reducing greenhouse gas emissions and production costs. It will become the sustainable and important pillar of the energy system in the future.

First, it is the wind. As a clean and renewable energy, wind energy can reduce the pollution from fossil fuels in the course of power generation. Large-scale

Table 8.5 Scale of China's wind power development in 2005–2009

Years	2005	2006	2007	2008	2009
Accumulative installed capacity (MW)	1267	2555	5866	12020	25805
Annual rate of growth (%)	66	106	126	106	115

Source China Energy Saving and environmental protection company Group: *2010 China Energy Industry Development Report—the road to explore low-carbon economy*, Beijing, China Water Conservancy and Hydropower Press, 2010

promotion of wind energy can conserve energy and reduce pollution effectively. China is rich in wind resources. According to a preliminary estimate, a theoretical calculation of the wind energy resource which is 10 meters from the ground is about 4.3 billion kW. Wind energy resources are mainly located in the northwest, north, northeast and coastal and nearby islands. There is about 380 million kW in about 200 thousand square kilometers that can be developed by technology.

As technology advances, the Chinese wind power development has expanded rapidly. According to preliminary statistics, by the end of 2009, total installed capacity of wind power has reached 25.8 million kW. In recent years, installed capacity of wind power continuous growing and wind power development reached a new level (see Table 8.5). Since 2005, China's total installed capacity of wind power has doubled in five consecutive years, and total installed capacity of wind power ranked first in the world. There are 24 provinces and cities constructing wind farms nationwide, national wind power bases in Hebei, Inner Mongolia, Gansu and other places entered growth period.

The second is solar energy. Solar energy is the most important basic energy among various of renewable energy. Biomass, wind energy, ocean energy and so on are all from solar energy. Broadly speaking, solar energy includes all above renewable resources. As one kind of renewable energies, solar energy refers to what can be conversed and used directly. Solar energy is different from oil, because it will not cause the "greenhouse effect" or environmental pollution. It is the cleanest, safest and most reliable energy and the ideal energy choice in the future. China's solar energy resources are very rich, with 2/3 of the regions get radiation more than 5.02 million kJ/m^2, and its annual hours of sunshine are more than 2000 h. Solar energy that Chinese land surface gets each year is equivalent to 170 billion tons of standard coal.

At present, applications of solar energy focuses on solar thermal and solar photovoltaic power generation. China has become the first in the production of photovoltaic power all over the world, and its products occupy a large share of the major markets in the world. In 2010, the installed capacity of the power generation mechanic has reached 600 thousand kW. The major companies have formed a complete industrial chain, and the annual output reached 8 million kW. According to *the* 11th *Five-Year Plan on the development of Renewable energy*, total installed capacity of solar water heaters has reached 150 million square meters by 2010. But there are also problems to concern and improve in China's solar industry, such as overcapacity, low conversion efficiency of photovoltaic power generation, high cost and other issues.

Column 8.1 China's Largest Solar Power Station Glows in Dunhuang

Dunhuang is located in the western end of Hexi Corridor, Gansu Province. It is an important city along the ancient "Silk Road", famous for the Mogao Grottoes. The unique geographical environment and climatic conditions in Dunhuang supplies all the necessary conditions for solar power generation, such as the annual average sunshine time of 3362 h, rainy days are less than 10 days throughout the year, the days with dust are few, topography is flat and open, so it can maintain stable power generation. Therefore, Dunhuang is called "China's gold zone of solar energy".

The global financial crisis in 2008 created the opportunity for development of solar cells. Because of the sharp decline in demand for solar cells in Europe, international prices of solar cells plummeted. With the improvement of profitability, China planned a proposal in which the scale of solar power stations expanded from 1800 c to 20000 MW by the end of 2020. That would be equivalent to the electricity generation made by 20 nuclear power plants.

In 2009, China's first large-scale solar power station which is officially approved to construct by Chinese government is located in Dunhuang. Its total generation capacity reached 20 MW, ranking first in China and third in Asia. National Investment and Development Corporation and the China Guangdong Nuclear Power Group constructed 10 MW-scale of the solar power station respectively. Dunhuang became a milestone in China's solar power station. On Dec. 27, 2010, formal access system of photovoltaic power plant in Dunhuang was constructed, and then all projects of PV power station in Dunhuang were completed. Because there is abundant sunlight resources and convenient power transmission conditions in Dunhuang, which can ensure great electrical load and grid-connected PV systems, it is the ideal region in China's "Optical Engineering in Desert" plan. Therefore, the impact of this project lies not only in Dunhuang and Gansu, but also it will lead other western provinces to develop solar power projects, so as to enhance China's overall level of the solar power industry.

In 2007, Dunhuang City was identified as the provincial sustainable development experimental zone by Gansu Science and Technology Department. In 2009, it applied for national sustainable development experimental zone. On January 15th and 16th 2010, 18 representatives of ministries including Ministry of Science, Development and Reform Commission, and other relevant departments held a joint review meeting on a national sustainable development. In this meeting, Dunhuang City was eventually identified as a national sustainable development experimental area.

Sources:

1. The world's largest solar power station is going to light in Dunhuang, http://club.business.sohu.com/read_elite.php b = zz0775& a = 1039148.
2. The Ministry of Science identified DunHuang as national sustainable development experimental zone.
http://www.sipo.gov.cn/dfzz/gansu/xwdt/ywdt/201004/t20100415_512392.htm.

Third is nuclear power. China is the seventh country which could self-design and build nuclear power plants in the world. In 2007, China's nuclear power generation was 1.92 % in proportion and installed capacity was 1.24 %, the lowest in all countries with nuclear power. In 2008, China's nuclear power installed capacity was 8.85 million kW, accounting for only 1.1 % of the total installed capacity, and nuclear power generating capacity was 68.4 billion kWh, accounting for 2 % of total power output. That is far lower than world's average nuclear power level of 17 %. Under the premise of safe operation of nuclear power plant, the domestic nuclear power still has much room to develop. Since 2005, government has approved 13 power projects including the Red river in Liaoning, Ningde in Fujian, Fuqing in Fujian and so on, with 34 units. During the period of

Table 8.6 China's nuclear power plant which is running (Unit: 10,000 kW)

Name of unit	Power	Time to open network	Name of unit	Power	Time to open network
Qinshan nuclear power station	30	1991-02-05	Ling'ao nuclear power station (unit two)	98.4	2002-12-15
Daya Bay nuclear power station (unit one)	90	1993-08-31	Qinshan (the third Stage; unit one)	72.8	2002-11-10
Daya Bay nuclear power station (unit two)	90	1994-02-07	Qinshan (the third stage; unit two)	72.8	2003-06-12
Qinshan (the second stage; unit one)	60	2002-02-01	Qinshan (the second stage; unit two)	60	2004-03-11
Ling'ao nuclear power station(unit one)	98.4	2002-04-05	Tianwan nuclear power station (unit one)	100	2006-05-12

Source China Energy Saving and environmental protection company Group: *2010 China Energy Industry Development Report—the road to explore low-carbon economy*, Beijing, China Water Conservancy and Hydropower Press, 2010

11th *Five-Year Plan*, 28 units with 30.97 million kilowatts were under construction. The scale under construction in China accounted for more than 40 % of global scales (Table 8.6).

8.1.2 Utilization of Water Resources

During the period of 11th *Five-Year Plan*, China made full efforts to construct water-saving society. The basic system of water resources management whose core is to control amounts and manage by quota was almost completed, and water management is greatly improved. The economic structure system was intensified with adaption to water carrying capacity, and industrial development accelerated the trend of water saving and emission reduction. Construction of water-saving society promoted the sustainable use of water resources effectively. And while economy is growing rapidly, water consumption kept a small degree of increase, efficiency and effectiveness of water utilization was further enhanced.

8.1.2.1 The Integrated Efficiency of Water Utilization

In the first four years of 11th *Five-Year Plan*, the country's population increased

Table 8.7 China's efficiency of water utilization

Years	GDP (100 million yuan)	Overall water consumption (100 million cubic meters)	Water consumption per ten thousand yuan of GDP (m^3)	Industrial added value (10,000 yuan)	Industrial water consumption (m^3)	Industrial water consumption per ten thousand (m^3)
2005	184937.4	5633	304.6	77230.78	1285.2	166.4
2006	216314.4	5795	267.9	91310.94	1343.8	147.2
2007	265810.3	5818.7	218.9	110534.9	1403.0	126.9
2008	314045.4	5910	188.2	130260.2	1397.1	107.3
2009	340506.9	5965.2	175.2	135239.9	1390.9	102.8

Source National Bureau of Statistics: *China Energy Statistical Yearbook 2010*, Beijing, China Statistics Press, 2011

from 1.314 billion to 1.335 billion, with an increase of 1.6 %, GDP increased from 22 to 34 trillion, and water consumption increased from 579.5 billion to 596.5 billion cubic meters. Growth of water consumption slowed down, while the integrated efficiency of water kept improving. In 2009, water consumption in ten thousand yuan of GDP was 43 % lower than that in 2005 (Table 8.7).

Table 8.8 Supply and utilization of water resources in China

Years	Overall water supply (100 million cubic meters)	Overall water consumption (100 million cubic meters)	Utilization of water				Water consumption per capita (cubic meter per capita)
			Agriculture	Industry	Life	Ecology	
2005	5,633.0	5,633.0	3,580.0	1,285.2	675.1	92.7	432.1
2006	5,795.0	5,795.0	3,664.4	1,343.8	693.8	93.0	442.0
2007	5,818.7	5,818.7	3,599.5	1,403.0	710.4	105.7	441.5
2008	5,910.0	5,910.0	3,663.5	1,397.1	729.3	120.2	446.2
2009	5,965.2	5,965.2	3,723.1	1,390.9	748.2	103.0	448.0

Source National Bureau of Statistics: *China Energy Statistical Yearbook 2010*, Beijing, China Statistics Press, 2011

8.1.2.2 Water Utilization in Agriculture

China's agricultural water consumption has been a major occupation of the water utilization for a long time, and its proportion is around 60 %. Because of extensive management and low efficiency of water utilization, China's water-saving project started from agricultural irrigation. During the period of 11th *Five-Year Plan*, agricultural water consumption was still growing, but effective utilization coefficient of agricultural irrigation water has increased to 0.49 (see Table 8.8).

8.1.2.3 The Condition of Saving Water in Industry and Urban Life

According to Table 8.8, we can see the condition of industrial water utilization from the unit water-use column. In 2009, water consumption per 10 thousand-yuan industrial added value was 102.8 cubic meters, 38 % lower than that in 2005. It achieved goals designed in the 11th *Five-Year Plan* ahead of schedule.

From Table 8.8, we can see the trend of life water consumption. Life water using increase as the population continued slowly growing. Due to the high level of urbanization, in 2005, it was 56.5 %, and increased to 58.8 % in 2008, the proportion of urban life water increased gradually, water consumption per capita in rural areas is lower than urban, but water consumption rate in rural life is higher than that in the urban. The main reason is the lack of drainage facilities in rural areas. In addition, public water and service water showed a consistent growing trend.

8.1.3 Changes of Forest and Land Resources

8.1.3.1 Changes in Forest Resources

Forests are the subject of terrestrial ecosystems and the material basis of economic and social development. It is the key guarantee to maintain ecological balance and improve the ecological environment, and it played an irreplaceable role in response to global climate change. Chinese Seventh National Forest Inventory (2004–2008) showed that: the national forest area is 195 million hectares, forest coverage is 20.36 % and forest stock volume is 13.721 billion cubic meters; forest plantation area is 62 million hectares, volume is 1.961 billion cubic meters; forest reserves of total carbon vegetation is 7.811 billion tons, the annual value of ecosystem services is 10.01 trillion yuan. China's forest area ranks fifth, following Russia, Brazil, Canada and the United States; forest reserves ranks sixth following Brazil, Russia, Canada, the United States and Congo. Compared with the results of the sixth inventory, China's forest resources show these three significant changes.

First, the accumulation of forest area continues to grow, and the forest coverage has increased steadily. There is a net increase of 20.543 million hectares in forest area, and national forest coverage rate increased from the 18.21 to 20.36 %, with an increase of 2.15 % points. There is a net increase of 1.128 billion cubic meters in standing stock, and a net increase of 1.123 billion cubic meters in forest reserves.

Second, the areas of the natural forest and project zones of natural forest protection increased significantly. There is a net increase of 3.9305 million hectares in natural forest area, and a net increase of 676 million cubic meters in natural forest accumulation. Compared to the sixth inventory, there is a net increase of 26.37 % in protected areas of natural forest, and the net accumulation of natural forests is about 2.23 times as much as that in the sixth inventory.

Third, plantation area grows rapidly, and the supporting forest resources shows the trend to increase. There is a net increase of 8.4311 million hectares in plantation area, and a net 447 million cubic meters in plantation accumulation. Immature forest area is 10.4618 million hectares of which tree species area is 6.3701 million hectares, with 30.17 % more than that in the sixth inventory.

Compared with the world's average level, China's total forest resources are still insufficient, with only 2/3 of the global average, ranking 139th in the world, and forest area per capita is 0.145 ha, less than 1/4 of the world consumption per capita.

8.1.3.2 Changes in Arable Land Resources

There are two main characteristics of China's cultivated land. First, the amount of cultivated land per capita is small, and farmland is in poor quality and with spacial imbalance. Second, there is a sharp decrease of arable land because the process of industrialization and urbanization accelerates. During the period of 11th *Five-Year Plan*, China has implemented the most stringent farmland protection system, keeping at least 1.8 billion mu of arable land unswervingly. It plays an important role in protecting national food security, economic development and social stability.

In a few years time, by taking various measures to strengthen the responsibility for farmland of local governments and key leaderships, and using economic measures to mobilize farmers to protect farmland, the momentum of rapid reduction of arable land has been effectively halted. The protection area of farmland keeps over 1.56 billion mu. 2006–2009, more than 16 million mu of land area are added nationwide, which is over 12.5 million mu more than what is occupied in construction in the same period. Except a few of national major projects, 97 % of the construction project realized "replenishment before occupation", and the Forced Mechanism on the occupied land formed gradually. Output of farm land increased by 10–20 % on the average, and production cost decreased by 5–15 %.

Column 8.2 Ecological Industrial Zone of Ancient Tea Trees in Pu'er, Yunnan Province

Pu'er Tea is the advantageous industry in Yunnan. It has a long cultural heritage, unique production processes, and has its own characteristics and health effects, so it is popular with domestic and foreign consumers. There are 1.36 million mu of tea plantations and ancient wild tea trees in Pu'er. Because these trees live mixed with forest, they maintain in original state. Besides, there is more than 0.9 million mu of modern tea plantations.

In recent years, consumers have a more intense demand for natural and non-polluting tea with high-quality. However, farming methods in most tea orchards is relatively backward, and the natural ecological environment is destroyed. Some tea orchards use the chemical fertilizers and pesticides, which affect the quality of tea more or less. These contribute to ecological environmental degradation in tea orchards and reduce the quality of tea. It seriously affects the farmers' income, and limits the development of tea industry.

In order to improve the quality of Pu'er Tea, Pu'er City strengthened the construction of ecological tea orchards bases. In 2008, Action Plan for Science-Pu'er was fully implemented. The plan regarded developing tea industry as the government's top priority, aiming to promote rapid and efficient development of tea industry. Ministry of Science made a national project supported by the 11th *Five-Year Plan* which is ecological tea garden and technologies research and demonstration. This project adopts technology of building three-dimensional composite tea garden ecosystem, technology of ecological management to tea garden pests, the technology of ecological fertilization, technology of recycling waste in tea garden, and technology of Tea Rejuvenation for aging tea orchards. In addition a certain technical effect, it produces good economic benefits.

On January 29, 2010, the city planning and ecological construction seminar was held to promote Pu'er Tea plantation, and Views on the construction of Pu'er ecological tea orchards' implementation was introduced, and tea farmers and cooperatives formed the leading office for the construction of ecological tea garden. More than 80 technicians in the office were trained on "how to set up farmers' cooperatives" and " master plan of ecological tea garden construction project, composition, and technology of intercropping trees cultivation". With the task of the office being launched, constructing ecological tea orchards reached the goal.

In order to improve the macro-environmental conditions of tea orchards, create a suitable environment for tea orchards, and form tea-based biomes, work on these four aspects which are the planning, the guidance of supporting species, the rational distribution of mountain, water, land, forest and road, the afforestation in steep sides, the peak, and other great gap which is not suitable for growing tea. Constructing this three-dimensional complex ecological tea garden, many good results have achieved. On the one hand, the tea industry in Pu'er City was developed, on the other hand, the ecological restoration area was expanded, a tea ecosystem service was effectively worked and ecological protection of arable land was protected. Therefore, the ability to fully enhance sustainable development of Pu'er was strengthened.

Sources:

1. Yang Li Xian, Pu'er enhances the quality of tea from the transformation of ecological tea orchards, the Chinese Corporation-Times, 2010 to -07-27.
2. Guogui Song, Zhu Kersee, An ecological economic analysis of key technologies for Pu'er Tea, Anhui Agricultural Science Bulletin, 2011 (5).

8.2 Regional Comparisons of Resources Support in China's Green Economic Development During 11th Five-Year Plan Period

There are vast territory and large population in China. Regional development is unbalanced and natural resource endowments of every region are strongly different. The cross-regional circulation of energy resources and water resources plays an important role in gap-narrowing and regional development, and supports the green economic development significantly.

8.2.1 West–East Gas Pipeline Project

China's West–East Pipeline Project is a large infrastructure arranged to be constructed by the national government during the 11th *Five-Year Plan*. It aims to transport the natural gas from Tarim Basin in Xinjiang to Yangtze River delta and other east-central regions, along through Xinjiang, Gansu, Ningxia, Shaanxi, Shanxi, Henan, Anhui, Jiangsu, Shanghai, Zhejiang. The huge project includes three parts. Those are exploration and development of natural gas in Tarim Basin, natural gas pipeline construction between Tarim and Shanghai, and facilities construction of downstream gas utilization. It is China's first major construction project since twenty first century, and a landmark project of China's western development, because it turns resource advantages in the west into economic advantages and transport clean energy to eastern and central regions which are short of energy. It is significant for adjusting China's energy structure.

West–East Gas Pipeline Project was approved by the State Council in August 2000, and its construction is started on July 4th, 2002. This project has the longest distance, the largest pipe diameter, the highest pressure, the largest gas volume, the highest technology in China by now. In 2010, the accumulated production of the Tarim Oilfield is more than 90 million tons, and became China's fourth largest major oil field and major gas production base. Its accumulated transported gas is more than 76 billion cubic meters, among which 15.8 billion cubic meters natural gas was transported. The Tarim Oilfield has kept producing for 3 years and has transported 10 billion cubic meters of natural gas. More than 300 million people in over 80 large and medium cities in China benefit from the project. It plays the most important role in supplying the economic circle of Yangtze River Delta and the Central Plains with natural gas, and complementing natural gas for Bohai Economic Rim and central regions.

Table 8.9 Coal balance in China's provinces (2009) (Unit: 10,000 tons)

Region (province)	Output	Consumption	The amount transferred in	The amount transferred out
East region	34832.49	138012.13	103182.09	−4732.02
Central region	123074.72	120713.14	34689.21	−50600.14
Western region	146620.74	97447.7	16290.6	−65604.61
Shandong	14377.72	36740.8	22543.26	−26.5
Hebei	8494.6	24050.59	18161.49	−2183.23
Jiangsu	2215.42	19465.22	16898.41	−318.85
Zhejiang	13.2	13076.91	12529.47	–
Duangdong	–	12508.9	9551.62	–
Shanxi	61535	32207.71	4612.77	−37845.83
Inner Mongolia	60058.45	24528.82	1400.49	−37900.49
Shan'xi	29611.13	9799.9	–	−18000

Source "China Energy Statistical Yearbook 2010"

8.2.2 West–East Coal Transportation Project

In China, the big differences of endowments of coal resource among different regions, the complex geographic transportation, the conditions of macroeconomic and regional economic development have determined the basic pattern of East–West Coal Transportation Project. East–West Coal Transportation refers to coal transporting from western regions to the eastern coastal areas. As can be seen from Table 8.9, the amount of coal transferred out from western regions reached 656 million tons in 2009, and the amount of coal transferred to eastern regions reached 1.032 billion tons. The feture of East–West Coal Transportation Project is very obvious.

The above sheet lists several provinces whose amount of transferred coal (transferred in or transferred out)is in the top five, they are Shandong, Hebei, Jiangsu, Zhejiang, Guangdong. Their total amount of coal that is transferred in is 797 million tons, accounting for 51.7 % of the whole amount transferred. The three major provinces which transferred coal out are Inner Mongolia, Shaanxi and Shanxi. Their amount transferred out accounts for 77.5 % of whole amount of coal transferred (see Table 8.9). Meanwhile, areas of coal transferred in are mostly in the eastern part of China. In these regions, their level of economy is relatively higher but storage of coal resources is fewer. Since there is not enough coal to meet needs of consumption, large part of their coal need to be transferred in from outside.

Areas which supply coal include Shanxi, Inner Mongolia in North China, and Shaanxi in northwestern China. Whole amount of coal resource storage is about 630 billion tons, accounting for 64 % of national reserves of coal resources. In these areas, coal resources are abundant, kinds of coal are various, and coal quality is good. Moreover, from the geographical point of view, they are relatively close to the regions where the coal is transferred in. Therefore, they have become China's most important bases of coal industry.

8.2.3 West–East Electricity Transmission Project

The year 2010 was the 10th anniversary of implementation of *Development of the West Regions* strategy. As an important part of the strategy of developing the western regions, West–East Electricity Transmission Project not only promoted China's power industry by leaps and bounds, but also made a profound impact on the pattern of the east–west economic and social development. This project is developing the power resources of Guizhou, Yunnan, Guangxi, Sichuan, Inner Mongolia, Shanxi and other western provinces, and transporting to Guangdong, Shanghai, Jiangsu, Zhejiang, and Beijing, Tianjin, Tangshan and other areas which are short of power energy. In the south areas, hydroelectric resource of Nanpan River, Beipan River and Hongshui River which are respectively at the junction between Wujiang and Guangxi, Wujiang and Yunnan, Wujiang and Guizhou provinces, and electricity of Kengkou Power station in Guizhou and Yunnan provinces, were developed and transported to Guangdong.

West–East Electricity Transmission Project is a typical projects in Development of the West Regions. Among the three remarkable projects, this is one with the largest investment and the largest engineering quantity. From the year 2000 to 2010, West–East Electricity Transmission Project has been invested over 526.5 billion, forming three-way transmission pattern of the north, central and southern. In the north line, electric power of Inner Mongolia, Shaanxi and other provinces (regions) is transported to the North China grid; in the central line, electric power of Sichuan and other provinces is transported to the Central China grid and East China grid; in the south line, electric power of Yunnan, Guizhou, Guangxi and other provinces is transported to South China. With the rapid development, the role of West-East Electricity Transmission Project has been not just adjusting inter-provincial power surplus and deficiency as before, but relates to the development of east–west power industry, security and stability of power grids, even becomes an important factor of economic and social development.

8.2.4 South-to-North Water Diversion Project

South-to-North Water Diversion Project is a remarkable project of cross-regional circulation of China's water resources, a strategic-type project alleviating severe conditions of water shortage in northern areas, and a major decision made by the central government to meet the need of economic and social development, because the south is flood-prone and the north is drought-prone. According to the reasonable distribution of cross-regional water resources, South-to-North Water Diversion Project alleviates severe conditions of water shortage in northern areas, and promotes the development of economy, society and population, resources, and environment. South-to-North Water Diversion Project is the largest water conservancy project in human history, with more engineering capacity and investment than that in world-famous project of Three Gorges Dam.

There are three water diversion lines: eastern line, central line and western line. West line is on Qinghai-Tibet Plateau, which can control the Northwest and North China geographically. Because of the limited water in higher reaches of Yangtze River, west line can only supply water for the Northwest and North China, which is located in upper-middle reaches of Yellow River. The central line abstracts water from the Yangtze River and its tributaries, and supplies the most parts of the North China Plain. Eastern line needs to pump water to supply the north because of low terrain. During the period of 11th *Five-Year Plan*, the first-stage construction of eastern and central lines has achieved positive results.

8.2.5 Afforestation Project

Aiming at improving the ecological environment and increasing carbon absorbing ability, China promotes reforestation and forest protection projects vigorously. In June 2009, the government held a national work conference on forestry, which is the first work conference on forestry held on behalf of central government since 60 years ago. The meeting clearly states: forestry has a special status in addressing climate change; in response to climate change, developing forestry should be regarded as a strategic option. In November 2009, the State Forestry Administration issued a *Forestry Action Plan on climate change*, determining the planning objectives and key actions of forestry development.

Firstly, continue implementation of the Three-North Shelterbelt and key shelter forests along the Yangtze River and other regions, cultivated land into forests engineering, protection project of natural forest, the Beijing-Tianjin Sandstorm Source Control Project, and fast-growing forest projects and others. First step, in August 2009, the State Council printed and distributed *Opinions to further promote the construction of the Three-North Shelterbelt*, requiring to further optimize layout of the construction, requiring to plan a number of key construction projects of different features and benefits based on the construction of one hundred acres or more plantation, requiring to build a number of centralized contiguous plantation bases with large-scale in the three northern regions and build a green barrier combined with points, lines and sides. Second step, from 2009 to 2010, forest engineering focused to improve the follow-up project planning and consolidate outputs of the construction. In typical ecologically fragile areas and important ecological zones, combined poverty alleviation with reservoir resettlement. To consolidate the results of returning farmland to forests, the long-term mechanism was established. Besides, the government also strengthened the investment and policy guidance, inspiring and motivating farmers to change the way of production and life in order to make efforts to solve the problem of poverty. From 1999 to 2009, the accumulated amount of returning farmland was 403 million mu; the implementation of the project covered more than 32 million households all over the country. Third step, promote Beijing and Tianjin sandstorm projects steadily. In the year 2009, nearly 15 million mu of the sandy land was treated and used.

Column 8.3 Ecological Construction in Youyu County, Shanxi Province: A Successful Practice of Building Oasis

Youyu County is located in northwest border of Shanxi province, in the forefront of Mu Us Desert, belonging to cold and arid areas. In the early days of New China, there was only 8,000 mu of defective forest in the county; the forest coverage rate is only 0.3 %, and desertification area formed by wind erosion and drought accounted for 76.2 % of the total land area. As the extreme deterioration of ecological environment, dry sand, soil erosion, frost, hail and other natural disasters took place frequently, agricultural development of the county is extremely slow, and the people lived in poverty.

After the founding of New China, Youyu county stated the guiding ideology "to be rich, must stop sandstorm; to stop sandstorm, plant more trees". One leadership after another stuck to the blueprint. It mainly reflected in three aspects: first, officials lead off to improve afforestation, and many township leaders become "tree leaders" and many leading cadres become "forest Directors"; second, one sector is responsible for one mountain and one organization is responsible for one bridge, distributing task by heads, surveillance by the masses; third, masses work together with one spade and two hands for every person, to create an artificial oasis by planting more and more trees year by year.

Firstly, the eco-efficiency has improved significantly. (1) To keep the sand on the ground. In recent years, the number of sandstorm days has been 50 % less than that in the early liberation, and the average wind speed has decreased by 29.7 % in the effective range of forest. (2) To maintain soil and water. As the vegetation increased year by year, sediment of runoffs and rivers decreased by 60 % than that before afforestation. There had been always floods, phenomenon of washing away farmlands and cross-flow sediment, but not any more since afforestation. (3) The conservation of water. Evaporation in field is 8.8 % annually decreased than that in wilderness.

Secondly, economic efficiency has improved significantly. As the ecological environment was improved, crop varieties which grow in the county's long growing season could be popularized in large scale, and the production and quality of pasture has been improved. Economic crop accounted for more than 50 %, green fodder and annual pasture accounted for 40 %, basically realizing the transition from traditional farming to high-quality, high-efficiency ecological agriculture.

Thirdly, benefits of tourism grew up significantly. With the continuous increase of ecological construction, the county built a number of eco-tourism attractions, such as mountain Park and forest Park, and eco-tourism was gradually becoming a major new industry and new economic growth point. According to an incomplete statistics, there were more than 0.7 million visitors to Youyu county in 2010. Output of tourism in Youyu county increased significantly, demonstrating the good results of ecological construction and enhancing the visibility and reputation. It also improved external image of Youyu county, and promote the local rapid development of service industry on the basis of the tourism industry.

Youyu County won many honorary titles by planting trees and improving ecological environment with unremitting efforts. These titles are the national advanced County of sand management, National Model Green County, National Advanced County for Three North Shelterbelt Construction, Outstanding County of ecological construction in Shanxi Province, Contribution Award for forestry construction in Shanxi Province, Forest Ecology County of Shaanxi Province. Youyu county became a national ecological demonstration zone and national experimental zone for sustainable development. Youyu had become today's Oasis through ecological construction. The successful experiences and practices have led an exemplary role in Shanxi Province and even in the whole country.

Source: Office of the Shanxi provincial government.

Secondly, implement carbon afforestation projects actively. carbon afforestation is one important measure of China in response to climate change. In 2010, National Forestry Administration enhanced further management of carbon afforestation. Enforcement record management system for existing carbon afforestation projects, and registration system for new projects; regulate management for carbon afforestation projects further to develop carbon afforestation healthily and orderly. For example, National Forestry Administration carried out afforestation pilot, guiding enterprises to increase voluntary donations for carbon afforestation. *Technology Formulates for Carbon Afforestation (Trial)* and *Inspection and Acceptance for Carbon Afforestation (Trial)* were established, which have made relevant provisions in the aspect of land eligibility, selection of planting, baseline surveys, job design, species selection, planting methods, planting of soil preparation, immature forest tending, inspection and acceptance, file management and so on, aiming to provide guidance to carry out afforestation pilots with standards.

Thirdly, carry out intensive urban afforestation. Afforestation work focuses on habitat ecological construction, mobilize all social forces to work for afforestation through the nationwide campaign of voluntary tree-planting and activity of "creating green homes, building wealth village", Much attention should be paid on afforestation along railways and highways and other channels, making efforts to build forest city, forest towns, forest villages, forest campus, Strengthening the green building in city center parks and squares, increasing green coverage and public green land per capita, establishing forest parks, and accelerating the construction of urban forest ecological barrier. By the end of 2009, National Urban green coverage reached 38.22 %, green land reached 34.13 % in built-up area, park and public green area per capita reached 10.66 m^2, and certain effects were achieved on increasing carbon sink.

8.3 Outlook of Resource Support to China's Green Economic Development in the 12th Five-Year Plan Period

The main features of China economy nowadays are high speed and extensive structure. In the next five years, China will face increasingly severe resource constraints, so *The* 12th *Five-Year Plan for National Economic and Social development of the People's Republic of China* emphasizes green development and building a resource-saving and environment-friendly society.

8.3.1 Energy Resources

According to the total energy consumption, China is the second largest energy consumer all over the world. With further development of economy, China may overtake the United States to become the first largest energy consumer. There is no doubt that economic growth will be severely constrained by energy resources during the period of 12th *Five-Year Plan.* In order to coordinate economic development and energy consumption, national energy strategies must be made, including three levels: national energy strategy, medium and long-term strategic development plan of national energy, energy plan of the *12th Five-Year Plan.* Energy plan of the 12th *Five-Year Plan* is positioned as the key to tackle energy issues. According to China's actual conditions, there should be six transformations of energy strategic concepts: from the emphasis on supporting supply to the scientific regulation of energy production and consumption, in order to promote economic development patterns by controlling energy consumption; from high reliance on coal resources to the green, diverse, low-carbon energy; from over-reliance on domestic energy supply to based on domestic energy supply and strengthen international cooperation; from the eco-environmental protection lagging behind environmental protection and energy development to the coordinated development of energy; from resource dependence development model to technological innovation-driven development model; from independent development of various energy to integrated complementary coordination development of all energy types. Among those transformations, the focus of development should include these four directions: insist on giving priority to energy conservation, implement the energy technology innovation vigorously, promote green production and clean use of coal, and develop Smart Energy Network with a core of electricity.

Looking forward to energy development of the 12th *Five-Year Plan*, it should focus on reducing the energy constraint, and progressively realize the overall target for energy development in which non-fossil fuels account for 11.4 % in primary energy consumption and energy consumption GDP per capita decreases by 16 %. There are three aspects in the specific point of view. First, optimize the energy structure further, realizing the energy structure transformation from "single" to "diversity". Namely, convert the supply structure from coal-based one to what bases on diversity species composed of coal, oil, natural gas, hydropower, nuclear power and new energy; Second, increase energy prices substantially. Owing to low energy prices, energy demand is unreasonable connivance in the past. Aluminum and small steel industry and other industries with high energy consumption, high pollution and high emission were able to develop abnormally. On the one hand, they waste resources and pollute the environment; on the other hand, they also intensified overcapacity of production in related industries, not conducive to sustainable development and industrial restructuring; Third, increase energy efficiency significantly. Increasing energy efficiency is the most effective way to alleviate current contradiction between energy supply and demand. It is by raising energy prices and the promotion of energy-saving technologies that correct unreasonable development of high energy-consuming industries because of low energy prices for a long time, and correct the phenomenon of low awareness of saving energy production in residents' life.

8.3.2 Water Resources

Looking forward to the 12th *Five-Year Plan*, the general idea of water conservation includes three key areas: controlling pollutant emissions, improving environmental quality, and preventing from risk. Actually, this is a transition from what bases on pollution control to what bases on the combination of pollution control and prevention; this is a transition from reducing pollutants to improving quality. During the period of 10th Five-Year Plan, total volume control was only guidance. There was no national assessment, so the effect was not very good. During the period of 11th *Five-Year Plan*, total volume control is binding, besides, the nation assessment is needed. In this situation, the total volume control has been well implemented. During the period of 12th *Five-Year Plan*, the total volume control is both constraint and quality-guided, the nation begins to focus on improvement of quality. Reduction of pollutants has an initial correspondence with improvement of quality. From 2015 to 2020, total volume control and quality improvement will be binding at the same time. From 2025 to 2030, the nation will begin to focus on controlling improvement of quality.

Looking forward to the 12th *Five-Year Plan*, the overall objective of water resources protection and development is to reduce water consumption per unit of industrial added value by 30 % and to increase by effective utilization coefficient of agricultural irrigation water to 0.53. Specifically, it is to decline the amount of water consumption and improve efficiency. Recycling is an effective solution to the issues. However, the key is to solve water pollution problems brought by heavy industry and manufacturing. Tasks mainly include these four aspects. The first is the control of emission. Discharge standards for industrial enterprises would be given severe treatment, especially in accordance with the requirements of regional total volume control. The second is the prevention of new contamination. It is

Table 8.10 Chinese arable land over the years (Unit: 10,000 ha)

Years	2005	2006	2007	2008
Agricultural average	13004	13004	12174	12172

Source National Bureau of Statistics: *China Statistical Yearbook 2005–2008,* Beijing, China Statistics Press, 2005–2008

mainly to determine which industries need to be encouraged, which need to limit, which need to eliminate according to industrial policy which is in accordance with the industrial structural adjustment, in order to solve the problems of new pollution and industrial structure optimization. The third is regional integrated management. Improve water quality further on the basis of emission attainment and the total volume control through the constructed wetlands, ecological restoration, intercept regional pollutant and other measures. The fourth is the construction of monitoring system of water environment. Currently, there are monitoring and pollution control approaches in China. The key of the 12th *Five-Year Plan* is to combine the two to speed up from pollution control to water quality monitoring.

8.3.3 Arable Land Resources

Arable land resources relates to food security, social security and environmental protection and other issues. Therefore, protection of arable land resources is related to social stability and national security. However, in recent years, with the accelerating process of urbanization, a large amount of suburban agricultural land is changed to urban construction land. That is bound to occupy much agricultural land, especially farmland, contributing to the tense situation of land resources. Data show that China's cultivated land area in 2008 decreased by nearly 9,000 ha compared with 2005(see Table 8.10), which makes situation of the land and food supply grimmer. In the aspect of supply, basis for food production is more and more fragile; in the aspect of demand, more and more farmers shift into grain consumers. With the rapid development of economy and society, especially the advancement of industrialization and urbanization, the contradiction between people and land is very prominent and pressure of farmland protection is heavy.

According to the current situation, there are 121.72 million hectares of arable land in China, the equivalent of 1.825 billion mu. Looking forward to the 12th *Five-Year Plan*, the bottom line of 1.8 billion mu of arable land should be hold. Referring to past empirical data and variation trend of Possession of arable land quantity, it is still an arduous task to realize the overall goals. One difficulty is that pollution problem becomes a major obstacle to protect arable land because pollution of soil, water and air will lead to lower quality of arable land inevitably. Another lies in the difficulty to make the statistics of arable land reduction, which brings more negative factors to achieve land protection. In the 11th National

People's Congress for the fourth meeting, Premier Wen Jiabao, in his government work report, states that China's urbanization rate would increase from 47.5 to 51.5 % in the 12th *Five-Year Plan*, therefore, intensifying efforts to protect arable land should be fit into China's key work and tasks in the 12th *Five-Year Plan*. The first is the strict implementation of basic farmland protection system, making sure to realize no reduction in the amount of basic farmland, no change of its function, no reduction of its quality. The second is to insist on the control system of land use, being strict in examination and approval of construction land. The third is to implement balance system of arable land strictly, fulfilling the responsibilities for additional land use. The fourth is to adhere to the execution of the system of requisition-compensation-resettlement, to ensure the legitimate rights and interests of farmers. The fifth is to increase the maintenance of land and complement farmland positively, making sure to realize the goals of farmland protection and development.

Chapter 9
Green Development and Environmental Carrying Capacity

Xianqiang Mao, Rui Hou and Peng Song

Earlier studies on Carrying Capacity are closely related to the development of Ecology. In 1921, Park and Burgess proposed the concept of Carrying Capacity in their work in Human Ecology, i.e. 'the maximum limit of the number of individual organisms in existence, in a particular environmental condition (mainly refers to the living space, nutrients, sunlight and other ecological factors in combination)'. This term was then used in environmental science academy in China, i.e. 'Environmental Carrying Capacity', reflecting the interaction between intensity of human activity and environment.[1]

'Environmental Carrying Capacity' in this chapter is focused on human activity, namely the threshold of the ability of environmental system to support human activities, which is a function of time, space and economic behavior. Focused on 'Green Development and Environmental Carrying Capacity', this chapter aims to discuss how to 'reasonably use the carrying capacity of environment, improving environmental carrying situations and conditions, and thereby improve the quality of the environment'.

In this chapter, we will use the indicator of 'Pollution-Economy Time Elasticity (PETE)', i.e. the ratio of the annual percentage variation of pollution indicators over the annual percentage variation of economy indicators in a specific country or region, to reflect the relationship between pollution and economy growth and to reveal the dynamics of pollution and economic development, thereby to reflect the environmental efficiency of economic growth.[2] 'Green Development and Environmental Carrying Capacity' in this chapter more refers to a 'flow' than a 'stock'. Therefore,

---

[1] GUO Xiurui, MAO Xianqiang, RAN Shenghong. Research Progress in Environmental Carrying Capacity in China [J]. China Population, Resource and Environment, 2000(3), 28–30.

[2] MAO Xianqiang, YANG Shuqian. The Concept, Method and Application of the Pollution-Economy Time Elasticity (PETE) [J]. China Population, Resource and Environment, 2010(11), 156–160.

X. Li and J. Pan (eds.), *China Green Development Index Report 2011*,
Current Chinese Economic Report Series, DOI: 10.1007/978-3-642-31597-8_9,

PETE is used to depict the dynamics of national or regional 'Green Development and Environmental Carrying Capacity'.

9.1 Environmental Carrying in China's Green Economic Development in the 11th Five-Year Plan Period

In the 'The 11th *Five-Year Plan* for National Economic and Social Development of the People's Republic of China' approved by the Fourth Session of the Tenth National People's Congress in March 2006, it was clearly advanced to reduce the energy consumption per unit GDP by 20 percent, and total emission of major pollutants [sulfur dioxide (SO_2) and chemical oxygen demand (COD)] by 10 %. This was identified as binding targets of the economic and social development in the 11th *Five-Year Plan* and was a milestone in the history of Chinese environmental protection.

9.1.1 Environmental Quality

In the period of 11th *Five-Year Plan*, China's economy developed stably with an annual GDP growth rate of 11.2 %, much higher than the world economy, and 1.4 % faster than the growth rate in the "10th Five-Year Plan" period, registered as one of the fastest periods since China's Opening and Reform. The continuous rapid growth of economy has brought great pressure on resources and the environment.[3] Under the leadership of the central government, in the 11th *Five-Year Plan* period, China had promoted the development of environmental protection through the construction of resource-saving and environment-friendly society to accelerate the transformation of economic development. In the last 5 year plan period, China's environmental protection efforts and pollution control investment maintained rapid increase. The total discharge of major pollutants has been under control; the prevention and control of environment pollution has obtained stage achievement; ecological protection has been strengthened comprehensively, and environmental quality has been improved substantively. In 2010, in the water quality monitoring sections in the seven major river systems, I-III water quality grades accounted for 59.6, 18.6 % higher than that in 2005; of the cities under air quality monitoring, 82.7 % met Grade II or above, a 22.4 % increase over 2005. At the end of 2010, the daily processing capacity of the city sewage treatment

[3] National Bureau of Statistics, Series report No.1 of the development achievement for the "'*11th Five-Year Plan*': New Development, New Leaps and New Chapter", March 1, 2011. http://www.stats.gov.cn/tjfx/ztfx/sywcj/t20110301_402706119.htm.

plant reached 102.62 million cubic meters, or increased by 79.2 % compared with that of the end of 2005. City sewage treatment rate reached 76.9 %, or increased by 24.9 %.[4]

The improvement of environmental quality is closely related to the increase of the investment in environmental protection. During the period of 11th *Five-Year Plan*, China has adopted a series of policy measures to further broaden the financing channel of investment in pollution control, and to strengthen the prevention and control of environmental pollution management. As a result, investment in pollution treatment has maintained rapid increase, especially in some key areas. In 2009, total investment in environmental pollution control was 452.5 billion Yuan, increased by 89.5 % over 2005. Environmental pollution control investment as a share of GDP has increased from 1.30 % in 2005 to 1.33 % in 2009. Among them, the city environmental infrastructure construction investment was 251.2 billion Yuan, increased by 94.8 % over 2005. The growth rate of the investment in city gas, central heating, drainage, landscaping and sanitation was 27.9, 67.4, 98.3, 122.4 and 114.1 % respectively. The investment on "three simultaneous (pollution prevention and control measures must be designed, constructed and put into use at the same time with the main project)" project was 157.07 billion Yuan, or 1.5 times more than that in 2005 (Table 9.1). In 2009, the investment in fixed assets was 135.1 billion Yuan, 1.9 times more than that in 2005, among which state investment was 71 billion Yuan, an increase of 101.4 %.[5]

In 2009, there were 418 emergency accidents of environmental pollution and damage, of which 116 water pollution and damage accidents, 130 air pollution and damage accidents, and 55 solid waste pollution and damage accidents, 2 marine and 115 other pollution accidents. The direct economic loss of pollution accidents reached to 433.544 million Yuan. Based on the data in recent years (see Figs. 9.1 and 9.2), although China's environmental pollution and damage accidents decreased distinctly, the caused direct economic loss remained at high level, which indicates that China's environmental risk early warning and processing system is in urgent need to be improved to guarantee a better environmental safety.

At present, environmental pollution in China is still an important issue in economic development. Economy and environment can complement each other and interact with each other. Under certain natural and social circumstance, the rapid development of the economy can be affected by environmental conditions and have an impact on the environment in turn. Then the changed environment will react to the material production and development of human beings. Therefore, in a certain environment carrying capacity condition, the pursuits of green

[4] National Bureau of Statistics, Series report No.1 of the development achievement for the "'*11th Five-Year Plan*': New Development, New Leaps and New Chapter", March 1, 2011. http://www.stats.gov.cn/tjfx/ztfx/sywcj/t20110301_402706119.htm

[5] National Bureau of Statistics, Series report No.14 of the development achievement for the "'*11th Five-Year Plan*': Environmental protection has made positive progress", March 10, 2011. http://www.stats.gov.cn/tjfx/ztfx/sywcj/t20110310_402709535.htm.

Table 9.1 Annual investment in environmental pollution control

Index	Unit	2005	2006	2007	2008	2009
Total investment in environmental pollution control	100 million Yuan	2388.0	2566.0	3387.3	4490.3	4525.3
Investment in urban environmental infrastructure	100 million Yuan	1289.7	1314.9	1467.5	1801.0	2512.0
Investment in treatment of industrial pollution sources	100 million Yuan	458.2	483.9	552.4	542.6	442.6
Environmental protection investment under the "three simultaneous" regulation	100 million Yuan	640.1	767.2	1367.4	2146.7	1570.7
Environmental pollution control investment/GDP	%	1.30	1.22	1.36	1.49	1.33

Source National Bureau of Statistics, "'11th Five-Year' series of reports on economic and social development achievements No. 14: Environmental protection has made positive progress", March 10, 2011

http://www.stats.gov.cn/tjfx/ztfx/sywcj/t20110310_402709535.htm

Note "Three Simultaneous" means the pollution abatement infrastructure/facilities of a project must be "designed, constructed and operated" simultaneously together with main body of the project

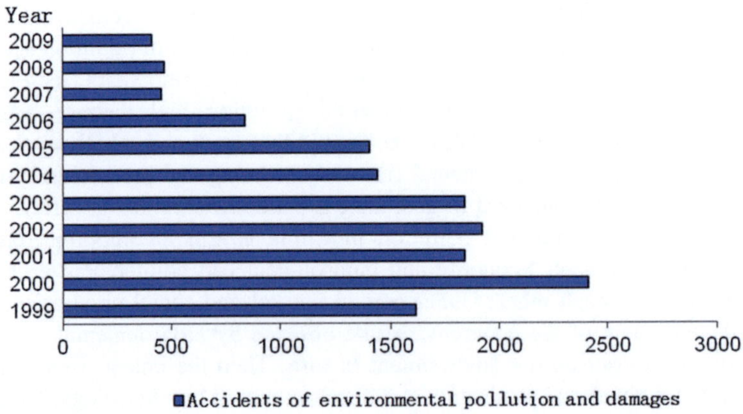

Fig. 9.1 Accidents of environmental pollution and damages in China (1999–2009). *Sources* National Bureau of statistics of China, "China Statistical Yearbook 2009", Beijing: China Statistics Press, 2009; National Bureau of statistics of China, "China Statistical Yearbook 2010", Beijing: China Statistics Press, 2010

development path, and efforts to reduce the pollutant emissions to improve environmental situation is of vital importance to promote sustainable economic and social development.

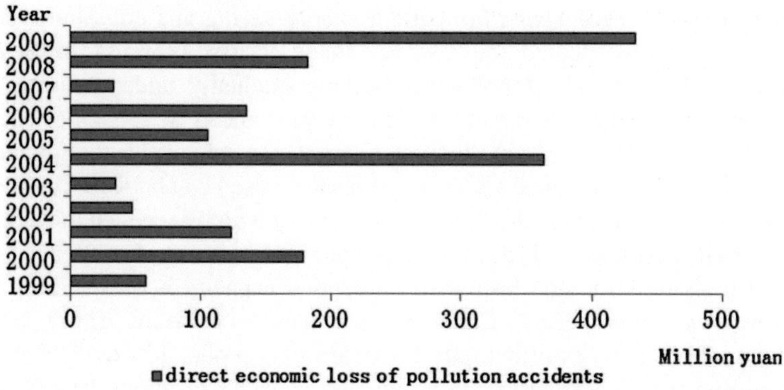

Fig. 9.2 China's direct economic loss of pollution accidents (1999–2009). *Sources* National Bureau of statistics of China, "China Statistical Yearbook 2009", Beijing: China Statistics Press, 2009; National Bureau of statistics of China, "China Statistical Yearbook 2010", Beijing: China Statistics Press, 2010

9.1.2 Pollutants Emission Reduction Progress

Since the "11th *Five-Year Plan* of National Economy and Social Development" put forward for the first time the targets of energy efficiency and environmental protection, energy saving and emission reduction has become a very important work for both the central and local governments. Governments at all levels attach great importance to this task, and set up a series of measures and policies to promote energy-saving and emission reduction work to be carried out smoothly. During the period of 11th *Five-Year Plan*, the main pollutants reduction had achieved remarkable effect, reflected by continue improvement for some environmental quality indicators (Table 9.2).

Table 9.2 Main environmental control targets of the 11th Five-Year Plan

Indices	2005	2010	Changes during the 11th Five-Year (%)
Total chemical oxygen demand emissions (10,000 tons)	1414.2	1272.8	−10
Total sulfur dioxide emissions (10,000 tons)	2549.4	2294.4	−10
Proportion of the national controlled surface water quality monitoring sections with water quality worse than grade V	26.1	<22	−4.1
Proportion of the national controlled surface water quality monitoring sections with water quality better than grade III in seven major water systems	41	>43	2
Proportion of key cities having air quality better than Class II standard for over 292 days per year	69.4	75	5.6

Source Central government portal, "Notice of the State Council on issuing the 11th *Five-Year Plan* for national environment protection" [No. 37 (2007) of the State Council], http://www.gov.cn/zwgk/2007-11/26/content_815498.htm

During the 11th *Five-Year Plan* period, energy-saving and emission reduction have obtained unprecedented emphasis and made positive progress. In the meantime, the total discharge of pollutants became gradually under control; sulfur dioxide emission reduction targets was met 1 year ahead of schedule; chemical oxygen demand (COD) emission reduction target was achieved 6 month ahead of schedule. In 2010, the national COD emissions decreased by about 12 % compared with 2005and sulfur dioxide dropped by 14 %, all of which exceeded the mandated emission reduction task in 11th *Five-Year Plan.* Furthermore, the effect of elimination of high emission and backward production capacity is outstanding. In the 11th *Five-Year Plan* period, China has eliminated backward iron production capacity of about 111.72 million tons, backward steel production capacity of about 66.83 million tons, backward coke production capacity of about 105.38 million tons, backward ferroalloy production capacity of about 6.63 million tons.[6]

9.1.2.1 Air Pollutant Emission Reduction

In the period of 11th *Five-Year*, the emissions of three kinds of main air pollutants, i.e. sulfur dioxide, soot and industrial dust all declined year after year. Up to 2009, sulfur dioxide, soot and industrial dust emissions decreased by 14.46, 22.19 and 35.23 % and their emissions were reduced respectively 3.74, 2.41 and 2.85 million tons. The effect of the major air pollutants emission reduction was significant (see Fig. 9.3).

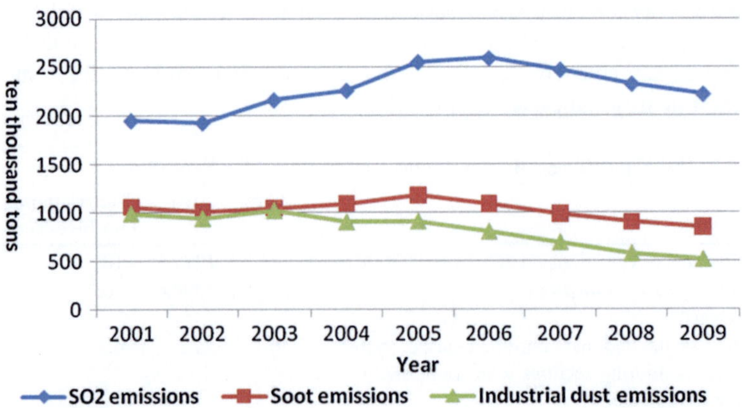

Fig. 9.3 Major air pollutants emissions. *Source* National Bureau of Statistics of China, "China Statistical Yearbook" (2002–2010), Beijing: China Statistics Press, 2002–2010

[6] National Bureau of Statistics, Series report No.1 of the development achievement for the "'11th *Five-Year Plan*': New Development, New Leaps and New Chapter", March 1, 2011. http://www.stats.gov.cn/tjfx/ztfx/sywcj/t20110301_402706119.htm.

Specifically, in 2008, the national total emissions of sulfur dioxide were 23.21 million tons, decreased by 5.95 % than that of 2007; among them, industrial sulfur dioxide emissions fell by 6.95 % from the previous year, and domestic sulfur dioxide emissions rose slightly by 0.55 %. In 2009, the sulfur dioxide emissions continued to decline, with a gross discharge of 22.14 million tons, or a decrease of 4.6 % from the previous year. Compared with 2005, sulfur dioxide emissions fell by 13.14 %, and overfilled the 11th *Five-Year Plan* emission reduction targets of 10 %.

9.1.2.2 Reduction of Water Pollutants

In the period of 11th *Five-Year*, the amount of wastewater discharge showed an increasing trend year after year, but COD discharge was declining, and the emission reduction targets of 10 % has been achieved 6 months earlier. According to the latest "China Statistical Abstract 2010–2011", in 2010, the national emissions of COD were 12.38 million tons, 12.45 % down compared with that of 2005, exceeding the emission reduction task. In addition, the ammonia emissions in the 11th *Five-Year* period did not change much in 2001–2009, showing the total emissions increased and then decreased in the trend (see Fig. 9.4).

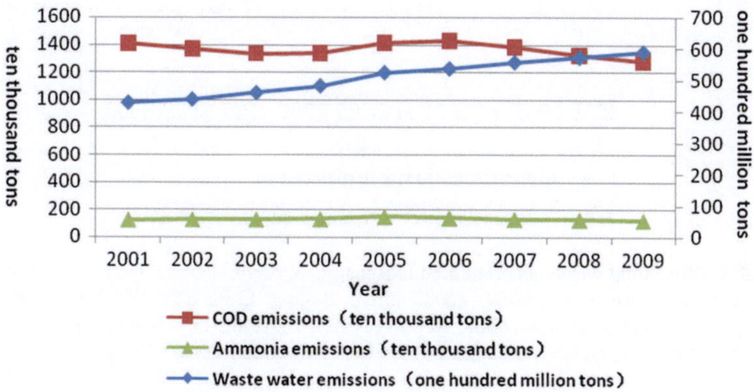

Fig. 9.4 Annual waste water and water pollutants discharge. *Sources* National Bureau of Statistics of China, "China Statistical Yearbook" (2002–2010), Beijing: China Statistics Press, 2002–2010. Ministry of environmental protection, "Report on the state of the environment in China 2009". http://www.zhb.gov.cn/gzfw/xzzx/wdxz/201006/P020100603551633387739.pdf

9.1.2.3 Solid Waste Reduction

In the period of 11th *Five-Year Plan*, China actively promoted the comprehensive treatment of solid waste, and "Solid waste pollution prevention information dissemination for national large and medium sized cities". Meanwhile, China continued to promote the implementation of "Hazardous waste operating licensing management approach", and released the regulations of "Guideline for recording and reporting the operation conditions of the hazardous waste operating unit" "Guideline for hazardous waste management review and licensing" and "Regulations on Administration of Collection and Disposal of Waste Electrical Appliances and Electronic Products". Additionally, strict waste import and export censorship system was set up and industrial solid waste conditions have been greatly improved.

In the period of 11th *Five-Year Plan*, the industrial solid waste generation grew year after year. In 2009, the amount of industrial solid waste was twice as that of 2001. In the same period, industrial solid waste treatment capacity (including utilization capacity, storage capacity and disposal capacity) continued to increase. Therefore, compared with the generation of the solid waste, discharges show the opposite trend. In 2009, discharges of industrial solid waste were less than one-third of that in 2001 (Fig. 9.5).

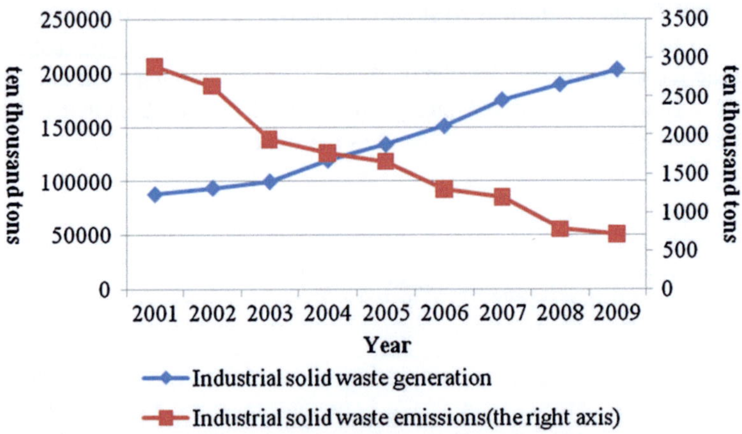

Fig. 9.5 Annual solid waste generation and discharge. *Sources* National Bureau of Statistics of China, "China Statistical Yearbook" (2002–2010), Beijing: China Statistics Press, 2002–2010. Ministry of environmental protection, "Report on the state of the environment in China 2009". http://www.zhb.gov.cn/gzfw/xzzx/wdxz/201006/P020100603551633387739.pdf

Column 9.1 Multi-Actions on Emission Reduction Achieved Remarkable Improvement—Enshi City in Hubei Province

During the 11th *Five-Year Plan* period, focusing on the reduction of major pollutants emission and aiming at improving the quality of ecological environment, Enshi has achieved significant emission reduction effects through Effective actions of reducing emission, increasing environmental protection investment, adjusting industrial structure, accelerating management of emission reduction projects, etc.

(A) Emission reduction targets were set to show government's determination

According to the agreement signed by Enshi city government with the prefecture government, Enshi had reduced 1877.56 tons of chemical oxygen demand (COD) and 2362.65 tons of sulfur dioxide (SO2) during the 11th *Five-Year Plan* period. In the context of high economic growth, Enshi actively took actions such as speeding up elimination of backward production capacity, construction of key energy-saving and emission reduction projects, increasing investment in environmental protection, implementing energy-saving system, to ensure the completion of Five-Year emission reduction target.

To make sure the emission reduction being carried out smoothly, Enshi had published local governmental documents to regulate local polluter's behavior, such as, 'Notice of Reduction Task of Total Amount of Major Pollutants', 'Notice of Time Limit for Environmental Treatment', 'The Provision of Environmental Treatment Projects and Environmental Protection Objectives Assessment and Reward for the Townships and sub-district Offices'. Meanwhile, the Enshi government allocated the task of emission reduction to specific enterprises, townships, and sub-district offices, in the form of signing responsible agreement of environment protection objectives. With environmental objectives assessment and strict supervision, the pollution reduction responsibilities could be ensured. Those who failed to meet the requirement within the time limit, would be given a veto, meaning that the leaders could be removed from their position.

(B) Strict environmental pollution source control

During the 11th *Five-Year Plan* period, Enshi enhanced environmental supervision, and shut down heavily polluting plants which could not meet the industrial policies and the environmental standard requirements. Those plants are as follows: 3 cement companies (Enshi Xintang Cement Company, Enshi Lianzhu Cement Co., Ltd. and Fumin Cement Co., Ltd.), a paper-making enterprise (Enshi Dongchang Paper-making Enterprise), and a chemical company (Enshi Balong Fertilizer Co., Ltd.). Just through those shutting down, Enshi achieved an annual reduction of 129.21 tons of COD and 92.52 tons of SO2. In 2008, Qiyangba Power Plant was cut off, resulting in more than 1,000 tons reduction of SO2 per year.

Meanwhile, Enshi took strict examination procedure to prevent new pollution sources. The newly built, rebuilt and expanded projects must strictly comply with 'Law of the People's Republic of China on Appraising of Environment Impacts', 'Law of the People's Republic of China on Appraising of Environment Impacts' and other laws or regulations as well as national industrial policies, and strictly abide by environmental function zoning, environmental quality objectives and total quantity control of pollutant requirements. The "threshold" of new construction projects was effectively raised.

(C) Supporting enterprises to construct pollution abatement projects

Because of technical and economical difficulties, Enshi Jiuzhou Livestock Co. was in lack of sewage treatment facility, and could not meet waste water discharging quality substandard. Enshi government actively sought the funding support with 600,000 Yuan from the 'national special environmental pollution control funds', to build a new sewage treatment facility. During the period of 2008–2009, the government raised more funding support to a new high-standard sewage treatment plant in Cuiba meat processing base, with treatment capacity of 700 tons per day using advanced UASB anaerobic wastewater treatment technology. The waste water discharged through the treatment can meet standard A of 'meat processing industry effluent discharge standards(GB13457-92)'.

With the strong support of Enshi government, Enshi Shanzhai Leather Co., Ltd., Hubei Jinhua Paper Co., Ltd., Hubei Shengfeng Pharmaceutical Co., Ltd., Hubei Chengcheng Pharmaceutical Co., Ltd., also managed to get help from 'national special environmental pollution control funds', with a total amount of 10 million Yuan. After these five companies' sewage treatment projects being constructed, the total sewage treatment capacity will reach 4,000 tons per day, so as to reduce 94 tons of COD per year.

(D) Sewage pipeline network construction

Enshi Shouchuang Water Co., Ltd, which lies in the downstream of Qingjiang river, can treat nearly 60,000 tons of domestic sewage per day, with its discharged water achieving emission standard B. Since the plant was put into trial operation, a sewage pipeline network of 35 km and more than 300 sewage outlets have been constructed to collect and transport urban sewage to the waste water treatment plant. During the 11th *Five-Year Plan* period, Enshi has spent 54.74 million Yuan on the construction of sewage pipe network. With the continuous extension and connection of the sewage pipe network, there will be more areas that could be accessed by sewage network during the 12th *Five-Year Plan* period, which will further improve water environmental quality in Enshi City.
Source: Enshi environmental protection bureau website
http://www.eshbj.com.cn/news/huanbaoxinwen/2010/1214/101214169323FG3F8F271AE1C434KK2.shtml.

9.1.3 Relationship Between Pollution and Economic Development

In this section, we will take use of the indicator, "Pollution-Economy Time Elasticity (PETE)", i.e. the ratio of annual growth rate (%) of the emissions of major pollutants over the annual growth rate (%) of China's economic growth, to discuss the relationship between the pollution and the economic development for the nation and the different regions.[7] The major pollutants indicators and economic

[7] MAO Xianqiang, YANG Shuqian. The Concept, Method and Application of the Pollution-Economy Time Elasticity (PETE) [J]. China Population, Resource and Environment, 2010(11), 156–160.

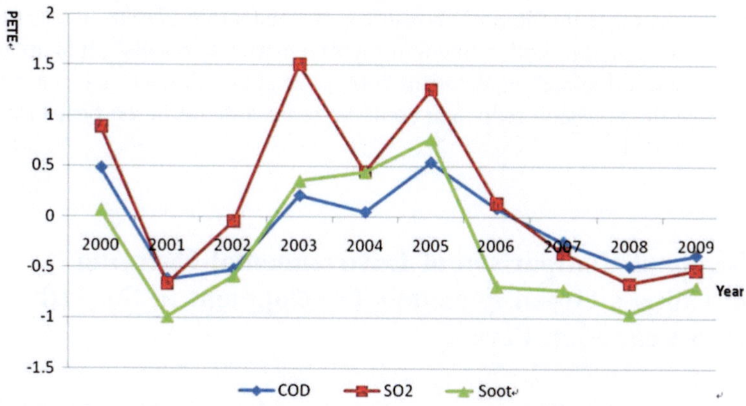

Fig. 9.6 Fluctuation of PETE of pollutant emissions in China (2000–2009)

indicators in 2000–2009 are employed to explore the variation of China's environmental loads in the periods of 10th *Five-year* and 11th Five-Year.

Considering the degree of social concern and data availability, in this section, chemical oxygen demand (COD), sulfur dioxide (SO_2) and soot are selected as the main pollutants indicators; and gross domestic product (GDP) as the economic indicator using the 1990 comparable price. The data used comes from "China Statistical Yearbook", "China Environment Yearbook" and "Compilation of Statistics for New China's 50 Years" and other information.

In 2000–2009, the total emission of COD, SO_2 and soot in China showed similar fluctuation characteristics (see Fig. 9.6).

Since 2000, China's GDP and industrial output value of enterprises has maintained growth trend, therefore, positive PETE means simultaneous growth of pollution and economy while negative PETE indicates an emission reduction over the previous year.

From Fig. 9.6, in the 10th *Five-Year Plan* period, PETE of the three pollutants fluctuated significantly. The values in 2001 and 2002 are negative, indicating a relative pollutant emissions reduction compared with the previous year, and the economic growth is relatively clean. The values in remaining years are positive, which indicates that the pollutant emissions increase along with the expansion of the economic scale year after year. In 2003 and 2005, the PETE of SO_2 is greater than 1, which means the rate of the emission increase is higher than the GDP growth rate, and the economic growth pattern is relatively "dirty". In that case, economic development pays bigger environmental cost and the environment bears larger pressure.

Since 2006, the PETE values of the three pollutants significantly decreased in different degree and turned negative in 2006 and 2007, while GDP has increased annually. This shows that, different from the 10th *Five-Year Plan* period, in the 11th *Five-Year Plan* period, the development direction of pollution and economy is opposite. Economy became "cleaner" and "clean and efficient" industries are replacing "dirty and inefficient" industries; or "economic output increase makes increased investment in pollution control with remarkable results". It is noticeable that, the PETE value has increased in 2009 over 2008, although still negative. It reminds that the emission reduction could be more difficult in the post 11th *Five-Year Plan* period.

9.2 Regional Comparison of Environmental Carrying in China's Green Economic Development in the 11th Five-Year Plan Period

To reflect the spatial differences and variation trends in environmental carrying in the process of China's economic development during 11th *Five-Year Plan* period, the relationship between pollution reduction and economic development has been analyzed at regional level.

9.2.1 Comparison of Regional Emission Reduction Progress

According to " Several Opinions of the Central Committee of the Communist Party of China (CCCPC) and the State Council on Boosting the Rise of the Central Region," " Directive by the State Council on the policies and measures of Western Development "and the spirit of the 16th Party Congress report, in order to scientifically reflect the regional difference in socio-economic development, and to provide the basis for regional development policy making by the central government, China has been divided into four economic regions, i.e. eastern, central, western and northeast regions. Eastern region includes Beijing, Tianjin, Hebei, Shanghai, Jiangsu, Zhejiang, Fujian, Shandong, Guangdong and Hainan. Central region includes Shanxi, Anhui, Jiangxi, Henan, Hubei and Hunan. Western region includes Inner Mongolia, Guangxi, Chongqing, Sichuan, Guizhou, Yunnan, Tibet, Shaanxi, Gansu, Qinghai, Ningxia and Xinjiang. Northeast region includes Liaoning, Jilin and Heilongjiang. Based on these four regions, in this section, we shall analyze the progress of Chinese emissions reduction and PETE indicator through regional comparison.

9.2.1.1 Comparison of Air Pollution Reduction

During 11th *Five-Year* period, China's three major air pollutants emission, i.e. sulfur dioxide, soot and industrial dust, decreased year by year.

Compared with 2005, total sulfur dioxide emission rate of the eastern region reached 19.02 % in 2009, 13.14 % higher than the national average. Shandong, Jiangsu and Hebei provinces ranked the top three in the eastern region. In the same

Table 9.3 Annual and cumulative air pollutants reduction rates by region (Unit %)

Sulfur dioxide emissions reduction rate					
	2005–2006 (%)	2006–2007 (%)	2007–2008 (%)	2008–2009 (%)	Cumulative reduction rate (2009 over 2005) (%)
National	−1.55	4.66	5.95	4.60	13.14
East	1.29	5.63	7.53	6.00	19.02
Northeast	−4.74	1.73	6.22	5.45	8.73
Central	−0.89	4.30	6.01	4.21	13.08
West	−3.69	4.51	4.44	3.45	8.65
Soot reduction rate					
	2005–2006 (%)	2006–2007 (%)	2007–2008 (%)	2008–2009 (%)	Cumulative reduction rate (2009 over 2005) (%)
National	7.95	9.36	8.62	5.99	28.32
East	2.89	12.80	3.91	4.63	22.40
Northeast	2.17	2.82	5.38	6.85	16.21
Central	7.71	11.44	13.59	8.17	35.14
West	14.16	7.96	9.24	4.68	31.64
Industrial dust reduction rate					
	2005–2006 (%)	2006–2007 (%)	2007–2008 (%)	2008–2009 (%)	Cumulative reduction rate (2009 over 2005) (%)
National	11.28	13.56	16.29	10.48	42.53
East	11.28	11.50	12.26	17.53	43.18
Northeast	5.74	3.61	29.79	13.49	44.81
Central	8.89	14.17	17.80	7.02	40.23
West	15.47	17.35	13.47	7.57	44.12

Source National Bureau of Statistics of China, "China Statistical Yearbook" (2002–2010), Beijing: China Statistics Press, 2002–2010

period, cumulative reduction rate in the western region was the lowest, only 66 % of the national average.

Regionally, the cumulative emission reductions of soot in the western and central regions were much higher than that in the east and northeast and the national average.

Cumulative emissions reduction of industrial dust of all regions decreased more significantly compared SO_2 and soot. The cumulative reduction rates in the eastern, northeastern and western regions from 2005 to 2009 were higher than the national average (Table 9.3).

9.2.1.2 Comparison of Water Pollution Reduction

During the 11th *Five-Year Plan* period, national waste water discharge showed increasing trend year by year, but the chemical oxygen demand emissions declined and achieved 10 % reduction target 6 months in advance. Regionally, from 2005 to 2009, cumulative emission reduction rate of chemical oxygen demand in the eastern emission reached 13.52 %, higher than the national average level of 9.66 %. For the same period, cumulative emission reduction rate of chemical oxygen demand in the western region was the lowest, only 63 % of the national average (Table 9.4).

The changes in ammonia emissions for different regions showed a declination year by year during 2006–2009. Ammonia emissions in the western region in 2008 showed an increase of 0.78 % compared with the previous year, following 2 years' consecutive decrease. While ammonia emissions in the eastern, northeastern and central regions achieved continuously decrease.

9.2.1.3 Comparison of Solid Waste Reduction

During 2001–2009, the amount of industrial solid waste generation increased year by year and was over twice of that in 2001. During this period the country's industrial solid waste treatment capacity (including capacity utilization, storage capacity and disposal capacity) continued to increase. Therefore, compared to industrial solid waste generation, solid waste emissions show an opposite trend. In 2009, emissions of industrial solid waste were less than one-third of that in 2001.

For the industrial solid waste emission reductions, in 2009, the top ten provinces and municipalities are: Shanxi, Hebei, Inner Mongolia, Hunan, Shaanxi, Sichuan, Ningxia, Henan, Zhejiang and Hubei. For the same period, industrial solid waste emissions in Chongqing, Shanxi and Xinjiang were more than 1million tons, 1.4986, 1.4157 and 1.052 million tons respectively.

For domestic solid waste, the municipal solid waste generation, waste treatment and disposal capacity, living garbage harmless treatment rate and processing capacity all increased during the 11th *Five-Year Plan* period.

Regionally, in 2009, garbage harmless treatment rate have significantly increased compared with 2008 both in overall level and provincial level. The top ten provinces are: Beijing, Zhejiang, Chongqing, Tianjin, Fujian, Jiangsu, Shandong, Guangxi,

Table 9.4 Annual and cumulative water pollutants reduction rates by region (Unit: %)

COD reduction rate

	2005–2006 (%)	2006–2007 (%)	2007–2008 (%)	2008–2009 (%)	Cumulative reduction rate (2009 over 2005) (%)
National	−0.99	3.25	4.42	3.27	9.66
East	0.75	3.64	5.44	4.37	13.52
Northeast	−0.06	2.58	5.36	3.42	10.90
Central	−1.92	2.62	3.86	3.10	7.54
West	−2.51	3.31	3.30	2.02	6.08

NH_3-N reduction rate

	2005–2006 (%)	2006–2007 (%)	2007–2008 (%)	2008–2009 (%)	Cumulative reduction rate (2009 over 2005) (%)
National	5.64	6.36	4.06	3.44	18.14
East	4.23	3.87	4.24	5.80	16.95
Northeast	10.33	7.74	3.77	4.04	23.60
Central	1.67	5.04	7.76	2.62	16.13
West	9.69	10.93	−0.78	0.53	19.36

Source National Bureau of Statistics of China, "China Statistical Yearbook" (2002–2010), Beijing: China Statistics Press, 2002–2010

Jiangxi and Sichuan. Among these 10 provinces, harmless treatment rates of 7 provinces were higher than 90 % and the others were higher than 83 %. In comparison, the national average rate was 71.39 %.

9.2.2 Regional Comparison of PETE Index

9.2.2.1 Air Pollution Elasticity

Sub-regional PETE for air pollution is detailed in Table 9.5.

Table 9.5 Comparison of regional PETE indices for air pollutants in 2006–2009

Sulfur dioxide				
	2006	2007	2008	2009
National	0.13	−0.36	−0.66	−0.53
East	−0.10	−0.40	−0.78	−0.66
Northeast	0.37	−0.12	−0.65	−0.60
Central	0.07	−0.30	−0.63	−0.46
West	0.29	−0.32	−0.46	−0.38
Soot				
	2006	2007	2008	2009
National	−0.69	−0.72	−0.96	−0.69
East	−0.23	−0.90	−0.41	−0.51
Northeast	−0.17	−0.20	−0.56	−0.75
Central	−0.61	−0.81	−1.42	−0.90
West	−1.11	−0.56	−0.96	−0.51
Industrial dust				
	2006	2007	2008	2009
National	−0.89	−0.96	−1.70	−1.15
East	−0.89	−0.81	−1.28	−1.93
Northeast	−0.45	−0.25	−3.10	−1.48
Central	−0.70	−1.00	−1.85	−0.77
West	−1.22	−1.22	−1.40	−0.83

Notes
1. In the calculation of the pollutant emission—economic time elasticity (PETE) in this form, we use the 2000 constant GDP in WDI database in World Bank
2. Data for pollutants comes from "China Statistic Yearbook 2006–2010"
3. Pollutant emissions-economic time elasticity is calculated as the ratio of annual emissions increase rate over annual GDP increase rate
4. In 2006–2009, China's economy has maintained rapid growth. So when pollution increases, PETE is positive, and the smaller the PETE, the less pollution cost of the country (or region, industry) to increase its GDP (or regional GDP, added value) is. PETE less than 1 indicates environmental deterioration rate less than the rate of economic growth; while PETE value greater than 1 indicates that the pollution increases at a rate higher than the rate of economic growth. When the PETE is negative, it is indicated that pollutant emission is reducing along with economic growth and the economic development is in a "clean" track

Regionally, in 2006 the PETE of sulfur dioxide emissions in the eastern region was negative, and those of the northeast, central and western regions were positive, i.e. 0.37, 0.07 and 0.29 respectively. Since 2007, along with rapid economic development, total sulfur dioxide emissions in the east, northeast, central and western regions have declined and PETE indices of sulfur dioxide were negative. In 2007, sulfur dioxide emissions reduction in eastern were better than the national average, western followed, while northeastern and central regions have same reduction. In 2008 and 2009, PETE indices of sulfur dioxide continued to decrease, and those of eastern and northeastern regions in two consecutive years were higher than national average, followed by the central and western. Especially, after achieving higher level of economic development, eastern region began to actively adjust the industrial structure, to develop cleaner industries, to increase investment in environmental governance, so as to achieve positive improvement of environmental quality.

During the 11th *Five-Year Plan* period, soot and industrial dust emissions maintained a constant decline trend, leading to consecutive negative PETE indices for both the country and each region from 2006 to 2009, indicating a "win–win" pattern of the emissions reduction along with economic growth.

9.2.2.2 Water Pollution Elasticity

During the 11th *Five-Year Plan* period, China had newly built a batch of wastewater treatment projects, as a result, the urban sewage treatment capacity of the country has been greatly improved. Although the quantity of industrial wastewater and domestic sewage increased, COD and ammonia emissions decreased year by year and water environment was also improved significantly (Table 9.6).

Table 9.6 Comparison of regional PETE indices for water pollutants (2006–2009)

COD				
	2006	2007	2008	2009
National	0.09	−0.25	−0.49	−0.38
East	−0.06	−0.26	−0.57	−0.48
Northeast	0.01	−0.18	−0.56	−0.38
Central	0.15	−0.18	−0.40	−0.34
West	0.20	−0.23	−0.34	−0.22
NH_3–N				
	2006	2007	2008	2009
National	−0.44	−0.45	−0.42	−0.38
East	−0.33	−0.27	−0.44	−0.64
Northeast	−0.81	−0.55	−0.39	−0.44
Central	−0.13	−0.35	−0.81	−0.29
West	−0.76	−0.77	0.08	−0.06

Note the same as Table 9.5

In 2006, the national PETE of COD emission was positive, and only in the eastern region the emissions of COD decreased along with economic development, while those in the northeast, central and western regions were increasing.

From 2007 to 2009, the PETE indices of national and the regional emissions of COD were all negative, showing the significant effects of regional COD emission reduction efforts.

PETE indices of NH_3–N in 2006–2009 kept negative value. In 2006 and 2007, NH_3–N emission reduction efforts in the northeast and western regions took better effects than the national average. In 2008, PETE indices of NH_3–N changed from negative to positive. In 2009, total NH_3–N emission reduction effect in the western region was not as obvious as other regions. In general, water pollution situation has been significantly improved along with the national and local economic development.

9.2.2.3 Solid Waste Emissions Elasticity

During the 11th *Five-Year Plan* period, the PETE indices of industrial solid waste for the nation and the regional mostly showed negative value. Exceptionally, in some year and in some regions, PETE indices showed positive values, e.g., northeast region in 2006 and 2009, east region in 2008, west region in 2009.

In comparison, PETE indices of municipal solid waste (MSW) were all negative in 2006, but showed positive values for the national average and the east and west regions in 2007–2009. This indicates that MSW could be a more challenging problem in the coming years (Table 9.7).

Table 9.7 Comparison of regional PETE indices for solid waste in 2006–2009

Industrial solid waste				
	2006	2007	2008	2009
National	−1.68	−0.57	−3.61	−1.00
East	−1.25	−1.04	4.14	−3.92
Northeast	11.50	−5.57	−5.11	6.21
Central	−2.21	−0.47	−4.09	−3.96
West	−1.46	−0.40	−3.89	1.46
Municipal solid waste				
	2006	2007	2008	2009
National	−0.27	0.07	0.16	0.21
East	−0.01	0.52	0.31	0.39
Northeast	−0.37	−0.23	−0.21	−0.06
Central	−0.89	−0.3	0.2	−0.02
West	−0.12	0.1	0.01	0.26

Note Municipal solid waste emission is calculated through dividing "municipal solid waste harmless treatment quantity" by "municipal solid waste harmless treatment rate"

Column 9.2 The Harvest of Harmonious Accomplishment of both Material Wealth and Environmental Wealth—Beilun district, Ningbo City, Zhejiang Province

Due to the unique advantages of the port, six port industries including petrochemical, energy, steel, paper, automobiles and ships, have been well developed in Beilun District. Beilun port now is becoming an important industrial base in Zhejiang province with the largest scale and the densest industries. During the 11th *Five-Year Plan* period, the regional economy in Beilun district has expanded by 134 % and container throughput has increased by 200 %. However, with the economic expansion, massive land occupation, water and energy consumption (coal consumption in Beilun district increased by 50 % during the 11th *Five-Year Plan* period), as well as large amounts of "three wastes" emissions, have threatened severely the environmental carrying capacity.

Although being confronted with such a tremendous environmental pressure of regional development, Beilun took the lead in passing through the "811" environment renovation examination in the province. Through a series of projects of environmental treatment, pollution reduction, environmental supervision and wisdom environmental protection, COD and SO2 were cut by 53 and 76 % in total, energy consumption per unit of GDP dropped by 23.8 %. The daily wastewater treatment capacity in the region has reached 420,000 tons and sewerage networks has expanded up to 329 km long, ranking the top in Ningbo, and achieving emission reduction target of 11th *Five-Year Plan* ahead of schedule. The production capacity of reclaimed water per day has amounted to 125,000 tons, which also ranks front row in the country. Beilun has constructed 210 km of clean water streams, and kept the drinking water sources' quality compliance rate at 100 %.

The "Forest Beilun" program has brought the green coverage rate of built-up areas up to 38.5 %, and the public green area per capita to 13.5 m^2, with a growth rate of 2.7 and 3.7 % compared with 2005 respectively. The proportion of class A air quality days in the five consecutive years has been kept above 90 %.

All these achievement enables Beilun being named as National Environmental Protection Model City and State-level ISO14000 Model District.

(A) Strengthened financial inputs into environmental treatment

Beilun district has successively invested 2.6 billion Yuan into completing 54 projects of industrial waste gas abatement, sewage pipeline network construction, and special pollution treatment projects for electroplating factories, printing factories, coal-fired boilers and foundries, to promote environmental management into "fast track".

(B) Emission Reduction

During the 11th *Five-Year Plan* Period, MEP (Ministry of Environmental Protection) had checked and ratified that Beilun had reduced COD and SO2 emissions 5,421 tons and 95,982 tons respectively, with reduction rates of 18.21 and 47.65 % of total quantities. Beilun had been commended as National Emission Reduction Advanced Model in 2007and 2009.

(C) Environmental Supervision

Beilun has enforced supervision and monitoring to key enterprises. A long-term integrated supervision and monitoring system of all-weather, full-time, comprehensive and multi-level, has been constructed. Special environmental protection programs, such as "Clear Water", "Environmental Safety", "Blue Sky" and "Ease of Radiation", were conducted to seek immediate effects. These experiences were reported on the front page of 'China Environmental News'.

(D) Smart Environmental Protection

To further enhance environmental governance capability, Beilun employed a "Smart Environmental Protection" project, to strengthen environmental information management focusing on basic data collection, monitoring and pre-warning, to achieve information integration and quantitative management for environment.

Source: Zhejiang environmental protection bureau website
http://www.zjepb.gov.cn/root14/auto506/201104/t20110419_95431.html.

9.3 The Prospect of the Environmental Carrying in China's Green Economic Development in the 12th Five-Year Plan Period

12th *Five-Year Plan* period is a critical period to open up the future and continue to implement the scientific development concept, and to vigorously develop green economy. Looking ahead, continuous efforts to promote pollution reduction, is still a crucial challenge to face. Through creative co-control mechanism to achieve multi-pollutants emission reduction is an important way to green development.

9.3.1 Multi-pollutants Emission Reduction Targets

In the 2009 UN General Assembly, President Hu Jintao promised to the world that, by 2020, China's carbon dioxide emissions per unit of GDP will be decreased by 40–45 % compared with 2005 and the share of renewable energy will be increased to about 15 %.[8] During Climate Change Conference in Copenhagen at the end of 2009, Premier Wen Jiabao further pledged to bring emission reduction targets as binding targets into long-term economic and social development planning, and to ensure its implementation receiving legal and public supervision. Government work report 2011 and the 12th *Five-Year Plan* established emission reduction

[8] See the press conference of the Fourth Session of Eleventh National People's Congress. http://gb.cri.cn/1321/2011/03/06/661s3174141.htm.

targets during the period 2011–2015: the ratio of non-fossil fuels in primary energy consumption being increased to 11.4 %, energy consumption and carbon dioxide emissions per unit of GDP being reduced by 16 and 17 % respectively, and the total discharge of major pollutants being decreased by 8–10 %, forest volume being increased by 6 billion cubic meters, with the forest coverage rate reaching 21.66 %.

After the release of 12th *Five-Year Plan* during the two conferences (i.e., the National People's Congress and the Chinese People's Political Consultative Conference) which proposed the binding control targets requirements to reduce the four major pollutants, the Ministry of Environmental Protection announced the 12th *Five-Year Plan* emissions reduction targets, i.e. sulfur dioxide and chemical oxygen demand emissions being decreased by 8 % compared with 2010, ammonia, nitrogen oxides emissions being reduced by 10 % (detailed in Table 9.8).

Table 9.8 Emission reduction targets during the 12th Five-Year Plan period

Indicator	Target
The ratio of Non-fossil fuels in primary energy consumption	Increase to 11.4 %
Energy consumption per unit of GDP	16 % reduction
CO_2 emissions per unit of GDP	17 % reduction
COD emissions	8 % reduction
SO_2 emissions	8 % reduction
Ammonia emissions	10 % reduction
NO_X emissions	10 % reduction
Forest volume	Increased by 6 billion cubic meters
Forest coverage rate	Reaching 21.66 %

Source Xinhua Net. The 12th "Five-Year Plan" of National Economic and Social Development of P. R. China
http://news.xinhuanet.com/politics/2011-03/16/c_121193916.htm

Column 9.3 Strengthening Circular Economy and Developing Green Industry—Miluo City, Hunan Province

In recent years, Miluo, a famous historical city,has developed its circular economy and fostered its green industry to step up to a resource-saving and eco-friendly way of development.

(A) Uprising industry developed from waste picking tradition

As early as the end of Qing dynasty, residents of Miluo began to pick up second-hand goods. To the 1990s, a 6 km recycling corridor had been formed from Xinshi town to Miluo city along the road. In recent years, Miluo City Committee and Government has successively invested 50 million Yuan into building Mid-South Renewable Resource Market which covers 50,000 m^2. The booming renewable resource industries have brought Miluo tremendous economic and social benefits. By 2010, the renewable resource industry has formed nearly 8,000 private entrepreneurs with assets over one million Yuan in Miluo, and created 120,000 job opportunities, and brought about wage income over 450 million Yuan.

In August 2010, the National Development and Reform Commission (NDRC), Ministry of Finance (MOF) jointly issued a special article to explicitly list Miluo's circular economy industrial park as one of first batch of seven "urban mining" demonstration bases. Miluo is facing an unprecedented development chance of circular economy.

(B) Inexhaustible "mine" created by green development

With numerous wastes streamed in from everywhere, there could be "secondary pollution" if those wastes are not properly handled. Therefore, Miluo set green development as the supreme concept. Through combining the development of renewable resource industry and environmental protection closely, Miluo finally explored a circular economy development way of "integrating the mine into the city". In 2010, Miluo reclaimed over 1.45 million tons of recycling waste materials from all over the country to produce 560,000 tons of recycled copper and aluminum. Miluo has become the country's major renewable material supply base of reclaimed copper, aluminum, stainless steel, plastic etc., which is equivalent to build a renewable mine worth of over 100 billion Yuan for each year.

(C) "Two-Orientation (resource-conserving and environment-friendly) Society" construction experiences explored

In October 2005, Miluo's market for renewable resources was listed in the first batch of national circular economy pilot scheme. In the past 5 years, Miluo conducted effective exploration using the opportunities and favorable national policies. The valuable experiences drawn for building "resource-conserving and environment-friendly society" are as follows:

1. Sound market systems and resource collection-distribution networks. Relying on the Mid-South Renewable Resource Market, Miluo gradually built three new networks, namely, renewable resource recycling network, logistic distribution network and electronic trading network, and three new markets, namely renewable resources market, renewable raw materials market and recycled product market. Currently, Miluo has 206 renewable resource recycling companies, more than 3,550 operating households and over 5,100 purchasing outlets distributed throughout the Country.
2. Solid industrial park platform and concentrated resources use. Miluo's circular economy industrial park has forged five processing sections of recycled copper, aluminum, steel, plastics and rubber, and there are 247 renewable resources processing companies with a total processing capacity of 500,000 tons per year.
3. Innovation and business development of high technology. By integrating production, education and research sectors, Miluo has constantly guided enterprises to upgrade technologies, so as to maximize utilization and productivity of renewable resources. For example, the oxygen-free copper rods of Xiangbei Copper Co, Ltd, and the recycled aluminum production technology of Zhongtian Aluminum Co, Ltd., have become the leading technologies in the country.

4. Standardized industry management. Renewable resources industry association, accounting-consulting company, asset management company, and SME credit guarantee company have been built up to form a suitable platform for enterprises' grown up.

(D) To build demonstration brand by optimization and upgrading

During the 12th *Five-Year Plan* period, with the aim of building Mid-South's largest and most influential "urban mining" pilot base, Miluo will take great efforts to promote "upgrading and improvement".

1. To upgrade and improve infrastructure. Miluo will invest about 2 billion Yuan to upgrade the industrial park road network, energy system, water system, fire-fighting, sanitation, and to build resettlement for residents so as to enhance the park's carrying capacity;
2. To upgrade and improve the environmental protection facilities. Wastewater collection pipe network, wastewater deep treatment plant, water reuse pipeline network, centralized industrial solid waste disposal center and other environmental projects will be constructed with investment of 700 million Yuan;
3. To upgrade and improve the recycling network. With Mid-South Renewable Resources Market and convenient transportation, professional recycling companies will be developed to expand its renewable resources reclamation capacity up to 4 million tons per year.
4. To upgrade and improve the industrial chains. 32 key industrial upgrading projects related to processing waste copper, aluminum, stainless steel, plastic and rubber, will be constructed to reach an annual renewable resources processing capacity of 2.4 million tons in 5 years.

Through 5 years' efforts, an "urban mining" pilot base with characteristics of advanced technology, standardization management and beautiful environment, will be constructed to promote resource recycling industries and finally to make new contributions to the construction of "resource-conserving and environment-friendly society".

9.3.2 Innovative Emission Reduction Mechanism: Co-Control

Reviewing the development of the 11th *Five-Year Plan* period, environmental protection and energy saving have been paid unprecedented attention, and made great progress.

However, emission reduction situation will be still severe during the 12th *Five-Year Plan* period. Compared with the 11th *Five-Year Plan* emission reduction targets, 12th *Five-Year Plan* period pays more attention to a comprehensive energy saving and emission reduction. Among them, two more pollutants were added to the list of pollutants emission reduction targets, i.e. the total ammonia nitrogen emission being reduced by 10 %, and the total nitrogen oxides emission being reduced by 10 %, by 2015. As the same time, the target for energy consumption per unit GDP during 12th *Five-Year Plan* period should be 16 % less and carbon dioxide emissions per unit of GDP should be reduced by 17 % by 2015. All these require more comprehensive thinking of using co-control measures to achieve multi-target.

Energy saving during 11th *Five-Year Plan* period mainly focused on pollution emission reduction and decrease of energy consumption per unit of GDP, but missed the synergistic reduction effect of various air pollutants and greenhouse gases. For example, environmental protection department only concentrated on the local air pollutant control target but not energy saving and CO_2 emission control, pollution emission management mainly focused on end-of-pipe control which is often at the cost of increased energy consumption and CO_2 emission. Also, when considering a variety of pollution control targets, end-of-pipe control measures or policies are not necessarily the most cost-effective optimization.

The challenging energy saving tasks of the 12th *Five-Year Plan* is calling on an urgent need for the co-control of both local and global pollutants and energy consumption.

In order to achieve the co-control effect of reducing emissions both for major air pollutants and greenhouse gases simultaneously, more measures of front-end-control and in-the-process control should be taken. As in the energy and industrial activities, SO_2, NOx and other air pollutants and greenhouse gas emissions are mostly generated from the burning of fossil energy, there is a considerable potential for co-control.

Co-control does not control a certain air pollutant or greenhouse gas, but apply the idea of integrated control in order to obtain maximum overall efficiency. Specifically, for a number of existing air pollutants and greenhouse gases, co-control means to choose and combine all those control measures from the source control, to the production process control, and to the end-of-pipe treatment (including virtual alternative control measures, macroeconomic policy measures, etc.), with the least unit emission reduction cost.

During the 12th *Five-Year Plan* period, there are three ways to implement co-control strategy for the major air pollutants and greenhouse gas emissions reduction based on the integrated consideration of efficiency effect, structural effect and scale effect:

1. To improve efficiency. Through strengthening management and technical progress, the efficiency of each links of the production chain and pollution abatement measures should be improved. Meanwhile, the emissions of carbon, sulfur, nitrogen and other pollutants per unit output will decrease. At the micro level, promoting efficiency means transforming from high-carbon, high-polluting production processes (or pollution abatement measures) to low-carbon and low-pollution production process, changing from the end-of-pipe control measures to the front-end and in-the-process control measures, to reduce energy consumption per unit output and pollutant emissions in various industry sectors.

2. Industrial restructuring. China will promote the adjustment of industrial and energy structure, and promote economic resources to be reallocated from high-carbon, high-polluting sectors and products to low-carbon, low-pollution sectors and products, and from energy-intensive industries to technology-intensive industries and services.
3. Optimization of production scale. Through economic development scale optimization and economic growth rate control, an acceptable rate of the discharge of pollutants will be achieved. Though China is still in the period of rapid economic development and development is still top priority, in order to coordinate the economic development and environment protection, keeping a moderate economic growth rate is still an effective optional strategy.

Chapter 10
Green Development and Ecological Construction

Yuan Jiang, Yongsheng Lin and Xiaopeng Sun

Ecological construction is an important part of green development which fully considers the effect and importance of ecosystem and environment. The Chinese government continued to focus on the protection and construction of ecosystem in the process of green development. During the 11th *Five-Year Plan* period, China had made important innovations and remarkable achievements on ecological policies and methods. Different scales of ecological engineering construction made the ecosystems all over China to get some degree of restoration and protection, and the area of nature reserve also increased largely. There were some new features in the ecological construction recently: first, ecological engineering construction turned from quantity increase to quality improvement; second, the connection of different ecological function regions improved the entire ecological functions; third, the economic functions of ecosystem appeared. These characteristics will be represented in the ecological construction in the 12th *Five-Year Plan*, which will bring the improvement and optimization of ecosystem quality and ecological service function to a new stage of the green development.

10.1 China's Achievements in Ecological Construction During the 11th Five-Year Plan Period

In "2011 China Green Development Index Report", six positive third class indicators were selected to measure the ecological construction in the process of China's green economic growth, namely, water resources per capita, forest area per capita, total standing stock volume per capita, forest coverage rate, the proportion of area of natural reserves in total area of a region and the proportion of area of wetlands in total area of a region. In other words, land greening, biodiversity and water conservation were important parts of China's ecological construction.

X. Li and J. Pan (eds.), *China Green Development Index Report 2011*,
Current Chinese Economic Report Series, DOI: 10.1007/978-3-642-31597-8_10,

During the past 5 years, under the guidance of correct principles and methods, a series of supporting policies and institutional arrangements were promulgated and implemented. In addition, members of the society participated in eco-environmental protection. Therefore, ecological construction had achieved great achievements during the 11th *Five-Year Plan* period.

In 2000, "*The National Programme for Ecological Environment Protection*"compiled by the State Council of China clearly stated that measures should be taken to prevent the environmental damage and degradation of ecological functions by establishing ecological function protected areas. In 2004, "Circular on Strengthening Biological Species Resources Protection and Management" compiled by General Office of the State Council demanded Ministry of Environmental Protection in conjunction with other departments concerned to develop the protection and utilization plan of national biological species resources. "The 11th *Five-Year Plan* for National Economic and Social Development of the People's Republic of China" and "*The Decision of the State Council on Implementing the Scientific Development View and Strengthening the Environmental Protection*" both pointed it out that the construction of the important ecological function protected areas should be regarded as one of most important missions contributed to the main function areas construction. In addition, the ecological function of the key ecological function protected areas should be maintained stable and the utilization of species resources should be reasonable and the prevention of its loss should be rigorous. In October, 2007, with the consent of the State Council, Ministry of Environmental Protection issued the "National Biological Species Resources Protection and Utilization Plan" (referred to as "Species Plan") and "The Outline of National Plan for Key Ecological Function Protected Areas" (referred to as "Ecological Function Protected Areas Plan") at the same time, which indicated China's major ecological principles and methods in the 11th *Five-Year Plan* period.

"Species Plan" proposed strategic thinking and tasks on protection and utilization of species resources. First, five principles on protection and utilization of species resources were determined, which were the principle of state sovereignty, scientific principle, the principle of conservation priorities, the principle of coordination between conservation and utilization and the principle of participatory. Second, propose the overall objective of national biological species resources protection and utilization in the next 15 years. Furthermore, the implementation would be divided into 3 five-year period. By 2010, the current dramatically reduction trend of biological species resources would be curbed. Furthermore, the loss of biological species resources would be controlled by 2015 and the biological species resources would be protected effectively by 2020. Third, the recent and medium-long term planning tasks covering the 12 critical areas were determined, including the traditional knowledge and exit-entry administration of animals, plants and microorganisms. Fourth, 10 priority actions and 55 priority projects were proposed during the national biological species resources protection and utilization during the 11th *Five-Year Plan* period, thus specifying the focus direction of fiscal funds.

"The Outline of National Plan for Key Ecological Function Protected Areas" defined the guiding ideology, basic principles and main tasks of construction of the ecological function protected areas based on the characteristics of ecological function protected areas. Firstly, ecological function protected areas were delineated regions with certain areas, which special protection were undertaken and development and construction were restricted due to its important ecological regulation effects, including water conservation, soil and water conservation, regulation of floods, sand—fixing, and maintenance of the biological diversity, etc. Secondly, key ecological function protected areas belonged to the restricted zone, which should be adhered to the principles of protection priority, development restraint, duplication avoidance and complementary during its construction. Under the principle of protection priority, these areas supposed to select development directions rationally, develop special advantage industries, strengthen ecological environmental protection and restoration, increase the intensity of ecological environment monitoring and protect and regenerate the regional ecological functions. Finally, local offices of ecological function protected areas should focus on the following three aspects in their work during the 11th *Five-Year Plan* period pointed by the "Plan Ecological Function Protected Areas". The first one was improving leadership in industry development reasonably, including developing ecological agriculture, ecological forestry and eco-tourism activity; limiting the high pollution, high energy consumption and high material consumption industries, which should be eliminated in accordance with law if they got serious pollution, serious damage to the regional ecology and serious resources waste; promoting the development of biogas, wind, small hydro, solar, geothermal and other clean energy industries activity to meet the rural energy needs and reduce the destruction to natural ecosystems based on the difference in resources. Secondly, protect and restore ecological functions following the principles of first urgent, focused, protection priority, active management, actions under local circumstances and damage, in conjunction with the implemented or planned ecological management projects there. Furthermore, increase the protection and regeneration efforts of region's natural ecosystem and improve the ability of water conservation, regeneration of soil and water conservation function, sand-fixing function, flood storage capacity and regional environmental quality. Thirdly, strengthen the environment regulatory. Via strengthening laws, regulations and monitoring capacity and strengthening monitoring and scientific research, the management would raise the level of ecological and environmental monitoring, prediction and early warning. Moreover, intensifying publicity and education could enhance the public's awareness of protecting the ecological safety of region and basin consciously.[1]

---

[1] Ministry of Environmental Protection: "National Biological Species Resources Protection and Utilization Plan" and "The Outline of National Plan for Key Ecological Function Protected Areas". http://news.xinhuanet.com/politics/2007-12/24/content_7304472_3.htm,2011-04-13.

In short, the 11th *Five-Year Plan* period got extraordinary development in the history, especially on eco-friendly construction of China. During the past 5 years, Chinese government focused on the construction of critical ecological function areas and protection of biodiversity, and took the emission of the main pollution reduction as the binding index of economic and social development planning and made great change on ecological protection from acknowledgement to practice. During the 11th *Five-Year Plan* period, China had intensified the work on ecological construction and environmental protection to make sure the water resources had been protected effectively and afforestation had been improved and the construction of nature reserves and wetlands had been pushed forward.

10.1.1 Effective Protection of Water Resources

In the early 2008, from the strategic perspective of construction of ecological civilization, Present Hu Jintao proposed that lakes and rivers should recuperate, which became the national guiding ideology of comprehensive management of water environment and water resources. Subsequently, the State Council transmitted “The Key Water Pollution Control Planning Appraisal of the Implementation of Interim Measures”. Meanwhile, the water quality evaluation system of crucial watershed based on trans-boundary section had been established.[2] By the end of 2009, the rate of completed pollution control project in planning had achieved 64.9 %. In addition, 80.1 % of water quality in the section was up to the standard. Under cooperation of three provinces, Heilongjiang, Jilin and Inner Mongolia autonomous region, Songhuajiang basin had achieved remarkable progress in comprehensive management. Water quality on 201 sections was assessed in seven hydrographic net of Hebei Province and the basin ecological compensation mechanism had been established. Jiangsu province improved the water quality of Taihu Lake by contributing to machine of appointing government officers as the leader of water resources commission. Shandong province adopted the approach of “rule, use, security” to restore the provincial key pollution controlled rivers, therefore, fishes regenerated in all rivers there. In 2009, permanganate index average concentration of the national state-controlled sections was 5.1 mg/L, decreased by 29 % compared with the year 2005. The rate of water quality which was better than Grade III of state-controlled sections in seven major river systems increased from 41 to 57 % since 2005.

In addition, during the 11th *Five-Year Plan* period, the government pushed the work of safeguarding the security of drinking water. Ministry of Environmental Protection organized governments at all levels to survey water resources area

[2] Zhou Shengxian, Rely firmly on the core of the new requirements of the theme and the main line, strive to create a new situation of environmental protection- Zhou Shengxian, Minister Speech at the 2011 National Conference on environmental protection. http://www.zhb.gov.cn/gkml/hbb/qt/201101/t20110120_200070.htm.

environment, assessed environmental conditions of 226 centralized drinking water source areas in key environmental protection cities and supervised the implementation of environmental protection requirements. "National Environmental Protection Plan of Urban Drinking Water Sources (2008–2020)" had been compiled and publicized, which proposed that the rate of national drinking water source areas with water quality standardized was no less than 90 % by 2015 and stabilized at 95 % by 2020.

10.1.2 The Constant Increase in Afforestation

On March 12, 2011, the office of the national afforestation committee released the "2010 China Land Greening Situation Bulletin". According to relevant data in the bulletin, during the 11th *Five-Year Plan* period, the afforestation increased constantly, mainly in the following areas. First, the national voluntary tree planting flourished. According to statistics, there were 590 million people participating in voluntary tree planting and 2.603 billion trees were planted in 2010. By the end of 2010, the number of people participated in voluntary tree planting reached to 12.7 billion. Meanwhile, 58.9 billion trees were planted. Second, key forestry afforestation projects got remarkable effect. Government gave priority to key forestry ecological engineering construction, taking reform of Collective Forest Right System as driving force, focusing on the mechanism innovation, vigorously carrying out afforestation in barren hills and nearby places and constructing the base of featured economic forests, carbon sink forests and energy forests. In addition, quality of native and rare species was strengthened to cultivate, consequently, completing the planting area to 5.9225 million hectares for the year 2010. Third, urban greening made steady progress. During the 11th *Five-Year Plan* period, that was, by the end of 2010, the urban greening covered an area of 1.4945 million hectares. Meanwhile, green areas reached to 1.3881 million hectares and park green area accounted for 401,600 ha. The proportion of the green coverage in the built-up region reached 38.22 % and the rate of green area arrived at 34.17 %. In addition, per capita park green area was 10.66 square. Fourth, departments carried out greening actively. Greening Committee of departments under the Party Central Committee and the direct subordinate organs of the Central State Organs carried out voluntary tree planting, garden virescence and beautification activities. In 2010, the government built over 120,000 squares new green space for the whole year and rebuilt more than 200,000 square green space. Via urban and rural areas hand in hand activities in voluntary tree planting, central government organs positively supported greening in suburbs of the capital and new rural construction. Fifth, the effect of the prairie construction was evident. During the 11th *Five-Year Plan* period, the government contributed to prairie protection and construction projects, including "Return Grazing to Grass", "Jing-Jin Sandstorm Source Control Project", grassland governance in karst mountainous area in southwest China and the Nomadic settlement projects. As a consequence, regeneration momentum of prairie vegetation was much better in these areas and the ecological environment ameliorated significantly, for instance, the vegetation coverage, height and biomass improved evidently. By the end of 2010, the national accumulative area of grass preservation was 22.6667 million ha., while grassland fencing area was 73.3333 million hectares and grass grazing, rotational grazing grassland was 0.107 billion hectares. Sixth, protection of forests and grassland was strengthened. During the 11th *Five-Year Plan* period, the prevention and control of national forests and grassland pests had made new progress, mainly on intensifying the prevention and control of forest pests, including pine wood nematode, American white moth, rat and rabbit damage, *Mikaniamicrantha* and other pests with significant risk, besides, the number of pine wood nematode disease in county-level epidemic points showed a decrease trend for the first time and the proliferation had been primary curbed. In 2010, the occurrence area of national major forestry pests was 11.9893 million hectares and the prevention area reached 8,296,700 ha. Moreover, the area of non-pollution control reached 6.548 million hectares. The area of controlling grassland rodents was 6.6667 million hectares and that of controlling insects was 400 ha grassland, by which direct economic loss was reduced by nearly 1.6 billion Yuan. Seventh, afforestation policy mechanism continued to be improved, especially in the reform of Collective Forest Right System which got significant benefit. By the end of 2010, 18 provinces, autonomous regions and municipalities directly under the jurisdiction of the central government all over the country had completed there form tasks of property rights clarification of Collective Forest Right System and land contracted to home. 162 million hectares offorest land was declared collective, which took 88.6 % of the total area, with 134 million hectares forest area and 72.6 million farmers issued warrants. Consequently, more than 300 million farmers benefited from it directly. Reform of Collective Forest Right System allowed peasants to become the owner of forests, which stimulated the peasants' enthusiasm in afforestation.

Column 10.1 Libo, Guizhou: An Emerald on the Belt of the Earth

Libo is known as an emerald on the belt of the earth. It is a minority mountain areas in the southern part of Guizhou province, which is located in 107°37′ ~ 108°18′E, 25.7° ~ 25.9°N. It is next to Congjiang and Rongjiang from Northeast, Huanjiang and Nandan from Southeast, Dushan from the west and Sandu County to the North. The county covers the area of 8459 ha and has a forest cover rate of 53.96 %. 6 towns, 11 villages (5 autonomous villages included), 170 villagers' committees and 4 residence committees were under the administration of Libo. Meanwhile, it is a typical example of cone karst topography existing in the transitional belt from Guizhou Plateau to Guangxi Basin, which has been considered to be a typical representative of South China Karst. Libo is the only green land on this latitude in the world, sparsely populated and full of strong vegetation, of which the rain thermal condition is superior, like a green gem isolated itself from the outside world. Libo was also one of the poorest 90,000 mountain counties supported by the state, where the transportation was inconvenient, culture was relatively backward, productivity level was low and peasants in rural communities and forest areas were under the line of extreme poverty for a long term. Since the national "87" poverty alleviation programme was implemented, Libo stuck to the road of "eco-development industrialization and ecological industry development" to create tourism-dominated city, with the goal of building "international professional tourist city". Libo was crowned with the World Natural Heritage title in 2007 for its fine example of Karst topography. The ecological system in Libo County has been well protected and the karst park here is now the most probable Chinese natural scenic spot that could be listed on the World Natural Heritage list. In 2008, its years of hospitality was 1.68 million, 71.1 % increased on the year; tourism direct revenue reached 19.98 million Yuan, an increase of 84.32 %; integrated tourism revenue was 0.326 billion Yuan, with an increase of 83.52 %; the county total retail sales of social consumer goods were up to 323.17 million Yuan, an increase of 26.24 %; average net income of peasants reached 2 618 Yuan, an increase of 21.27 %. "World Heritage" not only effectively promoting leap-forward development of the LIBO County Tourism, but also effectively promote the growth of services and the improvement of people's livelihood, meanwhile, Libo have markedly improved and repaired the ecological environment.

On March 1, 2011, "Libo County National Ecological County Construction Planning" (referred to as "Planning") reviewed in the meeting held by Guizhou Province Environmental Office, which clearly pointed out that : 2011–2018, Libo County would be expected to invest 1.7 billion Yuan in 5 key areas, which were ecological industry system construction, ecological environmental protection and construction, project of municipal facilities and the environmental pollution control, eco-residential construction, and integrated capacity-building, besides, 86 key projects would be under construction to reach national ecological County standard basically. It implies that Guizhou province takes a great step forward to the construction of ecological civilization with the characteristic of development, prosperity, and ecological soundness, with ecological county as a carrier.

Sources:

1. Baidu encyclopedia. http://baike.baidu.com/view/339007.htm,2011-05-12.
2. PeiYu, LIBO, road to "heritage",Guizhou ethnic minority report, 2011-03-21.
3. Reviewing on Libo County National Ecological County Construction Planning, official website of the Office of environmental protection in Guizhou province, 2011-05-13.

10.1.3 Nature Reserves and Wetland Construction Continued Promoting

China continued to promote the construction of nature reserves and wetlands, aimed at strengthening the construction of key ecological functional area and the protection of biological diversity. In 2010, China established the National

Committee for international year of biodiversity under the leadership of Chairman LiKeqiang and released "2010 International Year of Biodiversity China Action Plan". China State Council Executive meeting deliberated to set up "The Biodiversity Conservation Strategy and Action Plan of 2011–2030", then carried out activities for the international year of biodiversity.

By the end of 2010, the number of national forest nature reserves, including forest ecosystem, wetland ecosystem, desert ecosystem and wildlife protection area, was 2035, covering 124 million hectares, accounting for 12.87 % of the whole national area. These nature reserves had ready protected 90 % types of terrestrial ecosystem, 85 % of wildlife populations and 65 % higher plant communities effectively; therefore they played a pivotal role in maintenance of biodiversity. In addition, the protection and restoration of wetland ecosystem moved forward, including the implementation of "Chinese Wetland Protection Engineering Planning 2005–2010" and establishing a special central financial fund for protection of wetland. Furthermore, establish wetland parks, take its obligations seriously under the "Convention on Wetlands of International Importance Especially as Waterfowl Habitat", improve the construction of protection system of wetland continuously, and modify the ecological crisis and degradation of wetland effectively. In 2010, 45 new parks were added. At present, the total number of national wetland park reached 145, of which 37 were wetlands of international importance, covering 18.2 million hectares. Consequently, 50.3 % of the wetlands were effectively protected.

During the 11th *Five-Year Plan* period, 192 all types of nature reserves were newly established. The area of terrestrial nature reserves accounted for 14.72 % of land area and China continued to improve the work system of ecological construction demonstration zone, building up 11 national ecological counties (cities). Government released "Guidelines of the National Key Ecological Function Protected Areas Planning", "The National Ecological Function Zoning" and other important policies. Some provinces with key ecological function continued to focus on strengthening the ecological security, for instance, Tibet Autonomous Region invested over 10 billion Yuan to construct the ecological security barrier project. Yunnan province set up a foundation for biodiversity conservation and Qinghai province strengthened the ecological monitoring and protection project of three-river sources vigorously.[3]

Column 10.2 Lalu Wetland: Lung of Lhasa

Lalu wetland is located in the northern part of the capital city of Lhasa in Tibet Autonomous Region. The total area of Lalu wetland is 12.2 km^2, of which the core area is 6.6 km^2, accounting for 11.7 % of the built-up area of Lhasa and its average elevation is 3,645 m. It is a typical Qinghai—Tibet plateau wetland,

[3] National Afforestation Committee. 2010 Communiqué on China Land Afforestation Status.

belonging to the sedge peat swamp. Lalu is the highest and largest natural wetland in urban areas in the world, since it was listed as a provincial nature reserve in the late 1990 of the twentieth century. It gradually becomes home to Huang Ya, Bar-headed Goose and other migratory birds, besides, black—necked crane and other protected birds perch here, which makes Lalu a "bird's nest" in a real city. On July 23, 2005, Lalu was officially approved to be a national nature reserve by the State Council.

Lalu has more than 300 species of plants, 43 species of terrestrial wildlife, over 30 species of aquatic wild animals and 101 species of insects, making it a real "gene pool" of plateau biological species. Vegetation covers more than 50 % of its core area, of which Phragmites and Cyperaceae are the main plants. Due to the lush plant growth, transpiration is 3 times of that in grassland vegetation communities; therefore, water in Lalu wetland can be transported into air through the Meadow plants in the light continuously to increase the content of moisture in air and the humidity in Lhasa. In this sense, Lalu is a natural "humidifier". Every year around 3,000 wild migrant birds fly back for winter, which becomes an ecological harmonious image of a scene vividly.

Lalu is the main supply source of oxygen for Lhasa. Grass here grows well, which can absorb 78,800 tons of carbon dioxide and producing 53,700 tons of oxygen through photosynthesis every year. In addition, known as "the Lung of Lhasa ", Lalu is agiant air purifier and a water filter, which can absorb 5,475 tons of dust in the air each year and handle more than 10 million tons of sewage.

The Lalu wetland is acclaimed as 'the Lung of Lhasa' for its significant role in protecting Lhasa's environment. Lalu Wetland has been listed as an autonomous region-level nature reserve. A preservation scheme with an investment of over 100 million Yuan (US$12 million) has been launched. Recently as part of this input, the local environmental protection department has set up an automatic monitoring site to record temperature and humidity in the wetland.

From the beginning of the Lhasa city planning, District Government decided to give priority to ecological and environmental protection by abandoning the planning of changing the Lalu superior location and fertile soil of the area into building land and agricultural land. During the construction period of Lalu wetland nature reserve, government overcomes all kinds of difficulties. For example, close the quarries around the wetlands and other sources of pollution. At present, Lalu wetland has become a propaganda base for environmental protection education in Tibet and the window to display achievements of regional ecological environmental protection.

There are over 40 nature reserves like Lalu in Tibet, of which there are 9 national ones and 6 provincial ones. During the 11th *Five-Year Plan* period, funding for environmental protection and ecological construction of Tibet Autonomous region reached 10.1 billion, which was more than 3 times than that of Tenth Five-Year period. As data showed that by 2010, Tibet had established 47 nature reserves, which covered an area of 413,700 km^2, nearly 35 % of land area in Tibet, which was second to none in the national territory, making a significant contribution to the state work of ecosystem protection. 125 species of national protected wildlife and

39 species of national key protected wild plants are protected well. In addition, Tibet takes the lead in starting the country grassland ecological protection incentive pilot, establishing compensation system for forest ecological benefits. More than 150 million acres in all the 65 districts were included in public welfare forest compensation system. Afforestation area reached 0.11473 million hectares, sand prevention and control 42,666.6 ha, conversion of cropland to forest 39000 ha, returning cropping land to forage land 3.94 million hectares. The state key ecological functional area transfer payment work has been done.

On March 1, 2011, "Wetlands Protection Ordinance" officially christened in Tibet, thereby it made Tibet the first region that had the legal protection of wetland resources in China's history, which was significant for Tibet, especially in terms of the largest wetland in China there. Protection of wetland resources is an important part of engineering system of ecological environment protection in Tibet. Tibet treasures both of protection and development and vigorously supports the "Barrier Protection and Construction of Ecological Security in Tibet" project; therefore, the ecological environment has been improved and will be better and better.
Sources:

1. Zhaxi, Tibet constructs the ecological security screen. http://www.chinadaily.com.cn/hqgj/jryw/2011-03-20/content_2069048.html,2011-03-20.
2. Deyongjian, The exploration to highest city "bird's nest"—Lalu wetland.http://www.chinadaily.com.cn/dfpd/xizang_hjbh/2011-04-14/content_2307360.html, 2011-04-14.

10.2 Regional Comparison of Ecological Construction in Green Economic Development in China's 11th Five-Year Plan Period

10.2.1 Region Comparison of Ecosystem Status

The 11th *Five-Year Plan* has been significantly modified in ecosystem conservation and restoration. In the perspective of region configuration, the restoration of ecological function regions, including natural forest protection zones and important water self-restraint zones, has been emphasized, which formed a regional ecological security patterns.

Water Ecosystem

Water pollution is one of the most important factors which affect water ecosystem function. The 11th *Five-Year Plan* focused on the water pollution control and water source conservation, which is a difficult task, because it is varied in different geographic locations and development characteristics. The "national science and technology key projects on water pollution control and management" were launched by the Chinese government during 11th *Five-Year Plan* period (2007), which eased the situation of water pollution to some extent. Quality of water ecosystem varied in different regions, which were related to the economic development conditions.

According to the proportion of the water which is worse than grade V, water ecological conditions varied significantly in different regions of China (Fig. 10.1). The proportion of highly polluted water which is worse than grade V without ecological functions could represent the healthy condition of water ecosystem in a basin. The water pollution in Haihe River area was most seriously in different basin areas in 2009, and the proportion of grade V-plus water was more than 50 %. All the proportions of grade V-plus water were over 20 % in Liao River, Yellow river, and Huai River areas, which connected with their industry developments. South eastern River, Yangtze River, and Pearl River areas were polluted lighter relatively. Because of slow growth rates of industry, the rivers in northwestern and south western river areas were polluted lighter. The water qualities of serious polluted basins were improved compared with the conditions in 2005 in some extent, and the most obvious case was Huai River area. The similar case can be seen in Pearl River area. However, water qualities in the southeastern, northwestern and southwestern river areas became worse slightly.

Lake is special among all the water ecosystems, because it is a closed system, which always challenges restoration after pollution. Water pollution always brings a serious damage to the ecosystem, and it is difficult to eliminate and control. As the monitoring data shows in Table 10.1, water quality in Dianchi Lake was still worse than grade V in 2009. The water quality in Chaohu Lake had already fallen down from grade IV–V to grade V-plus. The water quality in Taihu Lake fell from grade III to grade V-plus, and the water quality in Baiyangdian Lake, Nansihu Lake, Namtso Lake, Poyanghu Lake and Honghu Lake was grade V or grade IV, which were belong to polluted water. The water quality in Qinghai Lake and Fuxian Lake were better, grade II. However, nearly half of the 15 lakes in our census were suffered from eutrophication. Eutrophication reveals the unbalance of ecological community in the lake ecosystems, which may be in danger of ecosystem breakdown.

Wetland has been recognized as the "kidney of the earth", which is one of the most important ecosystem types in the world for its powerful functions. Wetland performs a crucial role in local ecosystem regulation and sustainability. 550 wetland nature reserves, 100 national parks, 120 local wetland parks and 37 international important wetlands were established in 2010. 18.2 million hectares natural wetlands were efficiently protected, which was about 50.3 % of all natural wetlands.[4] The wetland coverage was more than 50 % in Shanghai due to the

[4] State Forestry Administration. Report on Economic Performance of Forestry Industry in 2010. http://www.forestry.gov.cn/portal/main/s/304/content-458701.html.

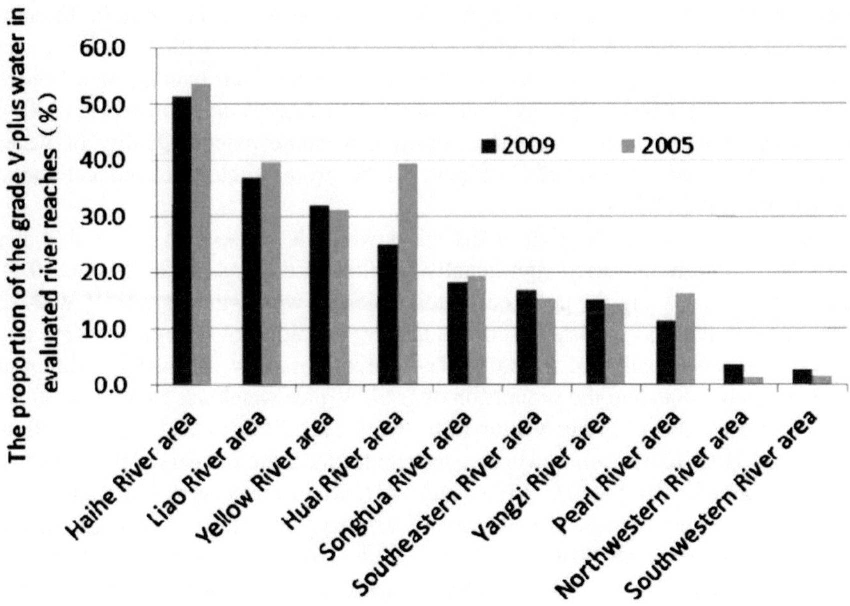

Fig. 10.1 Water qualities in different basin areas in China. (Grade V-plus water belongs to seriously polluted water which does not have ecological and environmental functions). *Sources* National Bureau of Statistics of China. China Statistical Yearbook of Environment China Statistical Yearbook of Environment 2006, China Statistical Yearbook of Environment 2010. Beijing:China Statistics Press, 2006, 2010

advantage of outfall, was more than 10 % in Jiangsu, Tianjin and Shandong, and was lowest in Guizhou, Chongqing, Yunnan, Xinjiang and Shaanxi, which located in southwest and northwest China. Considering the absolute value, the largest wetlands were in Tibet (5,232,000 ha.), Heilongjiang (4,314,800 ha.), Inner Mongolia (4,245,100 ha.), Qinghai (4,126,000 ha.) and Shandong (1,784,100 ha.); while the smallest wetlands were in Beijing (34,400 ha.), Chongqing (43,200 ha.), Guizhou (79,400 ha.), Tianjin (171,800 ha.) and Yunnan (235,300 ha.).[5] These wetland ecosystems, integrating with 201 wetland conservation and restoration projects during the 11th *Five-Year Plan* period, constructed the wetland conservation network primarily and would perform their holistic roles progressively[6] (Fig. 10.2).

[5] National Bureau of Statistics of China.China Statistical Yearbook 2010.

[6] National Bureau of Statistics of China. http://www.stats.gov.cn/tjfx/ztfx/sywcj/t20110310_402709535.htm.

Table 10.1 Water qualities of large lake ecosystems in China (2009)

Lake name	Districts	Evaluation area (km^2)	Water quality grade		Nutrition state
			2009	Contrast 2005	
Baiyangdian	Hebei	113.5	V		Moderate eutrophication
Taihu	Jiangsu, Zhejiang	2 338.0	V-plus	III	Moderate eutrophication
Hongzehu	Jiangsu	2 152.0	III		Moderate eutrophication
Luomahu	Jiangsu	625.0	III		Mild eutrophication
Chaohu	Anhui	110.0	V-plus	IV-V	Moderate eutrophication
Poyanghu	Jiangxi	2 184.0	IV		Moderatenutrients
Nansihu	Shandong	1 266.0	V		Mild eutrophication
Honghu	Hubei	395.4	IV		Mesotrophy
Dianchi	Yunnan	300.0	V-plus	V-plus	Moderate eutrophication
Fuxianhu	Yunnan	212.0	II		Mesotrophy
Erhai	Yunnan	250.0	III	II	Mesotrophy
Namtso	Tibet	1 920.0	V		Mesotrophy
Pumo Yum Tso	Tibet	284.0	III		Mesotrophy
Qinghai	Qinghai	4 247.0	II		Mild eutrophication
Bosten	Xinjiang	251.0	III	IV-V-plus	Mild eutrophication

Notes

[1] The selected lakes were large ones whose evaluation areas were more than 100 km^2 in 2009. Some data in 2005 was missing. Taihu Lake contained Wulihu in Shanghai

[2] Environmental quality standards for surface water(GB 3838–2002): Grade I, mainly for source water and national nature reserves; Grade II, mainly for Grade 1 protection areas for centralized potable water source, rare aquatic habitats, fish and shrimp production field, juvenile fish feeding ground, etc.; Grade III, mainly for Grade 2 protection areas for centralized potable water source, fishing waters for fish and shrimp wintering ground, migration pathway and aquaculture areas and swimming areas; Grade IV, mainly for general industrial water areas and entertainment water areas non-directly touched by body; Grade V, mainly for water areas for agriculture and general landscape requirement; Grade V-plus, seriously polluted water which does not have ecological and environmental functions

[3] According to the lake (reservoir) eutrophication evaluation method and classification technology regulation, lake (reservoir) eutrophication evaluation index uses chla, TP, TN, SD and CODMn as water comprehensive nutrition state evaluation index

Source National Bureau of Statistics of China. China Statistical Yearbook of Environment 2006, China Statistical Yearbook of Environment 2010. Beijing:China Statistics Press, 2006, 2010

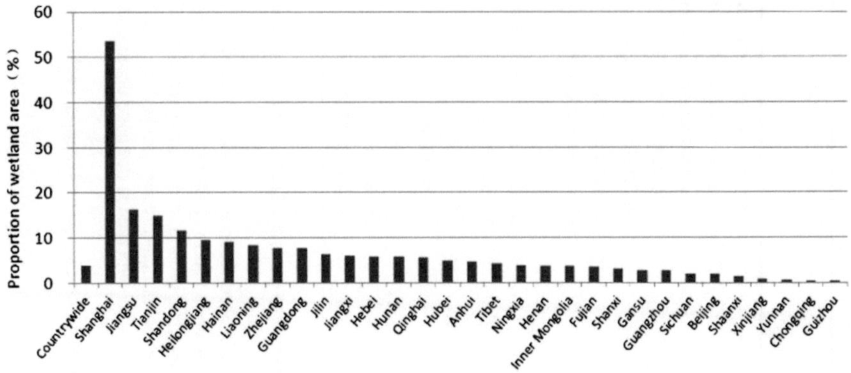

Fig. 10.2 Inter-provincial comparison of the wetland area proportion in China *Note* this chart was drawn according to China's first wetland survey (1995–2003) material, not including Taiwan, Hong Kong and Macao. Wetland area did not included paddies wetland. *Source* National Bureau of Statistics of China. China Statistical Yearbook 2010. Beijing:China Statistics Press, 2010

Forest Ecosystem

As the most important terrestrial ecosystem type, forest has significant ecological functions, including headwater and soil conservation, carbon fixating and oxygen releasing, organics accumulating, air cleansing and biodiversity conservation. As the 7th National Forest Inventory (NFI) (2004–2008) showed, the forest status during the 11th *Five-Year Plan* period were: forest coverage rate was not less than 20 % in 21 ones among all 31 provinces, was more than 50 % in only 5 provinces, including Fujian (63.10 %), Jiangxi (58.32 %), Zhejiang (57.41 %), Guangxi (52.71 %) and Hainan (51.98 %), was less than 10 % in 5 provinces, including Xinjiang (4.02 %), Qinghai (4.57 %), Tianjin (8.24 %), Shanghai (9.41 %) and Ningxia (9.84 %) (Fig. 10.3). The most regions with low forest coverage rate were located in arid region of northwest China and some circumlittoral developed regions, which illustrated that the forest distribution was affected by climate factors and economic development conditions.

The regional characters of forest ecosystem change in the 11th *Five-Year Plan* period could be realized by comparing 7th National Forest Inventory (NFI) (2004–2008) with the 6th one(1993–2004) (Fig. 10.4). The largest net growth of forest area was 23,442,600 ha of Inner Mongolia, about 10 % of national net growth. Based on the monitoring results, both desertification and sandy land areas declined in Inner Mongolia during the 11th *Five-Year Plan* period. The desertification expansion had been under control in key control areas around the 5 large deserts (Badainjaran, Tenger, Kubuqi, Ulanbuh and Bajanondor), and the desert area was relatively stable. The forest and grass coverage in all the 5 large sandy lands (Mu Us, Otindag, Horqin, Hulunbeir and Ujimqin) was increased, and the sandy area had shrunk. The ecological condition in Horqinsandy land and Mu Ussandy land reversed fortunately. 16 million hectares sandy land and 10 million

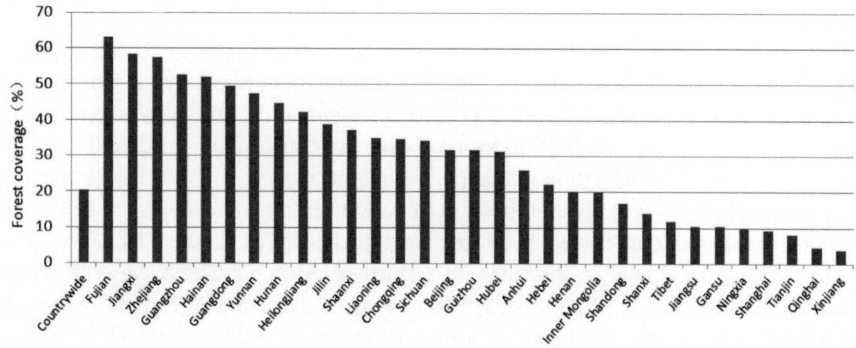

Fig. 10.3 Inter-provincial comparison of forest coverage rate in China. *Note* this data was from the 7th National Forest Inventory (NFI) (2004–2008). The countrywide count included Taiwan, Hong Kong and Macao. *Source* National Bureau of Statistics 19361212 of China. China Statistical Yearbook 2010. Beijing:China Statistics Press, 2010

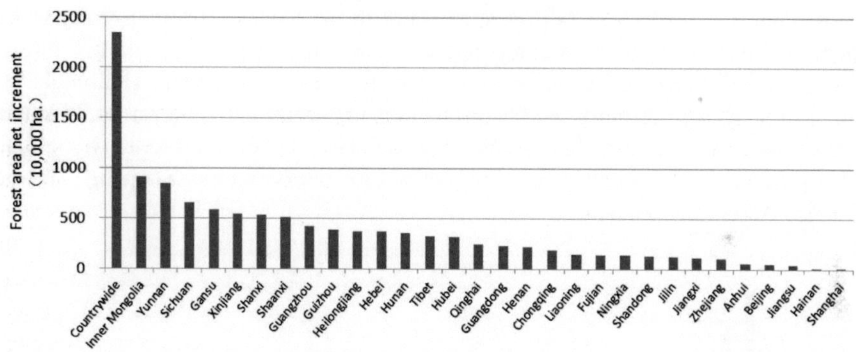

Fig. 10.4 Inter-provincial comparison of net increment of forest area between two forest resources inventories in China *Notes* this data was from the 6th (1993–2004) and 7th (2004–2008) National Forest Inventory (NFI). The countrywide count included Taiwan, Hong Kong and Macao. *Sources* National Bureau of Statistics of China. China Statistical Yearbook 2006, China Statistical Yearbook 2010. Beijing:China Statistics Press, 2006, 2010

hectares soil erosion area were under control, 4 million hectares farmland and 5.3 million hectares pasture and meadow was well protected by forest network.[7] In addition, forest coverage rate was also highly increased in the provinces where large potential area could be used to reforestation because of low vegetation coverage or serious ecosystem damage, such as Yunnan (9,160,800 ha.), Sichuan (8,473,200 ha.), Gansu (6,558,100 ha.), Xinjiang (5,825,000 ha.), Shanxi (5,463,900 ha.), Shaanxi (5,354,100 ha.) and Guangxi (5,126,200 ha.).Because of high vegetation coverage in Hainan, Zhejiang and Jiangxi and high development in Beijing, Shanghai and Tianjin, their potential forestation area was relatively limited, and the forest net growth was not obvious compared with other regions.

7 Inner Mongolia Broadcast Net. http://www.nmrb.cn/Item/240455.aspx.

Nature Reserve

The major tasks of nature reserves included national nature reserve establishment and improvement, critically endangered species salvation and conservation, and important nature reserves (e.g. three-river-source) construction in the 11th *Five-Year Plan* period.[8] Under this guidance, the nature reserve distribution characters were formed (Fig. 10.5). The proportion of nature reserve area was 15 % in China, which already beyond the 13 % as one of the objectives in the 11th *Five-Year Plan.*[9] The two largest proportions of nature reserves in China were in Tibet (34.51 %) and Qinghai (30.28 %), which suggested that the government paid high attention to protection of the river source ecosystems. Taking three-river-source nature reserve as an example, the area of major lakes increased by 245 km^2, and desert ecosystem area decreased by 95.63 km^2 from 2004 to 2009.Vegetation coverage was increased by 23.2 % in average in grassland desertification control area, and was increased by 80 %in Heitutan control area. Vegetation coverage reached 90 % in returning cultivated land to grassland area. Three-river-source ecosystem provided 16.89 billion $m^3$of water to the downstream area every year, increased by 4.46 billion m^3 compared with the count of 12.43 billion m^3 during 1975–2004.[10] Other nature reserves located in Sichuan, Shanghai, Heilongjiang, and Xinjiang were established for protecting important ecosystems, while some were located in some fragile regions, such as Gansu, Tianjin, and Inner Mongolia. The provinces with small proportions of nature reserves were Zhejiang, Hebei, Fujian, Anhui, and Henan.

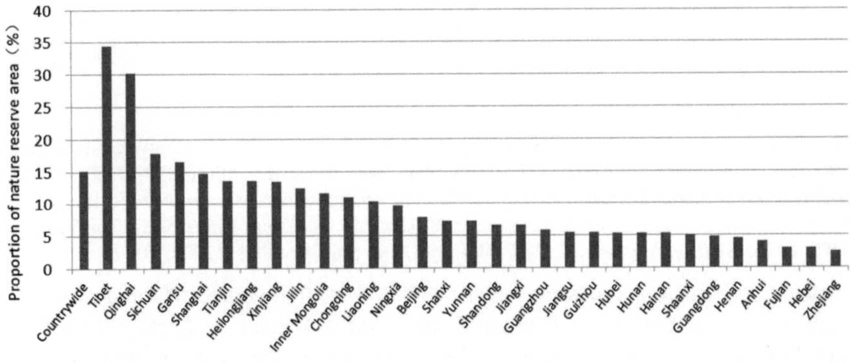

Fig. 10.5 Inter-provincial comparison of proportion of nature reserve area in China (2008). *Source* National Bureau of Statistics of China. China Statistical Yearbook 2010. Beijing: China Statistics Press, 2010

[8] The *12th Five-Year Plan* for National Economic and Social Development of the People's Republic of China.

[9] State Forestry Administration of China.11th Five-year and medium-and-long-term Plan on National Forestry Development, 2006.

[10] Qinghai News Net. http://www.qhnews.com/2010zt/system/2011/01/08/010267811.shtml.

10.2.2 Key Ecological Projects and Major Regions

Major ecological engineering projects were propelled in the 11th *Five-Year Plan* period, including Natural Forest Protection Project, Returning Farmland to Forest Project, Three-North and Yangtze River Basin Shelterbelt Project, Jing-Jin Sandstorm Source Control Project, etc. Due to the varied ecosystem characters in different regions, these projects ensured ecological functions performing well by increasing ecological resources quantity and quality. These projects obtained significant achievements, and the accumulative afforestation area was 16.87 million hectares from 2006 to 2009 (Fig. 10.6).

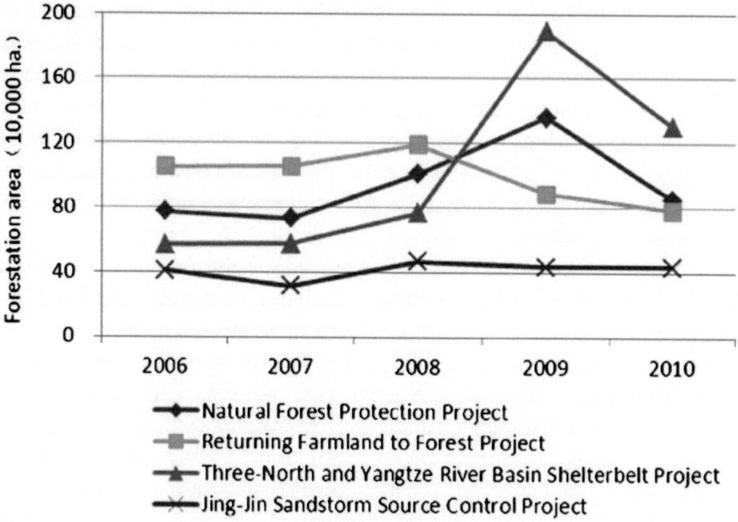

Fig. 10.6 Forestation areas of major ecological engineering projects in the 11th Five-Year Plan period. *Sources* (1) National Bureau of Statistics of China. China Statistical Yearbook 2010. Beijing:China Statistics Press, 2010. (2) National Afforestation Committee. 2010 Communiqué on China Land Afforestation Status

The purpose of Natural Forest Protection Project was to conserve and tend forest ecosystem that was one of the most important terrestrial ecosystems. The project included three aspects: stopping nature forest logging in the upstream area of Yangtze River and the middle-downstream area of Yellow river, reducing the forestry products in major state-owned forest in northeast China and Inner Mongolia, taking charge other natural forests by local governments. These tasks were further boosted in the 11th *Five-Year Plan* period. Therefore, 94.18 million hectares natural and other kinds of forests in the projects were effectively conserved, and forestation and planting grass processes were sped up in suitable barren land. Forestation area in Natural Forest Protection Project was about 4.73

million hectares cumulatively from 2006 to 2010,[11] which lead to both quantity and quality increase of important forest ecosystems in the upstream area of major rivers and important state-owned forest areas.

Column 10.3 AoHan County is Forging the Green Growth Pole to Realize the Sustainable Development

According to the requirement of circular economy development, Aohan County in Chifeng city in Inner Mongolia followed a "green growth" way to speed up the transformation of the economic development mode and to stick to the conservation of resources and environmental protection. The enterprises with small production scale and resources wasting had been closed, and the government supported the ones with large production scale and high technology, which attracted a large number of low carbon enterprises, including oil shale refining, chemical, machinery manufacturing, garment processing, wind power, etc. Due to the resource integration administration in 2011, the iron ore resources enterprises were integrated from about 80 to 20, and the gold miners were integrated from about 20 to 10 factories integration.

Aohan owned "Global 500 Roll of Honors for Environmental Achievement" award of UN with rich tourism resources. Aohan County planned tourism development as a key of the modern service industry. Insisting the government leading, departments cooperation, social participation, market operation, planning first and project support, they constructed Xinglongwa relics as a representative of the "village tour", Huangyangwa pasture as a representative of the "green tour", Aohan spa as the destination of the "spring tour", Chengzishan relics as the destination of the "heaven tour", which formed the tourism pattern in Aohan. Tourism has been becoming a special product and service as a part of the great tourism circle in Chifeng, even Lioaxi. This is a strong engine of the economic growth.

The government has been reforming the old block, developing the new block, completing the facilities and improving the livable quality. In the near future, Aohan will be a new city with complete functions, good ecosystem and environment, suitable living, prosperous industry and outstanding characteristics.

Source: Ministry of Finance of People's Republic of China: http://www.mof.gov.cn/xinwenlianbo/neimengcaizhengxinxilianbo/201007/t20100706_325859.html.

Returning Farmland to Forest Project was a great strategic decision of China against increasingly aggravated soil erosion. According to the plan, soil erosion area in 2010 would be retained under 22.7 million hectares, wind-breaking and sand-fixing area would be under 26.7 million hectares, and average quantity of sediment flowing into Yangtze River and Yellow River would be reduced by 260

---

[11] China Statistical Yearbook 2010, 2010 Communiqué on China Land Afforestation Status.

million ton. The project was emphasized in the upstream area of Yangtze River, the up-middle stream area of Yellow River, Jing-Jin wind and sandstorm source area, southern key rocky desertification area, water source area in the middle route of South-to-North Water Transfer Project, western and northeastern sandy area, etc. In addition, regions with higher than 25° of slope were considered as the important section in the project. Forestation area of the Returning Farmland to Forest Project was about 4,966,900 ha cumulatively from 2006 to 2010, and 99 % area had been maintained,[12] which reduced the soil erosion and sand threat in the project area, increased forest resources and recovered and restored some important ecosystems.

The Three-North and Yangtze River Basin Shelterbelt Project covered largest area among all the shelterbelt projects in China, including the shelterbelt project in Three-North, the coastal region, Pearl River, HuaiRiver, Taihang Mountain, the plain area, Dongting Lake, Poyang Lake, and the middle-downstream of Yangtze River basin. According to the plan, forestation would be 22.7 million hectares, and 71.87 million hectares forest would be well protected. This project strengthened the administration area and vegetation restoration mode to increase forest coverage rate by artificial and aerial seeding afforestation. The major section of the shelterbelt project included West Coast, Yangtze River Delta, Pearl River Delta, Bohaibay, Hainan Island, and BeibuGulf. The shelterbelt project in these areas advanced the establishment of shelterbelt system, built up lots of trunk shelterbelts, protected and restored plenty of mangrove wetlands, and built a series of demonstration areas. Forestation area of this project was 5,106,700 hectares cumulatively from 2006 to 2010.[13] Forestation area of Three-North Shelterbelt Project was 857,900 ha in 2010, and it increased from 5.05 % (1977) to 12.4 %.The forestation area of the Costal Shelterbelt Project was 231,500 ha, and the ability of defensing storm and storm surge would keep improving.[14]

Jing-Jin Sandstorm Source Control Project is an important project for Beijing, because of its special geographic location and necessitous situation. This project aimed to solve the sandstorm problem in and around Beijing. It promoted the region administration and vegetation restoration mode, and focused on returning farmland to forest, forestation in suitable barren land, artificial afforestation, aerial seeding afforestation, closing sandy land for upbringing vegetation, grassland control, small watershed management, water facilities construction and ecological migration during the 11th *Five-Year Plan* period. Forestation area of this project was 2,065,500 ha cumulatively from 2006 to 2010. Vegetation quality in and around Beijing was greatly improved, and sandy weather had been reduced significantly.

---

[12] 2010 Communiqué on China Land Afforestation Status.

[13] Source: China Statistical Yearbook 2010.

[14] 2010 Communiqué on China Land Afforestation Status.

10.3 Prospects for China's Ecological Construction in Green Economic Development During the 12th Five-Year Plan Period

10.3.1 Building the Strategic Pattern for National Ecological Security

During the 11th *Five-Year Plan* period, China's ecological construction was focused on some specific regions and remarkable achievements had been made. In the 12th *Five-Year Plan* period, the ecological construction has come to the stage of joint construction among different regions. Important ecological functional regions would be connected through scientific conservation and management measures, to form a new strategic pattern of the national ecological security and provide guarantee for *green development* nationwide.

Ecological security refers to a non-threatened condition for people's living, health, well-being, fundamental rights, life security sources, necessary resources, social orders and adaptability to environmental changes, etc., which constitutes a composite artificial ecological security system covering natural-, economic- and social-ecological security.[15] National-scale ecological security pattern is an important support for sustainable development and the proper functioning of national ecosystems. It is an interconnected and organic ecological network comprised by forests, grasslands, rivers, wetlands and other natural ecosystems, which has various important functions of conserving biodiversity, offering ecological resources, providing good environment support for economic development, and improving the quality of people's living conditions, etc. Studies have shown that the ecological security area at the lowest level of safety standard (the highest level of protection) accounts for 35.7 % of China's total land area.[16]

Key ecological functional regions are important components of ecological security pattern, which are areas of vital role in water source preservation, water and soil conservation, flood regulation, wind-break and sand-fixation, and biological diversity maintenance. Conservation and management of key ecological functional regions possess essential significance for mitigating and preventing natural disasters, coordinating watershed-and-regional ecological preservation and economic and social development, and ensuring national and local ecological security. *Plan for National Key Ecological Functional Conservation Zones* (No.165 document of SEPA in 2007) released by the former State Environmental Protection Administration of China pointed out that, major ecological functional areas in China were severely damaged, with the overall degradation or even

---

[15] MA Ke-Ming, FU Bo-Jie, LI Xiao-Ya, GUAN Wen-Bin.The Regional Pattern for Ecological Security (RPES): the Concept and Theoretical Basis [J]. *Acta Ecologica Sinica*, 2004, 24(4).

[16] YU Kong-Jian; LI Hai-Long; LI Di-Hua; QIAO Qing; XI Xue-Song. National-scale Ecological Security Pattern. *Acta Ecologica Sinica*, 2009, 29(10).

complete loss of ecological functions in some areas, which had become a serious threat to national and regional ecological security. The damages were mainly reflected by the following aspects: degradation of ecological functions and decline of water source preservation function in the headwater regions of major rivers threatening the ecological safety of downstream regions; vegetation deterioration and oasis shrink in some important sand-fixing regions of Northern China intensifying the threat from sand storms; shrinking of wetlands, rivers and lakes and degradation of these ecosystems reducing their functions of flood regulation; aggravated soil erosion in some areas menacing the regional sustainable development; damages to the offshore ecosystems leading to the production decline in major fishery waters; increasing degradation of natural habitats for some resource-intensive areas causing the decline of biodiversity maintenance function. These problems stem not only deep-rooted contradictions between economic development and environmental protection, but also from some historical reasons and management issues.

Thus it can be seen that, during the 12th *Five-Year Plan* period, to build a strategic pattern for national ecological security is faced with dual tasks. On one hand, the existing achievements of key ecological functional region constructions must be consolidated and developed, accompanied by strengthening ecological and environmental protection and improving management level, to ensure the proper exertion of such ecological functions like water source preservation, water and soil conservation, wind-break and sand-fixation, biodiversity conservation, etc. On the other hand, the ecological functional units need to be connected and managed integrally, which has been stated in the 12th *Five-Year Plan* as follows: constructing the ecological security strategy pattern based on Qinghai-Tibet Plateau ecological barrier, Loess Plateau-Sichuan-Yunnan ecological barrier, northeastern forest zone, northern sand-control belt, southern hilly and mountainous areas and major river systems, supported by other key ecological functional regions, and also comprised by punctate-distributional national prohibited development zones.

10.3.2 Strengthening Ecological Protection and Management

China's nature reserves experienced rapid development. The number of nature reserves of different levels nationwide reached 2541 at the end of 2009, with an increase of 192 in comparison with 2005. The area of nature reserves was altogether 147,750,000 ha in 2009, accounting for about 14.7 % of national land area.[17] Currently, there are some problems with the development of nature

[17] National Bureau of Statistics of China,*ReportSeries of Economic and Social Development Achievements during 11th Five-Year Plan period (No.14):PositiveProgresses in Environmental Protection. In* http://www.stats.gov.cn/tjfx/ztfx/sywcj/t20110310_402709535.htm, *2011-03-10.*

reserves as follows: unreasonable construction and expansion make for wrong compartmentation and decline of management quality in some nature reserves[18]; the step of protection could not catch up with the speed of destruction; the periphery area of some reserves are severely interfered by the anthropogenic factors, which is unfavorable to their protection; reserves mostly exist in isolation, lack of necessary inter-connection hindering the migration of species and restoration of biodiversity. Therefore, during the 12th *Five-Year Plan* period, the supervision of nature reserves construction needs to be strengthened, the mechanism and daily management of reserve construction need to be standardized, and the organic links among reserves should be established to gradually form the nationwide network system of nature reserves, putting ecological protection into an overall consideration with the economic development.

Biodiversity is an important part of ecosystem quality and also a basis for sustainable economic and social development as well as a guarantee for ecological security and food security. *China Biodiversity Protection Strategy and Action Plan(2011–2030)* (No. 106 document of SEPA in 2010)issued by the Ministry of Environmental Protection mentioned that, the legal and policy systems of biodiversity protection are still incomplete, the monitoring and early warning system has not been established yet, the input for biodiversity protection is insufficient, the management and protection level needs to be further improved, the basic scientific research ability is weak, and the ability to deal with new issues on biodiversity conservation is scarce. In addition, biodiversity is increasingly affected by growing urbanization and accelerated industrialization, habitats of species are threatened, and the environmental releases of invasive alien species as well as genetically modified organisms (GMOs) increase the pressure on bio-security. It also set objectives to strive to effectively curb the trend of biodiversity decline in key areas by 2015, which include: strengthen in situ conservation to maintain the ratio of total land nature reserves area to total land area being around 15 % and enable 90 % of national key protected species and typical ecosystem types to be well protected; carry out reasonable ex-situ conservation to obtain effective protection for over 80 % of those threatened species to which the in situ conservation is inadequate and whose wild population is quite small; in addition, bio-security management should be reinforced, the conservation and management of biological species resources should be intensified, species resources reduction and loss should be effectively prevented, the mechanism of monitoring and access control of species should be improved, and the alien species invasion should be positively prevented.

Although the national forest coverage had increased to 20.36 % during the 11th *Five-Year Plan* period, a lot of problems still exist as follows:

1. The total forest volume was not enough. The forest coverage was only 2/3 of the world's average level, ranking the 139th.

---

[18] *LEI Guang-Chun. Problems and* Challenges Faced with the Nature Reserves of China [J]. *World Environment*. 2008(4).

2. The reversal trend had been growing. *The* 7th *National Forest Inventory Data* showed that, during the 5 years' inventory interval, the area of non-forest land turned from forest land reached 8.32 million hectare due to deforestation and cultivation, natural disasters and project construction, 85 % of which had been changed reversely to other agricultural use.
3. The forest quality and productivity are very low. Existing forests present several negative phenomena: artificial pure forests are more than mixed forests, single-layer forests are more than multi-layer forests, immature and young forests are more than mature and old forests. The collectively-owned forest land, which accounts for 60 % of total forest land, has low stock volume per unit area of only 53 m^3/ha.

Forest lands deteriorate significantly. Waste mountains, unreclaimed lands and bare sand lands suitable for afforestation account for about 14.5 %. Forest land deterioration was quite critical especially those of western and southern regions as well as collective forest regions.[19]

Therefore, during the 12th *Five-Year Plan* period, it's essential, on one hand, to accelerate the process of ecological restoration and afforestation and the launch of a series of key forestry projects such as natural forest protection, shelter forest networks including northeastern-northern-northwestern (known as "Three-North Shelter Forests"), Yangtze River basin and coastal region shelter forests, etc., wildlife protection and nature reserves construction, to speed up some key ecological projects such as anti-desertification, comprehensive management of Karst rocky desertification in some southern Karst regions as well as forest protection and restoration, to launch the follow-up major ecological projects, to accelerate the afforestation pace of suitable lands, and to make use of forest resources sustainably. On the other hand, it's necessary to reinforce the ecological management of the degraded forest lands in key regions, focusing on vegetation recovery and ecological restoration, to significantly improve the ecological conditions in key ecological management areas as soon as possible.

10.3.3 Establishing and Optimizing Eco-Compensation Mechanism

The attempt of eco-compensation mechanism is a major highlight of ecological construction during the 11th *Five-Year Plan* period. Depending on economic measures, the eco-compensation mechanism maximizes the protection of existing ecological resources and ensures the effective functioning of ecosystem services. China started a pilot eco-compensation from the year 2005 and nearly 400 key ecological areas benefited from eco-compensation during the subsequent 5 years.

---

[19] See *Program Outline of National Forest land Utilization and Conservation (2010–2020).*

In the next 5 years, China will accelerate the establishment of eco-compensation mechanism to improve the level of ecological civilization.[20] Domestic eco-compensation can be divided into the following aspects: regional compensation (compensation from the eastern to the western region), basin compensation (compensation among basins inter-provincial or within local administrative areas), ecosystem compensation (compensation to services offered by ecosystems like forests, grasslands, wetlands, ocean, farmlands and other ecosystems), resources exploitation compensation (mineral resources exploitation, land reclamation and vegetation restoration, etc.).[21] The 12th *Five-Year Plan* clearly proposes to speed up the establishment of eco-compensation mechanism, following the principles of that those who conduct the explorations should take the responsibility of protection accordingly and those who get benefits should make corresponding compensations.

The current problems of eco-compensation can be summarized as follows: the theory of eco-compensation mechanism is far from complete to still in an exploratory stage. Eco-compensations vary widely among regions, and the standard of compensation and other aspects remains controversial, lacking of mature optional solutions; legal system of eco-compensation is incomplete which leads to subjective and arbitrary behaviors; compensation amount is generally too low to guarantee the successful implementation of eco-compensation mechanism; compensation relies excessively on transfer payments, increasing the fiscal burden while hindering the market to play its role; the relationship with economic development is not well handled, limiting the development of compensated regions to some extent.

To solve the problems stated above, two issues should be paid more attention to when carrying out eco-compensation: determine eco-compensation amounts based on the ecosystem service value theory, and introduce market mechanism into eco-compensation. For each compensation type, the range of compensation needs to be clearly set to avoid waste of money; stakeholders in the compensation mechanism should be precisely indentified to avoid social conflicts; reasonable quantitative criteria must be formulated to ensure the effective operation of compensation funds; a sound assessment system of eco-compensation efficiency needs to be set up to evaluate the effects of eco-compensation and correct arising problems in a timely manner. Moreover, for some areas in financial difficulties, it should be considered to increase financial support from the government and improve the compensation amounts. It's necessary to change the current situation of government finance payment being the sole source for eco-compensation. Enterprises and

[20] National Development and Reform Commission. *Taking Use of Economic Levers to Control Pollution Shows Full Effectiveness of the Eco-compensation Mechanism.* In http://www.sdpc.gov.cn/ajgx/t20101224_387531.htm, 2010-12-24.

[21] Li Wen-hua, Liu Mou-cheng. Several Strategic Thoughts on China's Eco-Compensation Mechanism [J]. *Resources Science*. 2010, 32(05).

individuals should be introduced to join the eco-compensation mechanism, and market mechanisms should be actively explored and encouraged. Ecosystem service should be increased through multiple ways. Eventually, the eco-compensation would be led to the track of enhancing the national ecological security.

Chapter 11
Measurement and Analysis of Carrying Potential of Natural Resources and Environment

Mingqing Jiang , Jiangxue Zhang and Yang Liu

To accelerate the transformation of the economic development pattern, it is necessary to coordinate the development of the economy, resources and environment. In the course of economic development, much attention should be paid to the resources, ecological protection, environmental stress and climate change. The Index of Carrying Potential of Natural Resources and Environment (ICPNRE) measures the Carrying Potential of resources, ecological protection, environmental stress and climate change that underlies future economic development and human activities in a region. As one of the important elements of the Green Development Index (GDI), ICPNRE reflects the conditions of natural resources and ecology, and the impacts of human activities on natural resources, environment, ecology and climate in that region.

From a regional comparison perspective, we adopted the evaluation and weight criterion of "China Province Green Development Index System" and "China City Green Development Index System" to calculate the ICPNRE of 30 provinces (autonomous regions and municipalities directly under the jurisdiction of the central government) and 34 provincial capital cities (including municipalities directly under the jurisdiction of the central government and specifically designated cities in the state plan) in this chapter. Basing on the calculated results, we analyzed and depicted the characteristics and basic configuration of two indicators in these regions: Resource Abundance Status and Ecological Conservation, and Environmental Stress and Climate Change. We also compared the ICPNRE among different regions and offered our suggestions.

11.1 Calculated Results of ICPNRE

According to the measurement system and weighing criterion of Green Degree of Economic Growth (GDEG) in the "China Province Green Development Index System" and "China City Green Development Index System", the Green Index

X. Li and J. Pan (eds.), *China Green Development Index Report 2011*,
Current Chinese Economic Report Series, DOI: 10.1007/978-3-642-31597-8_11,

Group (GIG) calculated the ICPNRE of 30 provinces (autonomous regions and municipalities directly under the jurisdiction of the central government) and 34 provincial capital cities (including municipalities directly under the jurisdiction of the central government and specifically designated cities in the state plan), and ranked all regions basing on the scores of two indicators: Resource Abundance Status and Ecological Conservation, and Environmental Stress and Climate Change.

11.1.1 Calculated Results of Provincial ICPNRE

The calculated results of ICPNRE of 30 provinces (autonomous regions and municipalities directly under the jurisdiction of the central government) were shown in the following Table 11.1 based on the measurement system and weighing criterion in the "China Province Green Development Index System".

As shown in Table 11.1, the top ten provinces in the ranking of ICPNRE were Qinghai, Guizhou, Yunnan, Hainan, Heilongjiang, Inner Mongolia, Sichuan, Gansu, Xinjiang and Jiangxi. The top ten provinces in the ranking of Resource Abundance Status and Ecological Conservation Indicator were Qinghai, Inner Mongolia, Heilongjiang, Yunnan, Hainan, Sichuan, Jilin, Shanghai, Jiangxi and Guangxi. The top ten provinces in the ranking of Environmental Stress and Climate Change Indicator were Qinghai, Guizhou, Yunnan, Hainan, Gansu, Xinjiang, Heilongjiang, Sichuan, Shanxi and Inner Mongolia.

Basing on the data in Table 11.1, we drew Fig. 11.1 to show the level of ICPNRE of different provinces. The horizontal axis represented the ICPNRE value, and the origin stood for the average ICPNRE of 30 provinces. The green bar represented the regions above the average level, and the higher the ICPNRE value, the longer the green bar. On the contrary, the white bar represented the regions below the average, and the lower the ICPNRE value, the longer the white bar.

From Table 11.1, we got the general characteristics of the ICPNREs of China's 30 provinces (autonomous regions and municipalities directly under the jurisdiction of the central government) as follows:

(1) There were a significant regional disparity of the ICPNRE

From a national perspective, there was a significant inter-provincial disparity of the ICPNRE. The score of the No. 1 province Qinghai was over twice that of the No. 2 Guizhou and the No. 3 Yunnan. Only three provinces scored more than 0.2. Nineteen provinces scored below the national average. The range of the 30 provinces was 4.5 times of their standard deviation.

From a regional perspective, a significant disparity of the ICPNREs existed among East, Central, Northeast and West China. As shown in Fig. 11.2, the ICPNRE level of the western region was higher, compared to the lower level of the eastern and central regions, and level of the northeastern region approximated the national average. Two reasons accounted for such a regional disparity. One was that the western and

Table 11.1 ICPNRE and rankings of 30 provinces in China in 2009

Indicator	Carrying potential of natural resources and environment		Second-class indicators			
			Resource abundance and ecological protection indicators		Environmental stress and climate change indicators	
Weight	100 %		30 %		70 %	
Province	Score	Ranking	Score	Ranking	Score	Ranking
Qinghai	0.556	1	0.143	1	0.413	1
Guizhou	0.269	2	−0.026	20	0.295	2
Yunnan	0.241	3	0.078	4	0.163	3
Hainan	0.160	4	0.060	5	0.100	4
Heilongjiang	0.156	5	0.121	3	0.035	7
Inner Mongolia	0.135	6	0.130	2	0.006	10
Sichuan	0.081	7	0.052	6	0.029	8
Gansu	0.070	8	−0.022	18	0.092	5
Xinjiang	0.056	9	0.011	12	0.045	6
Jiangxi	0.019	10	0.033	9	−0.014	13
Jilin	0.011	11	0.050	7	−0.039	15
Shaanxi	−0.009	12	−0.026	19	0.016	9
Guangxi	−0.031	13	0.027	10	−0.058	20
Fujian	−0.037	14	0.023	11	−0.059	21
Hunan	−0.047	15	−0.005	13	−0.042	17
Anhui	−0.054	16	−0.058	23	0.004	11
Chongqing	−0.055	17	−0.021	17	−0.033	14
Beijing	−0.063	18	−0.062	26	−0.001	12
Zhejiang	−0.092	19	−0.007	14	−0.085	26
Hubei	−0.104	20	−0.038	21	−0.065	23
Tianjin	−0.105	21	−0.046	22	−0.059	22
Shandong	−0.108	22	−0.062	24	−0.047	19
Guangdong	−0.117	23	−0.013	15	−0.104	29
Shanghai	−0.117	24	0.040	8	−0.157	30
Hebei	−0.118	25	−0.077	29	−0.041	16
Henan	−0.121	26	−0.079	30	−0.042	18
Liaoning	−0.123	27	−0.019	16	−0.103	28
Ningxia	−0.137	28	−0.071	27	−0.066	24
Shanxi	−0.158	29	−0.074	28	−0.084	25
Jiangsu	−0.159	30	−0.062	25	−0.097	27

Notes

1. Figures in this table are calculated based on data of each indicator for 2008 and 2009 in accordance with the indicator system of CCNRE embedded in the Province Measurement System
2. Index of each province in this table is ranked in descending order

Sources China Statistical Yearbook 2010, China Statistical Yearbook 2009, Desert and Its Treatment in China, Annual Statistical Report on Environment in China 2009, and China Environmental Statistical Yearbook 2010

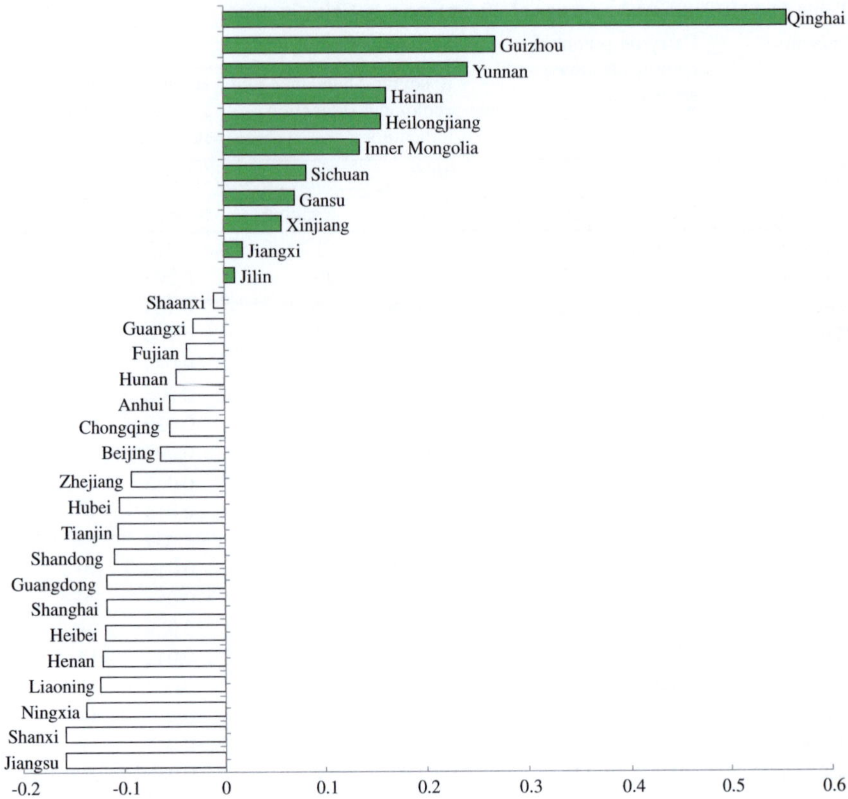

Fig. 11.1 Inter-provincial comparison of ICPNRE rankings

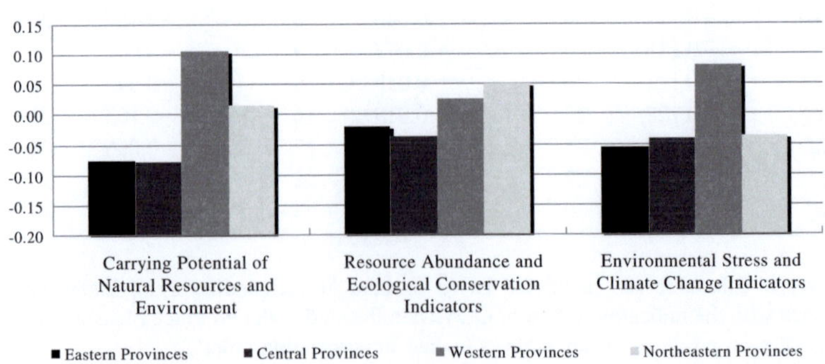

Fig. 11.2 Comparison of ICPNRE among 4 major areas in China

northeastern regions had far more resources and better ecology than the eastern and central regions. The other was that the western region was relatively better able to cope with environmental stress and climate change than the rest of the country.

From the perspective of provinces (autonomous regions and municipalities directly under the jurisdiction of the central government) inside the four major regions, we found that, among the ten eastern provinces, except the No. 4 Hainan, Fujian, Beijing and Zhejiang all ranked in the middle as No. 14, 18 and 19; Shandong, Guangdong, Hebei, Tianjin, Jiangsu and Shanghai all ranked behind, among No. 21–29, and Jiangsu even ranked at the bottom of the list. Among the six central provinces, Jiangxi ranked forward as No. 10. Hunan, Anhui, Hubei were in the middle, ranking No. 15, 16, and 20. Henan and Shanxi ranked behind as No. 26 and 29. Among the 11 western provinces (except Tibet), Qinghai, Guizhou, Yunnan, Inner Mongolia, Sichuan, Gansu and Xinjiang all ranked among the top ten. Moreover, Qinghai, Guizhou and Yunnan ranked at top 3 and Qinghai was ahead of all provinces with the score of 0.5561. In addition, No. 12 Shanxi (−0.009) and No. 13 Guangxi (−0.031) scored slightly below the national average. Chongqing and Ningxia ranked behind as No. 17 and 28 (or the last third place). Of the three northeastern provinces, Heilongjiang was better and ranked No. 5, the No. 11 Jilin was slightly above the national average, but Liaoning lagged behind, ranking No. 27.

(2) The level of ICPNRE mainly depended on the score of Environmental Stress and Climate Change Indicator

Table 11.1 indicated that the level of ICPNRE mainly depended on the score of Environmental Stress and Climate Change Indicator. Deeper statistical analysis showed a correlation of 0.935 between the level of ICPNRE and Environmental Stress and Climate Change Indicator, higher than that between ICPNRE and Resource Abundance Status and Ecological Conservation Indicator (0.745). The key reason for such a high correlation was that the indicator reflected lots of human activities such as methods of production, pollutant discharge and treatment.

(3) There was a negative correlation to a certain degree between the level of ICPNRE and that of economic development

GDP per capita was one of the most important indexes for measuring the level of regional economic development. Through analyzing the correlation between the level of ICPNRE and that of economic development, we found a negative correlation between the two: regions with high GDP per capita had low ICPNRE levels, and vice versa (See Table 11.2).

From Table 11.2, the top five provinces with highest GDP per capita in 2009 were Shanghai, Beijing, Tianjin, Jiangsu and Zhejiang, meanwhile their level of ICPNRE were much lower than the national average. Among them, the highest one was Beijing, ranking No.18, while Jiangsu ranked at the bottom of the list. Zhejiang, Tianjin and Shanghai ranked No. 19, 21 and 24. The last ten provinces with lowest GDP per capita were Guizhou, Gansu, Yunnan, Guangxi, Anhui, Jiangxi, Sichuan, Hainan, Qinghai and Xinjiang, but eight of them ranked among the top ten in terms of the ICPNRE level except Guangxi and Anhui. What's more, Qinghai, Guizhou, Yunnan and Hainan took the first four places.

Table 11.2 Inter-provincial comparison of ranking gap between GDP per capita and ICPNRE

Province	Ranking of GDP per capita (1)	Ranking of carrying potential of natural resources and environment (2)	Ranking gap (1)–(2)	Province	Ranking of GDP per capita (1)	Ranking of carrying potential of natural resources and environment (2)	Ranking Gap (1)–(2)
Shanghai	1	24	−23	Ningxia	16	28	−12
Beijing	2	18	−16	Shaanxi	17	12	5
Tianjin	3	21	−18	Shanxi	18	29	−11
Jiangsu	4	30	−26	Henan	19	26	−7
Zhejiang	5	19	−14	Hunan	20	15	5
Guangdong	6	23	−17	Xinjiang	21	9	12
Inner Mongolia	7	6	1	Qinghai	22	1	21
Shandong	8	22	−14	Hainan	23	4	19
Liaoning	9	27	−18	Sichuan	24	7	17
Fujian	10	14	−4	Jiangxi	25	10	15
Jilin	11	11	0	Anhui	26	16	10
Hebei	12	25	−13	Guangxi	27	13	14
Chongqing	13	17	−4	Yunnan	28	3	25
Hubei	14	20	−6	Gansu	29	8	21
Heilongjiang	15	5	10	Guizhou	30	2	28

Out of the 30 provinces (autonomous regions and municipalities directly under the jurisdiction of the central government), 24 experienced a ranking gap of ten or more places between the per capita GDP and ICPNRE, 13 experienced a ranking gap of 15 or more places, and six experienced a ranking gap of 20 or more places. The larger the gap, the stronger the negative correlation. Statistics showed the correlation coefficient between per capita GDP and ICPNRE was −0.5791. In other words, ICPNRE had a reverse gradient relationship with per capita GDP.

(4) There was a slight serial correlation between ICPNRE and Green Development Index (GDI)

Most of the 30 provinces (autonomous regions and municipalities directly under the jurisdiction of the central government) also experienced a ranking gap between the ranking of GDI and that of ICPNRE. 14 experienced a ranking gap of ten or more places, and they were Beijing, Tianjin, Zhejiang, Jiangsu, Guangdong, Shandong, Guizhou, Heilongjiang, Sichuan, Jilin, Guangxi, Hunan and Gansu, seven from the eastern region, one from the central region, four from the western region and two from the northeastern region (See Table 11.3).

Our analysis revealed a significant level of 0.205 between GDI and ICPNRE, far above 0.05. Hence a slight serial correlation between the two. Provinces with low

Table 11.3 Inter-provincial comparison of ranking gap between GDI and ICPNRE

Province	Ranking of GDI (1)	Ranking of ICPNRE (2)	Ranking Gap (1)–(2)	Province	Ranking of GDI (1)	Ranking of ICPNRE (2)	Ranking gap (1)–(2)
Beijing	1	18	−17	Xinjiang	16	9	7
Shanghai	2	24	−22	Jiangxi	17	10	7
Qinghai	3	1	2	Hebei	18	25	−7
Tianjin	4	21	−17	Sichuan	19	7	12
Hainan	5	4	1	Anhui	20	16	4
Zhejiang	6	19	−13	Chongqing	21	17	4
Yunnan	7	3	4	Hubei	22	20	2
Fujian	8	14	−6	Jilin	23	11	12
Jiangsu	9	30	−21	Guangxi	24	13	11
Guangdong	10	23	−13	Liaoning	25	27	−2
Inner Mongolia	11	6	5	Hunan	26	15	11
Shandong	12	22	−10	Ningxia	27	28	−1
Guizhou	13	2	11	Shanxi	28	29	−1
Shaanxi	14	12	2	Gansu	29	8	21
Heilongjiang	15	5	10	Henan	30	26	4

ICPNRE but high GDI are mostly from East China, such as Shanghai, Jiangsu, Beijing, Tianjin, Zhejiang, Guangdong and Shandong, where GDEG and Support Degree of Government Policies were both high. Provinces with high ICPNRE but low GDI such as Gansu, Sichuan, Jilin, Guizhou, Guangxi, Hunan and Heilongjiang experienced low GDEG and low Support Degree of Government Policies.

The different GDI and ICPNRE rankings of one place was due to the unique economic development pattern and domestic conditions of China. East China enjoyed a high GDI as a result of high economic development level and high GDEG and government policy support, but the region was low in ICPNRE. Part of Central, West and Northeast China enjoyed a high ICPNRE contributed by rich resources and well-preserved ecology, but the backward industrial structure, economic inefficiency and low GDEG of these regions pulled down the GDI.

Therefore, different regions had different priorities in transforming the economic development pattern. We should take actions that suited local circumstances, maximized favorable factors and minimized unfavorable ones. Regions with low GDI and high ICPNRE should focus on green economic growth and give more preferential policies in regard with resources and environmental issues. On the contrary, regions with low ICPNRE, especially the developed eastern region, should speed up the transformation of the development pattern and economic restructuring, and reduce environment damage and excessive reliance on resources, so as to improve green development.

11.1.2 Calculated Results of City ICPNRE

The calculated results of ICPNRE of 34 provincial capital cities (including municipalities directly under the jurisdiction of the central government and specifically designated cities in the state plan) were shown in the following Table 11.4 based on the measurement system and weighing criterion in the "China City Green Development Index System".

As shown in Table 11.4, the top ten cities in the ranking of ICPNRE were Kunming, Haikou, Harbin, Nanning, Fuzhou, Guiyang, Changchun, Chongqing and Changsha. The top ten cities in the ranking of Resource Abundance Status and Ecological Conservation Indicator were Lanzhou, Hangzhou, Nanning, Chongqing, Kunming, Guiyang, Changsha, Ningbo, Shenzhen and Guangzhou. The top ten provinces in the ranking of Environmental Stress and Climate Change Indicator were Kunming, Haikou, Harbin, Nanning, Hefei, Fuzhou, Changchun, Guiyang, Chongqing and Changsha.

Basing on the data in Table 11.4, we drew Fig. 11.3 to show the level of ICPNRE of different cities. The horizontal axis represented the ICPNRE value, and the origin stood for the average ICPNRE of 34 cities. The green bar represented the cities above the average level, and the higher the ICPNRE value, the longer the green bar. On the contrary, the white bar represented the cities below the average, and the lower the ICPNRE value, the longer the white bar.

From Table 11.4 and Fig. 11.3, we got the general characteristics of the ICPNRE of China's 34 cities as follows:

(1) Inter-city disparity of the ICPNRE was also significant, and the level of western and northeastern cities was much higher than that of eastern and central cities

The disparity of 34 cities in ICPNRE was far more significant than that of the 30 provinces. The score of No. 1 Kunming was respectively 1.5, 2.7, 2.7 and 3.3 times that of No. 2 Haikou, No. 3 Harbin, No. 4 Nanning and No. 5 Hefei. 32 cities scored within −0.2 and 0.2. The range of the 34 cities was 4.7 times their standard deviation.

If we look at the four major regions the 34 cities belong to: the eastern, central, northeastern and western regions, we would find that the western and northeastern cities had much higher ICPNRE levels than the eastern and central cities. The northeastern region scored 0.06 on average, western region 0.04, but the eastern and central regions scored below the national average. Such a regional disparity was caused by the difference in Environmental Stress and Climate Change among the four regions, whose Resource Abundance and Ecological Conservation did not differ much (Fig. 11.4).

From a regional perspective, among the ten calculated cities from West China, four ranked among top ten: Kunming, Nanning, Guiyang and Chongqing. The No. 1 Kunming scored 0.546 with an overwhelmingly superiority. Hohhot was slightly above the national average while Yinchuan and Lanzhou were slightly below it. Other western cities ranked behind. For example, Xining lied third from the

Table 11.4 Indexes of ICPNRE and rankings of 34 cities in China in 2009

Indicator	Carrying potential of natural resources and environment		Second-class indicators			
			Resource abundance and ecological protection indicators		Environmental stress and climate change indicators	
Weight	100 (%)		5 (%)		95 %	
City	Score	Ranking	Score	Ranking	Score	Ranking
Kunming	0.546	1	0.003	5	0.543	1
Haikou	0.368	2	−0.009	33	0.377	2
Harbin	0.202	3	−0.001	12	0.204	3
Nanning	0.201	4	0.010	3	0.190	4
Hefei	0.167	5	−0.006	22	0.173	5
Fuzhou	0.087	6	−0.005	18	0.092	6
Guiyang	0.073	7	0.003	6	0.071	8
Changchun	0.072	8	−0.006	21	0.078	7
Chongqing	0.060	9	0.008	4	0.052	9
Changsha	0.042	10	0.003	7	0.040	10
Dalian	0.029	11	−0.006	23	0.036	11
Hohhot	0.029	12	−0.004	16	0.034	12
Shenzhen	0.029	13	0.002	9	0.027	13
Qingdao	0.011	14	−0.006	24	0.017	14
Ningbo	0.004	15	0.002	8	0.002	15
Nanchang	−0.014	16	0.001	11	−0.015	17
Yinchuan	−0.020	17	−0.009	34	−0.011	16
Zhengzhou	−0.047	18	−0.008	31	−0.039	18
Shenyang	−0.064	19	−0.006	19	−0.059	19
Lanzhou	−0.067	20	0.091	1	−0.158	32
Guangzhou	−0.068	21	0.002	10	−0.070	20
Hangzhou	−0.071	22	0.013	2	−0.084	22
Shijiazhuang	−0.091	23	−0.007	28	−0.084	21
Beijing	−0.101	24	−0.007	27	−0.094	24
Taiyuan	−0.102	25	−0.008	32	−0.093	23
Xiamen	−0.107	26	−0.002	13	−0.105	26
Jinan	−0.110	27	−0.007	26	−0.103	25
Xi'an	−0.131	28	−0.007	25	−0.124	27
Chengdu	−0.135	29	−0.004	15	−0.131	29
Tianjin	−0.136	30	−0.008	30	−0.129	28
Nanjing	−0.152	31	−0.006	20	−0.146	30
Xining	−0.159	32	−0.004	14	−0.155	31
Shanghai	−0.166	33	−0.008	29	−0.158	33
Wuhan	−0.183	34	−0.005	17	−0.178	34

Lhasa and Urumqi. were unmeasured due to lack of basic data

Notes

1. Figures in this table are calculated based on data of each indicator for2098 and 2009 in accordance with the indicator system of ICPNRE embedded in the City Measurement System
2. Index of each province in this table is ranked in descending order

Sources China City Statistical Yearbook 2008, *China City Statistical Yearbook* 2009, *China Urban Life and Price Statistical Yearbook* 2010, *Annual Statistical Report on Environment in China* 2008, *Annual Statistical Report on Environment in China* 2009, and Data Center Network of the Ministry of Environmental Protection

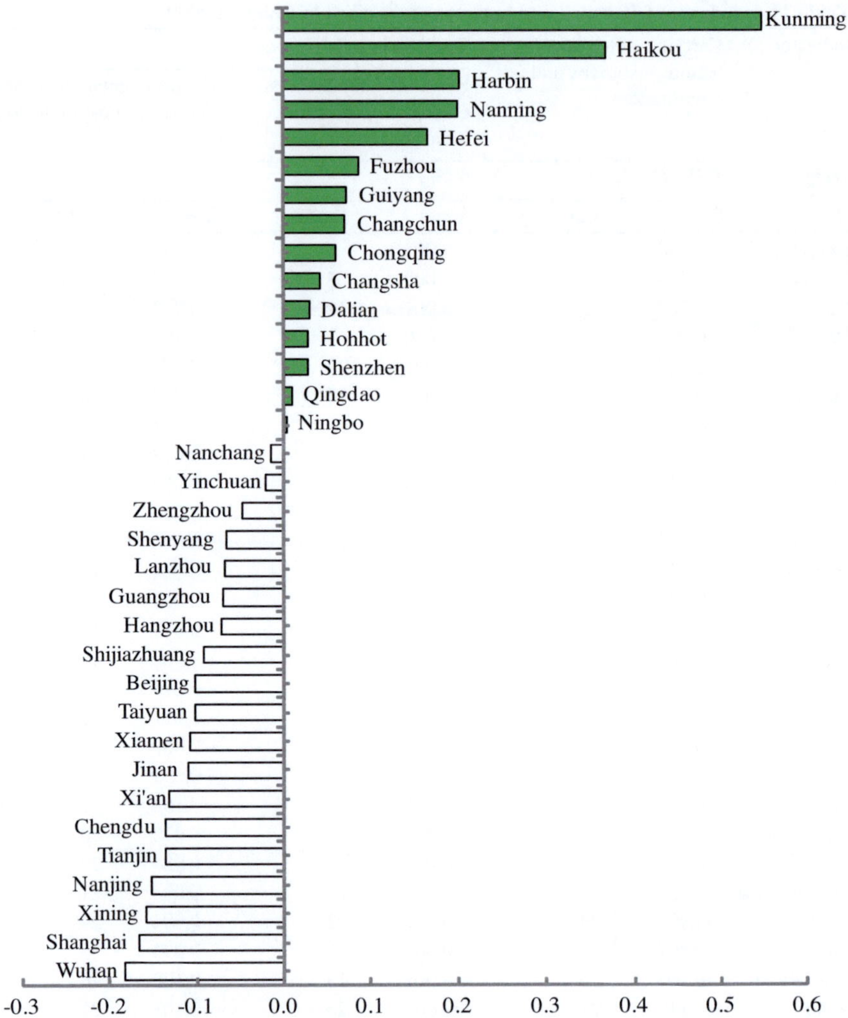

Fig. 11.3 Inter-city comparison of ICPNRE rankings

bottom. Among the four northeastern cities, two ranked among top ten: No. 3 Harbin and No. 8 Changchun. No. 11 Dalian and No. 19 Shenyang also ranked forward. Among the fourteen eastern cities, the highest-ranking city was No. 2 Haikou. Most cities in this region such as Beijing, Tianjin, Shijiazhuang and Nanjing ranked in the middle or the second half of the list, and Shanghai even lied second from the bottom. Two of the six central cities ranked among top ten: No. 5 Hefei and No. 10 Changsha. In particular, Wuhan was the tail-end.

(2) The level of ICPNRE mainly depended on the score of Environmental Stress and Climate Change Indicator

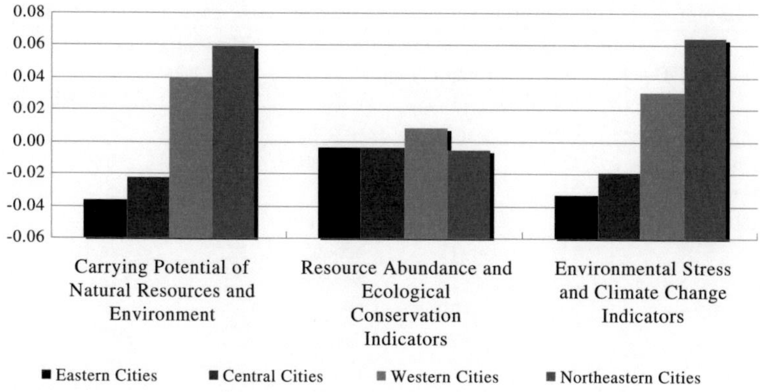

Fig. 11.4 Inter- city comparison of ICPNRE among 4 major areas in China

Table 11.4 showed that the level of city ICPNRE depended on the score of Environmental Stress and Climate Change Indicator to a great extent. Analysis demonstrated a correlation coefficient of 0.994 between the level of ICPNRE and Environmental Stress and Climate Change Indicator, distinctly higher than that between ICPNRE and Resource Abundance Status and Ecological Conservation Indicator (0.023). This revealed a huge impact of one city's Environmental Stress and Climate Change Indicator on its ICPNRE. As the indicator was closely related to the methods of production, pollutant discharge and control in a city, more favorable policies should be given in this field.

(3) There was a negative serial correlation to a certain degree between the level of ICPNRE and that of economic development

The ranking showed a negative serial correlation the level of ICPNRE and that of economic development in 34 provincial capital cities (including municipalities directly under the jurisdiction of the central government and specifically designated cities in the state plan). Cities with high levels of economic development had low levels of ICPNRE, and vice versa. Details were shown in Table 11.5.

From Table 11.5, the top five cities in terms of GDP per capita in 2009 were Shenzhen, Guangzhou, Shanghai, Dalian and Beijing, but these cities ranked low in terms of ICPNRE. The highest of them was No. 11 Dalian, and the lowest was No. 33 Shanghai. Shenzhen, Guangzhou and Beijing ranked No. 13, 21 and 24. The bottom ten cities in terms of GDP per capita were Xi'an, Harbin, Shijiazhuang, Lanzhou, Haikou, Kunming, Guiyang, Chongqing, Xining and Nanning. However, six of them ranked among top ten in terms of ICPNRE, four of them lied at top 4, and Kunming ranked even No.1.

In general, the 34 cities experienced a great ranking gap between the ranking of GDP per capita and that of ICPNRE. 19 cities experienced a ranking gap of ten or more places, 14 ones experienced a gap of 15 or more places, and nine experienced a gap of 20 or more places. Two cities even experienced a gap of 30 places. Further analysis revealed that the correlation coefficient between the rankings of

Table 11.5 Inter-city comparison of ranking gap between GDP per capita and ICPNRE

City	Ranking of GDP per capita (1)	Ranking of ICPNRE (2)	Ranking gap (1)–(2)	City	Ranking of GDP per capita (1)	Ranking of ICPNRE (2)	Ranking gap (1)–(2)
Shenzhen	1	13	−12	Zhengzhou	18	18	0
Guangzhou	2	21	−19	Hefei	19	5	14
Shanghai	3	33	−30	Nanchang	20	16	4
Dalian	4	11	−7	Fuzhou	21	6	15
Beijing	5	24	−19	Changchun	22	8	14
Xiamen	6	26	−20	Chengdu	23	29	−6
Nanjing	7	31	−24	Yinchuan	24	17	7
Hangzhou	8	22	−14	Xi'an	25	28	−3
Tianjin	9	30	−21	Harbin	26	3	23
Hohhot	10	12	−2	Shijiazhuang	27	23	4
Ningbo	11	15	−4	Lanzhou	28	20	8
Qingdao	12	14	−2	Haikou	29	2	27
Changsha	13	10	3	Kunming	30	1	29
Shenyang	14	19	−5	Guiyang	31	7	24
Wuhan	15	34	−19	Chongqing	32	9	23
Jinan	16	27	−11	Xining	33	32	1
Taiyuan	17	25	−8	Nanning	34	4	30

GDP per capita and ICPNRE was only −0.3665, a loose reverse gradient relationship.

(4) There was a significant ranking gap between the rankings of ICPNRE and GDI on the city level

The calculated results revealed that most cities experienced a significant ranking gap between the rankings of ICPNRE and GDI. 12, or 35 % of the 34 cities experienced a ranking gap of ten or more places: Beijing, Hohhot, Changchun, Harbin, Shanghai, Nanjing, Xiamen, Guangzhou, Shenzhen, Chongqing, Guiyang and Lanzhou. Six of them were from the eastern region, four from the western region and two from the northeastern region (See Table 11.6).

The data above showed that, due to high level of economic development, high GDEG and strong policy support from the government, some eastern cities such as Shenzhen, Nanjing, Shanghai, Guangzhou, Xiamen and Beijing ranked forward in terms of GDI. However, because of huge discharge of pollutants which led to great stress on environment, these cities ranked low in terms of ICPNRE. As a result, the ranking gap between GDI and ICPNRE was significant. As to some underdeveloped cities such as Chongqing, Guiyang, Lanzhou and Hohhot, despite the advantage in ICPNRE ranking, they still had problems with the GDEG and financial support from the government. Therefore, the GDI levels of these cities were relatively low.

Next, we worked out the reason why Beijing and Chongqing had the largest ranking gap. Beijing ranked No. 24 in terms of ICPNRE. As to the two Second-

Table 11.6 Inter-city comparison of ranking gap between GDI and ICPNRE

City	Ranking of GDI (1)	Ranking of ICPNRE (2)	Ranking gap (1)–(2)	City	Ranking of GDI (1)	Ranking of ICPNRE (2)	Ranking gap (1)–(2)
Shenzhen	1	13	−12	Nanjing	18	31	−13
Haikou	2	2	0	Shanghai	19	33	−14
Kunming	3	1	2	Changchun	20	8	12
Beijing	4	24	−20	Jinan	21	27	−6
Hefei	5	5	0	Yinchuan	22	17	5
Guangzhou	6	21	−15	Nanchang	23	16	7
Dalian	7	11	−4	Hohhot	24	12	12
Qingdao	8	14	−6	Zhengzhou	25	18	7
Changsha	9	10	−1	Guiyang	26	7	19
Fuzhou	10	6	4	Taiyuan	27	25	2
Xiamen	11	26	−15	Tianjin	28	30	−2
Nanning	12	4	8	Chongqing	29	9	20
Ningbo	13	15	−2	Xi'an	30	28	2
Shenyang	14	19	−5	Wuhan	31	34	−3
Harbin	15	3	12	Chengdu	32	29	3
Shijiazhuang	16	23	−7	Lanzhou	33	20	13
Hangzhou	17	22	−5	Xining	34	32	2

Level Indicators-Resource Abundance Status and Ecological Conservation Indicator and Environmental Stress and Climate Change Indicator, Beijing ranked No. 27 and 24. By analyzing the Third-Level Indicators under Environmental Stress and Climate Change, we discovered that Beijing ranked only No. 27 in terms of water resources per capita. What's more, Beijing ranked behind in terms of pollutant discharge per square and in the middle in terms of pollutant discharge per capita. However, due to the highly developed economy and preferential policies, Beijing ranked second in terms of GDEG and Support Degree of Government Policies. These advantages supplemented the deficiency of resources and environment. As a result, Beijing still led the country in terms of GDI. Chongqing ranked No. 33 in terms of GDEG and No. 30 in terms of Support Degree of Government Policies, but it ranked ninth in terms of ICPNRE. Hence a huge gap between the rankings of ICPNRE and GDI.

11.2 Comparative Analysis of ICPNRE on the Provincial Level

Basing on "China Province Green Development Index System", we analyzed the calculated results of Resource Abundance Status and Ecological Conservation and Environmental Stress and Climate Change Indicators of 30 provinces (autonomous regions and municipalities), in order to fully compare the different levels of ICPNRE of different regions.

11.2.1 Measurement Results and Analysis of Resource Abundance Status and Ecological Conservation Indicator on the Provincial Level

Under the "China Province Green Development Index System", Resource Abundance Status and Ecological Conservation Indicator consisted of six Third-Level Indicators including water resources per capita, forest area per capita, forest coverage rate, proportion of area of natural reserves in the total area of a region, proportion of area of wetlands in the total area of a region and total standing stock volume per capita. The weights and indications of the Third-Level Indicators were shown in Table 11.7.

Table 11.7 Third-class indicators, their weights and attributes of inter-provincial resource abundance and ecological conservation

Sequence number	Indicator	Weight (%)	Attribute
1	Water resources per capita	2.00	Positively correlated
2	Forest area per capita	2.00	Positively correlated
3	Forest coverage rate	2.00	Positively correlated
4	Proportion of area of natural reserves in total area of a region	2.00	Positively correlated
5	Proportion of area of wetlands in total area of a region	2.00	Positively correlated
6	Total standing stock volume per capita	2.00	Positively correlated

Note The contents in this table are determined by the task force after having conducted a lot of expert workshops

Basing on data normalization and the weights of the Third-Level Indicators, we worked out the calculated results of Resource Abundance Status and Ecological Conservation Indicator of different provinces (autonomous regions and municipalities) (See Table 11.8).

Generally speaking, from Table 11.8 and Fig. 11.5, provinces varied little in terms of Resource Abundance Status and Ecological Conservation, the score of which ranged from −0.08 to 0.08. Only Qinghai, Inner Mongolia and Heilongjiang scored slightly higher.

The top ten provinces were Qinghai, Inner Mongolia, Heilongjiang, Yunnan, Hainan, Sichuan, Jilin, Shanghai, Jiangxi and Guangxi, five including the top two

Table 11.8 Indexes of resource abundance and ecological conservation indicators and rankings of 30 provinces

Indicator	Resource abundance and ecological conservation		Indicator	Resource abundance and ecological conservation	
Province	Score	Ranking	Province	Score	Ranking
Qinghai	0.143	1	Liaoning	−0.019	16
Inner Mongolia	0.130	2	Chongqing	−0.021	17
Heilongjiang	0.121	3	Gansu	−0.022	18
Yunnan	0.078	4	Shaanxi	−0.026	19
Hainan	0.060	5	Guizhou	−0.026	20
Sichuan	0.052	6	Hubei	−0.038	21
Jilin	0.050	7	Tianjin	−0.046	22
Shanghai	0.040	8	Anhui	−0.058	23
Jiangxi	0.033	9	Shandong	−0.062	24
Guangxi	0.027	10	Jiangsu	−0.062	25
Fujian	0.023	11	Beijing	−0.062	26
Xinjiang	0.011	12	Ningxia	−0.071	27
Hunan	−0.005	13	Shanxi	−0.074	28
Zhejiang	−0.007	14	Hebei	−0.077	29
Guangdong	−0.013	15	Henan	−0.079	30

Sources China Statistical Yearbook 2010, *China Statistical Yearbook* 2009, *Desertification and Its Control in China, Annual Statistical Report on Environment in China* 2009, and *China Environmental Statistical Yearbook* 2010

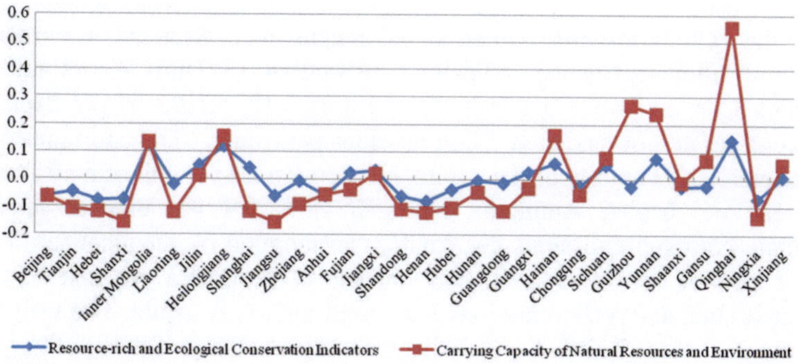

Fig. 11.5 Inter-provincial comparison of resource abundance and ecological conservation index and ICPNRE

from West China, two from Northeast China, one from Central China and two from East China.

From a regional perspective, among the No. 11–No. 20 provinces, half were from West China and ranked behind, including No. 12 Xinjiang, No. 17 Chongqing, No. 18 Gansu, No. 19 Shanxi and No. 20 Guizhou; outside the region were three eastern provinces, one central province and one northeastern province.

Among the bottom ten provinces, five were from East China, including No. 22 Tianjin, No. 24 Shandong, No. 25 Jiangsu, No.26 Beijing and No. 29 Hebei; four were from Central China, and one were from West China.

Among the calculated provinces, the No.1 province was Qinghai whose score was much higher than that of the rest. It ranked first in terms of two-Third-Level Indicators: the water resources per capita and proportion of area of natural reserves in its total area. Its forest area per capita ranked second, but the forest coverage ranked last but one. The No. 1 province in terms of forest area per capita and total standing stock volume per capita was Inner Mongolia, and this was probably due to major policy decisions like the Three North Shelter Forest Engineering Construction and Natural Forest Protection Plan.

As to eastern provinces such as Beijing, Shanghai and Jiangsu, except a relatively high ranking in terms of proportion of area of natural reserves in the total area, they ranked low in terms of water resources per capita, forest area per capita and forest coverage. Four out of the six central provinces ranked among the bottom ten. These data reflected a huge stress on resources confronted by the eastern and central regions. In addition, the only one western province in the bottom ten was Ningxia which ranked No. 27 in terms of Resource Abundance Status and Ecological Conservation, far behind the No. 1 Qinghai.

11.2.2 Measurement Results and Analysis of Environmental Stress and Climate Change Indicator on the Provincial Level

Under the "China Province Green Development Index System", Environmental Stress and Climate Change Indicator consisted of 13 Third -Level Indicators, including CO_2 emissions per unit of land area, CO_2 emissions per capita, SO_2 emissions per unit of land area, SO_2 emissions per capita, COD emissions per unit of land area, COD emissions per capita, ***NOx*** emissions per unit of land area, ***NOx*** emissions per capita, Ammonia Nitrogen emissions per unit of land area, Ammonia Nitrogen emissions per capita, consumption of chemical fertilizer per unit of area of cultivated land, consumption of pesticides per unit of area of cultivated land, and ***NOx*** emissions from road traffic per capita. The weights and indications of these Third -Level Indicators were shown in Table 11.9.

Basing on data normalization and weights of Third-Level Indicators, we worked out and ranked the calculated results of Environmental Stress and Climate Change Indicator of different provinces (autonomous regions and municipalities directly under the jurisdiction of the central government) (See Table 11.10).

From Table 11.10 and Fig. 11.6, the scores of Environmental Stress and Climate Change Indicator varied slightly among 26 provinces, ranging from −0.1 to 0.1 (including Liaoning and Guangdong). However, Qinghai, Guizhou and Yunnan scored much higher than the others. Shanghai scored rather low. On the whole, the range of the 30 provinces was 3.5 times higher than the standard deviation, indicating a significant disparity among provinces.

Table 11.9 Third-class indicators, their weights and attributes of inter-provincial environmental stress and climate change

Sequence number	Indicator	Weight (%)	Attribute
1	CO_2 emissions per unit of land area	2.45	Negatively correlated
2	CO_2 emissions per capita	2.45	Negatively correlated
3	SO_2 emissions per unit of land area	2.10	Negatively correlated
4	SO_2 emissions per capita	2.10	Negatively correlated
5	COD emissions per unit of land area	2.10	Negatively correlated
6	COD emissions per capita	2.10	Negatively correlated
7	Nitrogen oxides emissions per unit of land area	2.10	Negatively correlated
8	Nitrogen oxides emissions per capita	2.10	Negatively correlated
9	Ammonia nitrogen emissions per unit of land area	2.10	Negatively correlated
10	Ammonia nitrogen emissions per capita	2.10	Negatively correlated
11	Consumption of chemical fertilizers per unit of area of cultivated land	2.10	Negatively correlated
12	Consumption of pesticides per unit of area of cultivated land	2.10	Negatively correlated
13	Nitrogen oxides emissions from road traffic per capita	2.10	Negatively correlated

Note The contents in this table are determined by the task force after having conducted a lot of expert workshops

From a national perspective, the western region scored much higher than the other three regions, and no huge disparity existed inside the three.

The top ten provinces were Qinghai, Guizhou, Yunnan, Hainan, Gansu, Xinjiang, Heilongjiang, Sichuan, Shanxi and Inner Mongolia. From a regional perspective, eight of top ten were western provinces: No. 1 Qinghai, No. 2 Guizhou, No. 3 Yunnan, No. 5 Gansu, No. 6 Xinjiang, No. 8 Sichuan, No. 9 Shanxi and No. 10 Inner Mongolia. The other two were No. 4 Hainan from East China and No. 7 Heilongjiang from Northeast China.

The provinces ranking from No. 11 to No. 20 were Anhui, Beijing, Jiangxi, Chongqing, Jilin, Hebei, Hunan, Henan, Shandong and Guangxi, all of which except Anhui were below the national average. There was a slight disparity. From a regional perspective, there were four central provinces: No. 11 Anhui, No. 13 Jiangxi, No. 17 Hunan and No.18 Henan; two western provinces were No. 14 Chongqing and No. 20 Guangxi; three eastern provinces were No. 12 Beijing, No. 16 Hebei and No. 19 Shandong. No. 15 Jilin was the only northeastern province.

Table 11.10 Indexes of environmental stress and climate change indicators and rankings of 30 provinces

Indicator	Environmental stress and climate change indicators		Indicator	Environmental stress and climate change indicators	
Province	Score	Ranking	Province	Score	Ranking
Qinghai	0.413	1	Hebei	−0.041	16
Guizhou	0.295	2	Hunan	−0.042	17
Yunnan	0.163	3	Henan	−0.042	18
Hainan	0.100	4	Shandong	−0.047	19
Gansu	0.092	5	Guangxi	−0.058	20
Xinjiang	0.045	6	Fujian	−0.059	21
Heilongjiang	0.035	7	Tianjin	−0.059	22
Sichuan	0.029	8	Hubei	−0.065	23
Shaanxi	0.016	9	Ningxia	−0.066	24
Inner Mongolia	0.006	10	Shanxi	−0.084	25
Anhui	0.004	11	Zhejiang	−0.085	26
Beijing	−0.001	12	Jiangsu	−0.097	27
Jiangxi	−0.014	13	Liaoning	−0.103	28
Chongqing	−0.033	14	Guangdong	−0.104	29
Jilin	−0.039	15	Shanghai	−0.157	30

Sources China Statistical Yearbook 2010, *China Statistical Yearbook* 2009, *Desertification and Its Control in China, Annual Statistical Report on Environment in China* 2009, and *China Environmental Statistical Yearbook* 2010

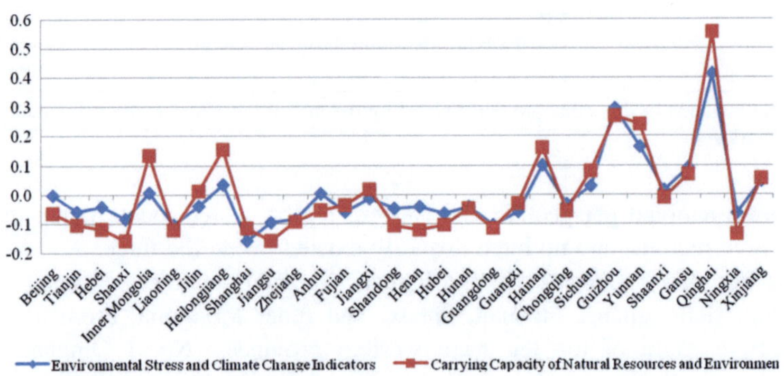

Fig. 11.6 Inter-provincial comparisons of environmental stress and climate change index and ICPNRE

The bottom ten provinces were Fujian, Tianjin, Hubei, Ningxia, Shanxi, Zhejiang, Jiangsu, Liaoning, Guangdong and Shanghai. From a regional perspective, there were six eastern provinces: No. 21 Fujian, No. 22 Tianjin, No. 26 Zhejiang, No. 27 Jiangsu, No. 29 Guangdong and No. 30 Shanghai; two central provinces were No. 23 Hubei and No. 25 Shanxi. No. 24 Ningxia and No. 28 Liaoning were the only western and northeastern provinces.

The top three provinces were Qinghai, Guizhou and Yunnan. Their high ranking was mainly contributed by low emissions of SO_2, COD, *NOx and* Ammonia Nitrogen per capita and per unit of land area.

Hainan ranked No. 4 as a result of its dominating service industry. In 2009, service accounted for 45.3 % of the local GDP, compared with 27.9 % contributed by secondary industry. In this regard, Hainan became the smallest emitter of SO_2 and *NOx* per capita across the country. Moreover, the total emission of pollutants of Hainan was lower than that of most provinces.

The bottom two provinces were Guangdong and Shanghai, as they were large emitters of pollutants per unit of land area. Shanghai was the largest emitter of SO_2, COD, *NOx and* Ammonia Nitrogen per unit of land area. Also one of the largest emitters in this regard, Guangdong was the second and third largest consumer of chemical fertilizer and pesticides per unit of area of cultivated land in China.

11.3 Comparative Analysis of ICPNRE on the City Level

Basing on "China City Green Development Index System", we analyzed the calculated results of Resource Abundance Status and Ecological Conservation and Environmental Stress and Climate Change Indicators of 34 provincial capital cities (including municipalities directly under the jurisdiction of the central government and specifically designated cities in the state plan), so as to fully compare their different levels of ICPNRE.

11.3.1 Measurement Results and Analysis of Resource Abundance Status and Ecological Conservation Indicator on the City Level

In the "China Province Green Development Index System", compared to Environmental Stress and Climate Change Indicator, Resource Abundance Status and Ecological Conservation Indicator was a smaller factor, whose weight was only 5 %. Below the indicator, there was only one-Third-Level positively correlated indicator-water resources per capita, whose weight in GDI was only 1.6 %.

Basing on data normalization and weights of Third-Level Indicators, we worked out and ranked the calculated results of Resource Abundance Status and Ecological Conservation Indicator of the 34 cities (See Table 11.11).

Table 11.11 showed that 33 cities varied from −0.009 to 0.013, a small disparity. Only Lanzhou was scored relatively higher (Fig. 11.7).

From a national perspective, 11 of the 34 calculated cities scored above the national average. From a regional perspective, the scores of the four regions varied slightly, the western region having the highest and the eastern region the lowest.

Table 11.11 Indexes of resource abundance and ecological conservation indicators and rankings of 34 cities

Indicator	Resource abundance and ecological conservation indicators		Indicator	Resource abundance and ecological conservation indicators	
City	Score	Ranking	City	Score	Ranking
Lanzhou	0.091	1	Fuzhou	−0.005	18
Hangzhou	0.013	2	Shenyang	−0.006	19
Nanning	0.010	3	Nanjing	−0.006	20
Chongqing	0.008	4	Changchun	−0.006	21
Kunming	0.003	5	Hefei	−0.006	22
Guiyang	0.003	6	Dalian	−0.006	23
Changsha	0.003	7	Qingdao	−0.006	24
Ningbo	0.002	8	Xi'an	−0.007	25
Shenzhen	0.002	9	Jinan	−0.007	26
Guangzhou	0.002	10	Beijing	−0.007	27
Nanchang	0.001	11	Shijiazhuang	−0.007	28
Harbin	−0.001	12	Shanghai	−0.008	29
Xiamen	−0.002	13	Tianjin	−0.008	30
Xining	−0.004	14	Zhengzhou	−0.008	31
Chengdu	−0.004	15	Taiyuan	−0.008	32
Hohhot	−0.004	16	Haikou	−0.009	33
Wuhan	−0.005	17	Yinchuan	−0.009	34

Sources China City Statistical Yearbook 2008, *China City Statistical Yearbook* 2009, *China Urban Life and Price Statistical Yearbook* 2010, *Annual Statistical Report on Environment in China* 2008, *Annual Statistical Report on Environment in China* 2009, and Data Center Network of the Ministry of Environmental Protection

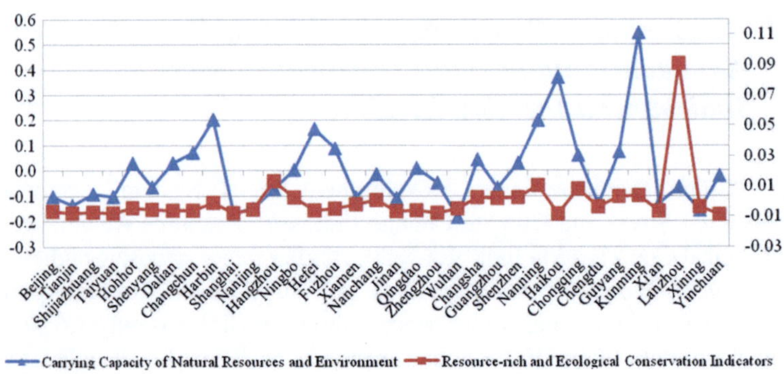

Fig. 11.7 Inter-city comparison of resource abundance and ecological conservation index and ICPNRE

The top ten cities were Lanzhou, Hangzhou, Nanning, Chongqing, Kunming, Guiyang, Changsha, Ningbo, Shenzhen and Guangzhou. Among them, five were western cities among the top six, one was from the central region and four from the western region.

Table 11.12 Difference in water resources per capita between Lanzhou and other cities (Unit: m^3/person)

Area	Water resources per capita	Difference between other cities and Lanzhou
Eastern cities	646.34	−9543.25
Central cities	577.93	−9611.65
Western cities	1865.07	−8324.52
Northeastern	489.34	−9700.25

Lanzhou ranked first with a leading advantage. The reason lied in its large amount of water resources as much as 10189.59 cubic meters per person. From table 11.12, the total water resource of Lanzhou was much higher than the average of the other three regions, which was less than 1/10 of Lanzhou's.

The bottom ten cities were Xian, Jinan, Beijing, Shijiazhuang, Shanghai, Tianjin, Zhengzhou, Taiyuan, Haikou and Yinchuan. Six of them were from the eastern region, taking up half of the calculated eastern cities. This meant the eastern cities behaved poorly as a whole. There were also two central cities and two western cities. Yinchuan lied at the end.

Four northeastern cities ranked in the middle and they were Harbin (No. 12), Shenyang (No. 19), Changchun (No. 21) and Dalian (No. 23). But they scored below the national average, indicating low storage of water resources in this region.

11.3.2 Measurement Results and Analysis of Environmental Stress and Climate Change Indicator on the City Level

In the "China Province Green Development Index System", Environmental Stress and Climate Change Indicator consisted of twelve Third-Level Indicators, including CO_2 emissions per unit of land area, CO_2 emissions per capita, SO_2 emissions per unit of land area, SO_2 emissions per capita, COD emissions per unit of land area, COD emissions per capita, *NOx* emissions per unit of land area, *NOx* emissions per capita, Ammonia Nitrogen emissions per unit of land area, Ammonia Nitrogen emissions per capita, ratio of days with air quality at and above level two to the whole year, ratio of days with principal pollutants as respirable particles to the whole year. The weights and indications of these indicators were shown in Table 11.13.

Basing on data normalization and weights of the Third-Level Indicators, we worked out and ranked the calculated results of Environmental Stress and Climate Change Indicator of the 34 cities (See Table 11.14).

Table 11.13 Third-class indicators, their weights and attributes of inter-city environmental stress and climate change

Sequence number	Indicator	Weight (%)	Attribute
1	CO_2 emissions per unit of land area	2.81	Negatively correlated
2	CO2 emissions per capita	2.81	Negatively correlated
3	SO2 emissions per unit of land area	2.67	Negatively correlated
4	SO2 emissions per capita	2.67	Negatively correlated
5	COD emissions per unit of land area	2.67	Negatively correlated
6	COD emissions per capita	2.67	Negatively correlated
7	Nitrogen oxides emissions per unit of land area	2.67	Negatively correlated
8	Nitrogen oxides emissions per capita	2.67	Negatively correlated
9	Ammonia nitrogen emissions per unit of land area	2.67	Negatively correlated
10	Ammonia nitrogen emissions per capita	2.67	Negatively correlated
11	Ratio of days with air quality at and above level 2 to the whole year	2.67	Positively correlated
12	Ratio of days with principal pollutants as respirable suspended particulate to the whole year	2.67	Negatively correlated

Note The contents in this table are determined by the task force after having conducted a lot of expert workshops

From Table 11.14 and Fig. 11.8, the scores of the 34 cities ranged from −0.178 to 0.543, a significant disparity. Some cities such as Kunming and Haikou scored much higher over others. Besides, since the weight of Environmental Stress and Climate Change Indicator was 95 %, the exponential curve of the indicator resembled that of ICPNRE. In other words, compared to Resource Abundance Status and Ecological Conservation Indicator, this indicator made a larger contribution to ICPNRE.

From a national perspective, northeastern and western cities scored much higher than other parts of China. The lowest were eastern cities, which approximated the scores of central cities.

The top ten cities were Kunming, Haikou, Harbin, Nanning, Hefei, Fuzhou, Changchun, Guiyang, Chongqing and Changsha. From a regional perspective, they covered all the regions. Among them were four western cities including No. 1 Kunming, No. 4 Nanning, No. 8 Guiyang and No. 9 Chongqing, two eastern cities including No. 2 Haikou and No. 6 Fuzhou, and two northeastern cities including

Table 11.14 Indexes of environmental stress and climate change indicators and rankings of 34 cities

Indicator	Environmental stress and climate change Indicators		Indicator	Environmental stress and climate change Indicators	
City	Score	Ranking	City	Score	Ranking
Kunming	0.543	1	Zhengzhou	−0.039	18
Haikou	0.377	2	Shenyang	−0.059	19
Harbin	0.204	3	Guangzhou	−0.070	20
Nanning	0.190	4	Shijiazhuang	−0.084	21
Hefei	0.173	5	Hangzhou	−0.084	22
Fuzhou	0.092	6	Taiyuan	−0.093	23
Changchun	0.078	7	Beijing	−0.094	24
Guiyang	0.071	8	Jinan	−0.103	25
Chongqing	0.052	9	Xiamen	−0.105	26
Changsha	0.040	10	Xi'an	−0.124	27
Dalian	0.036	11	Tianjin	−0.129	28
Hohhot	0.034	12	Chengdu	−0.131	29
Shenzhen	0.027	13	Nanjing	−0.146	30
Qingdao	0.017	14	Xining	−0.155	31
Ningbo	0.002	15	Lanzhou	−0.158	32
Yinchuan	−0.011	16	Shanghai	−0.158	33
Nanchang	−0.015	17	Wuhan	−0.178	34

Sources China City Statistical Yearbook 2008, *China City Statistical Yearbook* 2009, *China Urban Life and Price Statistical Yearbook* 2010, *Annual Statistical Report on Environment in China* 2008, *Annual Statistical Report on Environment in China* 2009, and Data Center Network of the Ministry of Environmental Protection

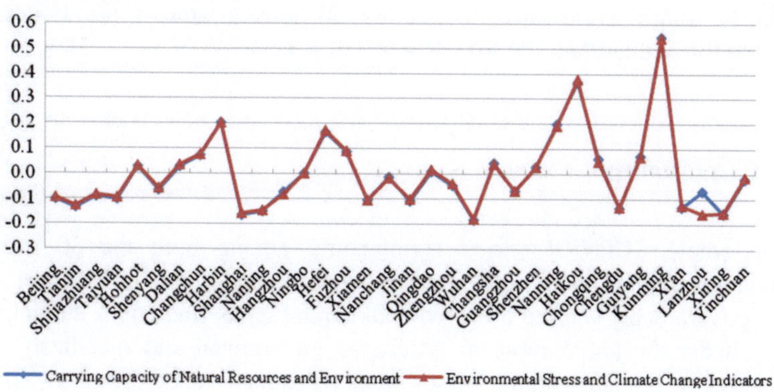

Fig. 11.8 Inter-city comparisons of environmental stress and climate change index and ICPNRE

No. 3 Harbin and No. 7 Changchun. The other two were central cities: No. 5 Hefei and No. 10 Changsha.

The cities ranking from No. 11 to No. 24 were Dalian, Hohhot, Shenzhen, Qingdao, Ningbo, Yinchuan, Nanchang, Zhengzhou, Shenyang, Guangzhou, Shijiazhuang, Hangzhou, Taiyuan and Beijing. Dalian, Hohhot, Shenzhen, Qingdao, and Ningbo scored higher than the national average. From a regional perspective, there were three central and two western cities, and seven eastern and two northeastern ones.

The bottom ten provinces were Jinan, Xiamen, Xian, Tianjin, Chengdu, Nanjing, Xining, Lanzhou, Shanghai and Wuhan, none of which was from the northeastern region. There were five eastern cities: No. 25 Jinan, No. 26 Xiamen, No. 28 Tianjin, No. 30 Nanjing and No. 33 Shanghai and four western cities: No. 27 Xian, No. 29 Chengdu, No. 31 Xining and No. 32 Lanzhou. The only central city Wuhan lied at the end of the list.

Viewed from the scores of the Third-Level Indicators, Kunming ranked first in terms of COD emissions per unit of land area and per capita, Ammonia Nitrogen emissions per unit of land area and per capita, ratio of days with air quality at and above level two to the whole year, which pushed the score of Environmental Stress and Climate Change Indicator to the first place. Other western cities like Nanning and Chongqing were low emitters of pollutants per square and per capita, mainly due to the low population in the vast region. However, Guiyang was a larger emitter in this regard, and the largest emitter of SO_2. In a word, the industrial structure varied greatly among western cities.

Tourism was the main economic driver of Hainan, a large province with a small population and a favorable natural environment. The province was the smallest emitter of SO_2 per unit of land area and per capita, and had the biggest number of days with air quality at and above level two in a year and the fewest days with principal pollutants as respirable particles in a year. That was why Hainan ranked second in terms of Environmental Stress and Climate Change Indicator.

The tail-end city of this indicator, Wuhan, except the emission of SO_2 per capita, the other Third-Level Indicators all ranked among the bottom ten. Undoubtedly, Wuhan had the lowest score in general.

11.4 Conclusion

On the whole, ICPNRE varied significantly among both the 30 provinces (autonomous regions and municipalities directly under the jurisdiction of the central government) and the 34 provincial capital cities (including municipalities directly under the jurisdiction of the central government and specifically designated cities in the state plan). West China was much better than East China in terms of ICPNRE and GDI. The disparity stemmed from the different economic development levels and natural endowments, but more importantly from different levels of Environmental Stress and Climate Change Indicator. Therefore, less consumption of resources and lower emission of pollutants were the right way to ease the stress on environment and improve ICPNRE.

Given the significant regional disparity of ICPNRE, we could reflect on our choices of the development pattern. We suggest all regions adjust measures to the local conditions, speed up economic restructuring and transforming the development pattern and strengthen cooperation among all sides in different fields to promote green development among themselves and the country. To be specific, underdeveloped regions should work out more innovative solutions towards green development and avoid the old track of low-efficient growth through high pollution and energy consumption. Developed regions should explore new ways in energy conservation and emission reduction and improving afforestation. Both West and East China should establish a compensation mechanism of environment and resources and a linkage mechanism of green development as soon as possible. The western region should vigorously promote sound and effective flow of resources and back the development of the eastern region. In return, the eastern region should help the western region develop by sharing experience, capital, technologies, talents, and market opportunities, etc.

Session III
Government Green Actions

Chapter 12
Government Actions in China's Green Development

Sanlin Jin, Jiancong Zhou and Fei Yang

During the 11th *Five-Year Plan* period, Chinese government paid great attention to green development and proposed ecological civilization for the first time, focusing on building a resource-conserving and environment-friendly society as a way to accelerate the transformation of the pattern of economic development. Over the past 5 years, governments at all levels continued to increase their investments in green development and acquired remarkable results, such as more improvements in infrastructure and environmental governance. During the 12th Five-Year Plan period, the Chinese government will promote a further integration of economic development with environmental protection in order to accelerate the pace of green development.

12.1 Green Economic Development and Government Function

12.1.1 The Economic Theory of Government Policies on Propelling Green Development

Green development is a development process characterized by conserving resources and protecting environment, in which "development" is the goal and "green" is both constraint and direction. The core is to correctly balance between development and environment. Environment is a public good with non-excludability and non-rivalry, whose property right is not easy to define. In the circumstance of no external restraints, manufacturers tend to excessively use the environment to make the private cost lower than the social cost to pursue extra profits at the expense of environmental pollution. Therefore, there are market failures in environmental protection which need to be remedied by government interventions in both micro and macro levels.

X. Li and J. Pan (eds.), *China Green Development Index Report 2011*,
Current Chinese Economic Report Series, DOI: 10.1007/978-3-642-31597-8_12,

12.1.1.1 Micro-Interventions Include Direct Administrative Regulation, Indirect Economic Instruments of Regulation, and the Emission Rights Trading Market

Direct administrative regulation is a compulsory emission standard set by the government. Enterprises must construct environmental protection facilities and use environmental protection technologies to ensure their products meeting the emission standard. Indirect economic instruments of regulation are mainly taxation and fees levied by the government for environmental pollution activities. When taxes exceed the marginal cost of controlling pollution emissions, enterprises will choose to eliminate pollution emissions to avoid taxes. The emission rights trading market is a market where the government issues tradable emission permits and allows them to be traded among enterprises of different pollution emissions levels.

Government should impose constraints and make the producers' external costs internalized, which is the core of micro-interventions. However, the restraints of the government are influenced by the government's efforts to increase regulation, its ability of law enforcement, the public sense in environmental protection, and the right of public acquisition for environmental protection. Generally speaking, because of information advantage, local governments play a key role in supervising environmental protection. If local governments' supervision is not strict, governmental failures may occur, and then the constraints on producers will be invalid.

12.1.1.2 Macro-Interventions Include Economic Policies that Affect Output Size, Industry Structure and Emission Efficiency

From the macro point of view, the lager the scale of economic activities is, the more resources will be consumed and the greater damage will be caused to the environment; the higher the proportion of low-pollution industries is and the better the industry structure is, the less damage will be caused to the environment. The higher the emission efficiency is, the less pollution will be produced per unit of output and the less pressure will be imposed on the environment.

Through fiscal, financial and industrial policies, the government can restrict the output scale of a particular industry or the economic scale in a specific area; it can restrict the development of high-pollution industries, encourage the development of low-pollution industries, and promote the optimization of the industrial structure; it can promote the adoption of clean technologies, clean management or environment-friendly products, reduce environmental depletion in unit inputs and outputs, and improve the discharge efficiency. The core of applying economic policies is to harmonize relationships among all relating parties' economic interests, to organically combine the goal of environmental protection and the behaviour of enterprises, and to prevent firms' economic activities from external diseconomies.

Generally speaking, a higher level of pollution in developing countries is mainly due to the higher implementation cost of environmental supervision and pollution standards, and the manufacturing industry's rising share in the process of industrialization. Therefore, the environmental protection in developing countries needs more governments' micro-regulation instruments combined with the use of macroeconomic policy.

12.1.2 Summary of Green Development Policies in Developed Countries

From global perspective, the understanding of development and environmental issues for international community has experienced a deepening process roughly through five stages, which can be summarized as follows. The first stage: taking the environmental pollution as a negative impact from economic growth. Polluters think that the environmental protection is an unnecessary measure taken by administrative authority fussily. Thus, enterprises have a resistance to the environmental regulation. At this stage, the measure of controlling pollution emission is the "End-of-Pipe Treatment", and may increase production cost from a general view. The second stage: taking the environmental pollution as one part of production costs. In this period, the polluters begin to realize that reducing pollution may be beneficial; therefore, they start to take measures to reduce the waste emissions in the production process. The third stage: taking the environment as one of the factors considered in decision-making process. When considering a new investment project, polluters must take environmental factors into account, and are forced to take measures to protect the environment in production and consumption process. The fourth stage: the environment is extremely important. When the polluters optimize the economic activities, they adopt different system designs, that is, to take measures to protect the environment systematically. The fifth stage: taking the environment as one of the goals of development. Environmental issues are considered as the goal of social and economic policies, which leads to the changes in production and consumption patterns, and changes people's attitudes to the environment.

At the same time, environmental protection policies are also changed from negative to positive policies. Many countries begin to take into account the environmental factors, economic factors and energy factors simultaneously, making long-term policies to develop the social economy and to protect the environment. Different countries have different situations and different concrete policies; however, the general trend is the same, which is to strengthen government supervision, and to emphasize more on introducing market incentives, public participation, and the role of justice. In addition, increasing government investment, encouraging social investment, and improving infrastructure condition are also important experiences of promoting green development in developed countries.

12.1.2.1 The Economic Regulation Instruments Adopted by Countries are Mainly Environmental Rules and Regulations, Environmental Taxes, Emission Rights Trading and Subsidies to Repeal or Application Mortgage, etc.

First, environmental rules and regulations, commonly used by various governments, mean that the governments set up relevant "standard", "ban", "license" and "quota" to control the pollution activities directly, which are common practice in all countries. Second, the environmental taxes (or fees on polluters) are important ways to internalize environmental costs and prevent "market failure". Currently, there are hundreds varieties of environmental taxes all over the world, such as taxes on petroleum and polluting products, etc. Third, the emission rights trading is another system designed to control the total amount of pollution emissions by using market mechanisms. Governments can set a limit on annual emissions or resource consumptions by issuing tradable emission permits, which can be traded in the emission rights trading market, and the price of which is determined by supply and demand. Fourth, the cancellation or the implementation of subsidies is to cancel the subsidies bad for environmental protection and implement subsidies on environment-friendly technologies. For example, subsidies on energy-saving products are very successful in the US. Fifth, application mortgage means that the government first collects guarantee deposits on the products that may cause pollution. It then returns the deposits when these products are brought back to the appointed place or treatment plant. Sixth, some countries try to accumulate funds for environmental protection through compensation, liability insurance and other methods.

12.1.2.2 In Developed Countries, Developing Circular Economy and Environment-Protection Industries and Promoting Enterprises' Environment-Protection Behaviors are also the Focuses of Government Actions

In the 1970s, enterprises began to explore new modes of production and vigorously developed the circular economy after OECD countries implemented the environmental protection policy "Polluters Pay". For example, since the late 1980s, Dupont has taken "Reduce", "Reuse" and "Recycle" as guiding principles to organize internal material recycling and created "3R Manufacturing Method" that greatly reduced waste emissions and costs. This has been supported and promoted by government with great efforts.

It is another important form of circular economy to build Eco-Industrial Parks, in which enterprises can connect to each other through waste exchange businesses. From the experiences in Canada, USA and other countries, we learn that comprehensive utilization of obsolete petroleum products and organics contribute to the development of ecological industry chain and Eco-Industrial Parks.

Many countries also advocate standard certifications of environmental management ISO14000. Lots of enterprises actively apply for the ISO14000 certification. Once approved, the enterprise will get high social reputation, and the competitiveness of their products can also be strengthened in the market.

All nations attach great importance to developing environmental protection industries. For example, in the USA and Canada, market with diversified investors plays an important role in environmental protection. The environmental industry has become an important rising industry.

12.1.2.3 Developed Countries have Established a Complete Legal System and Independent Monitoring System to Guarantee Policy Effects and Advocated the Public Participation

In countries like US and Canada, etc., governments at all levels have built a complete legal system in environmental protection, in which the laws can not only be applied to enterprises, but also used to guarantee the independence of monitoring institutions and the right of public acquisition for environmental protection. Take "Environmental Protection Law" in Canada as an example, it says that the heads in environmental protection agencies are nominated by government but not controlled by government. He has independent power of law enforcement. Moreover, there is a special law *Ontario Environmental Bill of Rights* to protect the right of public acquisition for environmental protection.

12.2 Government Actions in Green Development During the 11th Five-Year Plan Period: Policies, Achievements and Problems

12.2.1 Major Actions in Promoting Green Development During the 11th Five-Year Plan Period

During the *11th Five-Year Plan* period, Chinese government took various kinds of measures to promote green development due to the serious eco-environment degradation.

12.2.1.1 Further Improve Laws and Regulations

By the end of the *11th Five-Year Plan* period, China had formed a complete legal system, which was based on *Constitution of the People's Republic of China* and took *Environmental Protection Law of the People's Republic of China* as the main body. It met the requirements of constructing ecological civilization and played an important role in promoting green development. During the *11th Five-Year Plan* period, Chinese government further completed relevant laws and regulations, which provided a better legal basis for building the resource-conserving and environment-friendly society.

First, Chinese government promulgated and amended a number of laws targeting the new situation. The government promulgated *Circular Economy Promotion Law of the People's Republic of China* in 2008, which clearly stated the principle of "reduction, reuse and recycling" and made a circular economy step into the stage of legal system. In 2007, the government promulgated *Urban and Rural Planning Act of the People's Republic of China*, and amended *Energy Conservation Law*, *Law of the People's Republic of China on the Prevention and Control of Water Pollution* and *Renewable Energy Law* to meet the new requirements in the green development.

Second, the government promulgated many rules and regulations, which mainly included *Regulations Concerning the Prevention and Cure of Pollution Damage of Marine Environment by Seashore*, *Regulation on National General Survey of Pollution Sources*, *Energy Saving Regulations for Civil Constructions*, *Planning Environmental Impact Assessment Regulations*, *Regulation on the Defense against Meteorological Disasters*, *Regulations on Administration of Collection and Disposal of Waste Electronic Products*, *Regulations on Energy Sufficiency of Public Institutions*, *Hydrology Regulation of the People's Republic of China*, etc.

Third, relevant authorities also promulgated a number of administrative regulations to normalize industry management. State Environmental Protection Administration promulgated *Standard for Sector-specific Eco-industrial Parks* (*on trial*) (2006), *Measures for the Supervision and Inspection of National Nature Reserves* (2006), *Measures for Supervision of Prevention and Control on Pollution Environment from Electron Wastes* (2007), *Environmental Monitoring Management Measures* (2007); Ministry of Commerce and other five ministries promulgated *Administrative Measures for the Recovery of Renewable Resources* (2007); Ministry of Housing and Urban–Rural Development of the People's Republic of China promulgated *Green Building Evaluation Mark Management Approach* (2007); National Development and Reform Commission promulgated *Energy Saving Emission Reduction Plan* (2007); State-owned Asset Supervision and Administration Commission promulgated *Interim Measures for the Supervision and Management Concerning Energy Conservation and Emission Reduction in Central Enterprises* (2007).

12.2.1.2 Increase Investments in the Treatment of Pollution, Especially in Some Key Areas

Green development highly depends on investment in environmental protection. During the 11th *Five-Year Plan* period, Chinese government took a series of measures to further broaden financing sources in environmental protection, which promoted investments in the treatment of pollution, especially in some key areas.

Table 12.1 Investments in the treatment of pollution

Indicator	Unit	2005	2006	2007	2008	2009
The total funds in the treatment of pollution	100 million	2,388.0	2,566.0	3,387.3	4,490.3	4,525.3
Investment in urban environmental infrastructure	100 million	1,289.7	1,314.9	1,467.5	1,801.0	2,512.0
Investment in the treatment of industrial pollution sources	100 million	458.2	483.9	552.4	542.6	442.6
Investment in "three simultaneous" project	100 million	640.1	767.2	1,367.4	2,146.7	1,570.7
Investment in pollution treatment to GDP	%	1.30	1.22	1.36	1.49	1.33

Note Investments in "Three Simultaneous" project are actual investment funds, and "Three Simultaneous" means that the facilities for pollution prevention and the treatment of pollution must be designed, constructed, and put into operation at the same time with the main project
Source National Bureau of Statistics of China (2011)

As Table 12.1 shows, in 2009 the total amount spent in the treatment of pollution was 452.5 billion, which increased by 89.5 % from 2005, and its share in GDP changed from 1.30 % in 2005 to 1.33 % in 2009. Among the total funds, investments in urban environmental infrastructure were 251.2 billion, which increased by 94.8 % from 2005; also investments in urban gas, central heating, drainage, landscaping and environmental sanitation increased by 27.9, 67.4, 98.3, 122.4 and 114.1 % respectively; actual investments in "Three Simultaneous" were 157.07 billion, which increased by 150 % from 2005.

12.2.1.3 Strengthen Law Enforcement and Government Regulation

During the 11th *Five-Year Plan* period, governments at all levels strengthened law enforcement and government regulation.

First, they took measures of "Regional Restrictions" and "Industry Restrictions" which effectively prevented environmental violations. Environmental protection departments made 813 illegal items, which were worth of over 2.9 trillion RMB, inadmissible, disapproved or suspended. They also set insurmountable "firewall" to items of "heavy energy consumption, heavy pollution and resource-related", low-level repeating building and overcapacity. Besides, they organized assessments on the key industries in energy chemical areas in Bohai Rim, the west coast of the strait, Beibu Gulf, chengdu-chongqing zone and the middle or upper reaches of the Yellow River.

Second, they strengthened pollution prevention efforts in key areas. Environmental protection departments and other departments took effective measures to further promote rehabilitation of rivers and lakes and comprehensively established water quality assessment system in key valleys of provincial sections. They also carried out vigorous supervisions on regional environmental law enforcement, and put great efforts to build a regional mechanism of joint prevention and control for air pollution, which included "unified planning, unified monitoring, unified supervising, unified assessing, unified coordinating," to achieve the air quality goals for the Shanghai World Expo and Guangzhou Asian Games.

Third, they were focusing on serious environmental issues that harmed people's health. Besides, they carried out environmental investigations on the source of drinking water in cities and towns above county level, drew up and published the planning for protecting source of drinking water in cities, and strengthened the preventions and inspections for heavy metal pollution. Since 2006, the total number of law-enforcement officers was over 11 million. More than 140,000 environmentally illegal enterprises were investigated and more than 2,000 enterprises that discharged sewage illegally was shut down, which were good examples of punishment on the environmental violations.

12.2.1.4 Improve Incentive and Restraint Policies

During the 11th *Five-Year Plan* period, the government introduced a series of incentive and restraint policies on industry, finance, tax, banking, consuming, etc., which gradually formed a policy system in favor of resource conservation and environmental protection.

In terms of industry policies, the government strengthened regulation and control on high energy-consuming and high pollution industries, like iron and steel, coal, electricity, chemicals, etc. Ministry of Land and Resources and National Development and Reform Commission jointly enacted *Limited Project Directory with Land* and *Banned Project Directory with Land*, which restrict or prohibit land use in those projects that are not conductive to resource conservation and environmental protection. National Development and Reform Commission issued official *Notice on Advocating the Restructuring in Electric Power Industry for Healthy Development*, *Notice on Shutting down Small Thermal Power Plants*, etc. to supervise energy saving and emission reduction in electric power enterprises.

In terms of tax policies, the government took incentive tax instruments to promote energy saving and emission reduction. The new Law on Corporate Income Tax, which was passed in 2007, clearly defined that the income tax of environment-friendly activities could be reduced or exempted. In 2006, the Ministry of Finance and the State Environmental Protection Administration issued official *Notice on Adjustments and Improvements in Consumption Tax Policy*, which added some new taxation items such as disposable wooden chopsticks and wooden floor. It also reduced tax on some low-emission passenger cars, and increased tax on large-emission passenger cars.

In terms of credit policy, the government guided enterprises to increase green investments through financial means. In 2007, the People's Bank of China announced *General Guidance of Improving Financial Services in Energy Saving and Environmental Protection Field*, which proposed that we should make more progress in providing financial services targeting energy-saving fields. In July of 2007, the State Environmental Protection Administration, People's Bank of China and China Banking Regulatory Commission joined together for the first time to announce *Guidance of Carrying Out Environmental Policies and Regulations and Avoiding Credit Risk*, which stated that environmental protection departments at all levels should investigate and punish behaviors of construction prior to approval or leapfrog behavior. During the 11th *Five-Year Plan* period, over 40,000 environmental illegal issues and over 7,000 projects information concerning environment assessments was entered into the banks' credit management system. And, more than 10 insurance companies introduced new insurance products with environmental pollution liability.

12.2.1.5 Implement Important Special Projects

The 11th *Five-Year Plan* on environmental protection proposed ten environmental protection projects, including project of environment monitoring capacity, project of hazardous waste and medical waste disposal, project of chromium slag pollution control, project of urban sewage treatment, project of water pollution prevention in key watersheds, project of urban waste disposal, project of gas desulfurization of sintering machines in coal-fired power plants and steel industry, project of important ecological functional region and nature reserve construction, project of nuclear and radiation safety, and project of environmental protection in rural areas. Other ministries and commissions also brought forward "connect every village" project, the rural drinking-water safety project and other important projects, which aimed to improve people's wellbeing through green actions.

Among these projects, the rural drinking-water safety project, which was implemented jointly by the National Development and Reform Commission, the Ministry of Water Resources and the Ministry of Health, was a key project during the 11th *Five-Year Plan* period. It aimed to solve the problem of drinking-water safety for 160 million rural residents (about 150,000 administrative villages), reduce half of the populations who had no safe drinking-water, increase the percentage of people who benefit from centralized water supply to 55 %, and improve water quality during the 11th *Five-Year Plan* period. To better implement the project, related ministries and local governments made many innovative mechanisms according to local circumstances to expedite the settlement of rural drinking-water safety problem, such as centralized water supply, decentralized water supply, urban water supply networks extending to the rural areas, etc. During the 5 years, central government spending on this project totaled 59 billion (5 times as much as the investments during the 10th Five-Year Plan period), local governments and private spending on it was 44.3 billion, and non-government funds were 1.22 billion, so the total funds spending on this project were 104 billion. 210

million rural people got safe drinking-water during the 5 years, which exceeded the target proposed by the 11th *Five-Year Plan* to ensure that 160 million have safe drinking-water.

12.2.1.6 Enhance International Cooperation

It needs international cooperation to protect environment and develop green economy, as there is no boundary for pollution. During the 11th *Five-Year Plan* period, the Chinese government actively took measures to enhance international cooperation to strengthen global environmental governance.

First, the Chinese government actively undertook international obligations and fulfilled environmental-protection commitments. In 2007, the Chinese government formulated *National Program on Response to Climate Change*, which stated clearly the specific objectives, basic principles, focus areas and policies for response to climate change till 2010. In September of 2009, Hu Jintao attended the United Nations Summit on Climate Change and delivered an important speech on "Join Hands to Address Climate Challenge". Wen Jiabao promised in the Copenhagen Climate Summit that carbon dioxide emissions per unit of GDP would decrease by 40–50 % in 2020.

Second, the Chinese government actively participated in international organizations and promoted global environmental cooperation. China's related authorities communicated and cooperated with more than ten international (global) inter-governmental organizations, like UNEP, GEF, and almost ten international non-government organizations, such as IU-CN, WWF and so on.

Third, the Chinese government vigorously promoted bilateral environmental cooperation between countries. Environmental cooperation between China and the United States was raised to the strategic level; China and Russia set up a mechanism to improve environmental cooperation. Through carrying out cooperation projects with Sweden, World Bank, Asian Development Bank and so on, China gradually established a legal system to assess project program's environmental impacts, and made positive achievements in practice.

12.2.2 The Major Achievements in Government Green Actions During the 11th Five-Year Plan Period

During the 11th *Five-Year Plan* period, China's green development took a new step, in which environmental protection was constantly intensified, the investment of pollution control kept relatively fast growth, infrastructure projects increased, the release of major pollutants was under control, and environmental pollution control made a progressive achievement.

Table 12.2 Major industrial pollutant emissions (2005–2009) (Unit: 10,000 tons)

Index	2005	2006	2007	2008	2009
Volume of SO_2 emission by industry	2,168.4	2,234.8	2,140	1,991.4	1,865.9
COD discharge from industrial waste water	554.7	541.5	511.1	457.58	439.68
Ammonia Nitrogen discharge from industrial waste water	52.5	42.5	34.1	29.69	27.35

Sources National Bureau of Statistics of China (2006–2010)

12.2.2.1 The Effectiveness of Environmental Governance was Enhanced Significantly

First, the task of reducing pollution and emission was over-fulfilled. In the circumstance of economic growth and energy consumption exceeding the schedule, the target of reducing SO_2 emission was reached a year ahead of schedule, and the goal of reducing chemical oxygen demand was achieved half a year ahead of the schedule, so the task of reducing pollution emission was over-fulfilled. According to an initial calculation, in 2010 the emission of chemical oxygen demand in China decreased by 12 % comparing to that in 2005, while the emission of SO_2 decreased by about 14 %, which both exceeded the emission target.

Second, the major industrial pollutant emissions were greatly reduced. Table 12.2 shows, in 2009 the SO_2 emissions were 1,865.9 million tons, which decreased by 302.5 million tons comparing to 2005, with a decrease of 13.95 % and an average annual rate of 3.68 %. There was a sharp decrease of chemical oxygen demand discharge from industrial waste water from 2005 to 2009. Chemical oxygen demand discharge from industrial waste water in 2009 was 439.68 million tons, which decreased by 115.02 million tons comparing with 2005, with a decrease of 20.74 % and an average annual rate of 5.64 %. Ammonia nitrogen discharge from industrial waste water also decreased year by year, which was 27.35 million tons in 2009, decreased by 25.15 million tons comparing with 2005, with a decrease of 47.91 % and an average annual rate of 15.04 %.

Third, there was a whole descending in the number of environmental accidents. As Fig. 12.1 shows, in 2009 the number of environmental accidents dropped to 418 times, reduced 988 accidents comparing with 2005, with a decrease of 70.27 % and an average annual rate of 26.16 %.

Fourth, the treatment efficiency of industrial "three wastes" was improved year by year. During the 11th *Five-Year Plan* period, the proportion of industry waste water meeting discharge standards, proportion of industry SO_2 meeting discharge standards, proportion of industry soot meeting discharge standards, proportion of industry dust meeting discharge standards, and ratio of industrial solid wastes utilized all showed an increasing trend year by year.

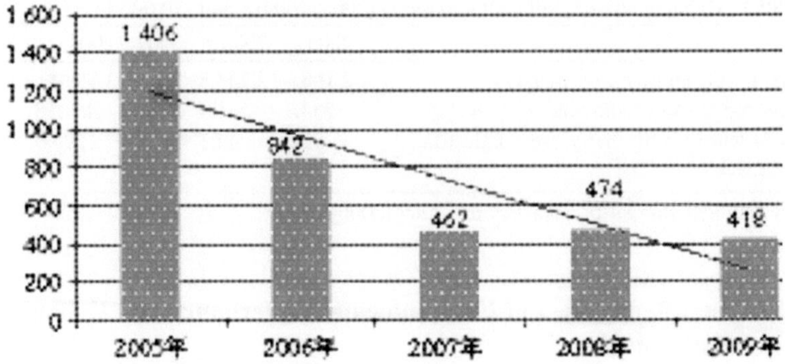

Fig. 12.1 Number of environmental accidents, 2005–2009. *Source* National Bureau of Statistics of China (2006–2010)

12.2.2.2 The Support Ability of Infrastructure to Green Development was Further Enhanced

First, urban green investment in fixed assets had a steady growth. Table 12.3 shows, from 2005 to 2009, the urban funds for maintaining construction in China increased by 12.3 %, with an average annual growth rate of 12.1 %, and the investment in fixed assets of municipal utilities and facilities increased from 560.22 to 1,064.15 billion Yuan, almost twice as much as before, with an average annual growth rate of 19.53 %. Among them, the investment in urban environmental infrastructure was 251.2 billion Yuan, increased by 94.8 % comparing with 2005, and the growth rate of investments in city gas, central heating, drainage, landscaping and city appearance and environmental sanitation were 27.9, 67.4, 98.3, 122.4 and 114.1 % respectively.

Second, urban water supply and wastewater treatment capacity were greatly improved. Table 12.4 shows by the end of 2009 urban water access rate in China reached 96.12 %, a 5.5 % increase over 2005, so the urban water supply capacity was improved significantly. At the same time, the treatment capacity per day of waste water treatment plants reached 90.52 million m^3, a 58.1 % increase over 2005. Besides, urban waste water treatment rate increased from 52.0 % in 2005 to 75.3 % in 2009.

Third, urban environment sanitation facilities in China have been improved constantly. Table 12.5 shows in 2009 the number of harmless treatment plants of urban garbage was 567 units, 96 units more than that in 2005, and the harmless treatment capacity per day of urban garbage reached 356,130 tons, while the urban proportion of harmless treated garbage was 71.4 %, which increased by 19.7 % over 2005. Besides, the green coverage rate improved significantly, from 32.6 % in 2005 to 38.2 % in 2009, and park green land per capita also increased by 2.7 %.

Fourth, urban road and public transportation facilities developed rapidly in China. Table 12.6 shows by the end of 2009 the length of operating routes of urban public transportation in China reached 209,249 km; and motor vehicles for public

Table 12.3 Investment in fixed assets in urban area in China (2005–2009) (Unit: 100 million)

Index	2005	2006	2007	2008	2009
Urban funds for maintaining construction	5,275.7	3,349.5	4,247.3	5,008.3	5,927.1
Investment in fixed assets of municipal utilities and facilities	5,602.2	5,765.1	6,418.9	7,368.2	10,641.5
Investment in urban environmental infrastructure	1,289.7	1,314.9	1,467.5	1,801.0	2,512.0

Sources Ministry of Housing and Urban–Rural Development of the People's Republic of China (2006–2010), National Bureau of Statistics of China (2011)

Table 12.4 Indicators of urban water supply and waste water treatment capacity in China (2005–2009)

Index	Unit	2005	2006	2007	2008	2009
Urban total water supply	100 million m^3	502.1	540.5	502.0	500.1	496.8
Urban water access rate	%	91.1	86.67	93.8	94.7	96.1
Urban population with access to water supply	10,000	32,723.4	32,304.1	34,766.5	35,086.7	36,214.2
Urban daily household water consumption per capita	Litre	204.1	188.3	178.4	178.2	176.6
Urban length of drainage pipes	10,000 km	8.5	26.1	29.2	31.5	34.4
Urban waste water discharged	100 million m^3	359.5	362.5	361.0	364.9	371.2
Urban waste water treatment plants	Unit	792	815	883	1,018	1,214
Treatment capacity	10,000 m^3/day	5,725.2	6,366	7,146	8,106	9,052
Quantity of waste water treated yearly	100 million m^3	186.8	202.6	227.0	256.0	279.4
Waste water treatment rate	%	52	55.7	62.9	70.2	75.3

Source National Bureau of Statistics of China (2006–2010)

transport per 10,000 population reached 11.1 standard units, an increase of 2.5 standard units over 2005; while volume of passenger transport in urban public transport reached 67.7 billion person-times, which was 39.92 times as much as that in 2005.

12.2.2.3 Rural Environmental Protection and Infrastructure Condition were Greatly Improved

First, toilet improvement in rural areas progressed smoothly. Table 12.7 shows, by the end of 2009, investment of toilet improvement in rural areas reached 11.44 billion Yuan, of which state investment was 4.25 billion Yuan, an increase of

Table 12.5 Environmental sanitation facilities construction in China (2005–2009)

Index	Unit	2005	2006	2007	2008	2009
Harmless treatment plants of urban garbage	Unit	471	419	460	509	567
Harmless treatment capacity of urban garbage	ton/day	256,312	258,048	271,791	315,153	356,130
Urban proportion of harmless treated garbage	%	51.7	52.2	62.0	66.8	71.4
Urban green coverage rate	%	32.6	35.1	35.3	37.4	38.2
Urban park green land per capita	m^2	7.9	8.3	9.0	9.7	10.7

Source National Bureau of Statistics of China (2006–2010)

Table 12.6 Urban road and public transportation facilities construction in China (2005–2009)

Index	Unit	2005	2006	2007	2008	2009
Length of operating routes	km	–	125,857	140,801	147,349	209,249
Number of public transport vehicles (year-end figure)	Unit	313,296	315,576	347,969	371,822	370,640
Motor vehicles for public transport per 10,000 population	Standard unit	8.6	9.05	10.2	11.1	11.1
Volume of passengers transport in public transport	10,000					
person-times		4,836,930	4,659,247	5,546,439	7,029,996	6,767,589
Operating number of rail transit (year-end figure)	Unit	2,364	2,764	3,480	4,530	5,479
Volume of passengers transport in rail transit	10,000					
person-times	–	181,599	220,582	337,390	365,770	

Source National Bureau of Statistics of China (2006–2010)

330 % over 2005. The access rate to rural sanitary toilets increased from 55.3 % in 2005 to 63.2 % in 2009, which made the households of using sanitary toilets reach 160.56 million.

Second, rural energy structure was constantly optimized, and the use of renewable energy increased significantly. Table 12.8 shows, in the end of 2009, the production of rural methane was 13.08 billion m^3, an increase of 79.4 % over 2005. At the same time, water heaters using solar energy was 4,997.1 m^2, an increase of 55.8 % over 2005, and the solar kitchen ranges was 1,484,271 units, an increase of 53.8 % over 2005.

Third, traffic conditions in rural China were further improved, and the "11th Five Year" goal about rural road construction was achieved. With the implementation of "village to village road project", by 2010, the length of highways in rural China (including the County Road, Township Road, Village Road) reached

Table 12.7 Toilet improvement in rural China (2005–2009)

Index	Unit	2005	2006	2007	2008	2009
Investment of rural toilet improvement	100 million Yuan	47.3	69.5	72.9	93.9	114.4
State investment	100 million Yuan	12.8	20.8	25.5	37.6	42.5
Proportion of state investment in total	%	27	30	35	40.1	37.2
Accumulative households using sanitary toilets	10,000 households	13,740	13,873	14,442	15,166	16,056
Access rate to sanitary toilets	%	55.3	55.0	57.0	59.7	63.2

Source National Bureau of Statistics of China (2006–2010)

3,506,600 km. And the township (town) with highways accounted for 99.97 % of total townships (town) in China, and the administrative villages with highways accounted for 99.21 % of total administrative villages in China, which increased by 6.33 % and by 22.3 % respectively compared to the end of the "10th Five Year". Basically, all the towns where conditions mature had asphalt (cement) roads, all the eastern and central administrative villages had asphalt (cement) roads, and some western administrative villages where conditions mature also had asphalt (cement) roads.

12.2.3 The Main Problems of China's Green Development and Its Reasons

12.2.3.1 The Main Problems of Green Development in China

First, the deterioration of environmental quality was curbed, but it was not completely stopped. During the second half of the "11th Five Year Plan", the momentum of improvement in environmental quality appeared, because of the increase in state investment, strengthened supervision, and the slowing down in economic growth. Nevertheless, during the "12th Five Year Plan" and even before 2020, the overall environmental situation is still "locally improved, but not overall curbed, and facing a continually increasing pressure". In China, the pressure on environment will be bigger than in any other country in the world, the resource and environmental issues will be more serious than in any other country, and the solutions to these issues will be more difficult. First of all, during this period, China's economy will still maintain a rapid growth, and the pressure of pollution emission is still large. Second, the power plant desulfurization and other relatively easy environmental protection measures were taken during the "11th Five Year Plan" period, so the environmental protection will be more difficult during the "12th Five Year Plan" Period. Table 12.9 shows, comparing to before, the proportion of industrial waste water meeting discharge standards, the treatment rate of

Table 12.8 Energy use in rural China (2005–2009)

Index	Unit	2005	2006	2007	2008	2009
Production of rural methane	100 million m^3	72.9	83.6	101.7	118.4	130.8
Water heaters using solar energy	10,000 m^2	3,205.6	3,941	4,286.4	4,758.7	4,997.1
Solar kitchen ranges	Unit	685,552	865,238	1,118,763	1,356,755	1,484,271

Source National Bureau of Statistics of China (2006–2010)

Table 12.9 Indicators relating to environmental protection technology in China (2000–2009) (Unit: %)

Years	Proportion of industrial waste water meeting discharge standards	Treatment rate of urban household waste water	Proportion of industry SO_2 meeting discharge standards from process of fuel burning	Proportion of industry SO2 meeting discharge standards from process of production
2000	82.1	–	–	–
2001	85.6	18.5	62.8	51.0
2002	88.3	22.3	72.9	55.1
2003	89.2	25.8	75.4	59.3
2004	90.7	32.3	78.6	59.4
2005	91.2	37.4	80.9	71.0
2006	92.1	43.8	82.3	81.0
2007	91.7	49.1	87.4	81.8
2008	92.4	57.4	89.3	86.5
2009	94.2	63.3	–	–

Source Collated according to the relevant years' *National Environmental Statistics Bulletin*

urban household waste water, and the proportion of industry SO_2 meeting discharge standards from process of fuel burning had been on a relative high level since 2006. Therefore, it is very difficult to make a sharp rise in the future. Besides, the industrial emissions started to decline absolutely or the growth rate was declining relatively, however, the emissions of urban residents are now increasing. The accelerating urbanization has an increasingly obvious impact on the environment. As the emissions of urban residents are still increasing, the emissions of waste water and other pollutants will continue to grow.

Second, small and medium enterprises become the focus of pollution treatment. During the "11th Five Year Plan" period, as the environmental supervision was more stringent in China, large enterprises' environmental protection facilities and investment were greatly promoted and their emissions decreased significantly. But, comparing with large enterprises, the proportion of emissions from small and medium enterprises was rising. Reasons for this situation are as follows. Small and medium enterprises developed fast, the proportion of which in the total economy increased rapidly. Besides, technical equipments of small and medium enterprises were on a relatively low level, while emissions per unit of output were high. With the increase of emission day by day, since the small and medium enterprises lag behind the large enterprises in pollution treatment capability, they had a rapid rise in the proportion of the whole industrial pollution. Currently, the key problems of improving small and medium enterprises' pollution treatment conditions are the cost of pollution treatment and environmental supervision. However, because the small and medium enterprises are too scattered and their treatment and supervision costs are too high, these two factors are also difficulties in environmental protection.

Third, environmental pollution are moving to central and western regions and rural areas. In the 1970s, China's environmental pollution showed dot distribution; in the 1980s, river pollution and air pollution were serious; in the 1990s, the pollution had a trend of regional expansion; and after the 21st century, the pollution had a trend of accelerating expansion. Especially labor-intensive and resource-intensive industries in coastal areas are transferred to the central and western regions quickly. Enterprises, located in provincial development areas, mostly experienced environmental assessment, whose pollutants meet the basic emission standards. Enterprises, located in county and township areas, had many serious pollution problems, which made rural pollution in central regions worsen and lead to some group events. As the environmental protection facilities in rural areas and farmers' environmental awareness are very poor, the pollution has been very serious, so the pressure on environmental protection in rural areas grows increasingly.

Fourth, more and more environmental pollutions are transferred to China through international trade and investment. International trade and investment affect the China's environment by influencing the domestic structure of production activities. A country which has less stringent environmental regulation may have the comparative advantage in polluting industries. Therefore, the pollution may be transferred from developed countries to less developed countries through international trade and investment. This trend also appears in China. Not only the high-pollution projects are accelerating the transfer to China, but also highly polluting commodities as well as environment waste are exported to China. Meanwhile, the export of domestic products is encountering more and more green barriers. Internationalization is adding new difficulties to China's environmental protection.

12.2.3.2 Deep Reasons for Restricting Green Development in China

From the macroscopic point of view, there are two reasons for environmental problems: the first is rapid economic growth, and the second is pollution emissions intensity (emissions per unit of output). From the micro perspective of view, environmental problems are mainly due to excessive emissions from enterprises, and the reason for excessive emissions is the weakness of the existing policies and institutions.

First, the objective basis for China's environmental problems is the particular stage of industrialization and urbanization.

Looking at the history of developed countries, environmental problems emerged because of industrialization, became serious because of the acceleration of industrialization and urbanization, and are being solved because of the great progress in industrialization and urbanization. For example, in the United States, from 1900s to 1970s, emission volume of sulfur dioxide per year increased from 9 to 28 million tons, but declined year by year after that. In 1990, the emission volume of sulfur dioxide in the United States was 20.935 million tons, which was 74.4 % of 1970. In 2005, it was 13.348 million tons, which was 47.5 % of 1970. Therefore, the process of industrialization and urbanization is the objective basis for the emergence and also the solution to environmental problems.

Generally speaking, since the "10th Five Year Plan" period, China entered a rapid development period of heavy industrialization and urbanization. Table 12.10 shows, after 2003, China's heavy industrialization accelerated significantly, the growth rate of heavy industry was obviously higher than that of light industry, and the pressure on environmental protection continued to increase. Looking at the data from the "10th Five Year Plan" period, waste water and chemical oxygen demand in waste water were mainly affected by urbanization, while sulfur dioxide and dust were mainly related to industrialization. In the discharge of waste water and chemical oxygen demand, the proportion of urban life had always been more than 50 % and increased year by year. In the sulfur dioxide emission, the proportion of industry had always been more than 80 % and continually increased, while declined slightly in 2008 and 2009. In the dust emission, the proportion of industry declined from 81.8 % in 2000 to 71.3 % in 2009, while the proportion of life emissions had been in ascendant trend.

Second, direct reasons for China's environmental problems are insufficient capital investment, backward technology, underdeveloped environmental protection industries and environmental protection market.

The first obstacle is insufficient investment in environmental protection. Investment in environmental protection in developed countries usually accounts for 2 % of GDP, while in China during the "7th Five Year Plan" period and the 8th *Five Year Plan* period the proportion was only 0.8 %. Although during the 9th *Five Year Plan* period Chinese government increased the investment in environmental protection, it only accounted for 1 % of GDP. During the 10th *Five Year Plan* period, the investment in environmental protection had an annual growth rate 17.6 %, but the highest proportion of investment in environmental protection of GDP was only 1.4 % (2004), an average of only 1.32 %, and the proportion was still very low. During the 11th *Five Year Plan* period, Chinese government increased the investment in environmental protection greatly, and the growth speed of investment in 2007 and 2008 were over 30 %. So, the proportion of investment in environmental protection in GDP was increasing. However, in 2009 the proportion was 1.33 %, not only lower than that in developed countries, but also in descendant trend.

The second obstacle is the relatively backward environmental technology. China lacks environmental protection facilities; most of which are at the international level between 1970s and 1980s, and environmental protection technology

Table 12.10 Industry distribution of main pollutant emissions in China (2000–2009) (Unit: %)

Years	Waste water		Chemical oxygen demand		Sulfur dioxide		Dust	
	Industry	Urban life	Industry	Urban life	Industry	Urban life	Industry	Urban life
2000	46.8	53.2	48.8	51.2	80.8	19.2	81.8	18.2
2001	46.9	53.1	43.2	56.8	80.4	19.6	79.4	20.6
2002	47.1	52.9	42.7	57.3	81.1	18.9	79.4	20.6
2003	46.2	53.8	38.4	61.6	83.0	17.0	80.7	19.3
2004	45.8	54.2	38.1	61.9	83.9	16.1	81.0	19.0
2005	46.3	53.7	39.2	60.8	85.1	14.9	80.2	19.8
2006	44.7	55.3	37.9	62.1	86.3	13.7	79.4	20.6
2007	44.3	55.7	37.0	63.0	86.7	13.3	78.2	21.8
2008	42.3	57.7	34.6	65.4	85.8	14.2	74.4	25.6
2009	39.8	60.2	34.4	65.6	84.3	15.7	71.3	28.7

Source Ministry of Environmental Protection (2010)

already falls behind. What's more, most enterprises are small, which have no capacity to develop environmental technology or to introduce new scientific and technological achievements. The professional environmental technology services are very weak and the low level of socialization results in poor transmission of technical and market information. Environmental technologies are not market-oriented and intermediate service agencies are not yet complete, which further restrict the improvement of China's environmental protection technology.

The third obstacle is that the environmental protection industry lags behind. China's environmental protection industry mainly focuses in two fields: one is producing environmental protection products; the other one is making use of "three wastes", while other fields, such as the production of low-pollution products, environmental technology services and ecological protection, are still very backward. Even in the traditional environmental protection industry, the level of development is low. For example, although the recycling rate of industrial water increased from 69.6 % in 2000 to 85.0 % in 2009, it is still lower than the average international level, and the growth rate is declining. Besides, the comprehensive utilization of industrial solid waste was only 67.0 % in 2009, far below the average international level. Meanwhile, the environmental protection industry in China develops highly unbalanced, and is very weak in central and western regions.

Third, the basic reason for China's environmental problems is lack of supervision and lax enforcement, which leads to the facto government failure and can't make the environmental costs fully internalized.

In theory, environmental protection particularly needs the government to play its role. The source of deterioration of environmental quality is the existing government failure, which shows in the following three aspects:

The first is lack of supervision. China's penalties for environmental violations are not severe and not introducing the criminal liability. More importantly, due to the influence of local protectionism, many local management institutions in

Table 12.11 Growth of investment in "three simultaneous" projects in China's environmental protection (2001–2009)

Years	Investment in pollution treatment projects		Investment in "three simultaneous" projects for environmental protection engineering		Proportion of investment in "three simultaneous" projects (%)
	Total (100 million Yuan)	Growth rate (%)	Total (100 million Yuan)	Growth rate (%)	
2001	1,106.6	–	336.4	–	30.4
2002	1,363.4	23.2	389.7	15.8	28.6
2003	1,627.3	19.4	333.5	−14.4	20.5
2004	1,908.6	17.3	460.5	38.1	24.1
2005	2,388.0	25.1	640.1	39.0	26.8
2006	2,567.8	7.5	767.2	19.9	29.9
2007	3,387.6	31.9	1,367.4	78.2	40.4
2008	4,490.3	32.6	2,146.7	57.0	47.8
2009	4,525.2	0.8	1,570.7	−26.8	34.7

Sources The relevant years' *National Environmental Statistics Bulletin*; Ministry of Environmental Protection (2010)

environmental protection have lax enforcement, even administrative omission, leaving the low cost of violating the law and high cost of obeying the law for the enterprises. In fact, this phenomenon encourages the enterprises to discharge more pollutants. For example, Under Chinese law, environmental protection facilities should be designed, constructed and put into operation simultaneously (three simultaneous) with the principal part of the project. But this institution has not been well implemented. Table 12.11 shows the investment in "three simultaneous" projects accounting for the investment in pollution treatment projects had been declining from 2001 to 2004. Although it had a rapid growth after that, it declined greatly in 2009 again, to only 34.7 %. Moreover, many investment projects in small and medium enterprises didn't accomplish the "three simultaneous".

The second is the policy deviation. China's present environmental policy is "Polluter Pays", but under fixed technical conditions, pollution treatment needs better economies of scale. The larger the pollution treatment is, the lower the cost is. To most individual small and medium-size enterprise (SMEs), there is a large gap between its pollution emission scale and the economy of scale for treatment. At the same time, SMEs don't have enough funds and it's very difficult and costly for them to raise funds to construct pollution treatment facilities. Therefore, pollution treatment is rather costly for SMEs, which has led to invalid pollution control. When someone comes over for inspection, they treat pollution only superficially or hand in some fines, in which there are always potential "rent-seeking" activities. Also, the price regulation and subsidy policies about waste disposal, water treatment and energy, are in fact encouraging enterprises and residents to discharge pollutants.

The third is strategic distortions. In China, taking the economic development as the center has been distorted as taking the GDP growth as the center. And the government has not paid much attention to environmental protection. Therefore, extensive ways of industrialization and urbanization haven't been improved, leaving much pollution emissions all the time. Some regions, whose ecological environment had been fragile already, still carried out industrial projects that exceeded the environmental capacity, even pollution-intensive heavy industry projects, which have made the environment worse.

12.3 Government Green Actions During the 12th Five-Year Plan Period: Planning and Prospect

12.3.1 The New Requirements for Government Green Actions During the 12th Five-Year Plan Period

The 12th *Five-Year Plan* puts forward that economic and social development during the 12th Five-Year period should take the scientific development as the theme, take accelerating the transformation of economic development pattern as the main thread, deepen the reform and opening up, ensure and improve people's wellbeing, consolidate and expand upon the success of our efforts to respond to the impact of the global financial crisis, promote long-term, steady and rapid economic development and social harmony and stability, and lay a solid foundation for building a moderately prosperous society in all respects. This guideline has new requirements for green development during the 12th *Five-Year Plan* period.

12.3.1.1 New Positions: New Requirements for Building a Resource-Conserving and Environment-Friendly Society

The 12th *Five-Year Plan* put forward clearly that taking building a resource-conserving and environment-friendly society as the focal point for accelerating the transformation of economic development mode. Thoroughly implement the basic national policy of resource conservation and environmental protection, save energy, reduce the intensity of greenhouse gas emissions, develop the circular economy, promote low-carbon technologies, actively respond to global climate change, and help the economy and society develop harmoniously with population, resource and environment in a sustainable way.

12.3.1.2 New Objective: New Requirements for Green Development's Results

The 12th *Five-Year Plan* put forward some specific objectives, such as "reducing energy consumption per unit of GDP by 16 % and reducing carbon dioxide emissions per unit of GDP by 17 %; reducing chemical oxygen demand and sulfur dioxide emissions respectively by 8 %; improving the forest coverage to 21.66 %, and increasing the forest volume by 600 million cubic meters".

12.3.1.3 New Deployment: New Requirements for Measures of Promoting Green Development

The 12th *Five-Year Plan* put forward some deployments for accelerating green development and building a resource-conserving and environment-friendly society from these aspects, such as positive response to global climate change, strengthening resource saving and management, vigorously developing circular economy, stepping up efforts for environmental protection, promoting ecological protection and restoration, etc. (Table 12.12).

12.3.2 Government Strategic Measures in Green Development During the 12th Five-Year Plan Period

Synthesizing the viewpoints of the 12th *Five-Year Plan* and relating departments' plan, the basic goal of government green action during the 12th Five-Year Plan period is to promote the integration of economic development and environmental protection: the government will, on one hand, continue to adopt the tight and stringent macro-environment policy, and on the other hand, implement the strategic guideline of environmental priority in special areas and fields.

12.3.2.1 Make Full Use of the Role of Environmental Protection in Accelerating the Transformation of the Pattern of Economic Development

Assess various *Plans*' environmental impacts actively, such as hydropower exploitation, development zones, and industrial parks; and apply these assessments' results in time, especially in the five regions of China. Deepen environmental assessment on construction projects, control strictly redundant projects of high energy-consuming and high emission, accelerate mergers and acquisitions, and eliminate backward production capacity. Launch the Special Action of Environmental Protection on remediating the illegal emission enterprises, and keep the supervision of environmental law enforcement in a high-pressure situation. Improve the mechanism of environmental risk prevention and emergence management, focus comprehensively on the treatment in chemical factories near rivers and lakes, and avoid frequent environmental accidents in chemical industries. Strengthen the overall capability to prevent emergency. Implement the

Table 12.12 Key projects for environment protection in the 12th Five-Year Plan

	Key projects for environment protection
01	*Projects for constructing treatment facilities for sewage and garbage in urban areas*
	Accelerate the construction of treatment facilities for sewage, sludge and garbage, and construct sewage collection pipe network, garbage removal facilities at the same time
02	*Projects for water environment treatment in major river valleys and regions*
	Strengthen the comprehensive treatment in "three rivers and three lakes", Songhua River, the Three Gorges reservoir area and upper reaches, Danjiangkou reservoir area and upper reaches, the upper and middle reaches of the Yellow River and other key marines, increase the water pollution prevention efforts in the middle and lower reaches of the Yangtze River, Pearl River and ecologically fragile highland lakes, and promote the comprehensive treatment in Bohai and other key sea areas
03	*Projects for desulfurization and denitrification*
	Build plants of desulfurization and denitrification for newly built coal-fired units, and install denitrification plants with efficiency no less than 60 % for newly built cement production lines, and denitrification plants for steel sintering machines and petrochemical industry
04	*Projects for heavy metal pollution prevention and control*
	Strengthen heavy metal pollution prevention and control in key areas, key industries and key enterprises; make key enterprises' emissions meet discharge standards; and make heavy metal pollution prevention and control in Xiangjiang River and other rivers or regions achieve significant results

Source "12th *Five-Year Plan* for National Economic and Social Development" http://www.gov.cn/test/2011-03/16/content_1825941_7.htm,2011-03-16

Heavy Metal Pollution Prevention Plan comprehensively, and put more efforts on remediating environmental law issues in key controlling areas and industries. Continue to promote environmental management of chemicals, implement the persistent organic pollutants (pops) and mercury pollution prevention plan, and complete indicators for standardized management of hazardous waste.

12.3.2.2 Take Comprehensive Measures to Actively Promote Pollution and Emission Reduction

Focus more on structural emissions reduction, continue to intensify emissions reduction from projects and the management of emissions, increase and implement strictly the emissions standard in industries like paper, textile, leather, chemicals and so on, comprehensively start the construction of sewage treatment plants in counties, vigorously carry out the project construction of agricultural water pollution abatement, continue to work on desulfurization and denitrification in coal-fired power plants, control tightly exhausting pollution of motor vehicles, carry on in-depth studies on denitrogenation in urban sewage treatment plants and other key technologies, promote an effective and stable operation of pollution control facilities, and vigorously conserve energy and reduce emissions in transportation sectors to achieve a 1.5 % decrease in emission volume of four major pollutants.

12.3.2.3 Step Up the Treatment of Water Pollution in Major Rivers and the Treatment of Sea Pollution

Implement comprehensively *Environmental Protection Plan for Drinking Water Sources in cities all over the country*, carry out assessments for drinking water sources in cities at prefecture level or above, and develop a technology and management system for assessing ecological safety in key lakes and rivers. Take the Beijing-Tianjin-Hebei Economic Band, the Circle of the Yangtze River Delta and the Pearl River Delta as the key areas, accelerate the new mechanism of joint prevention and control in all areas, put more efforts in pollution prevention for particles and volatile organics, and control tightly exhausting emissions from motor vehicles. Strengthen the management of urban environment and the prevention for industrial pollution, improve the verification system of public firms for environmental protection, and deepen the building of national environment protection model cities.

12.3.2.4 Focus on Solving Environment Problems Concerning the People's Wellbeing

Pursue the idea of putting people first and protecting environment for people, ensure the construction of drinking water projects; strengthen the treatment of heavy metal pollution, hazardous wastes pollution and soil pollution to safeguard people's interests. Improve the fundamental work and tighten the supervision of enterprise environment; increase the law reinforcement and take tight measures to punish illegal enterprises; implement the regulation of information publicity and accept people's supervision. At the same time, prevent environmental risks actively and ensure the safety of economy and environment to make all the people have clean drinking water, clean air, safe food and beautiful living environment.

12.3.2.5 Vigorously Promote Green Development in Rural Areas

Deepen the policy of "governing by rewards" constantly, and carry out contiguous improvement in rural environment and environmental evaluations. Promote the trials of goal responsibility system for comprehensive improvement of the rural environment, and prepare well for soil pollution investigations in key areas all over the country. Protect and recover biological diversity; and intensify supervision for activities concerning developing and constructing natural reserves. Build green energy villages and accelerate developing renewable energy in rural areas. Strengthen investments in rural infrastructure and direct the focus from "construction" to "both construction and management".

12.3.2.6 Deepen the Reform of Investment and Financing System, Increasing Investment in Green Development

Let the industrial policy play an important role in guiding investments to areas such as ecological development, environmental protection and resource conservation, etc.; Grasp the implementation of government investment regulations, enterprise investment project approval and filing regulations, and figure out management regulations for direct investment projects with central budget as soon as possible, further simplify the approving process of corporate bonds. Strengthen policy supports in planning guidelines, finance and tax, accelerate the construction of urban public facilities, prevent and treat "city diseases". We should continue to strengthen the foundations in monitoring, early warning, emergency and information to improve government management ability in green economy.

12.3.2.7 Further Improve Environmental Taxation and Market Policies

Consumer tax should be further improved to include products that can easily produce environmental pollution and consume large amounts of resources and some high-end luxury goods. The management system should be completed for collection and usage of pollution charges. The rate of pollution charges should be increased, and the trials of setting "Environmental Protection Tax" can be launched in an appropriate time and be expanded gradually. The trading mechanism of environmental property right should be established and improved to promote the compensable use and trading of emission rights.

References

Classic China-glorious 60 years: environmental protection achievements (in Chinese). Available at: http://env.people.com.cn/GB/146189/168051/

Ministry of Environmental Protection (2010) Annual statistical report on environment in China 2009. China Environmental Science Press, Beijing (in Chinese)

Ministry of Housing and Urban–Rural Development of the People's Republic of China (2006–2010) China urban construction statistical yearbook 2006–2010. China Building Industry Press, Beijing (in Chinese)

National Bureau of Statistics, Ministry of Environment Protection (2006–2010) China statistical yearbook of environment 2006–2010. China Statistical Press, Beijing (in Chinese)

National Bureau of Statistics of China (2006–2010) China statistical yearbook 2006–2010. China Statistics Press, Beijing (in Chinese)

National Bureau of Statistics of China (2009): Achievements report 17 in economic and social development during the 11th *five-year plan* period: great progress has been made in environmental protection (in Chinese). Available at: http://www.stats.gov.cn/tjfx/ztfx/qzxzgcl60zn/t20090928_402590846.htm

National Bureau of Statistics of China (2011) Achievements report 14 in economic and social development during the *11th five-year plan* period: great progress has been made in

environmental protection (in Chinese). Available at: http://www.stats.gov.cn/tjfx/ztfx/sywcj/t20110310_402709535.htm
Shi Y-d, Li C-y (2010) An analysis on the reform of chinese urban infrastructure investment and financing mechanism. J Cent Univ Finan Econ 2010(7) (in Chinese)
The State Council of the People's Republic of China (2011) The twelfth five-year plan for national economic and social development (in Chinese). Available at: http://news.xinhuanet.com/politics/2011-03/16/c_121193916.htm
Xia G (2010) Concept innovation on environmental protection planning in *11th five-year plan*. China Economic Times, 2010-05-19 (in Chinese)
Xia G (2010) Environmental protection and green development in *11th five-year plan*: target and policy. China Economic Times, 2011-03-22 (in Chinese)
Zhang S-h (2010) Establish and perfect the legal system of energy saving and emission reduction. Science-Technology and Law 2010(8) (in Chinese)
Zhou S (2010) China will actively explore new ways to protect environment during the 11th *five-year plan* period (in Chinese). Available at: http://news.ifeng.com/gundong/detail_2010_11/27/3247041_0.shtml
Zhou S (2010) Review of the 11th *five-year plan*—outlook on the twelfth five-year plan (in Chinese). Available at: http://news.xinmin.cn/rollnews/2010/11/29/7985458.html
Zhou S (2011) In the new starting point: accelerate the environment protection (in Chinese). Available at: http://www.gov.cn/gzdt/2011-01/07/content_1780051.htm
Zhou S (2011) Efforts to achieve a good start in environment protection in the first year of twelfth five-year plan (in Chinese). Available at: http://www.gov.cn/gzdt/2011-01/17/content_1786489.htm

Chapter 13
Government Green Investment

Desheng Lai, Ning Cai and Tingting Rong

Government green investment is a pre-condition and foundation for the whole country's green development. It provides guidance to corporate sector's green production, people's green consumption, and ecological environment protection. During the 11th *Five-Year Plan*, under the guidance of scientific concept of development, all levels of government, aiming to "save energy and reduce pollution", increased green investment to promote the construction of a resource-conserving and environment-friendly society, in order to push the structural adjustment and transform of the national economy.

This chapter focuses on the green investment activities at all levels of Chinese governments, and makes cross-regional comparisons by analyzing government green investment activities during the 11th *Five-Year Plan*. This chapter also looks into the future prospects of government green investment in the 12th Five-Year Plan, and tries to provide valuable reference for the governments in their future green investment activities.

13.1 Overview of Chinese Government Green Investment during the 11th Five-Year Plan Period

During the 11th *Five-Year Plan,* under the guidance of "scientific concept of development" promoted by the central government, all levels of governments in China, both central and local, put more emphasis on green investment and made great achievements. However, there are still some problems with China's government green investment, including: limited channels of investment, imperfect long-term mechanism for investment, etc. These problems need to be solved in future work.

D. Lai (✉) · N. Cai · T. Rong

X. Li and J. Pan (eds.), *China Green Development Index Report 2011*,
Current Chinese Economic Report Series, DOI: 10.1007/978-3-642-31597-8_13,

13.1.1 Main Achievements of China's Government Green Investment During the 11th Five-Year Plan Period

Chinese government has always paid high attention to green investment. In the theoretical exploration and empirical work during the 11th *Five-Year Plan*, the Chinese government become more determined on the idea of green investment; total amount of funding for green investment increased steadily; green investment projects become more efficient; the legal system in support of green investment become more complete; and many other achievements.

13.1.1.1 Government more Determined About Green Investment

The idea of green investment has been evolving and developing in the Chinese government for a long time. During the 11th *Five-Year Plan*, the government had a more thorough understanding of the concept. Before 2006, the traditional idea of government green investment refers to expenditures on environment protection and investment in pollution treatment. In 2006, against the background of establishing a resource-saving and environment-friendly society, it was proposed for the government to fully use its green investment as guidance for the whole society, to increase investment in green development, and to incorporate various activities to the system of green investment (including expenditures on natural ecological system protection, natural forestry protection, transforming of plant area back to forest, alternative energies research, comprehensive treatment of rural environment, etc.). The objective is for the government to expand its traditional concept of green investment into a broader scope.

In addition, before the 11th *Five-Year Plan*, environment protection expenditure, which is an important part of government green investment, was not included in an independent category of government expenditure, but was put under the entry of "expenditure on environment protection and urban water resources construction". In Feb 2006, considering the actual situation in the functions of Chinese governments and fiscal management, and borrowing the practice from foreign governments in their classification of government functions, the Ministry of Finance published the "*Reform Plan for Classification of Government Revenue and Expenditure*". According to the Plan, since 2007 "environment protection" will be a major category of government expenditure, under which there is administration of environment protection, environment monitoring and supervision, prevention and treatment of pollution, desertization treatment, transform of shepherd land back to grassland, and other entries. Expenditure and revenue in these entries are calculated separately, so that the planning and implementation of government work in environment protection become more regulated and reasonable.

After the 17th Plenary Conference of the CPC, government green investment has become a more and more important part of national green development. It has become a critical part of implementing the scientific concept of development and the construction of a harmonious society. It is now an integral path to promote faster and more efficient development of the Chinese economy. Up till now, a new structure of China's government green investment has come into being, with the central government in leadership, the central Ministries doing the appraisal and supervision, and the local governments pushing forward the projects. All governments have now a deeper and clearer understanding of the idea of green investment.

Column 13.1 Scientific Promotion of Green Development in Tongyu, Jilin Province

Tongyu District is located on the Horqin Grassland, in the northwest of Jilin Province. There are more than 200,000 ha of cultivated lands, 267,000 ha of grasslands and 173,000 ha of forestlands in the area. Tongyu ranks top in terms of cultivated land area per capita, grassland area per capita and forestland per capita in Jilin. Xianghai in Tongyu is a National Nature Reserve District and an A-class wetland. Moreover, home to 7 of the world's 15 crane species, Tongyu is called the Hometown of Cranes.

Guided by local government and with strong supports from local citizens, green agriculture, green energy industry and green eco-tourism have gradually become the main industries in Tongyu recently. Hence, the green development of Tongyu achieved a remarkable success.

Local government has made much effort to promote green agriculture in Tongyu in the 11th *Five-Year Plan* period. The government developed 100,000 ha of pollution-free agricultural production bases, accounting for nearly half of the cultivated land area in the district. Based on the green agricultural products, the government helped introduce investment from influential green enterprises to enlarge and strengthen local agricultural product markets. Besides, they increased subsidies to agriculture machinery and improved the level of agricultural mechanization. Meanwhile, 75 modern farms were established to reduce the pollution of agricultural production.

As to the green energy industry, with favorable national energy policies and supports from the provincial government, Tongyu was built to become the biggest wind power generation base in China and thc biggest wind power equipment manufacturing base in Jilin province. Whilst inviting leading companies in the industry like Huaneng International Power Development Corporation (HIPDC) and China Longyuan Power Group Co. Ltd to invest in Tongyu, the government also helped develop local wind power generation enterprises and promoted the industrilisation of wind power generation. In addition, Tongyu actively developed solar energy and bio-energy, making contribution to the exploitation and utilization of green energy in China.

Moreover, Tongyu put a great amount of manpower, material and financial resources into the green eco-tourism industry, sparing no effort to construct an

Eco-tourism District. It takes full use of the reputation of National Nature Reserve, A-class wetland and the Hometown of Crane to attract tourists, and spends most of the income on ecological environment protection.

In promoting scientific green development, Tongyu has found a green way for both economic growth and environmental protection.

Sources:

1. Ministry of Agriculture of the P.R. China (2010). *The Green Agriculture Development of Tongyu.* Retrieved April 20, 2011, from Ministry of Agriculture of the People's Republic of China website: http://www.moa.gov.cn/fwllm/qgxxlb/jl/201006/t20100625_1588014.htm.
2. The Government of Jilin Province (2010). *Developing Green Economic in County by Science.* Retrieved April 20, 2011, from Tong Yu Sheng Tai Lv You website: http://www.jl.gov.cn/zt/lygj/tystly/zjty/fzcj/201008/t20100820_768564.html.
3. Jilin Daily (2007). *The Green Development of Tongyu.* Retrieved April 20, 2011, from cpc.pople.com.cn website: http://cpc.people.com.cn/GB/64093/64387/6687515.html.

13.1.1.2 Steady Growth of Funding for Government Green Investment

During the 11th *Five-Year Plan*, at the same time of clearer understanding of green investment by China's governments, the size of funding has also been steadily increasing, year after year. Table 13.1 shows the growth of funding for China's government green investment during 2005–2009.

In terms of environmental protection, Chinese government invested a total of RMB99.6bn in 2007, about 2 % of fiscal expenditure. In 2007, the amount increased to RMB145.1bn, about 2.32 % of fiscal expenditure. In 2009, the amount further increased to close to RMB200bn, up RMB48.3bn over 2008, and the share in total government expenditure increased by 0.21 ppt. Within 3 years, total investment in environmental protection almost doubled at an average annual growth rate of 40 %.

In terms of pollution treatment, during the 11th *Five-Year Plan*, government investment had been increasing at an annual average rate of 17.3 %. Although the growth rate was relatively slow in certain years, the overall growth momentum was strong. Total investment amount increased from RMB238.8bn in 2005 to RMB452.5bn in 2009, with a positive growth rate every year. Specifically, in 2008 there was the largest amount of increase at RMB110.3bn, while in 2009 there was the smallest amount of increase at RMB3.5bn. The share in national fiscal expenditure decreased a bit in 2006 and 2009, but was in steady increase in the other 3 years.

Forestry construction and rural environment comprehensive treatment investment account for relatively smaller shares in total government green investment. But still in the 11th *Five-Year Plan*, their growth rates reached 31 and 26.5 %,

Table 13.1 Government green investment during the 11th Five-Year Plan (Units: RMB100mn, %)

	2005	2006	2007	2008	2009
National fiscal expenditure	33,930	40,423	49,781	62,593	76,300
Environment protection investment	–	–	996	1,451	1,934
Share of national fiscal expenditure	–	–	2.00	2.32	2.53
Investment in pollution treatment	2,388	2,566	3,387	4,490	4,525
Share of national fiscal expenditure	7.04	6.35	6.80	7.17	5.93
Investment in forestry construction	459	478	622	837	1,351
Share of national fiscal expenditure	1.35	1.18	1.25	1.34	1.77
Investment in comprehensive treatment of rural environment	144	198	209	282	369
Share of national fiscal expenditure	0.42	0.49	0.42	0.45	0.48

Note Because of changes in statistical caliber, environment protection data after 2006 are not comparable with later data, therefore the data start from 2007 in the above table
Source National Bureau of Statistics, *China Statistical Yearbook* 2007–2010, Beijing, China Statistics Publishing House, 2007–2010; National Bureau of Statistics, Ministry of Environment Protection: *China Statistical Yearbook of Environment* 2007–2010, Beijing, China Statistics Publishing House, 2007–2010

respectively, both above the average growth rate of national fiscal expenditure. Specifically, investment in forestry construction reached RMB135.1bn in 2009, about 2.94 times that of the final year of the 10th *Five-Year Plan*; investment in rural environment comprehensive treatment reached RMB36.9bn, increasing by RMB22.5bn over the final year of the 10th *Five-Year Plan*. Both categories of government green investment saw substantial growth.

In the 11th *Five-Year Plan*, the Chinese government started to pay more and more emphasis on green development. Various categories of government green investment saw high growth rate each year, contributing a lot to the grand cause of green development in China.

13.1.1.3 Substantial Achievement of Government Green Investment

In order to promote green development and to realize fast and high-quality transformation of economic structure, the Chinese government focuses on the target of green economic growth, and takes the coordinated green development of the economy and the environment as a priority task. It has made great achievement in terms of environment pollution treatment, forestry development, rural environment comprehensive treatment, and clean energy development.

In *the* 11th *Five-Year Plan for the National Economic and Social Development of the People's Republic of China*, a target was set to decrease the energy consumption per unit of GDP by 20 %, and to decrease the total amount of pollution by 10 %. Both are constraining targets. By 2010, the first target was pretty much achieved, and the second target was achieved ahead of schedule with two constraining indicators of SO_2 emission and chemical oxygen demand emission down by more than 10 %.

There was also great achievement in forestry construction in China during the 11th *Five-Year Plan*, especially in the several major forestry construction projects. During 2006–2009, the natural forestry protection project, the transforming of plant land back to forests project, the shelter forests construction project in northern China and the Yangtze River area, and the sand sources treatment project for Beijing and Tianjin have finished forestry construction of ha3.88mn, ha4.18mn, ha3.80mn and ha 1.63mn. In particular, the natural forestry protection project and the protection forests construction project in northern China and the Yangtze River area have both achieved more land areas than during the 10th *Five-Year Plan* (ha0.32mn and ha0.64mn, respectively). And the land areas of the other two projects have achieved targets very close to those during the 10th Five-Year Plan. According to the 7th National Forestry Resources Survey (2004–2008), in the past 5 years, the area of forests in China increased by ha 2054.3; forest area per capita increased by ha0.13 ha; and the forest coverage ratio increased by 2.15 ppts.

Rural environment protection is an important part of China's target to build a harmonious society, and is a critical target in the building of "socialistic new rural areas". The Chinese government made great achievements in this area during the 11th *Five-Year Plan*. In 2005, a total of 8,889 mn people benefited from rural water system reform, and the amount was 903 mn in 2009, up a total of 13.6 mn in 4 years. The share of people who benefited from rural water system reform increased from 94.1 % in 2005 to 94.3 % in 2009, up 0.2 ppt. In the first 4 years during the 11th *Five-Year Plan*, there were totally 23.2 mn households in China that started to use clean toilets, increasing by average 4 % annually. The ratio of clean toilets using increased by 7.9 ppts during the period, up 2 % each year on average. The use of pesticide and plastic membrane decreased over the past years, so that arable land pollution was cut at the same time of ensuring sufficient agricultural production.[1]

The government has made grand achievement in green investment during the 11th *Five-Year Plan*, and has paved a solid basis for the next stage of development. In the 12th *Five-Year Plan*, the Chinese government will be able to build on this basis, and realize even greater targets in the cause of green development.

Column 13.2 Dongshan District: The Oasis Near East China Sea

In the south of Fujian Province, Dongshan District locates at the intersection of East China Sea and South China Sea. It lies between Xiamen City and Shantou City, and only a small distance away from Taiwan. Dongshan is an island district and it is poor in natural resources like arable land and freshwater. The ecological environment of Dongshan used to be fragile since the Ming Dynasty. However, a green revolution has made great differences after the new P.R. China was founded.

[1] National Bureau of Statistics, Ministry of Environment Protection: "China Statistical Yearbook of Environment 2010", Beijing, China Statistics Publishing House, 2010.

In order to promote sustainable development and to build an environment-friendly society, Dongshan tried its best to enhance local ecological environment. To protect the limited cultivated lands, measures were taken to prohibit farmers from changing their farmlands to fishing grounds. Local government also planted a large number of trees to construct a windbreak belt, preserving the soils and water. Meanwhile, Dongshan banned sand dredging along the coast to support the development of clean production.

Dongshan saw many actions carried out to implement environment protection supervision during the 11th *Five-Year Plan* period. The government introduced many high technologies to supervise industrial production, reducing the marine pollution. They strictly complied with environment evaluation system and enhanced the management of new projects. Besides, they introduced a third party to monitor the enterprises, collecting the evidences of environment pollution.

Additionally, laws and regulations have been issued to ensure the green development in Dongshan. For example, provincial government and local government published the "*Construction Plan of Dongshan District as National Ecological Demonstration Zone in Fujian Province and the Coast Pollution Rectification Rules of Dongshan District*".

Dongshan became a National Sustainable Development Zone in 2002 and was awarded as one of the National Ecological Demonstration Districts in 2008. With extremely high forest coverage, Dongshan has become the Oasis of East China Sea today.

Source: This column is edited from materials provided by the National Sustainable Development Office of Dongshan District, Fujian Province.

13.1.1.4 Establishment of the Legal System for Government Green Investment

After several years of research and exploration, a comprehensive legal system for government green investment has been gradually established in China. The system is based on "*Constitution of the People's Republic of China*", and includes several special laws, administrative rules and local regulations. Many areas are covered in the legal system, including funding management of green investment projects, financial planning for green investment, financial auditing for green investment, etc.

In May 2007, the National Environment Protection Bureau and the Ministry of Finance published the "*Temporary Rules on the Funding Management of Central Fiscal Expenditure on Special Projects of Major Pollutant Reduction*". The Rules mainly focus on the fiscal funding management by the central government on projects of major pollutant reduction. The goal is to ensure that the targets of major pollutant reduction and related supervision and appraisal system can operate smoothly, so that the major pollutant reduction targets set in the 11th *Five-Year Plan* could be realized. In April 2009, the Ministry of Environment Protection and the Ministry of Finance published the "*Temporary Rules on the Funding Management of Central Fiscal Expenditure on Special Comprehensive Treatment Projects of*

Rural Environment Protection". The Rules provide new strategies to establish socialistic new rural area, to solve the increasingly important rural environment issue, and to improve the environment conditions in rural China. In May 2009, the Ministry of Environment Protection published the "*Notice on Strengthening the Audit of the Central Fiscal Special Funds for Environment Protection*". The Notice is a strong support to the strengthening and regulation of the management in central fiscal special funds for environment protection, to the improvement in efficiency of fund usage, to the improvement in project quality, and to the strengthening of project supervision.

Meanwhile, the central government, various Ministries and local governments also published a wide range of regulations on financial planning of investment projects and the management of "under-cover coffer". For example, there are "*Main Work Points for Financial Planning in the Environment Protection System in China*", "*Notice on the Further Work on the Management of 'Under-cover Coffer'*", "*Notice on the Establishment of Long-term Mechanism for the Management of 'Under-cover Coffer' by the Ministry of*", etc. These regulations actively promoted the strengthening of funding planning of government's fiscal expenditure on green development, the improvement of efficiency in the using of green investment fund, and the perfection of the supervision mechanism on green investment.

13.1.2 Main Problems with China Governments' Green Investment

The Chinese government put great emphasis on green investment in the course of green development, actively designed and perfected investment policies, increased funding for government investment, and made great contribution to China's green development. However, in meantime we must clearly recognize that with the challenging social and economic background nowadays, there are certain problems with green investment, which need to be dealt with during the next Five Year Plan.

13.1.2.1 Limited Funding Sources of Governments' Green Investment Projects

Although the size of funding for governments' green investment has been increasing steadily in recent years, compared with other types of investment, governments' green investment still relies on limited funding sources, and we still need to establish the market system for green investment. Since the targets of green investment (like environment protection, pollution treatment, forestry construction, rural environment comprehensive treatment, etc.) are all public goods, up till now most of the funding for green development is from the central and various local-level governments. There is only a small share of funding for green investment coming from the market system.

However, the development process of the socialistic market economy demonstrates that highly efficient and highly stable investment projects should get their funding from multiple sources and from the market system. The market should decide where the funding shall come from and invest into. Therefore, in the future work of the government, more emphasis should be put on how to fully utilize the function of markets in green investment, to perfect the market system in green investment, to utilize the market's function of resources allocation, to find new funding sources for China's green development, and to inject new energies to the whole cause of green development. In practice, the clarification of property rights by way of market mechanisms like auction and bidding could provide valuable experiences for green investment. And the future direction of green investment is to establish a financial market system of green finance and green bank credit.

13.1.2.2 Unbalanced Allocation of Funding

Since 2006, with the high attention from the central government, the size of funding for China governments' green investment has been growing steadily, and the results of these investments are showing up clearly. But in the specific using of green investment funding, because of various reason, there is the problem of unbalanced allocation. Every year the government will allocate 10 % of fiscal expenditure on green development. Within this 10 %, a large chunk is spent on pollution treatment, which has a substantially larger share than other categories of green investment, or 6 % out of the total 10 %. The share of investment in environment protection is also high at 2 %, and the ratio has been steadily increasing. In contrast, the shares for forestry construction and rural environment comprehensive treatment are relatively small at about 1.5 and 2 %, respectively. Such an allocation structure of funding resulted in the problem that every year pollution treatment and environment protection take up too much funding and the efficiency of fund using is not very high. At the same time, the amounting issue of forestry construction and rural pollution can not get enough funding and thus progress on these aspects have been slow.

13.1.2.3 Legal System for Government Green Investment Needs Improvement

There has been substantial improvement in China's legal system for government green investment, but the legal system today is still not yet good enough to meet the requirements of China's government green investment in practice, and thus is a constrain on China's course of green development. First, the legal system for supervision and appraisal of government green investment needs improvement. Because of the lack of supervision mechanism, there is lack of transparency in some local fund usage, and sometimes green investment fund is discounted in usage or some special project money is used in other areas. Because of the lack of appraisal mechanism, some of China's green investment funds are put to construction projects purely for the sake of showing-off, and such funds are not used for their more important purpose. Second, some laws and regulations overlap each other. Different laws and regulations are set by different government bodies based on their own interest. In the absence of long-term and coordinative system mechanism, there is an overlap of targets under various laws and regulations, which make the implementation of these laws and regulations unclear. Law-enforcing government bodies can not implement the laws even if the laws are already there, and this substantially lowers the efficiency of the laws and constrained green development in China. Third, the implementation of laws and regulations lag behind, and some of them are not practical. A considerable part of laws related to green investment are no longer suitable for modern green investment development. The strategy to develop the legal system for green investment should adjust the emphasis from ex post punishment to ex ante prevention, and from limitation and constraint to support and guidance, from being vague and overlapping to being clear and transparent. The objective is to absorb the idea of green investment into the legal system and regulation, and to protect the smooth development of green investment in China.

13.2 Regional Comparison of China Governments' Green Investment During the 11th Five-Year Plan Period

During the 11th *Five-Year Plan,* the Chinese economy witnessed fast growth and the overall size of China's economy become the second largest in the world. At the same time, economic development is constrained by both the shortage in resources and the environment pollution. Therefore, all levels of government try to transform the economic development pattern, to promote green development and to support green investment. According to the 2011 green development index system, this chapter tries to compare government green investment from the perspective of provincial-level and major cities comparison, in order to provide strong data and theoretical support of China governments' efforts to further promoted green development and green investment in the next 5 years.

13.2.1 Provincial Comparison of China Governments' Green Investment

In 2009, various levels of government made huge efforts in green development and lent strong support to green investment, in aspects like environment protection, pollution treatment, garden and park construction, rural water system reform, and toilets construction, etc. According to the indicators system, we selected some indicators and make comparisons of local green investment in the eastern, middle, western and north-eastern regions. We try to differentiate among these four regions in terms of their special characters and differences in green investment.

13.2.1.1 Green Investment in Eastern China

In recent years, the economic growth rate in eastern China regions has always been higher than that in middle, western and north-eastern China. With the fast growth of the economy, there come the issues of high pollution, high energy consumption and other serious environment problems in eastern China. Because of this, provincial governments in eastern China made great effort in pushing green development and strengthening green investment. Details about green investment in 10 eastern provinces and cities in 2009 are listed in Table 13.2.

As shown in the above chart, in the eastern region, Jiangsu and Shandong province had the largest amount of green investment in 2009, or RMB67.88bn and RMB67.66bn, respectively. The amount equaled to 16.9 and 20.71 % of the provinces' annual total fiscal expenditure, respectively. In contrast, Hainan province lagged behind with total investment amount of RMB4.51bn, or only about 6 % of the amount in the afore-mentioned two provinces, and which equaled to 9.27 % of Hainan's annual fiscal expenditure. In terms of the share of green investment in the local governments' fiscal expenditure, Shanghai, Fujian, Guangdong and Hainan all had ratios below 10, or 6.75, 9.01, 8.41 and 9.16 %, respectively, lower than most other eastern provinces. The seven more developed regions in eastern China (Beijing, Tianjin, Shanghai, Jiangsu, Zhejiang, Fujian and Shandong) spent more than 10 % of their fiscal expenditure on green investment. The ratio in Shandong was the highest at 20.71 %.

For investment in specific categories, Jiangsu had the largest amount of investment in environment protection, with a total amount of RMB14.76bn, accounting for 21.28 % of total such investment by the whole eastern China region. As for investment in pollution treatment, Shandong had the largest amount of investment at RMB45.95bn, which was about 67.91 % of total green investment by Shandong province, or 19.97 % of all such investment in eastern China. In terms of garden and park construction, Jiangsu was No. 1 way ahead with total investment of RMB13.94, while Guangdong had the smallest amount of investment at only RMB0.99bn. In the field of rural water system reform and toilets construction, Zhejiang, Guangdong and Jiangsu ranked the top 3 in terms of investment amount, at RMB2.69bn, RMB2.35bn and RMB2.19bn, respectively, which accounted for 20.84, 18.16 and 16.97 % of all such investment in eastern China, respectively.

The eastern region provides important leadership and guidance to the whole nation's economic development, and it has always led the country in terms of transformation of economic development pattern, economic structure adjustment and original innovation. The above characters of green investment in the eastern region are closely related to the economic development level, intensity of resources extraction and government policies implementation in eastern provinces.

Table 13.2 Green investment in Eastern China in 2009 (units: RMB100mn, %)

Province	Green investment projects					
	Environment protection investment	Pollution treatment investment	Garden and park construction investment	Rural water system reform and toilets construction investment	Total amount of government green investment	Share as total fiscal expenditure
Beijing	54.1	208.7	44.4	6.8	314.0	13.54
Tianjin	13.4	103.7	18.1	4.2	139.3	12.39
Hebei	104.2	248.6	72.6	4.2	429.6	18.30
Shanghai	34.0	160.1	7.8	7.6	209.5	7.01
Jiangsu	147.6	369.9	139.4	21.9	678.8	16.90
Zhejiang	55.4	198.0	34.3	26.9	314.6	11.86
Fujian	33.8	87.2	6.9	6.3	134.2	9.51
Shandong	76.2	459.5	125.2	15.8	676.6	20.71
Guangdong	100.8	240.1	9.9	23.5	374.3	8.63
Hainan	18.5	19.7	2.4	4.5	45.1	9.27

Sources National Bureau of Statistics: "*China Statistical Yearbook* 2010", Beijing, China Statistics Publishing Hose, 2010; Ministry of Environment Protection: "*Annual Statistic Report on Environment in China 2009*", Beijing, China Environment Science Publishing House, 2010; National Bureau of Statistics, Ministry of Environment Protection: "*China Statistical Yearbook of Environment 2010*", Beijing, China Statistics Publishing House, 2010

In 2009, the size of economies in Jiangsu and Shandong province ranked No. 2 and No. 3 in China. With strong growth in the economy, there comes more intensive resources extraction, and more environment pollution and industrial pollution. In face of bottle necks in development, Jiangsu province quickens its steps on economic development transformation and implements a series of investment measurements, e.g. the promotion of "green steel", establishment of high-end green and ecological water ways, strengthening the monitoring of waster water collective treatment facilities, treatment of particular environment issues in pollution treatment construction projects, etc. Therefore Jiangsu had relatively larger investment in environment protection and pollution treatment. In the area of construction of socialistic new rural area, Zhejiang, Guangdong and Jiangsu continues to quicken their steps of urbanization, and thus made large investment in rural water system reform and toilets construction. In particular, rural economic development in Zhejiang is a leading example in China. In 2009, Zhejiang provincial government published the "*Implementation Plan in Dinghai District of Rural Toilets Construction as Major Public Hygiene Projects in Zhejiang 2009*", which promoted the hygiene enhancement of rural environment, and focused on water system reform and toilets construction. Besides, Zhejiang also required that newly built houses in rural area must build waster water treatment facilities at the same time, and that old houses must quicken steps in such construction and renovations. Therefore, Zhejiang had quite large investment in these areas.

Table 13.3 Green investment in Middle China regions in 2009 (Units: RMB100mn, %)

Province	Green investment projects					
	Environment protection investment	Pollution treatment investment	Garden and park construction investment	Rural water system reform and toilets construction investment	Total amount of government green investment	Share as total fiscal expenditure
Shanxi	70.6	157.8	14.6	5.1	248.2	15.89
Anhui	59.3	139.2	52.8	12.9	264.2	12.33
Jiangxi	43.1	70.4	29.7	6.8	150.0	9.60
Henan	93.0	121.3	17.4	17.0	248.7	8.56
Hubei	74.2	150.6	21.2	21.6	267.6	12.80
Hunan	73.6	146.4	32.9	13.3	266.2	12.04

Sources National Bureau of Statistics: "*China Statistical Yearbook* 2010", Beijing, China Statistics Publishing Hose, 2010; Ministry of Environment Protection: "*Annual Statistic Report on Environment in China* 2009", Beijing, China Environment Science Publishing House, 2010; National Bureau of Statistics, Ministry of Environment Protection: "*China Statistical Yearbook of Environment* 2010", Beijing, China Statistics Publishing House, 2010

13.2.1.2 Green Investment in Middle China Regions

Middle China regions have always been the important production base for China's agricultural and husbandry products, energies and raw materials. Overall economic power in these regions is strong, but industrialization is generally at low levels. The heavy industries usually account for a large share, and the middle China regions mostly have their comparative advantages in some traditional industries, which have low technology content, low value-added, heavy pollution, and intensive use of resources and energy. Therefore, the middle regions should increase their efforts on government green investment, push on the cause of green development, and try to pursue a new road of industrialization. Detailed information about green investment in the 6 provinces and municipalities in middle China in 2009 is listed in Table 13.3.

As shown in the above chart, among middle China regions, Hubei and Hunan provinces had the largest amount of green investment in 2009, at RMB26.76bn and RMB26.62bn, respectively, which were equal to 12.8 and 12.04 % of each province's total fiscal expenditure that year. In contrast, Jiangxi province was lagged behind, whose total investment was only RMB14.5bn (9.6 % of its total provincial fiscal expenditure), or 56 % of the amount in the afore-mentioned two provinces. Besides, the share of green investment in total fiscal expenditure in Henan province was also low at 8.56 %.

In terms of specific types of investment, Henan province made the largest amount of investment in environment protection at RMB9.3bn, accounting for 17.8 % of total such investment in middle China. In terms of pollution treatment, Shanxi made the largest amount of investment at RMB15.78bn, about 64.78 % of total green investment in Shanxi, equaling to 16.44 % of total such investment in middle China regions. In terms of garden and park construction, Anhui made the largest amount of investment at RMB5.28bn, while Shanxi made the smallest amount of investment at RMB1.46bn. In terms of rural water system reform and toilets construction, Hebei and Hunan ranked top 2, with investment amounts of RMB2.16bn and RMB1.7bn, respectively, accounting for 24.19 and 19.07 % of total such investment in middle China regions.

There are huge differences of resources endowment among provinces in middle China, and the economic development stages in each province are different. In 2009, the economies of Hubei and Hunan were relatively more developed in middle China, and with economic development, they naturally tend to spend more on green investment. Shanxi is a major producer of coal, and because coal is heavily polluting and intensive in energy usage, Shanxi province has a more serious problem of environment pollution, thus spends more on pollution treatment. Anhui is a leading province in middle China in terms of garden and park construction, and has always been spending a lot on urban park and green land design and construction, as well as on the improvement of the public management system of urban parks and green land. In the field of rural water system reform and toilets construction, Hubei has investment large amounts. By 2009, Hubei has solved the clean drinking water issue for 10.82 million peasants, and 68 % of rural households now have their own toilets.

13.2.1.3 Green Investment in Western China

After more than ten years of "western China development" strategy, there have been substantial economic development and social progress in western China. Large achievements have been made in various fields, which strongly supported balanced regional economic development. In the future, the western region should grasp opportunities, hasten on the course to transform the economic development pattern, and put emphasis on improving the quality of development. The details of green development and green investment in western China in 2009 with its 11 provinces and municipalities are listed in Table 13.4.

As shown in Table 13.4, in western China in 2009, Sichuan province made the largest amount of green investment of RMB31.49bn, about 8.77 % of its annual fiscal expenditure. Hainan province made the smallest amount of green investment at RMB4.37bn, about 14 % the amount in Sichuan. Besides, Ningxia also made a small amount of green investment at RMB5.79bn. In terms of the share of government green investment in fiscal expenditure, Chongqing, Ningxia and Inner Mongolia all had ratios around 14 %, while Sichuan, Guizhou, Yunnan, Gansu and Qinghai all had ratios below 10 %. The ratio in Guizhou was the lowest at 6.28 %.

In terms of specific types of investment, Sichuan made the largest amount of investment in environment protection at RMB11.45bn, about 17.07 % of total such investment in western China. In the field of pollution treatment investment, Inner Mongolia ranked No. 1, with total investment of RMB15.52bn and Qinghai ranked the last with total investment of only RMB1.23bn. As for garden and park construction, Inner Mongolia ranked No. 1 with investment of RMB4.93, while

Table 13.4 Green investment in Western China in 2009 (Units: RMB100mn, %)

Province	Green investment projects					
	Environment protection investment	Pollution treatment investment	Garden and park construction investment	Rural water system reform and toilets construction investment	Total amount of government green investment	Share as total fiscal expenditure
Sichuan	114.5	103.5	27.6	69.3	314.9	8.77
Chongqing	50.1	109.7	42.1	10.9	212.8	16.47
Guizhou	55.3	21.2	0.6	9.1	86.2	6.28
Yunnan	82.2	79.6	7.0	8.6	177.4	9.09
Shaanxi	79.5	119.1	33.7	10.4	242.7	13.18
Gansu	53.2	44.4	2.8	7.2	107.5	8.63
Qinghai	29.0	12.3	1.2	1.2	43.7	8.97
Ningxia	22.6	28.7	4.3	2.4	57.9	13.39
Xinjiang	36.4	78.2	16.1	6.8	137.5	10.21
Guangxi	49.9	132.3	41.0	15.4	238.7	14.72
Inner Mongolia	97.9	155.2	49.3	7.0	309.3	16.05

Sources National Bureau of Statistics: "*China Statistical Yearbook* 2010", Beijing, China Statistics Publishing Hose, 2010; Ministry of Environment Protection: "*Annual Statistic Report on Environment in China 2009*", Beijing, China Environment Science Publishing House, 2010; National Bureau of Statistics, Ministry of Environment Protection: "*China Statistical Yearbook of Environment 2010*", Beijing, China Statistics Publishing House, 2010

Guizhou, Qinghai, Gansu and Ningxia made investments of RMB60mn, RMB120mn, RMB280mn and RMB430mn, respectively. In terms of rural water system reform and toilets construction, Sichuan ranked No. 1 with investment of RMB6.93bn, accounting for 46.76 % of total such investment in western China, while Qinghai and Ningxia only invested RMB120mn and RMB240mn, about 1.67 and 3.39 % the amount invested in Sichuan, respectively.

For a long time, there are huge difference in the stages of economic development, resources endowments and utilization capacities among various provinces in western China. Therefore, these provinces made largely different levels of green investment. In 2009, Sichuan province ranked high within this region in terms of economic development, and it strongly pushed forward infrastructure investment in both urban and rural areas, especially during the period of intensive construction activities after the 2008 earthquake. Therefore, Sichuan made huge amount of green investment in environment protection, rural water system reform and toilets construction, etc. In contrast, Qinghai and Ningxia made relatively small amounts of investment because (1) they have relatively less advanced economies and thus have less pollution and (2) they have not paid enough attention to strengthen comprehensive environment treatment. Specifically, Qinghai needs to pay more attention to major ecological projects like the nature protection zone of "three rivers sources". Wuhai and Erdos of Inner Mongolia have always been the largest resources-rich regions in northern China, and are the major production bases of coal, electricity, metallurgy, chemical products and construction materials in western China. Because of the old-fashioned strategy of energy-consumption and industrial development patterns, there are serious problems of environment pollution and resources wasting in the area. Therefore, Inner Mongolia made large efforts in aborting redundant production capacities and in comprehensive treatment of industrial wastes, and made large investment in pollution treatment.

Column 13.3 Guilin: Scientific Protection of Lijiang River to Achieve Green Development

Guilin is in the northeast of Guangxi Zhuang Autonomous Region, which is located in the southern end of the corridor between Hunan and Guangxi. It is not only a major tourist city in China, but also a historical and cultural city of China. In Guilin, mountains are beautiful, landscape is exotic. Above all the Lijiang River scenery enjoys the reputation of the most beautiful one in the country.

In 1973, Deng Xiaoping visited Lijiang River and said: "If we destruct the environment of the Lijiang River in order to develop the production, then that is not worth it." Over the years, with the words from Deng Xiaoping as guidance and under the leadership of municipal government, Guilin people explore a scientific way to protect Lijiang River with the basic idea of green development.

During the 11th Five-Year Period, Guilin proposed to focus on urban development strategy, such as creating a modern international tourist city, historical and cultural city and ecological landscape city. Meanwhile, the government put a lot of manpower, material and financial resources to shut down or relocate the dozens of factories and enterprises with heavy pollution and poor environment-friendliness along the Lijiang River. Through the pollution control and environment protection to Lijiang River and its tributaries, water quality has greatly improved and the urban water environment has also been greatly enhanced. Meanwhile, in order to reduce the effect to air quality of the Lijiang River from emissions of gases such as by motor vehicle, Guilin vigorously support new energy development and invest in building the electric car charging stations, using science and technology to protect the Lijiang River. In addition, Guilin has invested heavily in the strict control of infrastructure projects within the scenic area along the river to vigorously cultivate green vegetation, and finally to enhance the level of ecological civilization of Guilin.

To change the impact on Guilin Lijiang River's landscape from the current urban development and industrial sites distribution, so as to minimize the negative effects from the process of urbanization and economic growth on the natural scenery, Guilin build a new model to invest hugely for planning and construction of a new area without the traditional pattern of building and developing along the major river. Thus the whole city and industrial center will be moved westward. "Protect the Lijiang River, push the development of Lingui, recreate a new Guilin", has been identified as the new development strategies and initiatives to protect Lijiang River. To fully implement the resource-saving and environment-friendly development strategy,Guilin will reduce the old city's population density, free up green space along the Lijiang River, in order to achieve the aim of harmonious development between Guilin urban area and ecological protection of Lijiang River. Today Lingui "new area" is in the full construction to promote environmental protection along Lijiang River to a new level.

On the road of the protection of the Lijiang River, Guilin also actively mobilizes people's power, so that the massive people can get involved and contribute to the protection of the ecological environment. At first Guilin set up a youth volunteer team to protect Mother River, and then held some special campaigns, such as Lijiang River Youth Forum.

Through the efforts of the Guilin government and people, now the Lijiang River's water is still crystal clear, still tree-lined along the two sides. Under the leadership of the municipal government of Lijiang River, local people protect Lijiang River in various scientific projects and means to achieve green development.

Sources: Guilin environmental network; Baidu Encyclopedia: Guilin; Green investment in north-eastern China

13.2.1.4 Green Investment in North-Eastern China

As an old industrial production base, many regions in north-eastern China have been suffering from high energy-consumption and heavy pollution. There are quite a bunch of cities where resources are close to exhaustion. Water and air pollution and ecological issues have become huge hurdles to fast economic development in north-eastern China. In order to promote the transformation of economic development pattern in the region, north-eastern China has been increasing its government green investment to promote green economic development. Details of green investment in the three provinces in north-eastern China in 2009 are listed in the Table 13.5.

As shown in the above table, in north-eastern China, Liaoning made the largest amount of green investment in 2009 at RMB29.8bn, about 11.11 % of total fiscal expenditure that year; Heilongjiang ranked No. 2 with total green investment of RMB18.93bn, or about 10.08 % of its total fiscal expenditure; Jilin made the smallest amount of green investment at RMB13.27bn, about 8.97 % of total fiscal expenditure, or about 45 % of the amount invested in Liaoning.

In terms of specific investment categories, there is not much difference among the three provinces in environment protection investment. Heilongjiang and Liaoning made a bit larger investment at RMB5.91bn and RMB5.57bn, respectively, while Jilin made investment of RMB4.95bn. In the field of pollution treatment, Liaoning made the largest amount of investment at RMB20.49bn, Heilongjiang made investment of RMB10.78bn and Jilin made the smallest amount of investment at RMB6.61bn, which is about 32 or 61 % of the amounts invested in Jilin or Heilongjiang, respectively. In terms of garden and park construction, Liaoning invested RMB2.99bn, ranking No. 1, while Heilongjiang invested RMB1.92bn, and Jilin invested RMB780mn, ranking the last. In the field of rural water system

Table 13.5 Green investment in north-eastern China in 2009 (Units: RMB100mn, %)

Province	Green investment projects					
	Environment protection investment	Pollution treatment investment	Garden and park construction investment	Rural water system reform and toilets construction investment	Total amount of government green investment	Share as total fiscal expenditure
Liaoning	55.7	204.9	29.9	7.5	298.0	11.11
Jilin	49.5	66.1	7.8	9.3	132.7	8.97
Heilongjiang	59.1	107.8	19.2	3.3	189.3	10.08

Sources National Bureau of Statistics: "*China Statistical Yearbook* 2010", Beijing, China Statistics Publishing Hose, 2010; Ministry of Environment Protection: "*Annual Statistic Report on Environment in China* 2009", Beijing, China Environment Science Publishing House, 2010; National Bureau of Statistics, Ministry of Environment Protection: "*China Statistical Yearbook of Environment* 2010", Beijing, China Statistics Publishing House, 2010

reform and toilets construction, Jilin made the largest amount of investment at RMB930mn, Liaoning made investment of RMB750mn and Heilongjiang made investment RMB330mn.

The three provinces in north-eastern China have different levels of economic development. In recent years, the provincial governments set clear targets to push the transformation of those areas where natural resources are getting exhausted, and to push the reform of long-dated industrial bases. Specifically, Liaoning continued to strengthen efforts in the transformation of energy-intensive and heavily polluting industries, in promotion of green development of the whole province, and thus made large amount of green investment. Jilin focused on the protection of ecological forests and economic development pattern reform, and on the protection and construction of "general ecological circle" of the whole north-eastern China region. Heilongjiang is rich in forest and grass land resources and made substantial efforts in protection the "black soil", wet land, forests and grass land, and thus made relatively large investment in garden and park construction. Besides, Heilongjiang, Jilin and Liaoning co-operated with Inner Mongolia in western China regions in terms of ecological projects, and made huge efforts to protect the Changbaishan forestry area and Xing'anling Mountains. The goal is to build the largest-sized forestry land in China.

13.2.1.5 Comparisons of Government Green Investment in Eastern, Middle, Western and North-Eastern China

For a long time, there have been large discrepancies in economic development levels, natural resources endowment, resources extraction etc. among different regions in China. Naturally, there are big different in government green investment for these 4 regions. Table 13.6 lists out those differences in 2009 in detail.

Table 13.6 Comparison of government green investment in eastern, middle, western and north-eastern China in 2009 (Units: RMB100mn, %)

Region	Green investment projects					
	Environment protection investment	Pollution treatment investment	Garden and park construction investment	Rural water system reform and toilets construction investment	Total amount of government green investment	Share as total fiscal expenditure
Eastern	693.6	2300.4	461.0	129.2	3584.2	14.36
Middle	522.3	959.6	168.7	89.2	1739.9	13.95
Western	670.5	884.2	225.7	148.3	1928.6	11.27
North-	eastern	164.3	378.8	56.8	20.1	620.0
10.27						

Sources National Bureau of Statistics: "*China Statistical Yearbook* 2010", Beijing, China Statistics Publishing Hose, 2010; Ministry of Environment Protection: "*Annual Statistic Report on Environment in China* 2009", Beijing, China Environment Science Publishing House, 2010; National Bureau of Statistics, Ministry of Environment Protection: "*China Statistical Yearbook of Environment* 2010", Beijing, China Statistics Publishing House, 2010

As shown in above table, the ranking of shares of green investment in total fiscal expenditure was eastern, middle, western, and north-eastern, at 14.36, 13.95, 11.27 and 10.27 %, respectively. Since there are abundant natural resources like forestry in the three provinces in north-eastern China, which requires large sum of investment, those provinces have larger investment in garden and park construction. Also because there are only 3 provinces in north-eastern China, green investment in other categories in this region are much smaller than those in eastern, middle and western regions, and are thus not quite comparable. Therefore, in this section, we mainly focus on the comparison between eastern, middle and western regions. In 2009, total amount of green investment in eastern China was far larger than those in middle and western regions, at RMB358.42bn, compared to RMB173.99bn and RMB192.86bn in middle and western China. The amounts in middle and western China were equal to about 48.53 and 53.80 % of that in eastern China.

In terms of specific investment categories, eastern China made the largest amount of green investment in environment protection at RMB69.36bn, accounting for 37.77 % of total such investment in the whole country. Middle China region had the smallest environment protection investment at RMB52.23bn, about 27.69 % of national total. In the field of pollution treatment investment, eastern China also was by far the No. 1 with investment of RMB230.04bn, about 55.51 % of national total, while middle and western regions had similar levels of investment at RMB95.96bn and RMB88.42bn, or 23.16 and 21.34 % of national total. In terms of garden and park construction, the ranking in amount of investment was eastern, western and middle, with amounts of RMB46.1bn, RMB22.57bn and RMB16.87bn, respectively, accounting each for 50.54, 24.73 and 18.50 % of national total. In terms of rural water system reform and toilets construction, the ranking in amount of investment was western, eastern and middle, with amounts of RMB14.83bn, RMB12.92bn and RMB8.92bn, respectively, accounting each for 40.43, 35.23 and 24.34 % of national total.

The above-mentioned characters of green investment in eastern, middle and western China are closely related not only to the different levels of economic development and intensities of resources utilization in these regions, but also to the important fact that these regions have different regional economic development policies and self-positioning of the region's function in the whole national economy. Eastern China region focuses on cultivating new industrial competitiveness, quickening the development of strategic new industries, enhancing efforts in environment pollution treatment, and dealing with the bottle-necks of resources and environment constraints in the process of development. The middle region provides a connection between the eastern and western regions, and is a key base for agricultural production for the country, a key base for energy and raw materials, a production bases for modern equipment manufacturing and high-tech industries, and also have many critical transportation nexus. The western region sticks to the strategy of "western China development", continues to emphasize infrastructure investment, to strengthen ecological environment protection, to strengthen efforts in the prevention and dealing with natural disasters, and to push forward the construction of key ecological function areas. Because of these different characters of regional economies and development policies, eastern China region makes more investment in government environment protection, especially in pollution treatment. Western region of China makes more investment in garden and park construction, rural water system reform and toilets construction, which is closely related to the natural resources endowment in the region and its policy orientation to focus on infrastructure construction and coordinated urban and rural development.

13.2.2 Government Green Investment in Major Cities

Green development is not only the responsibility of the central and provincial government. All levels of local governments should participate. During the 11th *Five-Year Plan*, many cities in China paid large attention to green development and gave strong support to green investment. They made big achievements in environment protection, waste water treatment, waste gas treatment, garden and park construction, urban environment and hygiene construction, etc. Bases on the geographical pattern of China's government administration, we selected four municipalities directly under the jurisdiction of the central government, four provincial capital cities and five specifically designated cities in the state plan for comparison. We try to analyze the characters and difference in the green investment among these cities.

13.2.2.1 Green Investment of Four Municipalities Directly Under the Jurisdiction of the Central Government

In China, municipalities directly under the jurisdiction of the central government are under direct leadership of the central government, and they are equal to the provincial level in terms of administrative hierarchy. These municipalities directly under the jurisdiction of the central government have the characters of both provinces and independent cities. They have special administrative powers that other normal cities do not have, and they can enjoy special policies and preferential treatment that are absent to other cities. Therefore, in the field of green investment, these municipalities directly under the jurisdiction of the central government have their own special characters, which are worth of our analysis. The details of green investment in the four municipalities directly under the jurisdiction of the central government directly under the center government in 2009 are listed in Table 13.7.

In 2009, among the four municipalities directly under the jurisdiction of the central government, Beijing and Chongqing made the largest amount of green investment of RMB14.3bn and RMB10.2bn, accounting for 6.15 and 7.73 % of each city's annual fiscal expenditure. Shanghai and Tianjin were lagged behind with investment amounts of only about 40 % as those made in Beijing and Chonqqing, or RMB4.6bn and RMB4.3bn, respectively, accounting for 1.54 and 3.84 % of their annual fiscal expenditure. Beijing made the largest amount of investment in urban environment and hygiene construction, with an amount of 42.05 % of total green investment. Tianjin made the largest amount of investment in garden and park construction with 41.86 % of its total green investment amount. Shanghai and Chongqing both made the largest amount of investment in environment protection, whose investment amount equaled to 54.3 and 52 % of their annual total green investment, respectively.

Specifically in terms of individual investment categories, Chongqing made the largest amount of investment in environment protection of RMB5.3bn, accounting for 42.53 % of total such investment in all the four municipalities directly under the jurisdiction of the central government. In the field of waste water treatment projects, Tianjin made the largest amount of investment at RMB407mn, accounting for 50.7 % of total such investment in all the four municipalities. In the field of waste gas treatment, Tianjin also ranked No. 1, with total investment of RMB759mn, or 42.2 % of total such investment in all the four municipalities. As for garden and park construction and urban environment and hygiene construction, Beijing made the largest amount of investment, at RMB4.4bn and RMB6.0bn, accounting for 49.5 and 84.6 % of total such investments in all four major municipalities, respectively.

The different characters of green investment in Beijing, Tianjin, Shanghai and Chongqing are related to the different characters of economic and social development in these cities. During the 11th *Five-Year Plan*, Beijing was planning to host the 2008 Olympic Games and to construct a world-class municipality, and put a lot of efforts in urban garden and park construction and urban environment

Table 13.7 Government green investment in 4 municipalities directly under the jurisdiction of the central government in 2009 (Units: RMB10th, %)

Green investment projects	Environment protection investment	Waste water treatment investment	Waste gas treatment investment	Garden and park investment	Urban environment and hygiene investment	Total amount of government green investment	Share in total fiscal expenditure
Beijing	354,688	1,205	25,718	444,321	599,432	1,425,365	6.15
Tianjin	109,789	40,867	75,921	180,727	24,398	431,702	3.84
Shanghai	250,758	9,703	40,746	77,791	82,800	98	1.54
Chongqing	529,297	28,813	37,522	420,780	1,982	1,018,394	7.73

Sources National Bureau of Statistics: "*China Statistical Yearbook* 2010", Beijing, China Statistics Publishing Hose, 2010; Ministry of Environment Protection: "*Annual Statistic Report on Environment in China* 2009", Beijing, China Environment Science Publishing House, 2010; National Bureau of Statistics, Ministry of Environment Protection: "*China Urban Construction Statistical Yearbook* 2009", Beijing, China Statistics Publishing House, 2010

planning and regulation, so it made relatively large amount of investment in these areas. Shanghai was planning to host the 2010 World Exposition, and focused on protection and improvement of urban environment, and allocated a large amount of human and monetary resources to these tasks. Chongqing is the youngest municipality among the four, with a large land area and a large rural region. It had a relatively lower urbanization ratio, and thus its government green investment has similar characters with those in Sichuan and Hunan provinces. In addition, Chongqing government has been paying special attention to green development and green investment, and strengthened its efforts on environment protection, garden and park construction, waste water and gas treatment, etc. and thus it has invested relatively large amount of money to green investment, with the highest ratio to annual fiscal expenditure among the four municipalities.

Column 13.4 Shanghai Urban Green Land Construction

Shanghai which is one of the four municipalities directly under the jurisdiction of the central government is the largest city in the mainland of China. It has an area of 63.4 km^2 (at the end of 2009), and the total urban population is 17,020,000 people (at the end of 2009). In recent years, Shanghai has been improving the urban ecological construction comprehensively. In 2009, Shanghai urban green land area covers 11.69 million hectares, ranking among the top five all over the country. Urban green area per capita is 87.81 ha/million, ranking top three. Under the guidance of the "One City, Nine Towns" development model, urban development in Shanghai has focused on more "living area", with the urban public "activity centers" and "collective living cycles" in a decentralized pattern, which make up the beautiful ecology landscape in Shanghai. All in all, the Shanghai Municipal Government has made important achievements to promote green investment.

1. Ten years ago: the two "theme parks"

In 2000, Shanghai has completed the construction of "Pudong Century Park" and "Hongqiao Park" in Changning. With these two parks, Shanghai has further enhanced the urban residents' green area per capita and improved residents' quality of life. The Pudong century park covers the total area of 140.3 ha with investment of RMB 10 billion. The main facilities in the park are: weather stations, large lawn, ginkgo avenue, greenhouse, streams, seasonal flower beds, and Montreal garden; Hongqiao park is not only a public open-style theme park to meet the citizen's entertainment and recreational needs, but also a large ornamental garden area,which covers an area of 20,000 m^2 with more than 13,000 m^2 of green area, about 4,000 kinds of trees and a total of 122 species.

2. Five years ago: more "urban green land" to form a healthy life style for Shanghai residents.

In 2006, under the guidance of "One City, Nine Towns" development strategy of the city, Shanghai constructed many urban green land, from the construction of

"Huang Xing green land" in Yangpu District to "Daning green land" in Zhabei District, and from the "Qibao Sports Park" in Minhang District to the "Gucun living eco-park " in Baoshan District. With the construction of these parks, Shanghai has been building the urban environment from the "inner ring" to the "outer ring", so that more people can enjoy the healthy eco-living life style brought by "green city".

3. Nowadays: rebuild the urban landscape

In 2009, the first stage of "Gucun living eco-park" has completed, making up the blanks in urban ecological space in the north of shanghai. "Gucun living eco-park" is located in the northwest of Shanghai, where Gu Town is in Baoshan District, which is at the most vibrant part of the city and also is an important node on the system around the city's ecological planning. The master plan for the "Gucun living eco-park" area covers 434.5 ha, a part of Shanghai's largest urban green space which equals to three "Pudong Century Parks". Among them, the first stage of the park covers an area of 180 ha of theme parks, such as country forest garden, exotic garden, barbecue site, forest walks, and other ornamental garden plants. Currently, the "Gucun living eco-park" has been the most new bright spot around the eco-city construction and development of the whole city of Shanghai. The main planning feature is: "an axis" i.e. Chen Fu Road Park landscape development axis; "an area" i.e. 200 m^2 of ecological function forest along the Outer Ring Road; "three places" i.e. the culture forest area, healthy forest area and forest conservation area.

Sources: Shanghai Municipal Government website; National Bureau of Statistics; The Ministry of Housing and Urban–Rural Construction.

13.2.2.2 Green Investment in Typical Provincial Capital Cities

Provincial capital cities are usually the political and economic centers of each province, and are the focus of economic development policies in each province. Bases on their special positions in each province, provincial capital cities have the best resources and most preferential policies, and thus have the fastest economic development. In the cause of green development, provincial capital cities usually are more aware of the importance of green investment, and always have more funding available. Therefore, provincial capital cities usually do very well in terms of green investment. To analyze the green investment activities of provincial capital cities in China, we select one typical provincial capital city from each of the eastern, middle, western and north-eastern China, and list out the details of their green investment in related fields in 2009, as shown in Table 13.8.

As shown in the above table, within the 4 provincial capital cities of Nanjing, Wuhan, Chengdu and Changchun, Nanjing as a provincial capital cities in eastern China made the largest amount of green investment of RMB4.4bn, or 9.6 % of its annual fiscal expenditure. No. 2 is Chengdu, a provincial capital cities in western China, with total green investment amount of RMB3.35bn, or 5.58 % of its annual fiscal expenditure. No. 3 is Wuhan, a provincial capital cities in middle China, with total investment of RMB3.35bn, or 6.3 % of its annual fiscal expenditure. Changchun, a provincial capital cities in north-eastern China made the smallest amount of green investment, which is no more than half of those made in the other 3 cities, or 5.29 % of its own annual fiscal expenditure. Nanjing, Wuhan and Chengdu all made large amount of investment in garden and park construction, accounting for 65.77, 39.73 and 65.42 % of each local government's annual total green investment. Changchun made its largest green investment in environment protection, which accounts for 41.22 % of its total green investment.

In individual investment categories, Chengdu made the largest amount of investment in environment protection of RMB882mn, or 34.49 % of total such investment in all 4 cities. In the field of waste water treatment, waste gas treatment and urban environment and hygiene construction, Wuhan ranked No. 1, accounting for 44.76, 46.82 and 50.26 % of total such investments in 4 cities, respectively. In the field of garden and park construction, Nanjing made the largest amount of investment of RMB2.9bn, or 43.59 % of total such investment in all 4 cities.

As the above analysis demonstrates, the pattern of green investment in provincial capital cities is closely related with its geographic location. In the 4 typical provincial cities of Nanjing, Wuhan, Chengdu and Changchun, the city in eastern China has more developed economy and better social development, and thus it makes more reasonable and balanced investments in all categories of green investment. The city in middle China is in the key stage of industrial progress to pull up the overall regional development, and has more serious problems of industrial pollution, and thus makes more investment in industrial pollution treatment. Western city faces serious challenges in urban environment, and thus makes more investment in environmental protection. The city in north-eastern China had started urban construction much earlier, and its urban infrastructure is more out-dated, and thus it makes more investment in urban environment and hygiene construction investment.

13.2.2.3 Green Investment in 5 Specifically Designated Cities in the State Plan by the Central Government

Specifically designated cities in the state plan by the central government in China have equal economic management power as provincial level government, but are lower than provincial government in political hierarchy. They are not capital cities of the provinces, but are considered to be at equal level with capital cities. There are five specifically designated cities in China, namely Dalian, Qingdao, Ningbo, Xiamen and Shenzhen. They, just like capital cities, have abundant resources and receive preferential policies, and have made big achievement in green development. The green investment of these five cities in 2009 are listed in Table 13.9.

Among the five cities, Shenzhen made the largest amount of green investment of RMB17.8bn, or 1.78 % of its annual fiscal expenditure. Dalian, Qingdao and Ningbo made similar amounts of green investment of between RMB1-1.5bn.

Table 13.8 Government green investment in typical provincial capital cities (Units: RMB10th, %)

Green investment projects	Environment protection investment	Waste water treatment investment	Waster gas treatment investment	Garden and park investment	Urban environment and hygiene investment	Total amount of government green investment	Share in total fiscal expenditure
Nanjing	70,751	16,818	24,872	291,105	39,082	442,628	9.60
Wuhan	30,063	24,475	46,496	125,966	90,091	317,091	6.30
Chengdu	88,179	9,185	7,985	219,282	10,546	335,176	5.58
Changchun	66,679	4,202	19,962	31,424	39,516	161,783	5.29

Sources National Bureau of Statistics: "*China Statistical Yearbook* 2010", Beijing, China Statistics Publishing Hose, 2010; Ministry of Environment Protection: "*Annual Statistic Report on Environment in China* 2009", Beijing, China Environment Science Publishing House, 2010; National Bureau of Statistics, Ministry of Environment Protection: "*China Urban Construction Statistical Yearbook* 2009", Beijing, China Statistics Publishing House, 2010

Xiamen government made the smallest amount of green investment at RMB716mn, or 2.67 % of its annual fiscal expenditure. Dalian, Ningbo and Shenzhen mainly focused on environment protection investment, amounting to 55.43, 59.62 and 72.05 % of their total green investment, respectively. Qingdao mainly focused on garden and park construction, with amount equaling to 55.21 % of its total green investment. Xiamen focused on urban environment and hygiene investment, amounting to 58.19 % of its total green investment.

In terms of specific investment categories, Shenzhen ranked No. 1 in environment protection and waste water treatment, with amount of RMB1.28bn and RMB169mn, accounting for 41.11 and 50.01 % of total such investments by all five cities. In the field of waste gas treatment, Ningbo ranked No. 1 with investment of RMB170mn, or 53.23 % of total such investment by all five cities. In the field of garden and park construction, Qingdao made the largest amount of investment at RMB827mn, or 52.63 % of total such investment in all five cities. In the field of urban environment and hygiene construction, Xiamen made the most investment with amounts equaling 49.46 % of total such investments in all five cities.

Among the green investment activities in the five major specifically designated cities in the state plan in mentioned above, Dalian, Qingdao and Xiamen are famous within China and overseas for their beautiful environment, and thus made relatively larger investments in environment protection, garden and park construction, and urban environment and hygiene construction. Ningbo has been putting a lot of efforts in developing its industries, and thus made large investments in industrial waste gas treatment. Shenzhen has relatively higher level of economic growth, and has bigger environmental issues, and thus made relatively larger investments in environment protection.

13.2.2.4 Conclusion

By the comparison of green investment activities in major cities in 2009, we can see that because of different functions, self-images and characteristics of different cities, their green investment also have different characteristics. Municipalities directly under the jurisdiction of the central government have clear advantages in terms of resources and policies because of their special political and economic positions, and so they tend to make the most efforts in green development and made the largest amount of green investment. In comparison, provincial capital cities and specifically designated cities in the state plan have lower and lower amount of resources, preferential policy treatment and funding sources, and thus have less and less amount of green investment. In 2009, the average amount of green investment for municipalities directly under the jurisdiction of the central government, provincial capital cities and specifically designated cities in the state plan were RMB8.34bn, RMB3.14bn and RMB1.24bn, respectively. The differences are obvious. The amount for provincial capital cities is less than half of the amount for

Table 13.9 Green investment of 5 major specifically designated cities in the state plan in China in 2009 (Units: RMB10th, %)

Green investment projects	Environment protection investment	Waste water treatment investment	Waster gas treatment investment	Garden and park investment	Urban environment and hygiene investment	Total amount of government green investment	Share in total fiscal expenditure
Dalian	65,699	3,401	10,913	28,501	10,007	118,521	2.52
Qingdao	38,706	5,975	2,021	82,654	20,349	149,706	3.45
Ningbo	60,231	4,979	17,036	17,998	785	101,029	2.00
Xiamen	18,804	2,554	1,398	7,159	41,641	71,556	2.67
Shenzhen	128,052	16,915	637	20,729	11,405	177,738	1.78

Sources National Bureau of Statistics: "*China Statistical Yearbook* 2010", Beijing, China Statistics Publishing Hose, 2010; Ministry of Environment Protection: "*Annual Statistic Report on Environment in China* 2009", Beijing, China Environment Science Publishing House, 2010; National Bureau of Statistics, Ministry of Environment Protection: "*China Urban Construction Statistical Yearbook* 2009", Beijing, China Statistics Publishing House, 2010

the municipalities directly under the jurisdiction of the central government, and the amount for specifically designated cities in the state plan is only 1/7 of the amount for the municipalities directly under the jurisdiction of the central government.

13.3 Prospects of Government Green Investment in China During the 12th Five-Year Plan Period

During the 12th *Five-Year Plan*, the Chinese government insists on its emphasis about building a resource-conserving & environment-friendly society as the way to quicken the steps on transforming economic development patterns. The government adopts as basic national strategies to implement policies of economic using of resource and environment protection; it continues to strengthen efforts on economic energy consumption and pollution reduction, on developing recycling economy, on dealing with global climate changes, on taking the right path on green investment, on aiming at right targets in green investment, on pushing forward green development work, and on making new progress on government green investment.

The basic strategies and directions of development are clearly stated in the "*The* 12th *Five-Year Plan for the National Economic and Social Development of the People's Republic of China*" (12th *Five-Year Plan*), and are listed in Table 13.10.

As shown in the above table, in the next 5 years, the Chinese government established specific targets for saving resources, environment protection, and pollution treatment, dealing with global climate change, developing the recycling economy, and cultivating strategic newly emerging industries. Hence, with a comprehensive analysis of domestic development prospects, ahead of us is an important strategic period for China's government green investment. The Chinese government should solidly grasp this opportunity, and push forward government green investment, be aware of the challenges, and be ready to deal with difficult situations. On the basis of this, all levels of governments should adopt a series of effective measure, to promote the fruitful development of government green investment in China.

13.3.1 Continue to Perfect the Government Green Investment System in China

Green investment system is the kind of investment system that is built for the objectives of environment protection, promoting the development of recycling economy, and establishment of a harmonious society for man and nature. It includes a wide range of investments for pollution prevention and treatment,

Table 13.10 China government green investment targets for the 12th Five-Year Plan period

Economic using of resources	Maintain mu1.818bn of farm land, reduce water usage of per unit of industrial value-added by 30 %, increase the efficiency ratio of agricultural irrigation to 0.53, promote efficient and economic usage of water in agriculture like pipe water and irrigation under membrane, increase mu 50mn with high-efficiency irrigation, reduce construction land using per unit of GDP by 30 %
Environment protection	Reduce energy usage of per unit of GDP by 16 %, reduce CO_2 emission of per unit of GDP by 17 %, improve forestry coverage ratio to 21.66 %, increase total stock of forestry reserve to 600mn cubic meters
Pollution treatment	Substantial reduce in major pollution emission, 8 % decrease in SO_2 emission and chemical oxygen demand, 10 % decrease in NO chemicals emission, 85 % of urban waste water treatment and 80 % of urban garbage non-polluting treatment
Dealing with climate changes	Control of green house gas emission, utilization of measures like forestry absorption of carbon, promote forestry construction, new forestry land area of ha12.5mn
Development of recycling economy	Promote recycling production methods, increase the comprehensive utilization rate of solid wastes to 72 %, improve resources production rate by 15 %

Source "*The* 12th *Five-Year Plan for the National Economic and Social Development of the People's Republic of China*", Beijing, People's Publishing House, 2011

ecological environment protection and improvement, etc. It is a necessary route to establish a resource-conserving and society with recycling economy. With the adoption of scientific concept of development, the Chinese government has announced its determination to promote the recycling economy, and the green investment system is becoming more and more important in the overall investment structure in China. In the next 5 years, the Chinese government must continue to perfect the government green investment system, and to ensure the smooth implementation and progress in green investment. First, we must establish a series of formal institutional arrangements including that for environment protection, scientific using of funding, pollution treatment (clean production) and recycling use of materials, etc. We should rely on institutional arrangements to ensure that relevant government bodies will fulfill their duties in green investment. In addition, we must strength the supervision on the using of green investment funds, to avoid improper using of those funds, and to ensure that those funds are spent on where they are most needed. Also, the government should educate the general public to pay attention to environment protection, economic using of resources, recycling use of materials, and green consumptions, etc. These are the informal part of social institutions. With wide-spread propaganda from the government to promote environment protection, green investment shall become a natural habit and social awareness in the general public.

13.3.2 More Government Support on Green Investment with Fiscal and Financial Resources

In the future, the Chinese government shall put up more efforts to design a range of fiscal and financial policies to support government green investment. On the one hand, government shall design fiscal policies to support green investment, e.g. tax preferential, subsidies to green investment projects, etc. to reduce the cost for investors and enhance their profit, and collection of resources tax and environment tax to increase the cost for resources users and environment polluters, so as to encourage investors to make green investment. On the other hand, the government should design a financial system to support green investment, e.g. the state-owned commercial banks shall introduce priority loan treatment to corporations' green investment, and the policy banks should focus on investments that suit the national industrial policies and sectors and industries that fit the need of social and economic development, and the government shall encourage establishment of private financial organizations and to let them fully use their financing capacities, and the government shall also continue to expand and further promote international financing channels, so that we can seek funding from both domestic and international markets, and use the money to support green investment in China.

13.3.3 More Efforts in the Promotion of Government Green Investment in Key Areas

On the condition that the green investment system in China continues to be improved, the government should clarify on the key areas of future green investment, and to ensure that funds for green investment are efficiently used.

First, the government should promote the development of green industries. On the one hand, the government should promote the development of modern ecological agriculture, and continue to quicken the transformation of agricultural development. The government should try to promote the comprehensive production capacity of agriculture, the risk resistance ability of agriculture and the market competitiveness of agriculture. The government should continue to promote scientific and technological innovation of agriculture, to establish a better public distribution system of agricultural technologies, to development modern planting industries, and to promote machinery utilization in agriculture. The government should promote the standardization, specialization, economy of scale, and intensity of agricultural production, so that China could develop a modern agricultural industry system with higher production, better quality, higher efficiency, ecological and safe. On the other hand, the government shall promote the development of modern ecological industries, and try to realize the green transformation of industries in China. The government should insist on the combination of the fundamental role of markets and the guiding role of the government, so the enterprises can continuously optimize and upgrade the traditional industries by themselves, and the government could continuously strengthen its fiscal and financial support. The government should actively cultivate strategic newly emerging industries, based on the specific situations in China and the technological and industrial basis in China. The focus for the current period is to cultivate and promote industries like energy-saving and environment protection, alternative energies, new materials, new energy automobiles, etc., so that a series of leading and pillar industries could be established quickly, and the core competence and economic returns of industries could be established. Besides, the government should promote the development of the tertiary industry to make it more environment and resources-friendly.

Second, the government should make more efforts in the construction of green parks. Green parks are parks that include several related enterprises into one park, under the guiding principle of industrial rotation and co-existence. In the parks, all enterprises cooperate with each other, and the production, logistics and waste treatment facilities are shared, so that land using and investment are most economic. The government should play the roles of policy guidance and encouragement to the enterprises in the construction and investments of the green parks. The government shall also strengthen its efforts to educate the enterprises about the idea. With the green parks, the government is also building a green city at the same time.

Last, the government should enhance investment in green cities. On the basis of green parks, and in order to pursue the goal of a society with recycling economy,

Chinese government is actively implementing green city investments. The green city should follow the 3R rules of recycling economy. The government should develop the recycling and using of renewable energies, so that materials and energies in the cities could circulate within themselves. Many cities play the roles of industry, commerce, service and living at the same time, and recycling style cities are important. In terms of infrastructure, the government should try to build green transportation and green buildings by way of green investment. In terms of urban green land, the government should make scientific planning, and build green land and parks that are suitable to the cities' urban ecological development, so as to build a healthy living environment.

Green industries investment, green parks and green cities investments are along the same line and organically integrated. Government involvement, support and guidance are indispensable for the development of green activities, although the market mechanism is also important. The Chinese government should let the two hands of "government macro-regulation" and "market self-adjustments" play their roles together, to promote the development of green investment, the sustainable development of the society and economic transformation.

References

Huang H, Sun T, Cao Y (2005) Establish green investment system and promote recycling economic development in China. Collection of outstanding papers on environment protection in China

Inner Mongolia finance and fiscal information website http://www.nmgcz.gov.cn/

Jing C (2008) Reflections on green investment development in China. Science and technology and industry, 2008(7)

Li X et al. (2010) China economic development during global financial crisis. China Grand Encyclopedia Publishing House, Beijing

Ma X, Meng Y (2005) Ideas and strategies of green investment development in China. Econ Manag 2005(12)

Meng Y (2005) Green investment for resource and environment protection and its development pattern. Finan Econ Issues Res 2005(7)

Ministry of Environment Protection (2006–2010) Annual statistic report on environment in China 2005–2009. China Environment Sciences Publishing House, Beijing

Ministry of Housing and Urban-rural Development (2006–2010) China Urban construction statistical yearbook 2005–2009. China Planning Publishing House, Beijing

Ministry of Environment Protection (2010) China environment situations public report 2009. 31 May 2010

Ministry of Health website http://www.moh.gov.cn/

Ministry of Environment Protection website http://www.zhb.gov.cn/

National Bureau of Statistics (2006–2010) China statistical yearbook 2006–2010. China Statistics Publishing House, Beijing

National Bureau of Statistics, Ministry of Environment Protection (2006–2010) China statistical yearbook of environment 2006–2010. China Statistics Publishing House, Beijing

National Bureau of Statistics website http://www.stats.gov.cn/

National Development and Reform Commission website http://www.sdpc.gov.cn/

Research center of scientific development and sustainable economic growth in Beijing normal university (2010) 2010 China green development annual report—provincial comparison. Beijing Normal University Publishing House, Beijing

Shi B (2005) Industrial development in middle China, current situations, characters and strategic choice. Zhong Zhou Xue Kan 2005(9)

Su Y, Zhou H (2005) Discussion on several issues about China's promotion of recycling economy development. Econ Res Ref 2005(10)

The 12th *Five-Year Plan* for the National Economic and Social Development of the People's Republic of China (2011) People's Publishing House, Beijing

Chapter 14
Urban Infrastructure Construction and Management in Green Development

Qi Zhang, Ying Wang and Lei Zhu

Urban infrastructure construction and management, as the backbone of urban economic and social development, has significant impacts on green economic development. During the 11th *Five-Year Plan* period, in the context of building a resource-conserving and environment-friendly society, governments at all levels have attached great importance to urban infrastructure construction and management and worked out a series of urban green development blueprints, gradually establishing and improving institutions for urban economic and social development. Meanwhile, the speed-up of urbanization expands the demand for urban infrastructure and new problems and conflicts keeps popping out. It has been an irreversible trend to follow the concept of green and low-carbon development, sticking to energy saving and emission reduction for sustainable development and upgrading urban green development level.

14.1 Urban Infrastructure Construction Under the Background of Green Development in the 11th Five-Year Plan Period

During the 11th *Five-Year Plan* period, China made outstanding achievements in urban infrastructure construction for promoting urban green development. Due to the success of Beijing Olympic Games, Shanghai EXPO, as well as the promulgation of several regional development plans, input in urban infrastructure construction has been expanded significantly and the urban green development has made remarkable stride.

X. Li and J. Pan (eds.), *China Green Development Index Report 2011*,
Current Chinese Economic Report Series, DOI: 10.1007/978-3-642-31597-8_14,

Table 14.1 Fixed asset investment in municipal utilities

Index	Unit	2006	2007	2008	2009
Completed amount of fixed asset investment in municipal utilities	Billion	576.51	641.89	736.82	1064.15
The social fixed assets investment completed	Billion	10,999.82	13,732.39	17,229.11	22,459.88
Ratio in the social fixed assets investment completed	%	5.25	4.68	4.28	4.74
Expenditure on urban maintenance	Billion	334.95	424.73	500.83	592.71

Notes 1. Expenditure on urban maintenance only includes financial capital, social financing excluded

2. Data of 2009 excludes city public transportation, only rail transportation included

Source National Statistics Bureau, China Urban Construction *Statistical Yearbook 2009* [R]. Beijing, China Planning Press, 2010

14.1.1 Accomplishments in China's Urban Infrastructure Construction During the 11th Five-Year Plan Period

During the 11th *Five-Year Plan* period, China's urban infrastructure construction has made great progress in municipal utilities input, water supply and drainage, energy supply, public transportation, and environmental sanitation, etc.

Firstly, input in municipal utilities and maintenance has been largely expanded. The sum of investments completed in fixed assets increases from 576.51 billion RMB in 2006 to 106.42 billion RMB in 2009, almost doubled in the past 3 years. From 2006 to 2009, expenditure on urban maintenance keeps growing constantly, with a growth rate of 77 % in 3 years, which demonstrated the government's growing commitment to urban development with ever increasing investment (See Table 14.1).

Secondly, water-saving capacity has been enhanced with improved water supply capacity. During the 11th Five-Year Plan period, China has achieved remarkable achievements on both aspects. Table 14.2 indicates urban water supply and water saving from 2006 to 2009. The overall water supply capacity has been strengthened. The ratio of access to water has expanded continually, from 86.67 % in 2006 to 96.12 % in 2009. Meanwhile, urban water-saving capacity keeps improving. The table tells that urban population keeps growing while water supply keeps declining. The household daily water consumption per capita falls to 176.6 L in 2009 from 188.3 L in 2006. The amount of repeated utilization of industrial water keeps going up, with a growth rate of 17.8 % in 2009 over 2006. In 2008, cities with water-saving management agencies have about 6.60 billion cubic meters water saved.

Thirdly, construction of urban drainage facilities speeded up and the sewage treatment capacity expanded greatly. Table 14.3 indicates that urban drainage construction has made remarkable progress during 11th *Five-Year Plan* period. The length of drainage pipe has expanded from 261,400 km in 2006 to 343,900 km in 2009, up by 31.6 %. Both the construction and capacity of the

Table 14.2 Urban water supply and water saving

Index	Unit	2006	2007	2008	2009
The overall production capacity of water supply	10,000 m^3/day	26,965.6	25,708.4	26,604.1	27,046.8
Access ratio to water supply	%	86.67	93.83	94.73	96.12
Water supply	100 million cubic meters	540.52	501.95	500.08	496.75
Urban population	10,000	32,304.1	34,766.5	35,086.7	36,214.2
The household daily water consumption per capita	Liter	188.3	178.4	178.2	176.6
Repeated utilization of industrial water	100 million cubic meters	553.55	602.60	654.73	652.31
Amount of water saved	100 million cubic meters	41.58	45.48	65.91	62.87

Notes 1. The denominator of access ratio to water supply is the sum of urban population and temporary population living in cities
2. Repeated utilization of industrial water and Amount of water saved only include cities with water saving management agencies
Source Ministry of Housing and Urban Development: *China Urban Construction Statistical Yearbook 2009*, Beijing, China planning press, 2010

Table 14.3 Urban drainage and sewage treatment

Index	Unit	2006	2007	2008	2009
Length of drainage pipes	10,000 km	26.14	29.19	31.52	34.39
Amount of sewage drained	100 million cubic meters	362.53	361.01	364.88	371.21
Sewage treatment plants		815	883	1018	1214
Capacity of sewage treatment plants	10,000 cubic meters per day	6,366	7,146	8,106	9,052
Total amount of sewage treated annually	100 million cubic meters	202.62	226.98	256.00	279.35
Sewage treatment ratio	%	55.67	62.87	70.16	75.25

Source National Statistics Bureau, *China Urban Construction Statistical Yearbook 2009* [R]. Beijing, China Planning Press, 2010

urban sewage treatment have been promoted to a large extent. The number of sewage treatment plants in cities has increased from 815 in 2006 to 1214 in 2009, 1.5 times more than 3 years ago. The sewage treatment capacity has also expanded from 63.66 million cubic meters per day in 2006 to 90.52 million cubic meters per day in 2009, up by 42.2 %. The aggregated sewage treatment capacity in 2009 reaches 27.94 billion cubic meters, with a treatment ratio of 75.25 %, realizing a highly efficient utilization of water recycling.

Fourthly, urban energy supply increased with gradually optimized energy supply structure. During the 11th Five-Year Plan period, the energy supply structure keeps adjusting. The supply of electricity and natural gas keeps rising constantly. The annual electricity consumption increases from 15.12 kW h in 2006 to 19.25 kW h in 2009. As is commonly recognized as green energy, the consumption of natural gas has increased from 24.48 billion cubic meters in 2006 to 40.51 billion cubic meters in 2009, up by 65.5 %. The relative ratio of natural gas in total gas supply stays at 13.56 % in 2006 and 19.23 % in 2009, up by 5.67 % points in the past 4 years, all of which reflect the constant optimization of gas supply structure in China. The population using natural gas in 2006 was 83.19 million, 100 million in 2007, and 145.44 million in 2009, 1.75 times more than that of 2006. The population using manufactured gas and the pipe length both declined and the same trend for Table 14.4.

For urban heat supply, both the capacity and total amount have largely increased. Compared to 2006, the capacity increases by 31.4 % in 2009 and 35.2 % for the total amount. Meanwhile, the energy structure of urban heat supply has been optimizing constantly. The ratio of hot water heating in 2006 was 46.11 %, rising to 83.46 % in 2009. In 2008, the length of urban hot water pipes exceeded 100,000 km, reaching 110,500 km in 2009. Compared with hot water heating which saves more energy, both the capacity and total amount of steam heating maintain at the same level. In addition, the coverage of urban centralized heating has also elevated, reaching 3.796 billion square meters in 2009, 1.43 times wider than 2006.

Fifthly, urban public transportation develops rapidly, especially rail transportation. In the 11th Five-Year Plan period, close attention from the government facilitates the development of public transportation. The number of public transportation vehicles in operation, being 315,576 in 2006, increased to 370,640 in 2009. The number of standard vehicles in operation in 2009 reached 418,883, 1.2 times more than that of 2006. The length of public transportation network in operation increases by 83,392 km from 2006 to 2009, up by 66.3 %. Accordingly, the total passenger capacity is growing constantly, being 67,676 billion persons in 2009, increased by 21.083 billion (See Table 14.5).

It also indicates in this table that the urban rail transportation has developed rapidly. In 2006, the amount of rail transportation vehicles in operation at year end stayed at 2,764, increasing to 5,479 in 2009, 1.98 times more than 2006. The aggregated carrying capacity of urban rail transportation has doubled from 2006 to 2009. By the end of 2009, 10 cities nationwide have set up 33 rail transportation lines, totaling 893 km, with 558 stations, 116 transfer stations and 4801 vehicles. Another 28 cities are now constructing 81 rail transportation lines, with a length of 1991 km, 1342 stations and 306 transfer stations.[1]

Sixthly, the harmless disposal capacity of household waste has reinforced with expanded construction of sanitation facilities. Both the amount and capacity of waste disposed have increased largely. There are 567 harmless waste treatment plants in 2009, 148 more than that of 2006. The daily harmless disposal capacity of household waste has been growing constantly. In 2009, the treated waste reach

[1] Ministry of Housing and Urban-Rural Development: China City Construction Yearbook 2009, Beijing, China Planning Press, 2010.

Table 14.4 Urban energy supply and consumption

Index	Unit	2006	2007	2008	2009
Annual electricity consumption	100 million kilowatt hour	15,118.80	16,979.80	17,812.66	19,246.62
Household electricity consumption	100 million kilowatt hour	1,778.43	1,947.13	2,078.87	2,434.45
Natural gas supply	100 million cubic meters	244.77	308.64	368.04	405.10
Ratio of natural gas in total gas supply	%	13.56	14.71	17.93	19.23
Population using natural gas	10,000	8,319	10,190	12,167	14,544
Length of natural gas pipeline	10,000 km	12.15	15.53	18.41	21.88
Manufactured gas supply	100 million cubic meters	296.45	322.35	355.83	361.55
Ratio of manufactured gas in total gas supply	%	16.42	15.37	17.33	17.16
Population using manufactured gas	10,000	4,067	4,022	3,370	2,971
Length of manufactured gas pipeline	10,000 km	5.05	4.86	4.52	4.04
Liquefied gasoline supply	100 million cubic meters	163.66	1,466.77	1,329.11	1,340.03
Liquefied gasoline supply in total gas supply	%	70.01	69.92	64.74	63.61
Population using liquefied gasoline	10,000	17,100	18,172	17,632	16,924
Length of liquefied gasoline pipeline	10,000 km	1.75	1.72	2.86	1.42
Steamed gas electricity supply capacity	10,000 t per hour	9.52	9.4	9.45	9.32
Hot water heating supply capacity	10,000 MW	21.77	22.47	30.57	28.61
Steamed gas heating supply	100 million Kyrgyzstan	6.78	6.64	6.91	6.31
Hot water heating amount	100 million Kyrgyzstan	14.8	15.86	18.75	20.01
Steamed gas heating ratio	%	31.42	29.51	26.93	23.97
Hot water heating ratio	%	47.11	53.74	69.63	83.46
Length of steamed gas pipe	10,000 km	1.4	1.41	1.6	1.43
Length of hot water pipe	10,000 km	7.99	8.89	10.46	11.05
Acreage of concentrated heat supply	100 million square meter	26.59	30.06	34.89	37.69

Note the annual electricity consumption and household electricity consumption in 2009 are relatively growing in all society electricity consumption and household electricity consumption in urban and rural regions

Source National Statistics Bureau, China city management yearbook 2007 [R], Beijing, China Statistics Press, 2007–2010

Table 14.5 Urban public transportation

Index	Unit	2006	2007	2008	2009
The number of public transportation vehicles in operation		315,576	347,969	371,822	370,640
Standard vehicles in operation		337,240	379,021	412,080	418,883
Length of public transportation network in operation	Kilometer	125,857	140,801	147,349	209,249
Total passenger capacity	100 million person	465.92	55,464	703.00	676.76
Public vehicles owned by 10,000 persons		9.05	10.2	11.1	11.1
Rail transit in operation		2,764	3,480	4,530	5,479
Total passenger capacity of rail transit	10,000 person	181,599	220,582	337,390	365,770

Source National Statistics Bureau, Ministry of Environment, *China Statistical Yearbook of Environment 2007–2010* [R], Beijing, China Statistics Press, 2007–2010

Table 14.6 Environmental sanitation facilities in cities

Index	Unit	2006	2007	2008	2009
Number of harmless waste treatment plants		419	460	509	567
Capacity of harmless waste treatment plants	10,000 t per day	25.80	27.18	31.52	35.61
Ratio of harmless waste treatment plants	%	52.2	62.03	66.76	71.4
Urban green coverage	%	35.11	35.29	37.37	38.22
Urban green park acreage per capita	m^2	8.30	8.98	9.71	10.66

Source National Statistics Bureau, Ministry of Environment, *China Statistical Yearbook of Environment* 2007–2010 [R], Beijing, China Statistics Press, 2007–2010

356,100 ton per day, the harmless disposal ratio of household waste being 71.4 %, both increased largely compared to 2006 (See Table 14.6).

The urban landscaping construction has achieved great progress. In 2009, the forest coverage ratio is 38.22 %, up by 3.11 % points over 2006. The public green park coverage per capita has increased from 8.30 m^2 to 10.66 m^2, up by 2.36 m^2.

14.1.2 Regional Comparison of China's Urban Infrastructure Construction

As a whole, China has made tremendous accomplishments during the 11th *Five-Year Plan* period in urban infrastructure construction under the background of green development. However, a conspicuous unbalance exists among different regions.[2] According to the usual partitioning method of dividing eastern, central,

[2] Easter areas include: Tianjin, Hebei, Shanghai, Jiangsu, Zhejiang, Fujian, Shandong, Guangdong, and Hainan; Central areas include: Shanxi, Anhui, Jiangxi, Henan, Hubei, and Hunan; Western areas include: Inner Mongolia, Guangxi, Chongqing, Sichuan, Guizhou, Yunnan, Tibet, Shanxi, Gansu, Ningxia, Qinghai, and Xinjiang; Northeaster areas include: Niaoning, Jilin, and Heilongjiang.

Table 14.7 Fixed asset investment in municipal utilities among regions

Index	Unit	Eastern areas	Central areas	Western areas	Northeastern areas
Regional GDP	100 million RMB	196,674.4	70,577.6	66,973.5	31,078.2
Fixed asset investment in urban maintenance	100 million RMB	2,004.99	890.60	653.90	263.05
Fixed asset investment in municipal utilities	100 million RMB	5,697.63	1,829.64	2,142.64	971.57
Ratio of fixed asset investment in urban maintenance among regional GDP	%	1.02	1.26	0.98	0.85
Ratio of fixed asset investment in municipal utilities among regional GDP	%	2.90	2.59	3.20	3.13

Note Regional GDP includes urban and rural GDP
Source National Statistics Bureau, *China Statistical Yearbook for Regional Economy 2010* [R], Beijing, China Statistics Press, 2010; Ministry of Housing and Urban Development: *China Urban Construction Statistical Yearbook 2010*, Beijing, China Planning Press, 2010

western, and northeastern area, comparisons have been done among different regions on such aspects like: municipal utilities investment, water supply and water saving, water drainage and sewage treatment, gas supply, as well as environmental sanitation.

Firstly, in terms of municipal maintenance and investment, central and western areas are more accomplished. Table 14.7 indicates that central areas have the largest ratio of fixed asset invested in municipal maintenance, taking up as high as 1.26 % of the regional GDP. The western areas have the highest level of fixed asset invested in municipal utilities, as high as 3.2 % of the regional GDP. Northeastern areas have the second high ratio of 3.13 %.

Secondly, eastern areas have the strongest capacity in urban water supply while the central areas have the most efficient water resource utilization. As shown in Table 14.8, the overall production capacity of water supply in eastern areas is as large as 138.90 million cubic meters per day, taking up about half of the total nationwide. The amount of daily water consumption per capita in eastern, central, western, and northeastern areas are respectively 410.93, 351.04, 317.98, and 366.57 L per day, indicating that eastern areas have the highest water consumption while western areas have the lowest level. On the aspect of water supply, eastern areas have the largest supply amount, more than 50 % of the aggregated amount of the four areas. As for the repeated utilization of water resource, central areas have the highest ratio of 82.71 %, eastern and northeastern areas falling behind, being 75.25 and 74.80 % respectively. The amount of water saved per capita in central areas is as large as 47.79 m^3, further exceeding the other three areas, showing that water utilization in central areas is relatively more efficient.

Thirdly, eastern areas have the largest quantities of water drainage and sewage treatment facilities, leading to the highest sewage treatment capacity.

Table 14.8 Water supply and water saving in different areas

Index	Unit	Eastern areas	Central areas	Western areas	Northeastern areas
Overall production capacity of water supply	10,000 cubic meters	13,889.50	6,214.60	4,023.20	2,919.50
Amount of water saved	100 million cubic meters	266.53	93.89	81.19	55.14
Urban population	10,000 persons	17,769.97	7,327.48	6,995.42	4,121.34
Water consumption per day	Liter per day	410.93	351.04	317.98	366.57
Repeated water utilization	100 million cubic meters	287.14	152.71	71.58	140.89
Amount of water saved	100 million cubic meters	17.46	29.70	9.96	5.76
Water repeated utilization ratio	%	75.25	82.71	68.74	74.80
Amount of water saved per capita	cubic meters per day	12.57	47.79	24.75	19.71

Source Ministry of Housing and Urban Development: China City Construction Yearbook 2009, Beijing, China Planning Press, 2010

In comparison, northeastern areas have the highest sewage treatment capacity per plant in average. As indicated in Table 14.9, the length of water drainage and sewage treatment pipes in eastern areas is on top of all four areas, being 202,300 and 73,200 km each. The number of sewage treatment plants in eastern areas is the largest, amount to 643, 52.97 % of nationwide level. Sewage treatment capacity for these plants reaches 50.20 million cubic meters per day, far ahead of other areas. Northeastern areas have the strongest sewage treatment capacity per plant in average, as well as the largest treatment amount, being 94,500 m^3 per day per plant and 29 million cubic meters per plant respectively. Next to the eastern areas, central areas are 78,100 m^3 per day per plant and 24 million cubic meters per plant for the two figures. Western areas have a relatively low capacity and amount for sewage treatment.

Fourthly, for the supply structure of natural gas, western areas have the highest level of gas cleanness, while the ratio of liquefied gasoline supply in all four areas are all high. As we can see in Table 14.10, western areas have the largest ratio for the supply of manufactured gas and the commonly known *green energy*, the natural gas, 25.79 and 28.64 % for each. In northeastern areas, the ratio of liquefied gasoline in total gas supply is as high as 77.30 %, highest in all four areas. Western areas have the highest level of gas cleanness. For all four areas, liquefied gasoline takes up the largest ratio in total gas supply, 45.57 % in western areas and more than 50 % for another three areas.

Fifthly, in terms of urban environmental sanitation, eastern areas have the highest levels on both waste treatment and urban gardening. Table 14.11 indicates the status quo of urban environmental sanitation in four areas. Eastern areas have

Table 14.9 Water drainage and sewage treatment in different areas

Index	Unit	Eastern areas	Central areas	Western areas	Northeastern areas
Length of water drainage pipe	10,000 km	20.23	5.94	5.41	2.81
Length of sewage pipe	10,000 km	7.32	1.79	2.30	0.69
Sewage treatment capacity in plants	10,000 cubic meters per day	5,019.9	1,730.9	1,488.8	812.6
Sewage treatment amount	100 million cubic meters	157.15	52.62	44.43	25.15
Number of sewage treatment plants		643	241	244	86
Sewage treatment capacity per plant	10,000 cubic meters per day per plant	7.81	7.18	6.10	9.45
Sewage treatment amount per plant	100 million cubic meters per plant	0.24	0.22	0.18	0.29

Source Ministry of Housing and Urban Construction: China City Construction Yearbook 2009, Beijing, China Planning Press, 2010

Table 14.10 Gas supply in different areas

Index	Unit	Eastern areas	Central areas	Western areas	Northeastern areas
Manufactured gas supply	100 million cubic meter	200.93	33.49	117.29	9.84
Manufactured gas supply in total gas supply	%	15.52	13.28	25.79	9.35
Natural gas supply	100 million cubic meter	206.92	53.89	130.23	14.06
Natural gas supply in total gas supply	%	15.98	21.37	28.64	13.35
Liquefied gasoline supply	100 million cubic meter	886.65	164.80	207.20	81.39
Liquefied gasoline supply in total gas supply	%	68.49	65.35	45.57	77.30

Source Ministry of Housing and Urban Development: China City Construction Yearbook 2009, Beijing, China Planning Press, 2010

the largest amount of household waste treated harmlessly and the most plants, being 61.16 million ton and 268 plants. The household waste treated harmlessly for each city in northeastern areas is as large as 239,900 ton per plant. As for urban gardening, eastern areas have the largest acreage of green gardens in built-up areas, as well as highest level of green coverage, being 7563.23 km^2 and 42.95 % respectively. Western areas have the lowest green coverage ratio, only 34.70 %.

Table 14.11 Urban environmental sanitation in different regions

Index	Unit	Eastern areas	Central areas	Western areas	Northeastern areas
Amount of household waste treated harmlessly	10,000 t	6116	1967.8	2176.9	959.6
Number of harmless household waste treatment plants		268	114	144	40
Amount of harmless household waste treatment per plant	10,000 t per plant	22.82	17.26	15.12	23.99
Green gardens in built-up areas	Square kilometer	7,563.23	2,950.63	2,734.88	1,696.12
Built-up areas	Square kilometer	17,610.44	7,825.72	7,880.95	4,790.15
Green garden coverage ratio in built-up areas	%	42.95	37.70	34.70	35.41

Note Green garden coverage ratio in built-up areas is calculated from the division of Green gardens in built-up areas by Green garden coverage ratio in built-up areas
Source Ministry of housing and urban–rural development: *China urban construction statistical yearbook 2010*, Beijing, China Planning Press, 2010

14.2 City Management Under the Background of Green Development in the 11th Five-Year Plan Period

How to manage a city in a scientific way is fundamental for promoting urban green development. During the 11th *Five-Year Plan* period, governments at all levels, guided by *Scientific Outlook on Development*, gave great emphasis to city management. They have been adopting green development policies constantly in order to build a resource-conserving and environment-friendly society by practicing water saving, land conservation, energy saving and emission reduction.

14.2.1 Green City Management Policies Reinforced in the 11th Five-Year Plan Period

It states clearly in *The* 11th *Five-Year Plan for the National Economic and Social Development of the People's Republic of China* (The 11th *Five-Year Plan*) that China will "implement the fundamental national policies of resource conserving and environment protection, build a national economic system and a resource-conserving and environment-friendly society with low input, high output, low energy consumption, low emission, recycling and sustainable development." The Outline also gives emphasis to developing recycling economy through protection and restoration of natural ecological environment, enhancement of resource management, and rational utilization of ocean and climate resources, which are all fundamental for building a resource-conserving and environment-friendly society. Cites have all been responding actively by drawing out concrete practice measures. For instance, Beijing municipal government promulgates 2006 *Action Plan for Accelerating Recycling Economy and the Building Resource-saving City*, requiring that each district and county should carry out this plan without hesitation. On November 3, 2006, the Standing Committee of Beijing Municipal People's Congress approved *Decisions about Developing Recycling Economy and Building Resource-saving City* for further promoting the recycling economy and building a resource-conserving and environment-friendly society for the capital of China. Many other cities nationwide also take similar measures in succession for striving towards this goal.

Firstly, governments at all levels have adopted a series of measures for water-saving in cities. National Bureau of Environmental Protection issued 2006 *Outline for Nationwide Environment Supervision*, calling for supervising water quality in urban centralized drinking water source areas. Guided by this *Outline*, Nanjing municipal government adopted *Water Resource Protection Regulation for Nanjing*, and Ningxia Hui Autonomous Region promulgated *Water Resource Management Rule for Yinchuan*. In addition, several cities have exploited measures for unified water management. Beijing, Shanghai and some others have all realized unified water management.

Secondly, in terms of ecological compensation, pilot projects should be initiated in the aim of setting up ecological compensation policies. In August 2007, Ministry of Environmental Protection released *Guidance on Implementing Ecological Compensation Pilot Projects*, pointing out that there should be coordination between government and market. Innovative approach should be employed to launch ecological pilot projects based on local condition, so that an environment sharing mechanism can be institutionalized in the long run. And in so doing, ecological compensation policies and regulations can be improved, laying solid foundation for forming overall ecological compensation mechanism.

Thirdly, in terms of green financing, there should be concessional fiscal, financial and taxation policies for environment-friendly industries. On June 19, 2007, Ministry of Finance and State Administration of Taxation jointly issued notice of lowering down the export tax rebate for some commodities, encouraging the development of environment-friendly industries. People's Bank of China released guidance for improving and enhancing financial services in resource-conserving and environment-friendly areas, requiring branches and urban commercial banks to strengthen communication and cooperation, so as to upgrade financial service quality in energy saving and environment-friendly areas. On August 13th, 2007, National Bureau of Environment Protection issued *Notice of Further Rectifying Supervision on the Financing Application from Enterprises in Heavily Polluting Industries*, ruling that there should be more serious environmental supervision on industries like thermal power generation, iron and steel, cement, electronic aluminum, and other heavily polluting industries. More severe control should be inducted to industries with heavy pollution and high environmental risks, leading to the sound development of environment-friendly industries. In 2008, National Bureau of Environment Protection cooperated with China Insurance Regulation Committee in setting up green insurance mechanism. Previously the main approach for guiding the development of environment-friendly industry relied more on administrative regulations, while now it applied more market measures. According to this new mechanism, several pilot projects of insurance on environmental pollution responsibility have been set up, which initiates the insurance mechanism on environmental pollution responsibility. Until 2009, nine provinces and municipal cities have already developed green insurance pilot projects while Hebei, Shenyang, and Shanghai have incorporated regulations on environmental pollution responsibility into provincial environment laws. In April 2010, National Development and Reform Committee (NDRC), People's Bank of China, China Banking Regulatory Commission, and China Securities Regulatory Committee jointly promulgated *Notice of Supportive Financing Measures for Recycling Economy Development*, in order to expand investment and financing on recycling economy. It also provides all-round guidance for enhancing coordination among government at all levels in facilitating recycling economy, and promoting the implementation of policies.

Fourthly, in terms of urban low-carbon development, efforts should be made to promote polices for the development of low-carbon industries and the building of low-carbon cities. In early 2008, Ministry of Housing and Urban–Rural Development, together with World Wide Fund, initiated pilot projects of low-carbon city in China, following which, low-carbon city has become a common goal sought after by all cities. In 2010, NDRC released Notice of Implementing Pilot Projects of Low-carbon Provinces and Cities, kicking off experiments in exploring economic development, coping with climate changes, and realizing green development.

14.2.2 Green Management Institutions Further Improved and Perfected in the 11th Five-Year Plan Period

China has been setting up and constantly improving legal systems when giving emphasis to city management policies. So far, a comprehensive system has formed into shape, where the National People's Congress supervises the enforcement, while government at all levels is responsible for the implementation, environmental protection authorities and relevant departments conduct monitoring and management based on laws and regulation.

During the 11th *Five-Year Plan* period, guided by *Scientific Outlook on Development*, Central government at all levels issued a series of legal documents (see Table 14.12), standardizing procedures, evaluation indicators, implementation measures, management mechanism, and legal responsibility for national environmental protection from all aspects, which plays a crucial role for facilitating sustainable development and green city development. There are several features for these laws and regulations.

Table 14.12 Green management institutions during the 11th Five-Year Plan period in parts of cities

Level	Legislation	Promulgating agency	Effective Date
National	Law of the People's Republic of China on the prevention and control of water pollution	Standing committee of NPC	2008-6-1
National	Circular economy promotion law of the people's Republic of China	Standing committee of NPC	2009-1-1
National	Renewable energy law of the People's Republic of China	Standing committee of NPC	2010-4-1
National	Regulation of water permission and water resource fee	State council	2006-4-15
National	Regulation on national general survey of pollution sources	State council	2007-10-9
National	Energy saving regulations for civil constructions	State council	2008-10-1
National	Regulations on energy sufficiency of public institutions	State council	2008-10-1
National	Regulations on the administration of urban gas	State council	2011-3-1
Provincial	Regulations on Hunan chang-zhu-tan Urban Agglomeration Regional Planning	Standing committee of NPC of Hunan province	2008-1-1
Provincial	Hainan province town landscape regulations	Standing committee of NPC of Hainan province	2009-1-1
Provincial	Several provisions on prevention and control of environmental pollution by solid waste in fujian	Standing committee of NPC of Fujian province	2010-1-1
Provincial	Regulations on city appearance and sanitation management	Guangxi Zhuang minority autonomous government	2008-1-1
Provincial	Hebei province environment pollution prevention supervision and management measures	Hebei provincial government	2008-3-1
Provincial	Energy conservation monitoring measures in Hebei province	Hebei provincial government	2008-3-1
Provincial	Rules on household waste management	Tianjin municipal government	2008-5-1
Provincial	rules on city management	Tianjin municipal government	2008-6-1
Provincial	Tianjin Eco-city management regulations	Tianjin municipal government	2008-9-28
Provincial	Hainan province town household waste collection and management measures	Hainan provincial government	2009-12-1

Source http://www.gov.cn

Firstly, the coverage of green development in legislation has been expanding constantly. The above table shows a wide coverage, ranging from energy saving, water supply, energy supply, urban sanitation, to urban gardening and environmental pollution evaluation, etc. Central and provincial governments pay high attention to the formulation of regulations on green development management. NPC and State Council make legislations based on common issues in green development nationwide, while provincial governments should take into consideration of local situation and features before setting up rules and regulations, so as to fulfill green development.

Column 14.1 Breakthrough in Water-Saving and Building Water-Saving Society in Dalian

Putting the goal of water-saving society on the top priority for guaranteeing water safety, Dalian municipal government and relevant departments attach great importance to water-saving, sets up a leading workgroup for building water-saving society, led by municipal leaders. It aims at managing water-saving in the whole city and enlarging the input on water-saving. Two billion RMB has been input into refitting projects of water supply facilities from 2008 to 2010. Meanwhile, the municipal government has enhanced legal regulation and institution framework, promulgating a series of water-saving regulations. The practice and experience in building a water-saving society in Dalian can be summed up as follows:

Firstly, water-saving system construction and public propaganda have been enhanced. During the 11th *Five-Year Plan* period, Dalian makes great progress in restructuring water-saving management teams, thus improving relevant management system through the establishment of a five-tier management network of water-saving. In order to raise public awareness, various activities are conducted such as World Water Day, Urban Water-saving Week, and Environment EXPO. The water-saving awareness among citizens has been enhanced through regulation and policy training classes, news media, broadcasting and bulletin boards in enterprises and companies.

Secondly, legislation system has been improved for guaranteeing water-saving implementation orderly. During the 11th Five-Year Plan period, Dalian adopted three municipal water utilization rules: *Rules on Water Resource Management, Rules on Urban Water Supply and Utilization*, and *Rules on Protecting Water Draining and Supply Projects*. Additionally, there are several administrative rules such as *Managing Measures on Reclaimed Water Facility Construction, Managing Water-saving and Water Utilization*. All in all, there are more than 50 regulations on 17 items such as underground water management, recycling water supply, ocean water utilization, deployment of sanitation facilities, water pressure and water saving.

Thirdly, the institution of initial water right has been set up and the water-saving project has been facilitated through water pricing mechanism. Dalian municipal government issues *water use permit*, using water consumption quota as a media to promote the setting up of preliminary water right, thus forming a sound water utilization management system which combines aggregation control and quota management. Meanwhile, by reforming water pricing mechanism, urban water pricing can be adjusted via charging water resource fee and regulating recycling water price, and in so doing, the role of price as leverage can be given into full play.

Fourthly, the industrial distribution has been optimized for promoting the water-saving intensity in industrial enterprises. During the 11th Five-Year Plan period, Dalian municipal government developed recycling economy and water-saving economy, expanded supportive measures towards high and new tech such as electronic information, biological engineering, and new materials, as well as highly efficient water-saving industries like tourism and exhibition, commerce and trading, and financial information. Concrete measures included water quota for industrial enterprises, actual water use index management, water balance test, etc., so that water-saving can be expanded in industrial enterprises extensively.

Fifthly, water supply range has been expanded for overall distribution of different water resources. Dalian also focuses on exploiting unconventional water resource utilization. In terms of ocean water utilization, the utilization amount per day in Dalian has reached 11,900 m^3. Meanwhile, the ocean water desalination has also achieved great progress. Five ocean water desalination plants have been set up, with a desalination capacity of 11,900 m^3 per day. As for polluted water recycling, there are 9 plants in main urban district, with a capacity of 495,000 m^3 per day. The polluted water treatment ratio in main urban district stays at 71 %.

Under the highlight from government of each level, Dalian makes great breakthrough on water-saving project. During 11th Five-Year Plan period, water utilization of 10,000 *yuan* regional GDP and 10,000 *yuan* industrial value added decline 9.8 and 6.4 % on average, to 44.6 and 26.4 m^3 per 10,000 RMB. Both two indicators are in the lead in China and Liaoning province. The coverage ratio of water-saving tools in the city and the counties rises by 28 and 25 %. The comprehensive water utilization level keeps in the lead in China and Liaoning.

Source: Summarization of Constructing Water-saving Society in Dalian During 11th FIVE-YEAR PLAN. http://www.mwr.gov.cn/ztpd/2010ztbd/2010_jszgx/ckzl/201011/t20101109_245165.html.

Secondly, the green development legislation should reflect the requirement of green development policies so that green policy can be standardized and institutionalized. The 11th *Five-Year Plan Outline* advocates the developing of recycling economy. Following the introduction of polices for recycling economy in Beijing, Dalian, and Shenzhen, in 2008, the fourth meeting of the Standing Committee, 11th National People's Congress approved *Circular Economy Promotion Law of the People's Republic of China*, regulating reduced resource consumption and waste production in processing, circulating, and consumption, facilitating the development of recycling economy through raising the efficiency of resource utilization, thus realizing sustainable development.

Thirdly, the development of green management institutions has accelerated, with the accumulation of experience in city management. Relatively few regulations were adopted during 2006–2007, only *Regulation on National General Survey of Pollution Sources*. In 2008, 10 legislations come into promulgation. Take Tianjin as an example, during May to September in 2008, three regulations on urban green development have been introduced: *Rules on Household Waste Management, Rules on City management,* and *Rule on Tianjin Eco-City Development.*

14.3 China's Urban Infrastructure Construction and Green Development of City management

During the 11th *Five-Year Plan* period, China has made great accomplishments in green city development, which will still be a hot topic during the 12th *Five-Year Plan*. According to *the* 12th *Five-Year Plan Outline*, fundamental national policy for energy conservation and environment protection should be carried out thoroughly, with reduced greenhouse gas emission, as well as the development of recycling economy and dissemination of low-carbon technology. China should actively fight with global climate change; promote the coordination among economic social development, population, resources, and environment, thus realizing sustainable development. In this sense, ecological environment and social development should be considered as a core issue through reinforcing urban infrastructure and improving urban green management, eventually realizing green development of the city where man and the nature can coexist in harmony.

14.3.1 New Trend for Urban Infrastructure Construction in the 12th Five-Year Plan Period

Urban infrastructure construction is the backbone for green development. China has achieved remarkable accomplishments and accumulated substantial experiences during the 11th Five-Year Period. As the starting point of 12th *Five-Year Plan*, China faces more challenges with the accelerating urbanization, expansion of urban population and size. As designed in the 12th *Five-Year Plan Outline*, at a time when China is at the doorstep of strategically crucial period of development and transformation, the blueprint of urban infrastructure construction in the new era should realize the coordination between ground and underground public transportation, with the upgrading of infrastructure in fields like public transportation, communication, supply of power, heat, gas and water, sewage and waste treatment.

It is at the top agenda for China in the future to promote the development of diversified and clean energy as well as energy transmission channels. With the rapid and sustainable development of Chinese economy, demands towards energy keep expanding all the time and the energy issue has become a bottleneck for sustainable development. As required by *the* 12th *Five-Year Plan Outline*, China should prioritize conservation, enhance environmental protection and international cooperation with local condition suited and diversified development, so as to adjust and optimize energy structure and finally establish a modern energy industrial system which is safe, stable, economic, and clean. Cities are the region where most energy is consumed, and therefore, priority should be given to cities in terms of reduction of energy consumption and promotion of clean energy development. The *Outline* points out that priority should be given to combined heat and power plants in large and medium-sized cities and industrial parks, efforts should be reinforced to develop comprehensive power plants such as large pit coal-fired power stations and gangue, so that we can facilitate the construction of hydropower plants, efficient development of nuclear power plants based on safety, and promote the application of decentralized energy system. In addition, power grid should be further improved, advanced UHV transmission technology promoted, urban electric network upgraded, as to strengthen both the optimization and supply capability of power grid.

Column 14.2 Tianjin Eco-City Planning on the Enhancement of Clean and Comprehensive Energy Utilization

Tianjin Eco-City is a strategic cooperation project between China and Singapore governments, with the overall plan approved in 2008. The goal of the planning gives equal emphasis to land utilization, integration of urban transportation, as well as balanced employment and housing supply. The project aims at promoting resources recycling through deploying a series of innovative technology and environmental friendly policies and investment.

In terms of comprehensive utilization of energy, the plan calls for dissemination of new energy technology, the cascade utilization of energy, energy conservation, efficient utilization, development of recycling economy, clean production, energy conservation and emission reduction, forbidding the use of non clean-coal, for the purpose of establishing a safe, efficient, and sustainable energy supply system.

According to the overall plan, Tianjin Eco-City plans to use 80 % traditional energy and 20 % renewable energy. Two thermal power plants are built to provide heat for the city in the form of hot water through two transmission pipes. Almost 16.32 million square meters in the eco-city require heating and cooling. Two thermal power plants could provide heat to an area of 12.31 million square meters, while local geothermal energy and heat pump can provide heat for the remaining area. Meanwhile, the thermal power plants transmit power to the State Grid, which in turn, provide most of the electricity for the eco-city. The natural gas network in the ecological city will cover the whole city and the natural gas is provided mainly by Gansu and Shanxi province. Heating, cooling, household hot water, and street lighting in the city are delivered mainly by renewable energy, among which solar energy takes up 30.7 %. The renewable energy accounts for 10 % of residential energy consumption, and 15 % of the total for public buildings.

In addition, there is also key performance indicators designed for directing the urban planning and development, as well as project progress evaluation. By 2020,

the carbon emission per unit GDP in the city will be controlled within 150 tons per million dollar GDP and the ratio of green buildings will reach 100 %. Renewable energy should take up at least 20 % in the total energy supply and the coverage ratio of centralized heat supply reach 100 %.
Sources:

1. Website of Tianjin Eco-City: http://www.eco-city.gov.cn/eco/shouye/main.html.
2. World Bank, *Tianjin Eco-City: Research of New Ecological City Cases in China*; Tianjin Eco-City, Urban Residential House, 2011.

It will become the basic requirement for urban infrastructure construction to boost energy conservation, emission reduction and promote development of green building. High energy and resource consumption urges China to transform its economic development pattern. It's a fundamental approach for China to disseminate advanced energy conservation and environment-friendly technology when striving towards the path of sustainable development through energy saving and emission reduction. During the 12th Five-Year Plan period, China will make great efforts in promoting the advance of energy saving and environment-friendly industries and facilitating the development of resource recycling industries. The 12th Five-Year Plan Outline makes energy saving and emission reduction a top priority by encouraging the implementation of projects in areas of heat and power cogeneration, systematic optimization of energy, gasoline conservation and alternatives, energy conservation in building, transportation and green lighting. It also increases subsidy towards highly energy efficient household appliances, vehicles, motors and lighting products. On April 21st, Ministry of Housing and Urban–Rural Development promulgated *Report on Supervision of Energy Conservation in Buildings under the Project of Energy Saving and Emission Reduction Supervision Nationwide in the Area of Housing and Urban–Rural Development*, promoting the implementation of energy saving project of warm house in key cities and counties. It determines to complete at least 35 % of heat metering and renovations for energy saving of old residential houses which is worthwhile for renovations by the year 2015. In addition, energy conservation conference in Hubei province has set up the goal for green building during the 12th Five-Year Plan period, that is, an increased energy saving capacity equivalent for 5.2 million tons of coal, a total utilization area of 25 million square meters for buildings with renewable energy supply, and a green building area of 100 million square meters.

It will become the prerequisite for China's urban infrastructure construction to develop recycling economy and boost environment protection efforts. Recycling economy combines cleaning production with waste utilization based on the principle of decrement, recycling and resource recovery for the purpose of making best use of resources and protecting the environment through appropriate measures. During the 11th Five-Year Plan period, China has achieved great progress in developing recycling economy and the 12th Five-Year Plan Outline put forward new requirement for the recycling economy in the new era. That is, to realize efficient land use, cascade use of energy, recycling of waste water, and centralized waste treatment, the overall utilization ratio of industrial solid waste reaching 72 %, and the resource output ratio increased by 15 %. As the urban population keeps growing, environmental pollution has been one of the daunting illnesses of cities. During the 12th *Five-Year Plan* period, China will take comprehensive measures to solve the problems of drinking water safety, air pollution, and soil contamination, and meanwhile, intensify the monitoring in terms of width and in-depth on environmental pollution and speed up the development of treatment facilities for household sewage, sludge, and other waste. By 2015, harmless treatment ratio household waste will surpass 80 % and harmless treatment for household waste will be realized in municipalities directly under the jurisdiction of the central government, provincial capital cities, and specifically designated cities in the state plan. At least one model city in each province will be set up for household waste classification. 50 % of community cities will basically realize classified collection and treatment for kitchen waste, ratio of resource utilization of household waste reaching 30 and 50 % for municipalities directly under the jurisdiction of the central government, provincial capital cities, and specifically designated cities in the state plan. 80 % of cities at prefectural levels have air quality standard of above class two, and the ratio of household waste water treatment and harmless treatment for household waste being 85 and 80 %.

It will become priority areas for China's urban infrastructure to develop urban public transportations and improve its service level. *The* 12th *Five-Year Plan Outline* calls for implementing the strategy of prioritized development for public transportation, advancing urban public transportation system and increasing the sharing ratio of public transportation. Accordingly, light rail, subway, tramcar and other forms of urban railway should be promoted. Meanwhile, the goal of urban rail transportation has been drew out, indicating that railway network will be built in Beijing, Shanghai, Guangzhou and Shenzhen during the 12th *Five-Year Plan* period; for Tianjin, Chongqing, Shenyang, Changchun, Wuhan, Xi'an, Hangzhou, Fuzhou, Nanchang, and Kunming, a basic framework for railway transportation shall come into being; and for Hefei, Guiyang, Shijiazhuang, Taiyuan, Jinan, and Urumqi, railway transportation backbone lines shall be planned. Efforts should be made to expand public transportation station coverage, develop BRT (urban rapid transit) system, regulate taxi operators, encourage use of non-motor vehicles, and promote ways of transportation that is energy saving and environment-friendly.

Column 14.3 Shenzhen Expands Input in Public Transportation and Wins the Title of National Public Transportation Model City

Facing with increasing challenges brought by public transportation, Shenzhen puts public transportation development on top priority and designs the guideline of *government leadership, market operation, scale operation, policy support, and prioritized development*, realizing a huge leap for urban public transportation through optimized government management, improved market mechanism, expanded infrastructure construction of urban public transportation and upgraded service level of public transportation. From the year 2006–2009, the number of public bus per 10,000 persons in Shenzhen increased from 99 to 103.02, much higher than other cities in China. There are mainly three measures carried out during the 11th Five-Year Plan period.

Firstly, improving urban public transportation planning and design specific plan documents like *Overall Planning for Public Transportation in Shenzhen (*3rd *version),* 111th *FYP for Public Transportation in Shenzhen, Recent Mid-term Construction Plan for Rail Transit in Shenzhen, Construction Plan for Bus Station in Near Term,* and *BRT Plan in Shenzhen,* etc. All of these guarantee full implementation of plans and sound development of public transportation.

Secondly, expanding financial input in public transportation and proactively develop infrastructure for public transportation. The 1st phase subway project in Shenzhen has an investment of 11.55 billion RMB, among which 70 % from direct governmental investment. All public stations are invested and constructed by the government. Meanwhile, in order to guarantee a lasting and stable funding source for prioritized public transportation development, Shenzhen has set up a special fund for public transportation disbursed by municipal finance. In terms of infrastructure construction of public transportation, Shenzhen advances the construction of bus lanes, port stations, and double platform stop, thus greatly improving the speed.

Thirdly, promoting constantly the operation innovation and improving market operation mechanism. Following the principle of *policy guidance, market operation, support for the strong, and resource integration*, Shenzhen enhanced concentration of public transportation through mergers and acquisitions. By implementing specialized operation in public transportation zone, Shenzhen advances the establishment of a market competition platform with appropriate scale, intensified operation and rational competition, and in so doing, market resources have been optimized and service level has also been upgraded constantly.

In 2007, together with Beijing, Shanghai and another nine cities, Shenzhen was awarded the title of *national public transportation model city* by Ministry of Housing and Urban–Rural Development.

Source: Optimizing policy system, guaranteeing public transportation on priority, Zhang Siping, Deputy Mayor of Shenzhen,

http://www.citieschina.org/ArticleShow.Asp?ID=5054&CatId=334;Shenzhen acquired the title of urban public transportation demonstration, Shenzhen News Internet, http://www.sznews.com/zhuanti/content/2007-02/14/content_873910.htm.

14.3.2 Outlook of China's City Management

The 12th Five-Year Plan period is crucial for China to build a well-off society in an all-round way, when urbanization level will be accelerating and cities are prompted to transform from extensive to intensive development with the aim of upgrading city management level. In the future, city management, guided by

Scientific Outlook on Development, shall promote institution innovation based on current experiences. All social forces should be mobilized to participate in city management with the implementation of the strategy of main functioning zone. City management should be advanced with specific focus with deepened scientific and democratic dimensions when fulfilling the idea of putting people first.

Institution innovation can be served to deepen institutional building of city management. In terms of urban green development, governments at various levels have designed a series of city management legislations and the green management system has taken into shape. These legislations guarantee rational implementation of law enforcement. During the 12th Five-Year Plan period, the legal framework of city management in China will be more comprehensive and inclusive, covering all aspects in city planning, building and management. It puts forward rigid requirement for the city management staff to study the legislation carefully and strictly, as well as conduct law enforcement, thus realizing institutionalized and standardized city management.

City management system should be innovated and all social forces should be mobilized to participate into city management, which can intensify city management. The 12th Five-Year Plan Outline points out that innovating social management should follow the principle of all party participation, co-governance and dynamic coordination. Meanwhile, Party Committee leadership should be improved, as well as the enhancement of government responsibility, society coordination and citizen participation, so as to push forward decentralized management with wide management participation from the whole society. City green development should allow combined action from the government and the society, by initiating a coordinated mechanism, finally achieving synergy among different government departments, as well as between government and social organizations. Functions of departments should be specified clearly. Via comprehensive governance, promotion and education, resources should be integrated by making full use of administration and market and encouraging enthusiasm of all the social parties, with a purpose of realizing the sound development of city management.

The strategy of main functioning zone should be implemented and city management should be advanced steadily with a specific focus. According to *Outline of Nationwide Main Functioning Zone*, there should be classified policies towards zones with development priority, targeted development, restricted development, and forbidden development when considering factors like history, natural conditions, resources and environment carrying capacity. Eco-city pilot projects should be advanced proactively. Pollution and emission control should be differentiated with different standards for different main functioning zones, by using various evaluation systems and incorporating indicators of ecological protection into the evaluation system and in so doing, the trend of environment deterioration can be reversed completely which follows the sustainable development for the society, economy, resource, and environment.

The idea of People First should be fulfilled so as to guarantee the scientific and democratic dimensions of future city management. The *Scientific Outlook on Development* requires all social parties to participate in city management. Modern city management is a huge systematic project and the whole society participation could cut down the cost of city management and satisfy different demands from various interest groups, meaning that city management could focus more on human and democratic dimensions. During the 12th Five-Year Plan period, government should plan, coordinate, organize, and supervise, as well as draw out major planning so that city management can be more systematic and democratic, finally realizing an equal and coordinated relationship among government, social organizations, and individuals. In this sense, management cost can be cut down, management efficiency can be upgraded and the demand of scientific management for city planning and urban development can be met.

References

Beijing Normal University (2010) *Scientific outlook on development* and sustainable development research base, etc. 2010 China green development index annual report: provincial comparison. Normal University Press, Beijing

Chen H (2011) New ideas for city management and theoretical breakthrough of city management, Shanghai city management (1)

Conclusion of Dalian constructing water-saving society during 11th *FIVE-YEAR PLAN* [EB/OL], website of Ministry of Water Resources

Huang J (2011) Analyzing city management legislation and institution reform, Legislation and Society (7)

Hubei province determines the goal for building energy saving during *12th FIVE-YEAR PLAN* [EB/OL], website of Ministry of Housing and Urban and Rural Construction

Inform by state council to circular the opinion from ministry of housing urban and rural construction on further strengthening the treatment of urban household waste [2011] No. 9

Lian Y (2010) Research report on core issues of 12th *FIVE-YEAR PLAN* in China. China time economy press, Beijing

Liu Wenjian (2011) Trend and strategy for city management City (2)

National Statistics Bureau (2010a) China city management yearbook 2007. China Statistics Press, Beijing

National Statistics Bureau (2007–2010) Ministry of Environment, China environment yearbook 2007, Beijing, China Statistics Press

National Statistics Bureau (2010b) China city management yearbook 2009. China Planning Press, Beijing

National Statistics Bureau (2010c) China regional economy yearbook, 2010. China Statistics Press, Beijing

National Statistics Bureau (2010) China urban life and price yearbook 2010, China Statistics Press

Shenzhen acquired the title of urban public transportation demonstration, Shenzhen News Internet, 2011-05-02

Special supervision and inspection on 2010 national housing, urban and rural construction field of energy saving and emission reduction [EB/OL], 2011-05-06

YueJinzhu (2010) Speeding up capital city management and environmental construction, City management and technology (4)

12th *FIVE-YEAR PLAN* Outline for National Economy and Social Development of China, Beijing, People's Press, 2011

http://www.gov.cn

http://www.eco-city.gov.cn/eco/shouye/main.html

Chapter 15
Environmental Governance in Green Development

Renwu Tang, Guanqiong Yang, Zheng Zhao and Yang Liu

Environmental issue is a major issue related to the national and regional sustainable development. Ever since China's adopting Reform and Opening-up, China's economy has got sustainable and rapid growth; industrialization and urbanization has enjoyed constant acceleration; comprehensive national power has been significantly advanced; people's material life has been generally improved. However, with the rapid economic growth, environmental problems become increasingly serious, which has apparently hampered China's sustainable developments of economy and society. During the period of the 11th *Five-Year Plan*, China's central government focused on building up resource-saving and environment-friendly society and accelerated the transformation of economic development. As a result, environmental governance has been constantly strengthened; the task of pollution reduction has been finished ahead of schedule; the task of environmental pollution control has made progress in the phased sense. Nevertheless, deficiencies still exist in the field of environmental governance, which is still unable to match the environmental demands of sustainable economic development. Facing the new situation of economic and social development, and increasingly enhanced resource and environmental constrains, there is an urgent need in the period of the 12th Five-Year Plan, through focus on the main theme of accelerating the transformation of economy development and the new requirements of improving the ecological civilization, to strengthen the capacity building of environmental governance, to explore a new approach of environmental governance, and to guarantee the long-term sustainable economic and social development of China.

X. Li and J. Pan (eds.), *China Green Development Index Report 2011*,
Current Chinese Economic Report Series, DOI: 10.1007/978-3-642-31597-8_15,

15.1 The Effects and Problems of China's Environmental Governance in the 11th Five-Year Plan Period

During the period of the 11th *Five-Year Plan*, China's environmental governance has played a more significant role in the national economic development: the legal system of environmental governance has gradually improved; public participation in environmental governance has increasingly risen; the effectiveness of environmental governance has enhanced remarkably. Thus, environmental governance has became not only the major content of implementing *Scientific Outlook on Development* and building the harmonious socialist society, but also the important approach and strong force to transform economic growth pattern. However, we should also be clearly aware of that China's current environmental situation is still rather grim and along with the expansion of economic scale and acceleration of economic growth pace, new environmental problems become more and more prominent, so for China's environmental governance, there are still many problems needing to be urgently solved in the aspects like system and mechanism.

15.1.1 The Main Achievements of Environmental Governance in the 11th Five-Year Plan Period

During the period of the 11th *Five-Year Plan*, China adhered to the important theme of exploring the new path of environmental governance, formulated and improved environmental policy actively, and strengthened the efforts on environmental policy implementation, which has led to that: pollution treatment investment has maintained relatively fast growth; the total amount of major pollutants discharge has been under control; environmental governance's macro strategic system, overall preventing and controlling system, regulation and policy standard system and public participation system that are adapted to fundamental realities of China have been established or improved; the environmental governance capability has increased constantly and environmental governance has already gained remarkable achievements.

15.1.1.1 The Constant Improvement of Law and Regulation and the Gradual Enhancement of Law Enforcement

During the period of the 11th *Five-Year Plan*, China's environmental governance legislation developed rapidly: "*Law of the People's Republic of China on the Prevention and Control of Water Pollution*" was amended and promulgated; "*Circular Economy Act*" was formulated and implemented; "*Regulations on Planning Environmental Impact Assessment*", "*Regulations on Administration of Collection and Disposal of Waste Electrical Appliances and Electronic Products*" and the other five environmental administrative regulations were promulgated in succession; "*Energy-Saving and Emission Reduction Plan*", "*National Program of Response to Climate Change*" and other legal documents were issued. The number of national environmental standards increased at the speed of 100 per year, which filled the blank of environmental quality standards, and China finished amending and revising more than 60 key industry pollution emission standards and carried out 1,050 amendments and revisions of national environmental standards. The quantity of current national environmental standard has reached 1,300, and it is 502 more than that of the 10th Five-Year Plan. The continuous improvements of legislation have basically realized that in the main fields of environmental governance there are laws to follow. The local People's Congresses and governments also worked out local environmental governance regulations and local government rules accordingly, and thus legal and standard systems of environmental governance which are adapted to market economy system have been formed initially. These laws and regulations have played an important role in promoting the restriction of the behaviors which do harm to resources and environments and the acceleration of pollution control process.

Based on ensuring environmental governance has legal support, China's governments further strengthened law enforcement: on the basis of reinforcing routine check, governments carried out a large number of special inspections on key issues such as heavy metal pollution, paper-making enterprises, sewage treatment plants and landfills, and thus the law enforcement system which combines centralized law enforcement inspection with daily supervision has gradually formed and the environmental law enforcement capability and level have also constantly increased. In the meanwhile, many regions learned from the experience and practices of other domestic and overseas regions actively to carry out the innovations and practices in the field of environmental justice; through establishing environmental protection tribunals and other forms, they safeguarded the authority of the environmental protection law enforcement; through integrating the powers of executive and judicial sectors, they provided judicial guarantee to combats against environmental illegal activities and restraints of environmental pollution and environmental crimes.

Column 15.1 Yanqing County Established Environmental Protection Tribunals to Strengthen the Environmental Law Enforcement

On November, 10, 2010, the first environmental protection tribunal in Beijing was established in the law Court of Yanqing County. The environmental protection tribunal mainly used the experience and practices of other domestic and overseas regions for reference, and exercised the pattern of "Three in One" and "Dual Track System". "Three in One" means integrating criminal, civil and administrative proceedings together; as long as the cases are related to environmental protection, they will all be judged by the environmental protection tribunal. The cases include environmental pollution tort cases, administrative litigation cases caused by administrative omission or private party's refusing to accept administrative institutions' specific administrative behaviors on eco-environmental protection and resource administrative management, executive compensation cases caused by specific administrative behaviors like implementing environmental policy and carrying out environmental governance, and criminal cases which involves destroying the ecological environment and protecting living environment. "Dual Track System" is to integrate executive and judicial powers through both administrative means and judicial means, to form the integration power in environmental protection. Meanwhile, after establishing the environmental protection tribunal, Yanqing will establish the liaison, joint session and information communication system, and build the environmental protection law enforcement coordination mechanism among environmental protection departments, environmental protection tribunals and other environmental protection law enforcement institutions. In addition to hearing cases related to environmental protection and strengthening the linkage with administrative institutions, the tribunal will also adopt a variety of measures to participate in the social management activity positively, use judicial advices to supervise and guide related departments to improve environmental protection measures, strengthen the prevention of environmental violations, and propagate environmental laws and regulations through interpreting cases in attractions, scenic spots, communities and rural areas to raise the consciousness of environmental protection in all classes and all groups.

The environmental protection tribunal in Yanqing is a reform and exploration of the trial mechanism within the existing legal framework, which aims to maximize the protection of ecological environment and legitimate interests of the public. The "Three in One" and "Dual Track System" have not only provided a new path for the application of judicial means in environmental protection, but also to some extent compensated the defect that environmental violation is difficult to be fundamentally curbed due to administrative penalties being unable to be mandatorily enforced and incapable of investigating litigant's criminal responsibility, which has adequately strengthened the special effect of judicial function on the environmental protection and provided judicial guarantee to the combat against environmental illegal activities and the restraints of environmental pollution and environmental crimes.

Source: *Beijing established the first environmental protection tribunal* on "Science and Technology Daily", November, 18, 2011.

15.1.1.2 The Ever-Increasingly Definite Governance Ideas and the Enhancement of Public Participation

Since the 11th *Five-Year Plan*, the Central Committee of the Party and the State Council have proposed a serial of new ideas and new perspectives on constructing ecological civilization, promoting the historical transformation of environmental protection and exploring the new path of China's environmental protection, particularly emphasized guiding practice with advanced concepts, put environmental governance in a prominent position in the overall arrangement, determined the objectives and tasks of environmental protection, and solved lots of major problems in the field of environmental protection. The awareness of the importance and urgency of environmental governance in each region has also obviously increased in the practice of economic and social development. Closely encircling the local development realities, many regions have done exploration and innovation actively: they combined ecological civilization with economic development, and specially worked out special plans on environmental governance and ecological protection; they also clarified and normalized the implementation of the concept of environmental governance from various aspects like guiding ideology, governance target, security measures and policy system, which further promoted the practice of environmental governance concept.

In the meantime, the concept of public participation in environmental governance has also got further definition and guarantee. During the period of the 11th *Five-Year Plan*, China's environmental governance attached more importance to the protection of public rights, especially to absorbing public participation and to public interest demands, emphasized the participation of all social sectors, and ensured the effective public participation through institution construction.

In recent years, China has issued "National Action Plan of Energy-Saving and Emission Reduction" to mobilize the whole society to participate in the energy-saving and emission reduction and to cope with the challenge of climate change, and thus formed the work pattern of energy-saving and emission reduction: governments function as the leader; enterprises functions as the main body; the whole society works together to promote. China has published the first department regulation which aims to propel public participation in environmental protection "*Interim Measures on Environmental Impact Assessment of Public Participation*", and announced the first department regulation "*Measures for the Disclosure of Environmental Information (for Trial Implementation)*" to ensure the implementation of "*Regulation of the People's Republic of China on the Disclosure of Government Information*". With particular emphasis on utilizing newspapers, websites, televisions and other news media and public channels, governments disseminate the knowledge of environmental governance, expose environmental violation behaviors, carry forward the culture of environmental protection, create a national environmental protection atmosphere, ensure the public's rights to know about, to participate, and to supervise environment, bring a strong impetus to the open, democratic and scientific decision-making process, and enhance environmental governance's public participation.

Column 15.2 Zhangjianggang City in Jiangsu Province Worked Out the First Ecological Civilization Construction Plan in China

While keeping a rapid economic development, Zhangjiazhang has attached great importance to and vigorously promoted environmental protection and ecological construction, and successively won the honorary titles of the First National Model

City of Environmental Protection, National Model Green City, China Habitat Environment Model Award, International Garden City, United Nations Habitat Award and the Top 100 Counties (Cities) in Chinese County Comprehensive Socio-economic Development Index System. In order to further strengthen the effectiveness of ecological construction and enhance sustainable development capability, Zhangjiagang City has in time proposed building ecological civilization, and worked out the first ecological civilization construction plan in China.

On the basis of using advanced experience and ideas of home and abroad for reference, "Zhangjiagang City Ecological Civilization Construction plan" takes boosting the harmony between human and nature and promoting sustainable development as the core. It has been formulated aimed at the features like Zhangjiagang City's environmental quality, resource consumption and economic coordination. Its objective evaluation system is comprised of five systems: ecological consciousness civilization, ecological behavior civilization, ecological institution civilization, ecological environment civilization and ecological habitat civilization. It has also proposed the main framework and feasible planning scheme of ecological civilization construction, and established the construction evaluation index system of ecological civilization. (1) Ecological consciousness civilization includes expanding the investment in education to build a learning-oriented Zhangjianggang, developing comprehensively ecological education to cultivate ecological culture among all citizens, propagating ecology widely to enhance ecological consciousness of all citizens, and vigorously carrying forward the culture of Yangtze River to build basin ecological civilization together. (2) Ecological behavior civilization includes speeding up the development of circular economy to build industrial systems with the civilization of saving and low-carbon, strengthening the high-end agriculture construction to create the multi-functional ecological agriculture system, accelerating the development of sanitary services to build the effective environmental protection service system, deepening ecological protection awareness to cultivate the consumption pattern of saving and environmental protection, and carrying out the "Three Reductions" action to advocate civilized and frugal life and work mode. (3) Ecological environment civilization includes optimizing river systems' functions to create rivers of life, following laws of nature to reasonably position regional ecological function, optimizing space group to establish regional ecological security pattern, relying on the advantage of waterway network to create the biological harmonious system. (4) Ecological habitat civilization includes implementing the regulating and guiding strategy to establish the optimum population system, optimizing the urban living environment to construct the green proper and liveable city, highlighting the constructing characteristics of villages to construct waterside villages with the style of the southern Jiangsu Province, and upholding the people-oriented concept to create the saving and efficient ecological community. (5) Ecological institution civilization includes constructing efficient and transparent governmental administrative system, constructing enterprise institutional civilization of coordination and order, and building efficient public participation mechanism.

Environmental protection and ecological construction is a complex social system project. "Zhangjiagang City Ecological Civilization Construction Plan" has proposed the relatively systematic and comprehensive planning and constructing system of ecological civilization for the first time, which not only has an important effect on and meaning to the ecological civilization construction of Zhangjiagang city, but also is an useful exploration of China's further deepening ecological civilization construction. It will also set an example to other regions in ecological civilization construction planning.

Source: Gao Jixi, Huang Qin, Nie Yihuang, Xu Meijiang. Regional Practice and Exploration of Ecological Civilization Construction: Zhangjiagang City Ecological Civilization Construction Plan. Beijing: China Environmental Science Press, 2010.

15.1.1.3 The Gradual Improvement of Management System and the Obvious Strengthening of Governance Capability

During the period of the 11th *Five-Year Plan*, China has gradually established a supervision management system, in which National People's Congress is in charge of legislation; governments of all levels are responsible for the implementation; environmental protection administrative departments are in charge of the unified supervision and management; all related departments carry out supervision and management according to laws. In the year 2008, the Ministry of Environmental Protection was approved to be set up, and becomes one of the composition departments of the State Council, which has significantly improved the importance of environmental protection sector. This governmental institution reform strengthened the functions of coordination, macro-control, supervision, law enforcement and public service, which provided effective organizational guarantee to the promotion of environmental governance's historical transformation. Along with the upgrading of the national environmental protection department, regional environmental protection departments are playing more and more prominent role and their environmental governance capability has also improved; many regional environmental protection departments adopted stringent measures of environmental governance when dealing with the tasks of preventing and regulating pollution and improving environment, which has decreased the tendency of environmental pollution and ecological damages, and ameliorated the environment of urban and rural areas.

Meanwhile, in the practice of environmental governance, China's environmental governance instruments have increased progressively, and the function of environmental economic policy has also become increasingly obvious. For example, the promulgation of the list of products with "high pollution" and "high environmental risk", the expansion of emission right trade, the pilot of ecological compensation, the announcement of policies of "green credit", "green insurance" and "green tax", have all strengthened the environmental governance capability effectively, improved the incentive and restraint mechanisms of environmental protection and sustainable development, supported and encouraged the environmental protection behavior of enterprises and the whole society. In addition, with the establishment of the national environmental protection key laboratory, the national environmental protection engineering technology center in succession, China's environmental monitoring and early warning system has gradually realized the coverage of sea, land and sky; the environmental emergency response system has increasingly improved; many provinces and cities have established special environmental emergency management institutions; a number of major environmental issues which have attracted great public concern have been properly settled; all of these have significantly enhanced the average capability and emergency response capability of China on environmental governance.

Column 15.3 Shizuishan City, Ningxia Province: the Road of Transformation From the National Pollution City to the Pollution Control Model City

Shizuishan city is located in the intersection of the Helan Mountain and the Yellow River. This city was built because of coal and also became prosperous because of coal. However, the development of coal industry brought a lot of ecological problems to Shizuishan city. In 2003, Shizuishan was on the blacklist of "National Top 10 Cities with Serious Air Pollution". In recent years, Shizuishan city has made great efforts to strengthen environmental governance to promote the transformation of resource-exhausted city and to explore a sustainable development path of excellent economic effectiveness, resource consumption decrease and environmental pollution reduction. Nowadays, the city's green coverage has reached 36 % and per capita public green area has reached 14 m^2. The main indicators of energy-saving and emission reduction rank the first in Ningxia Province, and thus Shizuishan City has won the honors of "China Habitat Environment Model Prize", "National Demonstration City of Soil and Water Conservation and Ecological Construction". This gorgeous historical transformation from the national pollution city to the pollution control model city has mainly benefited from the environmental governance practice of the "ironhanded" pollution control and strict environmental supervision and monitoring.

On the one hand, Shizuishan city controlled pollution with "ironhanded" measures. Since 2004, Shizuishan has made great efforts to implement "Three Wastes Management", and achieved positive outcome. In the control of air pollution, the city shut down 98 small polluting enterprises in the urban area at a single swoop, relocated 182 coal processing enterprises around the urban area, banned 64 enterprises with backward production capacity and technology by law, and stopped 5 railway stations transporting coal by the deadline. In dealing with industrial wastes, the city invested nearly 1 billion yuan to bulldoze, cover and green the coal gangue hills formed by the coal mining and coal washing in the mining area, and backfilled the subsidence caused by coal mining to form a unique coal mining subsidence (geology) ecological park. In dealing with industrial and urban sewage, the city invested 750 million yuan to carry out the recycling use of sewage and the construction of city green project, and built a number of water treatment projects including recycled water plants.

On the other hand, the city implemented strict regulatory monitoring and strengthened environmental governance. Firstly, the city carried out strict investigation and approval of construction project and ensured the first approval right of environmental protection. The municipal environmental protection department would resolutely disapprove the projects which could not meet the requirements of national and regional industrial development policies, or the projects with immature pollution control technology. The implementation rates of the whole city's environmental impact evaluation of construction project and the "Three Simultaneities" of major construction project both reached 100 %. Secondly, Shizuishan city increased the environmental enforcement inspection of construction project to effectively prevent the generation of new pollution source. At the same time, the city particularly strengthened the environmental supervision of the enterprises with major pollution sources, and urged the normal operation of pollution control facilities to ensure their pollutants' discharge could meet the standard. Finally, Shizuishan city strengthened environmental monitoring. The city facilitated the national special funds of the state-controlled key pollution source automatic monitoring capacity building project to complete the construction of pollution source control center and 21 state—controlled key pollution source online monitoring systems: through using advanced monitoring measures to realize information sharing of the local pollution source data of atmosphere, water and so on, the city could master the related information of the city's pollution sources and pollutant emission more scientifically, accurately and timely and thus realize the real-time monitoring of the pollutant emission of key pollution sources.

Source: S*hizuishan City in Ningxia Province: the Strategy of National Pollution Control Model City.* http://env.people.com.cn/GB/10942510.html.

15.1.1.4 Over-Fulfilled Emission Reduction Task and Remarkable Environmental Governance Effects

During the period of the 11th *Five-Year Plan*, China set the reduction of major pollution emission as the binding indicator of national economic and social development plan, creatively put forward the principle of diluting cardinal number, counting increment, and verifying thc decrement to grope a set of effective methods of accounting and verifying, strengthened structural emission reduction and project emission reduction and managed emission reduction measure, and implemented strict assessment accountability to form a work pattern with vertical linkage and horizontal coordination. When both economic growth speed and energy consumption exceed the planning expectation, the SO_2 emission reduction target was fulfilled a year earlier than the schedule, and the COD emission reduction target was fulfilled half year earlier than the schedule, which can be showed in the Table 15.1.

Table 15.1 Major pollutant emission and sewage treatment capacity

Indictors	Unit	2005	2006	2007	2008	2009
COD emission	10,000 t	1,414.2	1,428.2	1381.8	1,320.7	1,277.5
SO_2 emission	10,000 t	2,549.4	2,588.8	2,468.1	2,321.2	2,214.4
Sewage treatment quantity	100 million cubic meters	–	202.62	226.98	256.00	279.35
Sewage treatment rate	%	–	55.67	62.87	70.16	75.25

Sources National Bureau of Statistics. *China Statistical Yearbook 2006–2010*. Beijing, China Statistical Press, 2006–2010. Ministry of Housing and Urban–Rural Development: *China City Statistical Yearbook 2009*. Beijing, China Planning Press, 2010

Through these data, we can find that from 2006 COD emission began to decline year by year: compared with 2005, COD emission in 2009 decreased 9.67 %. SO_2 emission in 2006 slightly increased compared to 2005, while in the years after 2006 it presented a downward trend: SO_2 emission in 2009 was only 86.9 % of that in 2005. Compared with the year 2005, COD emission in 2010 decreased 12.45 %, while SO_2 emission decreased 14.29 %, and both of these two indicators over fulfilled the emission reduction task.[1] In addition, China's sewage treatment capacity has been strengthened year by year; the annual sewage treatment quantity and sewage treatment rate have both increased year by year: sewage treatment quantity increased from 20.262 billion cubic meters in 2006 to 27.953 billion cubic meters in 2009; sewage treatment rate increased rapidly from 55.67 % in 2006 to 75.25 % in 2009. At the same time, China also paid attention to the construction of environmental governance infrastructure. During the 11th *Five-Year Plan*, China built up and operated 500 million kilowatts desulphurization facilities in coal-fired power plants, and desulphurization unit rate increased from 12 to 80 %.[2] The management acreage of water loss and soil erosion to a relatively large extent increased; the construction of desertification prevention and treatment was pushed forward solidly: the desertification land area reduced averagely 1,717 square kilometer per year, and provinces of desertification land reduced to 29.[3]

Corresponding to environmental governance in the national level, various regions have also gained remarkable effectiveness of environmental governance. During the 11th Five-Year Plan, the concept of environmental governance was widely reflected in the economic and social development practice in different regions in China. The importance of environment get constantly realized and

[1] Ministry of Environmental Protection: *The task of major pollutant emission in* 2010 *has been over fulfilled.* On www.cri.cn, June, 10, 2011.

[2] "*Closely encircling the new demand of theme and main line, strive for new situation of environmental protection*", the speech of Minister Zhou Shengxian at the 2011 national environmental protection conference. On http://www.zhb.gov.cn/zhxx/hjyw/201101/t20110120_200062.htm.

[3] National Bureau of Statics : *New development, new transcendence, new chapter: a report of economic and social development achievements during the period of "the* 11th *Five-Year Plan"* On http://www.stats.gov.cn/tjfx/ztfx/sywcj/t20110301_402706119.htm.

developed and ecological civilization is growing in popularity. To strengthen environmental protection, to carry out environmental governance, and to create favorable environment have became the main theme of and key to many regional development in China. In order to strengthen environmental protection, and improve environmental quality, many regions in China have adopted many environmental governance measures and carried out a series of environmental governance activities like special treatment and comprehensive management, increasingly strengthened environmental governance efforts, and achieved positive results.

Column 15.4 Haikou City, Hainan Province: Creating High Quality Water Environment Through Comprehensive Governance

The river network of Haikou City is rather developed and locates in a crisscross pattern. The network presents the "Two Horizontal and Five Vertical" distribution pattern around the main city. However, in the processes of urbanization, the Wuyuan River, the eastern part of Nandu River and the upper reaches of Meishe River are still natural water system, but the remaining river systems (basin) have all been subject to human transformation and reshaping: the river system has been dismembered; the lakes has been filled and occupied; the water quality has degraded. Facing the increasingly severe water pollution problems, Haikou city based on the reality, carried out comprehensive treatment from six aspects, and gained positive results.

Firstly, the city implemented the water network power engineering project in the city center. After the completion of this project, the annual water supply to urban rivers and lakes is about 916.8 million cubic meters, which has contributed to water flow, circulation and replacement, and enhanced the river network power and water self-purification ability in the city center.

Secondly, the city implemented the sewage outfall closure and network integration project, which intensively investigated 13 sewage outfalls of brooks and lakes in the city center and brought more than 300 pollutant outfalls respectively into the closure and network integration project of the center city, and the remaining sewage outfall closure and network integration project of Meishe River.

Thirdly, the city carried out the Meishe River diversion canal project, which started from Shapo reservoir spillway, and ended at the intersection of Ding Village, with the total length of 3.2 km.

Fourthly, the city carried out the governance project of Baisha River and the clear-return project of Yaweixi-Wuxi Road open channel and water body. The main contents include closure of sewage outfalls along the river, dredging remediation, and construction of water-pumping from Henghegou, which made the channels change from the stagnant and flow again, and realize the water body return clear.

Fifthly, the city implemented the decontamination project of Jinniu Lake. This project is collecting schemes, project establishment and researches all over China, which is aimed at improving the ecosystem of Jinniu Lake.

Sixthly, the city carried out the water source pollution control projects. It comprehensively started the basic environmental investigation and evaluation of drinking water source, regularly monitored and assessed the water quality of Longtang reservoir, Yongzhuang reservoir and other drinking water source protection areas. In the drinking water protection areas, the strictest environmental admittance and environmental protection approval institution were carried out: the project which cannot meet the environmental requirements will definitely not be approved and allowed to be constructed; the municipal environmental protection bureau, water affairs bureau, land and resources bureau, commerce and industry bureau jointly launched investigation on the illegal buildings, structures and pollution sources in the water source protection areas.

After the comprehensive governance, the water environment of Haikou City has been significantly improved. The quality of the city's rivers, lakes, reservoirs and other surface water has reached corresponding functional areas standards; the qualified rate of water quality in drinking water source has reached 100 %; public awareness of environmental protection has grown constantly. The city has realized the multi-win situation: water environmental governance keeps pace with the improvement of disaster prevention and reduction, the acceleration of urbanization process and the improvement of living environment.

Source: Six Comprehensive Measures were Carried Out on the Water Environment Governance in Haikou, on China Daylily, April, 11, 2011.

15.1.2 The Major Problems Existing in Environmental Governance at the Present Stage

During the period of the 11th *Five-Year Plan*, China attached great importance to environmental issues in economic development, actively developed and improved environmental policies, strengthened implementation efforts on environmental policies, and has achieved remarkable and phased outcomes in the aspects like pollution source regulation, discharge standards, comprehensive environmental governance and major pollutant quantity control. However, we should realize that China's environmental pollution situation is still grim. New environmental problems become more and more prominent along with the expansion of economic scale and the acceleration of economic growth; the mechanism, capacity construction and other aspects of environmental governance still cannot satisfy the demands of environmental protection and there are still some environmental problems needing solving.

15.1.2.1 Environmental Governance Legal System Needs Improving

In the period of the 11th *Five-Year Plan*, the construction of China's environmental governance legal system has made great progress. However, China's environmental legislation and environmental protection is far from meeting the actual demand of environmental governance, and has restraint environmental governance and the economic and social sustainable development. On the one hand, between the introduction and improvement of environmental governance laws and the practical need of environmental governance there still exists a gap. For example, there is still no law and regulation in the fields like green finance. The existing system of environmental governance regulations and policies primarily focuses on pollution regulation like pollutant emission standards and emphasizes on punishment; the governance means are relatively homogeneous and cannot solve problems completely; the environmental protection effect is relatively poor with high economic cost and low environmental effects; all these are not conducive to the fundamental improvement of environmental quality. On the other hand, there are still problems existing in the field of environmental law enforcement. Affected by local interest and departmental interest, some local governments and environmental law enforcement departments took relatively tolerant attitude to the environmental violation behaviors of the pillar industries and key enterprises which are strongly connected to local economic development. The phenomenon of lax enforcement and impunity still exists in environmental governance.

15.1.2.2 Environmental Governance Mechanisms Need Improving

At present, China's environmental governance mechanism still has an obvious color of "strong government", and the transformation from administrative management to pluralistic governance is still very tough. In the environmental governance practices, although the public plays an increasingly important role in environmental governance, most specific measures in the formulation and implementation of environmental policy are carried out through the power of administrative system, and they are operated by different levels of governmental departments directly and carried out as through governmental system an administrative behavior. Public participation from the private sector has written legality and rationality, but it is lack of specific provisions in operation procedures and rights and interests safeguard and workable guarantee mechanisms; the channels and paths for public participation in environmental governance are not sufficient and smooth; participation modes are usually the participation at the bottom and at the same time face with many practical obstacles. The existing environmental governance mechanism has restraint the exertion of social resource and power to a large extent, and impacted the enthusiasm and creativity of public participation. Hence it is in urgent need to innovate the institutions, to optimize environmental governance structure, to reinforce the basis for public participation into environmental governance fundamentally, and to construct a pluralistic environmental governance mechanism.

15.1.2.3 Environmental Governance Instruments Need Enriching

In the process of environmental governance, environmental economic policy instruments have the advantages like promoting the innovation of environmental protection technology, strengthening market competitiveness, reducing environmental governance cost and administrative monitoring cost. During the period of the 11th *Five-Year Plan*, China introduced a lot of environmental economic policy instruments and methods into environmental governance. The implementation and deepening of environmental economic policies such as green credit, green insurance, green trade, green tax and other policy instruments enriched the national macro-control measures and reduced the environmental costs of economic growth. However, on the whole, the main body of China's environmental policy is still environmental policies and orders under the direct control of government, and the environmental policy instruments which will be used more frequently is the command-control. In the premise of each region and each enterprise pursuing profits, direct regulation policies will meet huge resistance in the practical implementation process because it cannot improve environment at the least cost. In the meanwhile, the environmental economic policy that focuses on the coordinated development of economic construction and environmental protection is lack of systematicness, and the environmental economic policy system which will be beneficial to the implementation of sustainable development hasn't been formed. When formulating environmental policies or adopting environmental policy instruments, governments considers more from the inner environmental management system, are short of the adjustment based on environmental resource interest aimed at various market subjects, and seldom make comprehensive environmental policies from the perspective of the governmental macro-control function conversion and coordination with other policies.

15.2 Regional Comparison of China's Environmental Governance in the 11th Five-Year Plan Period

During the period of the 11th *Five-Year Plan*, China's economy has continued to maintain strong growth momentum, and the total economy quantity has exceeded Japan and thus ranked the second place in the world. However, rapid development of economy has brought tremendous pressure to the environment; resource scarcity and environmental pollution have become the bottleneck which constrains economic development. Therefore, different levels of governments put more attention on strengthening environmental governance, accelerated the transformation of economic growth, and promoted the development of green economy. Based on the evaluation results of "China Provincial Green Development Index System" and "China City Green Development Index System", this section intends to do comparative analysis of governmental environmental governance in 2009 from the angle of provinces and major cities separately, which was supposed to provide strong data and theoretical support to Chinese government's vigorously promoting environmental governance capacity and green development in the 12th *Five-Year Plan*.

15.2.1 Interprovincial Comparison of China's Environmental Governance

In order to strengthen China's environmental governance capacity, all levels of governments have made great efforts on environmental protection, pollution control and other aspects. Based on "China Provincial Green Development Index System", we selected five positive indicators and one inverse indicator[4]: Incremental area of afforestation per capita, Removing rate of industrial SO_2 emissions, Removing rate of COD in industrial waste water, Removing rate of industrial NOx emissions, Removing rate of ammonia nitrogen in industrial waste water and Sudden accidents effecting environment (inverse), analyzed the environmental governance capacity of the eastern region, central region, western region and north-eastern region respectively, and further processed difference comparison of this four regions in terms of environmental governance characteristics.

15.2.1.1 Environmental Governance in Eastern Region

Based on the geographical advantage, the economic growth speed of eastern region has been in a leading position for a long time. With the rapid economic development and the continuously advancing urbanization, the environment in eastern region also suffers great pressure. For the regional sustainable development, strengthening environmental governance is in urgent need. Provinces in eastern region promoted the development of environmental governance technology, which has boosted the environmental governance quantity and efficiency. The situation of environmental governance of 10 provinces in eastern China in 2009 is shown on Table 15.2.

On the whole, eastern provinces have done relatively good performance on the indicators of "Removing rate of industrial SO_2 emissions" and "Removing rate of

[4] The bigger the value of positive indicator is, the stronger the environmental governance capacity is; the bigger the value of inverse indicator is, the weaker the environmental governance capacity is.

Table 15.2 Environmental governance of 10 provinces in eastern China in 2009 (Units: hectares/10,000 people, %, times)

Provinces	Environmental governance indicators					
	Incremental area of afforestation per capita	Removing rate of industrial SO_2 emissions	Removing rate of COD in industrial waste water	Removing rate of industrial NOx emissions	Removing rate of ammonia nitrogen in industrial waste water	Sudden accidents effecting environment
Beijing	10.2	65.3	86.2	36.3	92.8	31
Tianjin	13.0	59.7	63.2	16.7	29.8	–
Hebei	43.7	54.3	15.2	4.7	60.5	4
Shanghai	1.1	61.6	36.8	6.1	77.3	118
Jiangsu	10.9	64.7	20.9	6.9	79.0	10
Zhejiang	5.3	65.9	8.1	3.2	81.7	50
Fujian	9.2	47.0	18.2	7.2	72.7	6
Shandong	19.3	66.7	0.9	0.2	90.0	19
Guangdong	2.1	64.9	19.3	6.0	83.1	10
Hainan	22.6	73.7	3.1	1.6	56.8	7
Average	13.7	62.4	27.2	8.9	72.4	25.5

Sources National Bureau of Statistics: *China Statistical Yearbook 2010.* Beijing, China Statistical Press, 2010. Ministry of Environmental Protection: *Annual Statistical Report on Environment in China 2010*, China Environmental Science Press, 2010. National Bureau of Statistics, Ministry of Environmental Protection: *China Statistical Yearbook of Environment 2010*, Beijing, China Statistical Press, 2010

ammonia nitrogen in industrial waste water", which reached 62.4 and 72.4 % respectively, while the indicators "Removing rate of COD in industrial waste water" and "Removing rate of industrial NOx emissions" was relative low, the average value of which was only 27.2 and 8.9 %. In addition, the indicators of "Incremental area of afforestation per capita" and "Sudden accidents effecting environment" presented great difference among eastern provinces.

Specifically speaking, on the positive indicators, there was big difference on the indicator "Incremental area of afforestation per capita" in eastern provinces, the top two provinces in the rank were Hebei and Hainan, and their value reached 43.7 ha/10,000 people and 22.6 ha/10,000 people respectively; the last two provinces in the rank were Shanghai and Guangdong, and their quantities were only 1.1 ha/10,000 people and 2.1 ha/10,000 people. The incremental areas of afforestation per capita of the other provinces was all between 5 and 20 ha/10,000 people. In regard to the four indicators of pollutant removing rate, eastern provinces presented little difference on the indicators of "Removing rate of industrial SO_2 emissions": the highest value was in Hainan, that is 73.7 %, and the lowest value 47.0 % was in Fujian; the removing rate of industrial SO_2 emissions of the other provinces was around 60 %. However, on the indicators of "Removing rate of COD in industrial waste water" and "Removing rate of industrial NOx emissions", the value of eastern provinces presented obvious disparities. The average value of "Removing rate of COD in industrial waste water" was 27.2 %;

the highest value was in Beijing, that is 86.2 %, and the lowest value 0.9 % was in Shandong. Besides that, Beijing's "Removing rate of industrial NOx emissions" reached 36.3 %, which was the also the highest, while the lowest was still in Shandong, that is 0.2 %. On the indicator of "Removing rate of ammonia nitrogen in industrial waste water", with the exception of that the value in Tianjin (29.8 %) was relatively low, the indicator value in the other provinces all exceeded 55 %, and in Beijing, the indicator value was even up to 92.8 %.

In terms of the inverse indicator "Sudden accidents effecting environment", the frequency of Hebei, Fujian, Hainan and Guangdong was less than 10 times (10 times including). The best performance came from Hebei, only 4 times; the frequency of the other provinces was all more than 10 times. Shanghai has done the worst performance in this aspect, and its frequency of "sudden accidents effecting environment" in 2009 reached 118. On the whole, the average annual sudden accidents effecting environment in eastern provinces was 25.5.

15.2.1.2 Environmental Governance in Central Region

In the overall strategy of China's regional development, the central areas acts as the role of "connecting east with west". Central region has always been the major agricultural and sideline product base of China. Shanxi, Jiangxi and other provinces have abundant coal resources. Therefore, the development of the central region will be beneficial to the improvement of food and energy security capabilities of China. However, there are problems such as low level of industrialization, serious pollution, huge resource and energy consumption in central region. Hence, strengthening the environmental governance in central region and taking a new road of industrialization are important factors that are crucial to China's economic and social development. The situation of environmental governance of 6 provinces in central China in 2009 is shown on Table 15.3.

Holistically speaking, on the indicators of "Incremental area of afforestation per capita" and "Sudden accidents effecting environment", with the exception of some provinces, the value of which was relatively prominent, there was little difference among most central provinces. Central provinces did well on the indicators of "Removing rate of industrial SO_2 emissions" and "Removing rate of ammonia nitrogen in industrial waste water", which respectively reached 64.0 and 56.5 %, while the indicator values of "Removing rate of COD in industrial waste water" and "Removing rate of industrial NOx emissions" was relatively low, the average value of which was only 17.1 and 6.9 %.

Specifically speaking, judged from the analysis of positive indicators, the average value of "Incremental area of afforestation per capita" among central provinces was 41.4 ha/10,000 people: the highest value was 95.5 ha/10,000 people in Shanxi, while the lowest value was in Anhui of 11.2 ha/10,000 people. The indicator value of "Removing rate of industrial SO_2 emissions" in central provinces was relatively high; the gap between the lowest in Henan (54.5 %) and the highest in Anhui (76.9 %) was only 12.3 %. In regard to the indicators "Removing

Table 15.3 Environmental governance of 6 provinces in central China in 2009 (Units: hectares/10,000 people, %, times)

Provinces	Environmental governance indicators					
	Incremental area of afforestation per capita	Removing rate of industrial SO_2 emissions	Removing rate of COD in industrial waste water	Removing rate of industrial NOx emissions	Removing rate of ammonia nitrogen in industrial waste water	Sudden accidents effecting environment
Shanxi	95.5	58.7	51.6	16.9	58.5	4
Anhui	11.2	76.9	30.3	11.7	83.8	22
Jiangxi	51.8	74.5	1.3	0.7	57.7	6
Henan	44.0	54.6	0.7	0.2	52.1	10
Hubei	26.1	62.6	7.9	3.1	47.2	11
Hubei	19.6	56.6	10.7	8.7	39.9	–
Average	41.4	64.0	17.1	6.9	56.5	8.8

Sources National Bureau of Statistics: *China Statistical Yearbook 2010*. Beijing, China Statistical Press, 2010. Ministry of Environmental Protection: *Annual Statistical Report on Environment in China 2010*, China Environmental Science Press, 2010. National Bureau of Statistics, Ministry of Environmental Protection: *China Statistical Yearbook of Environment 2010,* Beijing, China Statistical Press, 2010

rate of COD in industrial waste water" and "Removing rate of ammonia nitrogen in industrial waste water", Jiangxi, Henan and Hubei performed relatively poor and ranked the last three place on the two indicators; the province that ranked the first place on these two indicators was Shanxi and there was a big gap between it and other provinces with low ranking. On the indicator of "Removing rate of ammonia nitrogen in industrial waste water", Anhui (83.8 %) was far ahead from the other provinces, and Hunan (39.6 %) was relatively low; the gap between the other provinces was not great and the indicator value was all around 56.5 %.

On the inverse indicator "Sudden accidents effecting environment", central provinces all performed well, with the average annual frequency of 8.8. With the exception of the highest frequency 22 times of Anhui, the frequency of the other provinces were all beneath 12 times.

15.2.1.3 Environmental Governance in Western Region

Western region is abundant in natural resources. For more than 10 years ever since the implementation of the western development strategy, western region has provided strong resources support for the development of eastern region. Meanwhile, the region has also gained a series of achievements in its own infrastructure construction and economic development. In the next five years, the western region will continue to accelerate the pace and obtain significant development in comprehensive economic strength, people's living standard, ecological environment

Table 15.4 Environmental governance of 11 provinces in western China in 2009 (Units: hectares/10,000 people, %, times)

Provinces	Environmental governance indicators					
	Incremental area of afforestation per capita	Removing rate of industrial SO_2 emissions	Removing rate of COD in industrial waste water	Removing rate of industrial NOx emissions	Removing rate of ammonia nitrogen in industrial waste water	Sudden accidents effecting environment
Sichuan	59.8	47.0	7.7	4.1	42.7	–
Chongqing	33.6	57.1	5.8	3.6	47.9	33
Guizhou	62.2	67.9	73.3	17.3	57.1	4
Yunnan	156.6	77.2	–	–	85.6	3
Shaanxi	119.3	50.1	9.5	2.9	47.2	10
Gansu	80.7	82.7	15.4	4.3	18.0	36
Qinghai	253.1	9.6	0.005	0.003	17.1	2
Ningxia	144.0	51.3	–	–	43.5	–
Xinjiang	160.2	5.4	2.3	1.1	32.6	–
Guangxi	28.8	51.2	3.7	9.7	39.1	11
NeiMongol	356.6	57.8	2.1	0.3	61.1	5
Average	132.2	50.7	10.9	3.9	44.7	9.5

Sources National Bureau of Statistics: *China Statistical Yearbook 2010*. Beijing, China Statistical Press, 2010. Ministry of Environmental Protection: *Annual Statistical Report on Environment in China 2010*, China Environmental Science Press, 2010. National Bureau of Statistics, Ministry of Environmental Protection: *China Statistical Yearbook of Environment 2010*, Beijing, China Statistical Press, 2010

protection and other aspects. The situation of environmental governance of 11 provinces in western China in 2009 is shown on Table 15.4.

On the whole, western provinces performed relatively well on "Incremental area of afforestation per capita", with the average value of 132.2 ha/10,000 people. They also had relative good performance on the indicator of "Removing rate of industrial SO_2 emissions" and "Removing rate of ammonia nitrogen in industrial waste water", the value of which reached 50.7 and 44.7 % separately, while the indicator value of "Removing rate of COD in industrial waste water" and "Removing rate of industrial NOx emissions" was relatively low, the average value of which was only 10.9 and 3.9 %. In terms of sudden accidents effecting environment, most western provinces did relatively well.

Specifically speaking, on the positive indicators, there were six provinces in western region, of which incremental area of afforestation per capita of western provinces exceeded 100 ha/10,000 people: Inner Mongolia, Qinghai, Xinjiang, Yunnan, Ningxia and Shanxi; the highest value was 356.6 ha/10,000 people of Inner Mongolia, while the lowest value was in Guangxi, which was just 8 % of that of Inner Mongolia.

On the indicator of "Removing rate of industrial SO_2 emissions", the value of Qinghai and Xinjiang was obviously lower than that of the other provinces, and was only 9.6 and 5.4 % respectively; Gansu performed well on this indicator, and its value reached 82.7 %. Guizhou ranked the first on both indicators of "Removing rate of COD in industrial waste water" and "Removing rate of industrial NOx emissions", the value of which was 73.3 and 17.3 % and it was far ahead from the average level of western provinces; these two indicator value in Qinghai was almost 0. In addition, there was a big discrepancy on the indicator of "Removing rate of ammonia nitrogen in industrial waste water" among western provinces: the highest value in Yunnan (85.6 %) was 65.8 % more than the lowest value in Qinghai (17.1 %).

On the whole, western provinces performed well on the indicator of "Sudden accidents effecting environment". Among the 8 provinces with estimated value, the frequency of sudden accidents effecting environment in 5 provinces was less than 10 times (10 times including). The frequency of Chongqing and Xinjiang was relatively high and reached 33 and 36 times respectively. Calculated from the whole, the average annual frequency of sudden accidents effecting environment was 9.5 times.

15.2.1.4 Environmental Governance in Northeastern Region

In the period of the 11th *Five-Year*, a series of strategic measures to revitalize old industrial bases like northeastern region were introduced in succession. After five years' efforts, northeastern region's economic structure has been further optimized; the independent innovation capacity has improved significantly; the opening up degree has increased remarkably; the condition of infrastructure has got upgraded; the outlook of urban and rural areas has undergone great changes. Meanwhile, the development of green economic development also requires northeastern region to continue constantly strengthening ecological construction, actively promote emission reduction and enhance environmental governance. The situation of environmental governance of 3 provinces in northeastern China in 2009 is shown on Table 15.5.

Holistically speaking, there existed certain gap among the provinces in northeastern region on the indicator of "Incremental area of afforestation per capita". On the four indicators of pollutant removing rate, "Removing rate of ammonia nitrogen in industrial waste water" presented relatively good condition, with the average value of 73.1 %, while the value of the other three indicators were relatively low. On the indicator of "Sudden accidents effecting environment", the frequency of the accidents was relatively small on the whole.

Specifically speaking, the indicator value of "Removing rate of industrial SO_2 emissions" and "Removing rate of ammonia nitrogen in industrial waste water" in Liaoning was comparatively high, and reached 52.7 and 76.5 % respectively, while on the indicator of "Removing rate of COD in industrial waste water", its value was relative low (only 2.3 %); the indicator value of "Incremental area of afforestation per capita" ranked the second among the three provinces, which is 30.1 ha/10,000 people. The indicator value of "Removing rate of ammonia nitrogen in industrial waste water" in Heilongjiang was the highest (80.1 %),

Table 15.5 Environmental governance of 3 provinces in northeastern China in 2009 (Units: hectares/10,000 people, %, times)

Provinces	Environmental governance indicators					
	Incremental area of afforestation per capita	Removing rate of industrial SO_2 emissions	Removing rate of COD in industrial waste water	Removing rate of industrial NOx emissions	Removing rate of ammonia nitrogen in industrial waste water	Sudden accidents effecting environment
Liaoning	30.1	52.7	7.1	2.3	76.5	6
Jilin	11.0	29.9	–	–	62.7	–
Heilongjiang	55.7	15.0	10.4	3.5	80.1	–
Average	32.3	32.5	5.8	1.9	73.1	2

Sources National Bureau of Statistics: *China Statistical Yearbook 2010*. Beijing, China Statistical Press, 2010. Ministry of Environmental Protection: *Annual Statistical Report on Environment in China* 2010, China Environmental Science Press, 2010. National Bureau of Statistics, Ministry of Environmental Protection: China Statistical Yearbook of Environment *2010*, Beijing, China Statistical Press, 2010

while its "Removing rate of industrial SO_2 emissions" was the lowest (only 15.0 %). The indicator value of "Incremental area of afforestation per capita" of Heilongjiang was the highest in these three provinces, which reached 55.7 ha/ 10,000 people. The data of Jilin has been lost on a relatively large scale, but its estimated data basically ranked the second place among the three provinces.

15.2.1.5 The Disparity Analysis on Environmental Governance in Eastern, Central, Western and Northeastern China

For a long time, the regional disparities have existed in China's economic development. In recent years, the gap between undeveloped provinces in western and central regions with the eastern coastal provinces is increasingly widened, which has further aroused the wide concern of China's government and all walks of life. Table 15.6 shows the disparity of environmental governance in eastern, central, western and northeast China.

On the whole, there are great disparities in terms of indicators among the four regions. On the indicator the incremental area of afforestation per capita, western region was much better than the other regions; the indicators of four pollutants removing rate in eastern region were obviously better than those of other regions, but eastern region had relatively high frequency of sudden accidents effecting environment. The indicator "Removing rate of industrial SO_2 emissions" in central region ranked the first among the four regions, while the other indicators remained in the midstream. Northeastern region was comparatively weak in the indicator ranking, and its three indicators ranked the last.

Table 15.6 Disparity of Environmental Governance in Eastern, Central, Western and North-eastern China (Units: hectares/10,000 people, %, times)

Provinces	Environmental governance indicators					
	Incremental area of afforestation per capita	Removing rate of industrial SO_2 emissions	Removing rate of COD in industrial waste water	Removing rate of industrial NOx emissions	Removing rate of ammonia nitrogen in industrial waste water	Sudden accidents effecting environment
Eastern region	13.7	62.4	27.2	8.9	72.4	25.5
Central areas	41.4	64.0	17.1	6.9	56.5	8.8
Western areas	132.2	50.7	10.9	3.9	44.7	9.5
Northeastern areas	32.3	32.5	5.8	1.9	73.1	2

Note all the indicator value is average value

Sources National Bureau of Statistics: *China Statistical Yearbook 2010*. Beijing, China Statistical Press, 2010. Ministry of Environmental Protection: *Annual Statistical Report on Environment in China 2010*, China Environmental Science Press, 2010. National Bureau of Statistics, Ministry of Environmental Protection: *China Statistical Yearbook of Environment 2010*, Beijing, China Statistical Press, 2010

Specifically speaking, western region had the largest incremental area of afforestation per capita, with a quantity of 132.2 ha/10,000 people, which was obviously higher than other three regions. The overall situation of removing rate of industrial SO_2 emissions was relatively high, while northeastern region in this aspect was relatively weak, with a percentage of 32.3 %, but other regions all exceeded 50 %. However, the overall situation of removing rate of ammonia nitrogen in industrial waste water and removing rate of COD in industrial waste water were low, and especially on the indicator of "Removing rate of ammonia nitrogen in industrial waste water", there was no region that could exceed 10 %; in northeastern region, the value was only 1.9 %. On the indicator "Removing rate of COD in industrial waste water", the situation of each region was optimistic: even the lowest value in western region reached 44.7 %; the highest value was in the eastern region, which was 72.4 %.

On the inverse indicator of "Sudden accidents effecting environment", the frequency in eastern region was smallest, that is 2 times; the frequency in central and western regions were comparatively high, that is 8.8 and 9.5 times; eastern region's performance on this indicator was comparatively poor and frequency reached 25.5 times.

Table 15.7 Environmental governance of four municipalities directly under the jurisdiction of the center government in 2009 (Unit: %)

Cites	Environmental governance indicators			
	Removing rate of industrial SO2 emissions	Removing rate of COD in industrial waste water	Removing rate of industrial NOx emissions	Removing rate of ammonia nitrogen in industrial waste water
Beijing	65.3	89.1	36.3	92.8
Tianjin	59.7	67.7	16.7	29.8
Shanghai	61.6	88.5	6.1	77.3
Chongqing	57.1	52.5	3.6	47.9
Average	60.9	74.4	15.7	61.9

Sources Ministry of Environmental Protection: *China Statistical Yearbook of Environment 2010*, Ministry of Environmental Protection: *China Statistical Yearbook of Environment 2009*, China Environmental Science Press, 2009. Ministry of Housing and Urban–Rural Development: *China City Statistical Yearbook 2009*. Beijing, China Planning Press, 2009

15.2.2 Analysis of China's Major Cities' Environmental Governance

In order to well complete environmental governance, each city in China has made a lot of efforts on environmental protection, pollution control and other aspects. Based on"China City Green Development Index System", we selected 4 positive indicators: "Removing rate of industrial SO_2 emissions", "Removing rate of COD in industrial waste water", "Removing rate of industrial NOx emissions" and "Removing rate of ammonia nitrogen in industrial waste water", and respectively analyzed environmental governance capacities of four municipalities directly under the jurisdiction of the central government, four typical provincial capitals and five speically designated cities, and further compared the disparities of environmental governance characteristics among these cities.

15.2.2.1 Environmental Governance of Four Municipalities Directly Under the Jurisdiction of the Central Government

Each municipality directly under the jurisdiction of the central government has its own typical characteristics of environmental governance launched by government. The environmental governance situation of four municipalities directly under the jurisdiction of the central government is shown on Table 15.7.

In terms of indicators, the four municipalities under the jurisdiction of the central government all achieved better performance on "Removing rate of industrial SO_2 emissions", "Removing rate of COD in industrial waste water" and "Removing rate of ammonia nitrogen in industrial waste water" than on "Removing rate of industrial NOx emissions". In addition, among the four municipalities there was almost no difference on the indicator of "Removing rate

Table 15.8 Environmental governance of four typical provincial capital cities in 2009 (Unit: %)

Cites	Environmental governance indicators			
	Removing rate of industrial SO2 emissions	Removing rate of COD in industrial waste water	Removing rate of industrial NOx emissions	Removing rate of ammonia nitrogen in industrial waste water
Chengdu	62.3	63.9	0.8	26.3
Shenyang	34.9	86.1	0.5	27.2
WUhan	48.4	85.7	0.2	59.5
Nanjing	78.7	76.0	15.9	90.7
Average	56.1	77.9	4.3	50.9

Sources Ministry of Environmental Protection: China Statistical Yearbook of Environment 2010, Ministry of Environmental Protection: China Statistical Yearbook of Environment 2009, China Environmental Science Press, 2009. Ministry of Housing and Urban–Rural Development: China City Statistical Yearbook 2009. Beijing, China Planning Press, 2009

of industrial SO_2 emissions", and all were around 60 %, while there were obvious disparities among the four cities on the other three indicators.

As far as city is concerned, Beijing ranked the first on the four indicators among the four cities, and its indicator value of "Removing rate of industrial NOx emissions" and "Removing rate of ammonia nitrogen in industrial waste water" has distinct advantages over the other three municipalities directly under the jurisdiction of the central government. Tianjin had the lowest "Removing rate of ammonia nitrogen in industrial waste water", which was 63 % lower than that of Beijing. Chongqing ranked the last on the indicators of "Removing rate of industrial SO_2 emissions", "Removing rate of industrial NOx emissions" and "Removing rate of COD in industrial waste water" among the four cites.

15.2.2.2 Environmental Governance of Four Typical Provincial Capital Cities

In order to investigate the environmental governance of provincial capital cities, we separately chose four representative provincial capitals from eastern region, central region, western region and northeastern region, and analyzed their environmental governance situation, which is showed on Table 15.8.

As far as indicator is concerned, "Removing rate of COD in industrial waste water" of the four cities was the highest among the four indicators, and the average value reached 77.9 %, while "Removing rate of ammonia nitrogen in industrial waste water" was the lowest, and was only 4.3 %. In addition, obvious disparities existed on the indicator of "Removing rate of industrial SO_2 emissions" among the four typical provincial capitals, and on the indicators of "Removing rate of industrial NOx emissions" and "Removing rate of ammonia nitrogen in industrial waste water", the value of some cities was distinctly higher than the average level.

As far as city is concerned, Nanjing had the best performance on environmental governance indicators among the four cites, and ranked the first on three

Table 15.9 Environmental governance of five specially designated cities in the state plan in 2009 Unit: %)

Cites	Environmental governance indicators			
	Removing rate of industrial SO2 emissions	Removing rate of COD in industrial waste water	Removing rate of industrial NOx emissions	Removing rate of ammonia nitrogen in industrial waste water
Dalian	74.8	64.5	0.3	96.2
Qingdao	66.4	85.1	0.9	89.5
Ningbo	84.7	91.0	7.6	97.7
Xiamen	48.6	96.1	52.3	84.3
Shenzhen	60.4	88.4	6.5	65.8
Average	67.0	85.0	13.5	86.7

Sources Ministry of Environmental Protection: *China Statistical Yearbook of Environment 2010*, Ministry of Environmental Protection: *China Statistical Yearbook of Environment 2009*, China Environmental Science Press, 2009. Ministry of Housing and Urban–Rural Development: *China City Statistical Yearbook 2009*. Beijing, China Planning Press, 2009

indicators, and its indicator value of "Removing rate of industrial NOx emissions" and "Removing rate of ammonia nitrogen in industrial waste water" was obviously higher than the other cities. "Removing rate of industrial SO_2 emissions" of Shenyang ranked the last among the four typical provincial capital cities, which was only 34.9 %, and this value was 43.8 % lower than that of Nanjing. With the exception of Nanjing, "Removing rate of industrial NOx emissions" of Chengdu, Shenyang and Wuhan was relatively low, and was 0.8, 0.5, and 0.2 % respectively. In addition, "Removing rate of ammonia nitrogen in industrial waste water" of Chengdu and Shenyang was far beneath the average level (50.9 %).

15.2.2.3 Environmental Governance of Five Specially Designated Cities in the State Plan

Five specially designated cities Dalian, Qingdao, Ningbo, Xiamen and Shenzhen have gained remarkable achievements on environmental governance, and the environmental governance situation of five specially designated cities in the state plan is showed on Table 15.9.

As far as indicator is concerned, "Removing rate of industrial SO_2 emissions", "Removing rate of COD in industrial waste water", and "Removing rate of ammonia nitrogen in industrial waste water" of these five cities were relatively high, and respectively reached 67.0, 85.0, and 86.7 %, while "Removing rate of industrial NOx emissions" was relatively low and was only 13.5 %, and there were also great disparities among these cities on this indicator: the highest was in Xiamen (52.3 %), while the lowest was in Dalian and Qingdao, which were only 0.3 and 0.9 % respectively.

As far as city is concerned, Xiamen ranked the first on two indicators "Removing rate of COD in industrial waste water" and "Removing rate of industrial NOx emissions", but also ranked the last on the indicator "Removing rate of industrial SO2 emissions"(48.6 %). Ningbo also ranked the first on two indicators "Removing rate of ammonia nitrogen in industrial waste water" (84.7 %) and "Removing rate of ammonia nitrogen in industrial waste water" (97.7 %). "Removing rate of COD in industrial waste water" and "Removing rate of industrial NOx emissions" of Dalian were the lowest in comparison with the other specially designated cities. The average value of "Removing rate of ammonia nitrogen in industrial waste water" of these five cities reached 86.7 %, among which that of Dalian and Ningbo exceeded 90 %.

15.3 Tendency and Outlook of China's Environmental Governance in the 12th Five-Year Plan Period

The 12th *Five-Year Plan* is the critical phase for China's comprehensively building a moderately prosperous society, and is also the important stage for China's transforming from the middle-income country to the moderately developed country. With the further acceleration of industrialization and urbanization, environmental problems in China will become more complex, and environmental governance will be destined to be a long and arduous task. Facing increasingly enhanced constraints of resource and environment, it can be inferred that, during the period of the 12th Five-Year Plan, the efforts put into China's environmental governance will continue to increase. Focusing on solving the prominent environmental problems which affect sustainable development, China will constantly improve related institutions and mechanisms, strengthen comprehensively the construction of environmental legal system, explore new environmental governance instruments actively and achieve sound and fast development. This will be the main orientation of China's environmental governance.

15.3.1 Pollution Prevention and Control in Key Fields will Get Further Strengthening

During the 12th *Five-Year Plan*, the pollution prevention and control in key fields will be still the top priority of China's environmental governance. Judged from the current situation, in the future China will continue to consolidate the phased outcomes of the 11th *Five-Year Plan*: China will focus on solving the prominent environmental problems which do harm to people's health such as drinking water pollution, air pollution, soil pollution and other environmental problems, take the pollution control of key industries and enterprises, environmental governance of drinking water source protection areas, and environmental management in rural and urban areas as the core, and enhance the pollution treatment of thermal power, steel, nonferrous metals, chemicals, and building materials industries to stabilize the effectiveness of standard discharge. In the meanwhile, China will strictly implement the protection institution of drinking water source, increase the efforts on environmental management and pollution prevention and control of major trans-boundary rivers, promote water pollution prevention and control of major basins and areas, control urban air pollution and noise pollution, increase the township treatment capacity of sewage and garbage, improve urban environment, and implement rural environmental contiguous treatment project to promote the improvement of rural environment. Through the pollution prevention and control of key industries and urban and rural regions, China will solve the key environmental problems which affect China sustainable development.

15.3.2 Multi-Subject Environmental Governance Pattern will be Further improved

During the 12th Five-Year Plan, with the deepening of environmental governance practices, the multi-subject environmental governance pattern will continuously get innovation and improvement. On the one hand, the public rights to know, to supervise and to discuss will be expanded. Besides the published public information like environmental condition communiqué and weekly report of air condition, pollutant discharge, settlement of pollution accidents and environmental decisional information will also be opened to public gradually to increase the transparency of environmental information. On the other hand, the threshold for public participation in environmental governance will be lowered; China will improve the operability of public participation and forming the mass basis and social basis of environmental protection will be the imperative demand of improving multi-subject environmental governance pattern. In the future, the guarantee mechanism of public participation in environmental governance will be improved, the path; path, form and procedure and other issues of public participation will be embodied in substantive law, procedural law and environmental management systems; the long-term positive interaction of environmental governance mechanisms between governmental supervision and public participation will tend to be improved.

15.3.3 The Legal System of Environmental Governance will be Further Strengthened

Ecological environment has the typical property of public goods, thus lots of "public domains" exist. In the practice of environmental governance, the improvement of legal system is beneficial to the better protection of people's fundamental interests, the better definition of proprietary rights, use rights, management rights and profit rights that are closely related to ecological environment, and the better maintaining of the normal order of market economy. During the period of 12th Five-Year Plan, the legal construction of environmental governance in China will get further strengthening: in order to adapt to the demand of environmental governance practice, China will introduce, amend environmental related laws, and add more content of ecological protection, resource recycling and efficient use to the existing laws, and in the meantime further coordinate individual environment and resource laws, and establish strict legal system and effective enforcement methods. On the other hand, increasing the efforts on environmental law enforcement will be the important measure of environmental governance in China. To overcome the interference of local interests and departmental interests, to process and correct environmental violation behaviors in time, to enhance the administrative supervision of environmental law enforcement, and to realize that "Once a law is put into force, it must be observed and strictly enforced and violators must be punished" will be the important part of improving environmental governance development system in China.

15.3.4 The Capacity Building of Environmental Governance will Get Further Enhancement

Environmental governance capacity is the basis of environmental governance and the important support of the environment-friendly society construction. Nowadays, China's environmental protection undertaking has entered into a new period. During the 12th *Five-Year Plan*, both gaining good environmental governance and solving the structural, compound and compressive environmental problems need the enhancement of environmental governance capacity. On the one hand, the capital investment in the environmental governance will be increased. According to the unified arrangement of the central government, all levels of governments will gradually establish environmental finance system, and absorb environmental governance funds into annual budget, and based on governments' financial condition and environmental protection task, government will have to truly allocate the 12th Five-Year Plan environmental protection fund and specifically decompose it into the pollution control projects. On the other hand, China's environmental governance capacity system will be improved. Special technological projects related with environmental governance will attract more attention; the process of environmental monitoring network, management programming, methods standardization, and monitoring automation will be accelerated; environmental standard system and technology management system will be gradually improved; general management and emergency management capacity will be continuously enhanced.

15.3.5 *Policy Instruments of Environmental Governance will Get Further Enrichment*

The key of enriching environmental governance policy instruments is to strengthen the function of economic instruments of environmental governance, to use market mechanism to internalize the economic subjects' external cost caused by the pollution in economic activities, to save and comprehensively utilize resources and to implement the policies which aim at pursuing social welfare maximization. In the period of the 12th *Five-Year Plan*, China will constantly improve and reduce command-control policy, and at the same time, China will build up environmental governance economic policy system based on market mechanism. On the one hand, China will implement the environmental economic policy which will benefit scientific development, comprehensively use economic measures like pricing, taxation, finance, credit, fees and insurance to regulate or influence the behavior of market subjects; while strengthening the "polluter pays" principle and avoiding the shift behavior and non sufficient payment, China will improve the environmental governance's taxation, credit, finance guiding mechanisms to encourage enterprises to carry out pollution control actively, establish the incentive and restraint mechanisms of protecting and sustainably using resources. In the meanwhile, China will enrich and improve green economic policies which will benefit the mutual promotion of environmental governance and economic construction, enhance the research and implementation of green credit, green insurance, green securities, green trade, green taxes, ecological compensation and other policies, to form relatively complete green economic policy system.

15.3.6 *The Collaborative Linkage Mechanism of Environmental Governance will Get Further Improvement*

Environmental governance is a systematic project with wide involvement and complex operation mechanism. Environment itself has the prosperities of public resources, externality and spatial extension, which determines environmental governance also has integrity. During the 12th Five-Year Plan, the cooperation mechanism of environmental governance will be paid more attention to. On the one hand, inter-regional cooperation of environmental governance will increase; different regions will put more consideration on integrity, balance and timeliness when making regional holistic environmental protection policy, implementing environmental standards and processing environmental supervision, to form the prevention and control pattern of exchange and cooperation, co-management, information sharing, and mutual promotion of win–win situation; regional environmental linkage mechanism will be improved. On the other hand, in the context of globalization, any country cannot carry out environmental governance without the collaboration and cooperation from the other countries. In the future,

China will continue the multilateral, bilateral and regional international cooperation in the field of desertification prevention and control, air pollution control, acid rain prevention and sewage disposal. The environmental governance experience communication and learning with international environmental organizations, developed countries and neighboring countries will be more frequent; international cooperation on environmental governance will continue to be strengthened, and the transnational collaborative linkage mechanism on environmental governance will get further improvement.

Chapter 16
Measurement and Analysis of the Support Degree of Government Policies

Junli Zhao, Xin Jiang and Tao Song

Green economic development has become the common choice of all countries to cope with global warming and climate changing realize sustainable development of humanity. For China, in particular, it is an important approach to practice the Scientific Concept of Development, transform the pattern of economic growth and coordinate the development of economy, society and environment in a sustainable way. As the policy orientation and implementation directly affect the level of green economic development in a region, the Index of Support Degree of Government Policies constitutes one of the three important sub-indexes under Green Development Index (GDI). Since 2009 was an important year of the *11th Five-Year Plan* period, governments at all levels had made efforts to save energy, reduce pollution, protect environment, build up ecological civilization, and guide the green development of socio-economy. These efforts were fruitful.

The Support Degree of Government Policies (SDGP) is a comprehensive evaluation of the participation of a government in local green development. In accordance with the measurement criterion of SDGP in China Province Green Development Index System (PGDIS) and China City Green Development Index System (CGDIS), basing on data of 2009, we measured the SDGP of 30 provinces (autonomous regions, and municipalities directly under the jurisdiction of the central government),[1] 34 large and mid-sized cities (including four municipalities directly under the jurisdiction of the central government, 25 provincial capital cities and five cities specifically designated cities in the state plan, except Lhasa and Urumqi) in China by analyzing three indicators: green investment, infrastructure and environmental control. Based on the results, we elaborated on the relationship between policy support and green economic development on a regional comparative basis and put forward policy suggestions for the government to act more effectively towards green development.

[1] For lack of key data, Tibet, Hong Kong and Chinese Taipei are excluded from the calculation.

X. Li and J. Pan (eds.), *China Green Development Index Report 2011*,
Current Chinese Economic Report Series, DOI: 10.1007/978-3-642-31597-8_16,

16.1 Calculated Results of SDGP

In order to fully reflect the characteristics of provinces and cities, we designed different Third-Level Indicators with different weights under SDGP in PGDIS and CGDIS. The calculated results of provinces and cities are as follows:

16.1.1 Calculated Results of Provincial SDGP

According to the measurement system and weight standards of PGDIS and CGDIS, we worked out the results of SDGP of 30 provinces (autonomous regions and municipalities directly under the jurisdiction of the central government) in China (See Table 16.1).

The results showed that these provinces varied little in SDGP, indicating that local governments are all making efforts (see Table 16.1). Beijing ranked first with a SDGP of 0.278, 28 % higher than the national average;The lowest was Gansu with − 0.165, 17 % lower than the national average. The index values of 15, or half of the measured provinces (autonomous regions and municipalities directly under the jurisdiction of the central government) were above the national average. The top ten provinces were Beijing, Jiangsu, Shanghai, Zhejiang, Shandong, Ningxia, Guangdong, Fujian, Yunnan and Hebei (see Fig. 16.1). A deeper look into the three indicators showed that, in terms of green investment, the top ten provinces were Ningxia, Beijing, Qinghai, Shanxi, Chongqing, Shaanxi, Gansu, Guangxi, Hebei and Xinjiang; in terms of infrastructure, the top ten provinces were Shanghai, Beijing, Jiangsu, Zhejiang, Shandong, Guangdong, Fujian, Chongqing, Jiangxi and Hebei; in terms of environmental control, the top ten provinces were Beijing, Inner Mongolia, Yunnan, Guizhou, Anhui, Jiangsu, Shandong, Guangdong, Shanxi and Fujian.

In 2009, the SDGP of 30 provinces (autonomous regions and municipalities directly under the jurisdiction of the central government) in China had the following characteristics:

16.1.1.1 SDGPs in Economically Developed Provinces were Relatively Higher

The calculated results showed a slight inter-provincial disparity, but economically developed provinces did have a higher SDGP. Among the four major regions[2]

[2] The eastern provinces included Beijing, Tianjin, Hebei, Shanghai, Jiangsu, Zhejiang, Fujian, Shandong, Guangdong and Hainan; the central provinces included Shanxi, Anhui, Jiangxi, Henan, Hubei and Hunan; the western provinces included Inner Mongolia, Guangxi, Chongqing, Sichuan, Guizhou, Yunnan, Tibet, Shaanxi, Gansu, Qinghai, Ningxia and Xinjiang; the northeastern provinces included Liaoning, Jilin and Heilongjiang.

Table 16.1 Indexes of support degree of government policies and rankings of 30 provinces in China in 2009

Indicator	Support degree of government policies		Green investment indicators		Infrastructure indicators		Environmental control indicators	
Weight	100 %		25 %		45 %		30 %	
Province	Score	Ranking	Score	Ranking	Score	Ranking	Score	Ranking
Beijing	0.278	1	0.052	2	0.137	2	0.089	1
Jiangsu	0.121	2	−0.018	20	0.11	3	0.029	6
Shanghai	0.107	3	−0.003	15	0.152	1	−0.043	26
Zhejiang	0.096	4	−0.003	16	0.1	4	0	15
Shandong	0.087	5	−0.022	22	0.083	5	0.027	7
Ningxia	0.081	6	0.068	1	0.019	12	−0.006	16
Guangdong	0.077	7	−0.028	26	0.081	6	0.024	8
Fujian	0.051	8	−0.034	28	0.065	7	0.019	10
Yunnan	0.049	9	0.001	13	−0.01	15	0.057	3
Hebei	0.043	10	0.01	9	0.025	10	0.007	12
Shanxi	0.036	11	0.038	4	−0.022	19	0.02	9
Chongqing	0.035	12	0.036	5	0.042	8	−0.044	28
Shaanxi	0.017	13	0.033	6	−0.008	14	−0.009	17
Jiangxi	0.01	14	−0.026	23	0.035	9	0.001	14
Anhui	0.004	15	−0.019	21	−0.014	18	0.037	5
Hubei	0	16	−0.008	19	0.019	13	−0.011	21
Tianjin	−0.009	17	−0.032	27	0.021	11	0.002	13
Guangxi	−0.03	18	0.015	8	−0.024	20	−0.021	24
Inner Mongolia	−0.03	19	0.003	12	−0.092	27	0.058	2
Hainan	−0.05	20	−0.002	14	−0.061	25	0.013	11
Liaoning	−0.061	21	−0.039	30	−0.012	17	−0.01	20
Xinjiang	−0.063	22	0.008	10	−0.01	16	−0.062	29
Sichuan	−0.074	23	−0.007	18	−0.048	22	−0.018	22
Hunan	−0.075	24	−0.004	17	−0.051	24	−0.02	23
Henan	−0.076	25	−0.027	25	−0.038	21	−0.01	19
Guizhou	−0.076	26	0.003	11	−0.127	29	0.048	4
Qinghai	−0.082	27	0.051	3	−0.05	23	−0.082	30
Heilongjiang	−0.147	28	−0.037	29	−0.101	28	−0.009	18
Jilin	−0.154	29	−0.027	24	−0.085	26	−0.042	25
Gansu	−0.165	30	0.017	7	−0.138	30	−0.044	27

Notes (1) Figures in this table are calculated based on data of each indicator for 2008 and 2009 in accordance with the indicator system of SDGP embedded in the Province Measurement System (2) Index of each province in this table is ranked in descending order

Sources China Statistical Yearbook 2010, *Annual Statistical Report on Environment in China* 2009, *China Environmental Statistical Yearbook* 2010, *China Industrial Economic Statistical Yearbook* 2010, and *China City Statistical Yearbook* 2010

(eastern, central, western and northeastern), the eastern provinces had the highest SDGP. Eight of the ten eastern provinces ranked among the top ten in the country, including Beijing (No.1), Jiangsu (No. 2), Shanghai (No. 3), Zhejiang (No. 4),

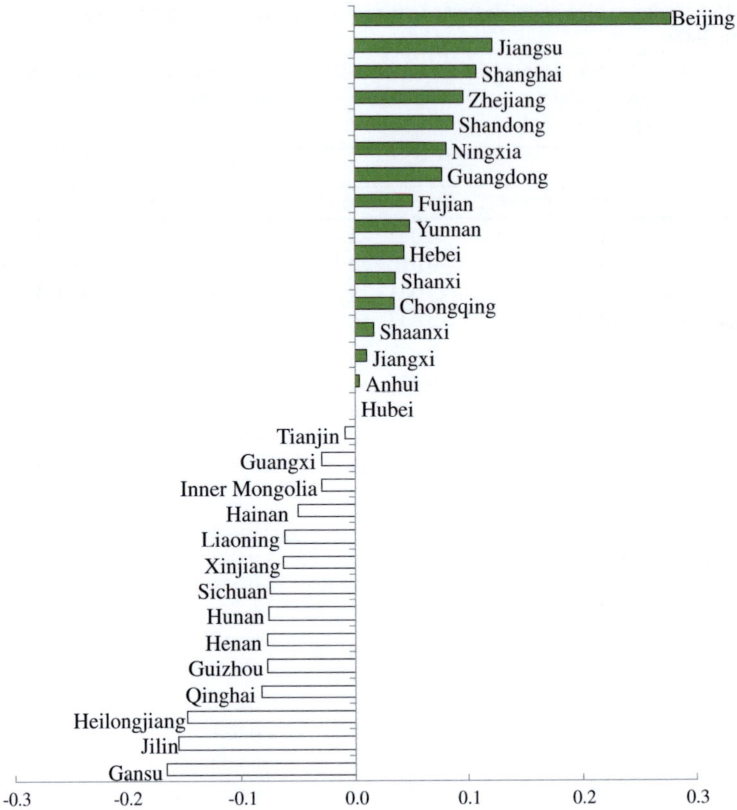

Fig. 16.1 Inter-provincial comparison of SDGP rankings

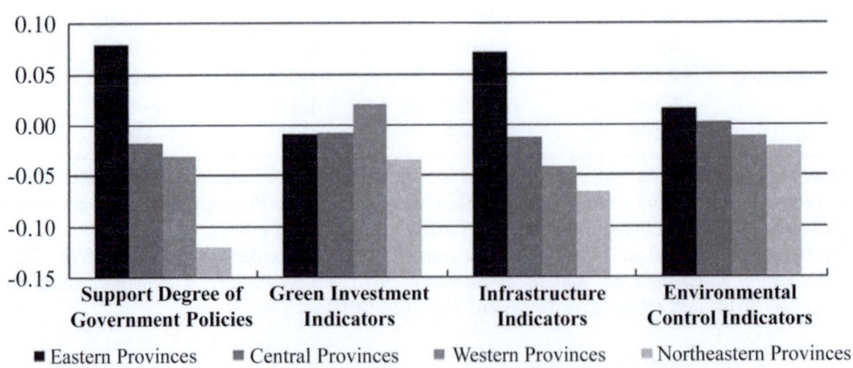

Fig. 16.2 Comparison of SDGP among 4 Major Areas in China. *Notes* Figures in the chart are the arithmetic mean of provinces within each of four major areas in China

Shandong (No. 5), Guangdong (No. 7), Fujian (No. 8), and Hebei (No.10). Only Tianjin and Hainan were slightly lower, ranking 17 and 20th. The central provinces had the second highest SDGP in general. Of the six central provinces, four ranked among No. 11 ~20. They were Shanxi (No. 11), Jiangxi (No. 14), Anhui (No. 15) and Hubei (No. 16); Hunan and Henan were slightly lower, ranking 24 and 25th. As to the western provinces, half were high, half were not. Six of the 11 provinces had high SDGP: Ningxia, (No. 6), Yunnan (No. 9), Chongqing (No. 12), Shaanxi (No. 13), Guangxi (No. 18) and Inner Mongolia (No. 19). However, the other five provinces ranked among the national bottom ten. The SDGP of northeast provinces were much lower. Liaoning (No. 21), Heilongjiang (No. 28) and Jilin (No. 29), below the national average, lied among the bottom ten.

16.1.1.2 Green Development of Local Governments Among Different Regions was Distinctive

Given the different natural endowments and economic development levels, all local governments should tailor their own policies towards green development to their specific conditions and needs. In terms of the three indicators, the provinces differ most in infrastructure, next in green investment, and least in environmental control. This meant different policy orientations and priorities for each province in green development.

The eastern provinces had the finest infrastructure and relatively effective environmental control but insufficient green investment. The much better infrastructure (see Fig. 16.3) provided well-equipped hardware for the government to carry out more favorable policies. The environmental control was slightly above the national average and not so much an advantage. The green investment, slightly below the national average and equal to that of the central region, was far below that of the West, but higher than that of the Northeast. Thus, the eastern provinces should invest more in environmental protection and scientific research to match their economic strength.

The central provinces had high environmental control and green investment but poor infrastructure. The environmental control level approximated the national average, second to the East with a slight gap; the green investment identified with that of the East, slightly below the national average; the infrastructure was slightly below the national average but significantly lower than that of the East. Improved infrastructure was the necessary step towards greener development. Sandwiched by the East and West, Central China was confronted with double trouble in the underdeveloped economy and shortage of natural resources. Thus more policy support was needed for greener development.

The western region made the greatest green investment, compared with poor infrastructure and insufficient environmental control. Though rich in resources, the West was confronted with dual challenges in economic development and environmental protection. The provinces had trouble in energy conservation and emission reduction. The region's greatest efforts into green investment indicated that policy support for Western Development was paying off. Yet the region needed to invest more in infrastructure and environmental control that were better than the Northeast but slightly below the national average.

The northeastern provinces were poor in all the three indicators that were significantly lower than the national average. Compared with other parts of China, the region was closest with them in environmental control and most distant in infrastructure. As the old industrial base of China, the region had no time to delay in transforming the economic development pattern. In this regard, policy support served as a strong backup Fig. 16.2.

16.1.1.3 Government Policy Support Offset Deficient Resources and Environment

By comparing the rankings of SDGP Index and GDI, we found 14, or half of provinces had changed their rankings by five places or less. Only seven provinces changed their rankings by ten places or more, namely Qinghai, Tianjin, Hainan, Guizhou, Heilongjiang, Ningxia and Shanxi (see Table 16.2).

Overall, the SDGP rankings of 18 provinces were higher than their GDI rankings. Of the top ten GDI provinces, seven ranked among the top ten SDGP ones (see Table 16.2), namely Beijing, Shanghai, Zhejiang, Yunnan, Fujian, Jiangsu and Guangdong. The eastern provinces lagged behind in terms of potential bearing capacity indexes of resources and environment, such as Beijing (No. 18), Shanghai (No. 24), Zhejiang (No. 19), Fujian (No. 14), Jiangsu (No. 30) and Guangdong (No. 23). Their strong policy support greatly pushed forward the green development. As was indicated, government policy support played an important role in offsetting the deficient resources and environment and safeguarding green development.

16.1.2 Calculated Results of City SDGP

Given the central role of cities in regional economy and the constantly improved urbanization in China, measuring urban SDGP was of great significance. The Third-Level Indicators of urban SDGP were different from those under the provincial measurement system, for they were more representative of the characteristics of urban socio-economic development. In this report, 13 Third-Level Indicators were selected and the results of SDGP in 34 cities were ranked as follows (see Table 16.3).

As can be seen from Table 16.3, the highest index value was that of Shenzhen, 0.319, 32 % higher than the average, while the lowest was that of Xining, −0.346, 35 % lower than the average. The index values of 18, or over half of the cities were above the average. The top ten SDGP cities were Shenzhen, Beijing, Xiamen, Guangzhou, Shijiazhuang, Ningbo, Nanjing, Qingdao, Dalian and Fuzhou

Table 16.2 Inter-provincial comparisons of ranking gap between GDI and SDGP

Province	Ranking of GDI (1)	Ranking of SDGP (2)	Ranking gap (1)–(2)	Province	Ranking of GDI (1)	Ranking of SDGP (2)	Ranking gap (1)–(2)
Beijing	1	1	0	Xinjiang	16	22	−6
Shanghai	2	3	−1	Jiangxi	17	14	3
Qinghai	3	27	−24	Hebei	18	10	8
Tianjin	4	17	−13	Sichuan	19	23	−4
Hainan	5	20	−15	Anhui	20	15	5
Zhejiang	6	4	2	Chongqing	21	12	9
Yunnan	7	9	−2	Hubei	22	16	6
Fujian	8	8	0	Jilin	23	29	−6
Jiangsu	9	2	7	Guangxi	24	18	6
Guangdong	10	7	3	Liaoning	25	21	4
Inner							

Note This table is derived from Table 16.1

(see Fig. 16.3). The top ten green investment cities were Shijiazhuang, Lanzhou, Fuzhou, Beijing, Harbin, Hangzhou, Yinchuan, Guangzhou, Shenzhen and Zhengzhou; the top ten infrastructure cities were Shenzhen, Beijing, Dalian, Xiamen, Qingdao, Nanjing, Hangzhou, Nanchang, Nanning, Shijiazhuang; and the top ten environmental control cities were Xiamen, Beijing, Ningbo, Guangzhou, Guiyang, Nanjing, Kunming, Taiyuan, Qingdao and Shanghai. The SDGP of all the cities was shown in Fig. 16.3.

To sum up, the inter-city SDGP in 2009 had the following characteristics:

16.1.2.1 There was a Significant Disparity Among Cities and Clear Advantages of the Eastern Cities

Deviation values of SDGP among cities were high, indicating a significant disparity of urban SDGP. The eastern cities[3] enjoyed greater support. Among the four regions in China, the eastern cities had the highest SDGP, higher than the average of the 34 cities and the SDGP of any city in the other three regions. Among the ten eastern cities, nine ranked among the top ten, namely Shenzhen (No. 1), Beijing (No. 2), Xiamen (No. 3), Guangzhou (No. 4), Shijiazhuang (No. 5), Ningbo (No. 6), Nanjing (No. 7), Qingdao (No. 8) and Fuzhou (No. 10). The SDGP of the central cities was relatively high, close to the city average. These cities were

[3] The eastern cities included Beijing, Tianjin, Shijiazhuang, Shanghai, Nanjing, Hangzhou, Ningbo, Fuzhou, Xiamen, Jinan, Qingdao, Guangzhou, Shenzhen and Haikou; the central cities included Taiyuan, Hefei, Nanchang, Zhengzhou, Wuhan and Changsha; the western cities included Hohhot, Nanning, Chongqing, Chengdu, Guiyang, Kunming, Xi'an, Lanzhou, Xining and Yinchuan; the northeastern cities included Shenyang, Dalian, Changchun and Harbin.

Table 16.3 Indexes of SDGP and rankings of 34 cities in China in 2009

City	Support degree of government policies		Green investment indicators		Infrastructure indicators		Environmental control indicators	
	100 %		25 %		45 %		30 %	
	Score	Ranking	Score	Ranking	Score	Ranking	Score	Ranking
Shenzhen	0.319	1	0.015	9	0.29	1	0.017	18
Beijing	0.222	2	0.035	4	0.08	2	0.109	2
Xiamen	0.168	3	−0.019	26	0.07	4	0.12	1
Guangzhou	0.141	4	0.027	8	0.03	13	0.086	4
Shijiazhuang	0.133	5	0.072	1	0.04	9	0.019	16
Ningbo	0.11	6	0.002	14	0.02	16	0.089	3
Nanjing	0.099	7	−0.019	24	0.05	6	0.065	6
Qingdao	0.083	8	−0.012	23	0.06	5	0.034	9
Dalian	0.065	9	−0.026	31	0.07	3	0.018	17
Fuzhou	0.048	10	0.043	3	0.01	17	−0.01	21
Kunming	0.047	11	−0.026	30	0.02	14	0.049	7
Haikou	0.047	12	0.004	13	0.02	15	0.022	15
Hangzhou	0.041	13	0.029	6	0.05	7	−0.036	25
Shanghai	0.03	14	−0.01	21	0.01	19	0.031	10
Yinchuan	0.024	15	0.028	7	−0.02	25	0.014	20
Taiyuan	0.009	16	0.013	11	−0.05	28	0.047	8
Jinan	0.008	17	−0.007	18	−0.01	23	0.027	14
Hefei	0.008	18	−0.038	32	0.03	11	0.015	19
Changsha	−0.011	19	−0.009	20	0.01	18	−0.015	22
Nanning	−0.021	20	−0.007	17	0.04	10	−0.05	27
Zhengzhou	−0.027	21	0.013	10	0	21	−0.045	26
Shenyang	−0.034	22	0.007	12	0.03	12	−0.071	29
Nanchang	−0.045	23	−0.006	16	0.04	8	−0.084	32
Guiyang	−0.057	24	−0.009	19	−0.12	32	0.071	5
Wuhan	−0.059	25	−0.025	28	−0.01	24	−0.021	23
Xi'an	−0.094	26	−0.019	25	0	22	−0.08	31
Hohhot	−0.098	27	−0.039	34	−0.09	30	0.03	11
Tianjin	−0.099	28	−0.011	22	−0.05	29	−0.035	24
Chengdu	−0.102	29	−0.039	33	0.01	20	−0.072	30
Chongqing	−0.121	30	−0.026	29	−0.03	26	−0.068	28
Harbin	−0.135	31	0.029	5	−0.19	33	0.03	12
Changchun	−0.163	32	−0.005	15	−0.03	27	−0.127	33
Lanzhou	−0.184	33	0.06	2	−0.27	34	0.028	13
Xining	−0.346	34	−0.02	27	−0.12	31	−0.208	34

Notes (1) Figures in this table are calculated based on data of each indicator for 2008 and 2009 in accordance with the indicator system of SDGP embedded in the City Measurement System
(2) Index of each province in this table is ranked in descending order

Sources China Statistical Yearbook 2010, *Annual Statistical Report on Environment in China* 2009, *China Environmental Statistical Yearbook* 2010, *China Industrial Economic Statistical Yearbook* 2010, and *China City Statistical Yearbook* 2010

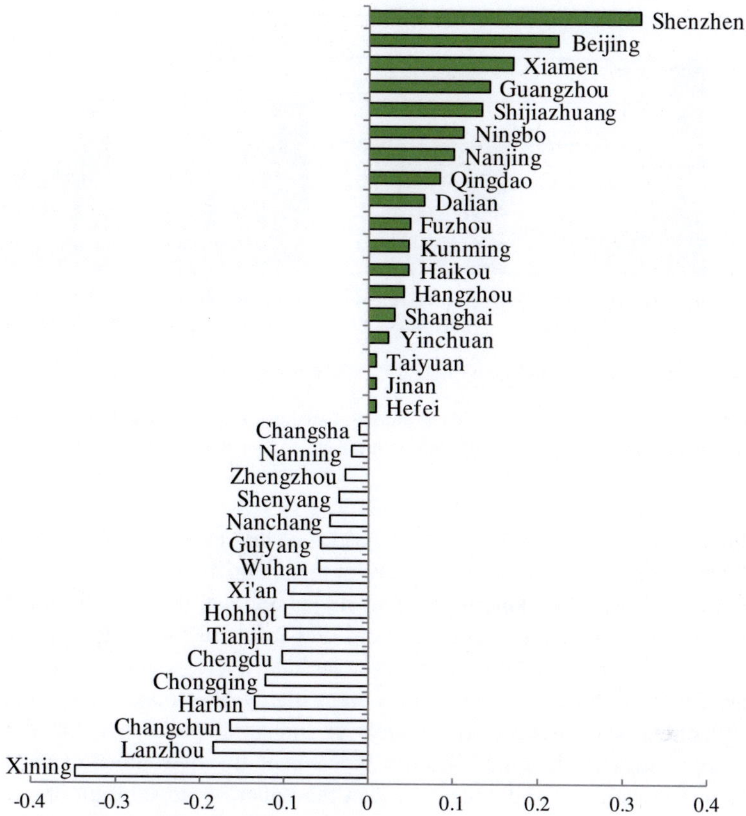

Fig. 16.3 Inter-city comparison of SDGP rankings

Taiyuan (No. 16), Hefei (No. 18), Changsha (No. 19), Zhengzhou (No. 21), Nanchang (No. 23), Wuhan (No. 25). The SDGP of the four northeastern cities were relatively low. They were Dalian (No. 9), Shenyang (No. 22), Harbin (No. 31) and Changchun (No. 32). The western cities had the lowest SDGP, for among the ten measured cities, eight were far below the national average except Kunming (No. 11) and Yinchuan (No. 15).

The measured results of cities differed slightly from those of provinces. On the provincial level, western provinces had higher SDGP than the northeastern ones, while on the city level, we got the reverse outcome.

16.1.2.2 Different Cities had Their Specific SDGPs

We measured the three indicators including Green Investment, Infrastructure and Environmental Control in order to measure SDGP under GDI. The results showed

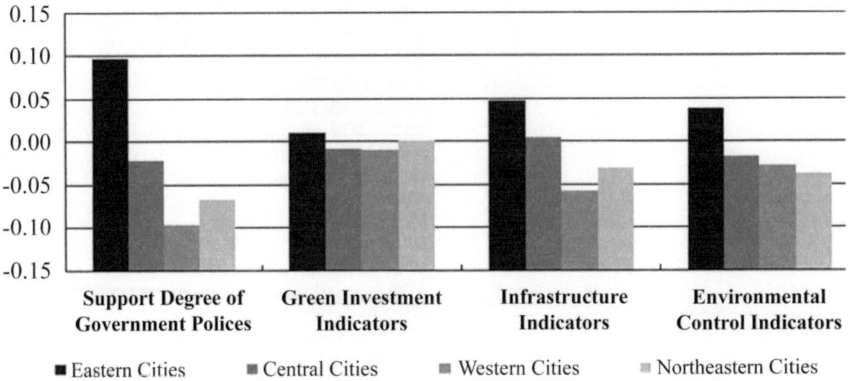

Fig. 16.4 Inter-city comparison of SDGP among 4 major areas in China. *Note* Figures in the chart are the arithmetic mean of provinces within each of four major areas in China

that different cities had different trends, priorities and extent of efforts in making and implementing policies on green development Fig. 16.4.

From the above table, among the four major regions, the cities in the eastern region had much higher green investment and infrastructure value and slightly higher environmental control value over the national average and cities in the other regions. On the one hand, eastern cities were highly developed in infrastructure, urban planning management and control of industrial pollutants. On the other hand, these cities made much heavier investment in environmental protection, science and technology and education than the underdeveloped regions.

The values of all indicators of central cities were far below those of eastern ones. In the central region, the infrastructure value was slightly above the national average and also higher than that of the West and Northeast. The environmental control value was above the West and Northeast, and the green investment value was equal to the West and slightly below the Northeast, but the value of both indicators were both lower than the national average. This revealed a solid hardware basis of Central China and greater input into environmental protection by provincial capitals and other major and mid-sized cities. As a result, the region was delivering a better environment for the people.

The three indicators of the western and northeastern cities were all low. However, the northeastern cities had more green investment and better infrastructure than the western cities. The green investment in the Northeast approached the city average and was second only to the East, but the region lagged far behind East and Central China in infrastructure. The western cities had better environmental control over the Northeast, but lagged behind the eastern and central cities. Mostly on the verge of resource exhaustion, the resource-relied western cities and the long-established industrial cities in Northeast China must transform their patterns of economic development. It was also imperative for local governments to make more preferential policies and implement the Scientific Approach to Development.

16.1.2.3 Government Policies Underpinned Urban Green Development

Policy Support from Government contributed to economic development. Among the top ten GDI cities, six were also among the top ten SDGP cities, namely Shenzhen (No. 1), Beijing (No. 2), Guangzhou (No. 4), Dalian (No. 9), Qingdao (No. 8), and Fuzhou (No. 10). Although Beijing, Guangzhou and Shanghai were poor in the potential carrying capacity of natural resources and environment, yet the robust local economy and strong policy support from government offset the disadvantage and enhanced green development. Among the last ten GDI cities, seven were also among the last ten SDGP cities, namely Tianjin (No. 28), Chongqing (No. 30), Xi'an (No. 26), Wuhan (No. 25), Chengdu (No. 29), Lanzhou (No. 33), and Xining (No. 34) (see Table 16.4).

Table 16.4 Inter-city comparison of ranking gap between GDI and SDGP

City	Ranking of GDI (1)	Ranking of SDGP (2)	Ranking gap (1)–(2)	City	Ranking of GDI (1)	Ranking of SDGP (2)	Ranking gap (1)–(2)
Shenzhen	1	1	0	Nanjing	18	7	11
Haikou	2	12	−10	Shanghai	19	14	5
Kunming	3	11	−8	Changchun	20	32	−12
Beijing	4	2	2	Jinan	21	17	4
Hefei	5	18	−13	Yinchuan	22	15	7
Guangzhou	6	4	2	Nanchang	23	23	0
Dalian	7	9	−2	Hohhot	24	27	−3
Qingdao	8	8	0	Zhengzhou	25	21	4
Changsha	9	19	−10	Guiyang	26	24	2
Fuzhou	10	10	0	Taiyuan	27	16	11
Xiamen	11	3	8	Tianjin	28	28	0
Nanning	12	20	−8	Chongqing	29	30	−1
Ningbo	13	6	7	Xi'an	30	26	4
Shenyang	14	22	−8	Wuhan	31	25	6
Harbin	15	31	−16	Chengdu	32	29	3
Shijiazhuang	16	5	11	Lanzhou	33	33	0
Hangzhou	17	13	4	Xining	34	34	0

Notes This table is derived from Tables 0.5 and 16.3

16.1.2.4 There was a Disparity in SDGP of a City and the Province Where it was Located in

Basing on a comparative analysis, we found that most cities had different SDGP rankings from those of the provinces they were located in (see Table 16.5 for details). The city rankings of four municipalities directly under the jurisdiction of the central government were lower than their provincial ranking. Beijing's city ranking was one place behind its provincial ranking, but the other three municipalities experienced a ranking gap by 11 $\sim$ 18 places. Such a ranking gap took place among all provincial capitals and their corresponding provinces. 14, or half of the cities dropped by five or more places. The five cities specifically designated in the state plan were among the top ten cities, higher than the rankings of their corresponding provinces and the capitals.

Table 16.5 Comparison of ranking gap of SDGP between cities and their corresponding provinces

Indicator	Support degree of government policies		Indicator	Support degree of government policies		Ranking gap (1)–(2)
Province	Score	Ranking (1)	City	Score	Ranking (2)	
Beijing	0.278	1	Beijing	0.222	2	−1
Jiangsu	0.121	2	Nanjing	0.099	7	−5
Shanghai	0.107	3	Shanghai	0.03	14	−11
Zhejiang	0.096	4	Hangzhou	0.041	13	−9
			Ningbo	0.11	6	−2
Shandong	0.087	5	Qingdao	0.083	8	−3
			Jinan	0.008	17	−12
Ningxia	0.081	6	Yinchuan	0.024	15	−9
Guangdong	0.077	7	Guangzhou	0.141	4	3
			Shenzhen	0.319	1	6
Fujian	0.051	8	Fuzhou	0.048	10	−2
			Xiamen	0.168	3	5
Yunnan	0.049	9	Kunming	0.047	11	−2
Hebei	0.043	10	Shijiazhuang	0.133	5	5
Shanxi	0.036	11	Taiyuan	0.009	16	−5
Chongqing	0.035	12	Chongqing	−0.121	30	−18
Shaanxi	0.017	13	Xi'an	−0.094	26	−13
Jiangxi	0.01	14	Nanchang	−0.045	23	−9
Anhui	0.004	15	Hefei	0.008	18	−3
Hubei	0	16	Wuhan	−0.059	25	−9
Tianjin	−0.009	17	Tianjin	−0.099	28	−11
Guangxi	−0.03	18	Nanjing	−0.021	20	−2
Inner Mongolia	−0.03	19	Hohhot	−0.098	27	−8
Hainan	−0.05	20	Haikou	0.047	12	8
Liaoning	−0.061	21	Shenyang	−0.034	22	−1
Xinjiang	−0.063	22	Dalian	0.065	9	12
Sichuan	−0.074	23	Chengdu	−0.102	29	−6
Hunan	−0.075	24	Changsha	−0.011	19	5
Henan	−0.076	25	Zhengzhou	−0.027	21	4
Guizhou	−0.076	26	Guiyang	−0.057	24	2
Qinghai	−0.082	27	Xining	−0.346	34	−7
Heilongjiang	−0.147	28	Harbin	−0.135	31	−3
Jilin	−0.154	29	Changchun	−0.163	32	−3
Gansu	−0.165	30	Lanzhou	−0.184	33	−3

Note This table is derived from Tables 16.1 and 16.3

Table 16.6 Third-class indicators, their weights and attributes of inter-provincial green investment

Sequence number	Indicator	Weight (%)	Attribute
1	Ratio of environmental protection expenditures to government expenditures	1.50	Positively correlated
2	Ratio of investment in the control of environmental pollution to GRP	1.50	Positively correlated
3	Per capita investment of water sanitation and toilet improvement in rural areas	1.50	Positively correlated
4	Investment in converting cultivated land into forests and grassland per unit of area of cultivated land	1.50	Positively correlated
5	Ratio of expenditures for science and technology, education, culture, and medical and health care to government expenditures	1.50	Positively correlated

Note The contents in this table are determined by the task force after having conducted a lot of expert workshops

16.2 Comparison of Inter-Provincial SDGP

Weighing 30 % of GDI, Inter-provincial SDGP was made up of 19 Third-Level Indicators, including 18 positively correlated indicators and one negatively correlated indicator. Each positively correlated indicator weighted 1.50 % and the negatively correlated one 1.69 %.

16.2.1 Measurement and Analysis of the Green Investment Indicator on the Provincial Level

According to the perspectives of Western organizations and scholars, green investment was an investment model based on environmental norms, social norms and economic return norms. Taking the three bottom lines of economy, society and environment into account, it was also known as "triple surplus" investment.

Green Investment was an investment strategy that conformed with the Scientific Approach to Development and sustainable development. It was an integration of economic, social, environmental and other factors. It encouraged investors to take due responsibilities in line with the profits and brought both investors and the society sustainable development. Government played the leading role as green investment was policy oriented and worked for public good.

Table 16.7 Indexes of green investment indicators and rankings of 30 provinces in China in 2009

Indicator	Green investment indicators		Indicator	Green investment indicators	
Province	Score	Ranking	Province	Score	Ranking
Ningxia	0.068	1	Zhejiang	−0.003	16
Beijing	0.052	2	Hunan	−0.004	17
Qinghai	0.051	3	Sichuan	−0.007	18
Shanxi	0.038	4	Hubei	−0.008	19
Chongqing	0.036	5	Jiangsu	−0.018	20
Shaanxi	0.033	6	Anhui	−0.019	21
Gansu	0.017	7	Shandong	−0.022	22
Guangxi	0.015	8	Jiangxi	−0.026	23
Hebei	0.010	9	Jilin	−0.027	24
Xinjiang	0.008	10	Henan	−0.027	25
Guizhou	0.003	11	Guangdong	−0.028	26
Inner Mongolia	0.003	12	Tianjin	−0.032	27
Yunnan	0.001	13	Fujian	−0.034	28
Hainan	−0.002	14	Heilongjiang	−0.037	29
Shanghai	−0.003	15	Liaoning	−0.039	30

Sources China Statistical Yearbook 2010, *Annual Statistical Report on Environment in China* 2009, *China Environmental Statistical Yearbook* 2010, *China Industrial Economic Statistical Yearbook* 2010, and *China City Statistical Yearbook* 2010

Weighing 25 % of SDGP, Green Investment Indicator (GII) consisted of five Third-Level Indicators, namely the ratio of environmental protection expenditures to government expenditures, ratio of investment in the control of environmental pollution to GRP, per capita investment of water sanitation and toilet improvement in rural areas, Investment in converting cultivated land into forests and grassland per unit of area of cultivated land, and the ratio of expenditures for science and technology, education, culture, and medical and health care to government expenditures. Those five indicators were currently the priorities of government green investment. Measuring them would be an objective and comprehensive evaluation of China's green investment. We adopted the averaging weight method (Table 16.6) and ranked the green investment levels of all provinces (see Table 16.7).

Calculated results showed a slight difference among the provincial GDI that ranged from 0.068 to −0.039. Ningxia ranked the first, 7 % higher than the national average; 13, or half of the provinces were above the national average level, which in descending order included Ningxia, Beijing, Qinghai, Shanxi, Chongqing, Shaanxi, Gansu, Guangxi, Hebei, Xinjiang, Guizhou, Inner Mongolia and Yunnan. The other provinces such as Hainan and Shanghai were below the average (Table 16.7).

From a regional perspective, West China had the highest GDI value, followed by equivalent East and Central China, and lastly Northeast China. Among the ten western provinces (see Fig. 16.6) that were above the national average, seven were among the national top ten, including Ningxia, Qinghai, Chongqing, Shaanxi,

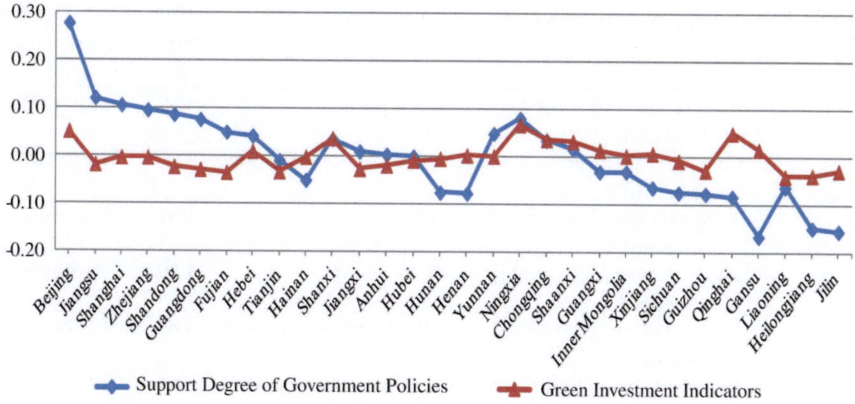

Fig. 16.5 Inter-provincial comparison of green investment index and SDGP index. *Notes* From the area division perspective of eastern, central, western and northeastern areas, this chart arranges these regions from *left–right* in terms of indexes of SDGP in descending order

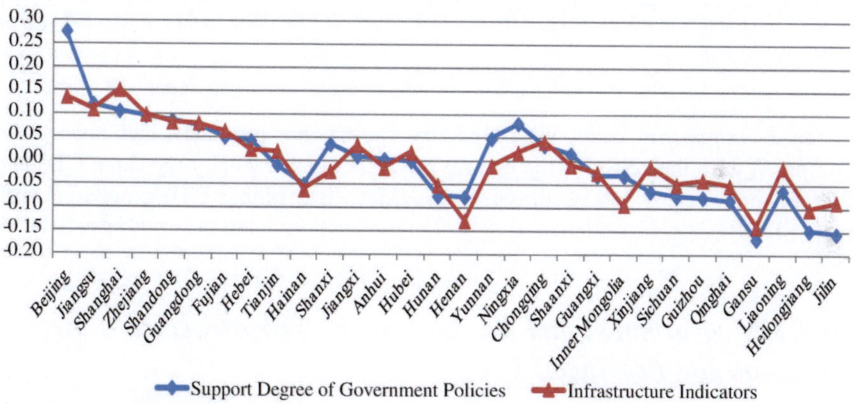

Fig. 16.6 Inter-provincial comparisons of infrastructure index and SDGP index. *Notes* From the area division perspective of eastern, central, western and northeastern areas, this chart arranges these regions from *left–right* in terms of indexes of SDGP in descending order

Gansu, Guangxi, Xinjiang, and the other three, namely Guizhou, Inner Mongolia and Yunnan respectively ranked 11th, 12th and 13th. Among the eastern provinces, only Beijing and Hebei ranked among the top ten, while Guizhou, Inner Mongolia, Yunnan, Zhejiang, Jiangsu were among the No. 11 ~ 20. Shandong, Guangdong, Tianjin and Fujian were among the bottom ten, whose GDI did not match their economic strength. Among the central provinces, Shanxi was the only one among the top ten, Hubei and Hunan among No. 11 ~ 20, and Anhui, Jiangxi and Henan among the bottom ten. Among the Northestern provinces, Jilin ranked 24th, and Heilongjiang 29th and Liaoning 30th.

Table 16.8 Third-class indicators, their weights and attributes of inter-provincial infrastructure

Sequence number	Indicator	Weight (%)	Attribute
6	Area of green land per capita in urban areas	1.69	Positively correlated
7	Coverage rate of urban population with access to tap water	1.69	Positively correlated
8	Treatment rate of urban waste water	1.69	Positively correlated
9	Ratio of urban consumption wastes treated	1.69	Positively correlated
10	Public transportation vehicles per 10000 urban population	1.69	Positively correlated
11	Per capita length of urban public transit operating routes	1.69	Positively correlated
12	Ratio of rural population benefiting from water improvement projects to total rural population	1.69	Positively correlated
13	Ratio of green covered area to completed urban area	1.69	Positively correlated

Notes The contents in this table are determined by the task force after having conducted a lot of expert workshops

On a comparison basis, we found that the eastern provinces had lower GDI compared with their SDGP; the central provinces had basically the same GDI and SDGP; the western and northeastern provinces had higher GDI compared with their SDGP.

16.2.2 Measurement and Analysis of the Infrastructure Indicator on the Provincial Level

Given the current situation of policy making and implementation in regard to infrastructure, we designated 45 % of weight to the Infrastructure Indicator in SDGP. Under the indicator were eight Third-Level Indicators, namely area of green land per capita in urban areas, coverage rate of urban population with access to tap water, treatment rate of urban waste water, ratio of urban consumption wastes treated, public transportation vehicles per 10,000 urban population, per capita length of urban public transit operating routes, ratio of rural population benefiting from water improvement projects to total rural population, and ratio of green covered area to completed urban area. The weights and indications of the Third-Level Indicators were shown in Table 16.8.

Calculated results (Table 16.9) showed that the value of provincial Infrastructure Indicator ranged from 0.152 to −0.138, with a larger disparity than green

Table 16.9 Indexes of infrastructure indicators and rankings of 30 provinces in China in 2009

Indicator	Infrastructure indicators		Indicator	Infrastructure indicators	
Province	Score	Ranking	Province	Score	Ranking
Shanghai	0.152	1	Xinjiang	−0.010	16
Beijing	0.137	2	Liaoning	−0.012	17
Jiangsu	0.110	3	Anhui	−0.014	18
Zhejiang	0.100	4	Shanxi	−0.022	19
Shandong	0.083	5	Guangxi	−0.024	20
Guangdong	0.081	6	Henan	−0.038	21
Fujian	0.065	7	Sichuan	−0.048	22
Chongqing	0.042	8	Qinghai	−0.050	23
Jiangxi	0.035	9	Hubei	−0.051	24
Hebei	0.025	10	Hainan	−0.061	25
Tianjin	0.021	11	Jilin	−0.085	26
Ningxia	0.019	12	Inner Mongolia	−0.092	27
Hubei	0.019	13	Heilongjiang	−0.101	28
Shaanxi	−0.008	14	Guizhou	−0.127	29
Yunnan	−0.010	15	Gansu	−0.138	30

Sources China Statistical Yearbook 2010, *Annual Statistical Report on Environment in China* 2009, *China Environmental Statistical Yearbook* 2010, *China Industrial Economic Statistical Yearbook* 2010, and *China City Statistical Yearbook* 2010

investment. Shanghai ranked first and Gansu the last. 13 provinces were above the national average. In descending order, they were Shanghai, Beijing, Jiangsu, Zhejiang, Shandong, Guangdong, Fujian, Chongqing, Jiangxi, Hebei, Tianjin, Ningxia and Hubei. The other 17 provinces such as Shaanxi and Gansu were below the national average.

Among the four major regions (see Fig. 16.7), there was a significant disparity. East China had much higher Infrastructure value, followed by Central China, West China and Northeast China in order. SDGP highly depended upon the Infrastructure Indicator.

Most of the eastern provinces had higher Infrastructure value than the national average, except Hainan. Eight provinces were among the top ten, namely Shanghai, Beijing, Jiangsu, Zhejiang, Shandong, Guangdong, Fujian, and Hebei. Tianjin ranked 11th and Hainan 25th. Highly urbanized with great social progress and civilization, this region had perfect infrastructure and scientific urban planning and management guided by relevant philosophy.

However, other regions had much lower Infrastructure values. Most provinces from the other three regions were below the national average. As economic development improved and urbanization sped up, local governments would invest more in infrastructure and embrace more scientific philosophy of urban planning and management. Improved infrastructure would be a solid basis underpinning green development.

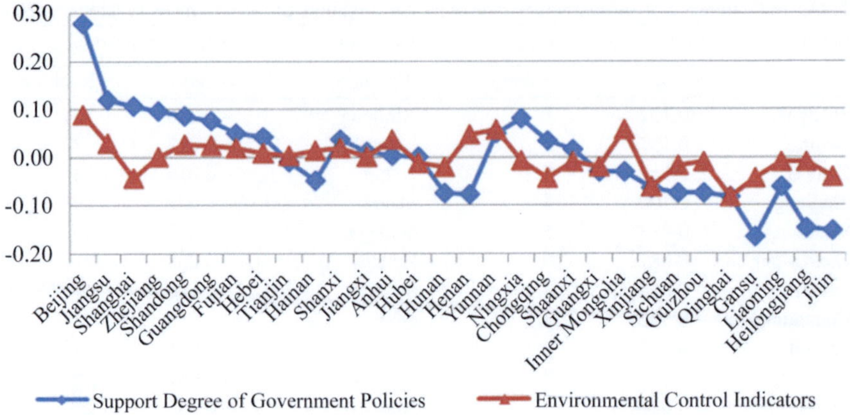

Fig. 16.7 Inter-provincial comparison of environmental control index and SDGP index. *Notes* From the area division perspective of eastern, central, western and northeastern areas, this chart arranges these regions from *left–right* in terms of indexes of SDGP in descending order

16.2.3 Measurement and Analysis of the Environmental Control Indicator on the Provincial Level

Given the current situation of policy making and implementation in regard to environmental protection and ecological management, we designated 30 % of weight to the Environmental Control Indicator in SDGP. Under the indicator were six Third-Level Indicators, namely area of afforestation per capita, removal rate of industrial sulfur dioxide emissions, removal rate of chemical oxygen demand in industrial waste water, removal rate of industrial nitrogen oxide emissions, removal rate of ammonia nitrogen in industrial waste water and the sudden accidents effecting environment. The weights and indications of all Third-Level Indicators were shown in Table 16.10.

Calculated results (Table 16.11) indicated that the values of the Environmental Control Indicator ranged from 0.089 to −0.082, with a slight provincial disparity. Beijing ranked first with the value of 0.089, 9 % higher than the national average. Another 14 provinces including Inner Mongolia, Yunnan, Guizhou, Anhui, Jiangsu, Shandong, Guangdong, Shanxi, Fujian, Hainan, Hebei, Tianjin, Jiangxi and Zhejiang were above the national average. However, the other sixteen including Ningxia, Shaanxi, Heilongjiang, Henan, Liaoning, Hubei, Sichuan, Hunan, Guangxi, Jilin, Shanghai, Gansu, Chongqing, Xinjiang and Qinghai were below the national average.

Calculated results showed a slight regional disparity of the Environmental Control Indicator while disparity within the regions was significant. The eastern region had the highest index values, significantly higher than other regions. The western and central regions ranked second and third whereas the northeast the last.

Table 16.10 Third-class indicators, their weights and attributes of inter-provincial environmental control

Sequence number	Indicator	Weight (%)	Attribute
14	Area of Afforestation per capita	1.50	Positively correlated
15	Removal rate of industrial SO2 emissions	1.50	Positively correlated
16	Removal rate of COD in industrial waste water	1.50	Positively correlated
17	Removal rate of industrial NOx emissions	1.50	Positively correlated
18	Removal rate of ammonia nitrogen in industrial waste water	1.50	Positively correlated
19	Sudden accidents effecting environment	1.50	Negatively correlated

Notes The contents in this table are determined by the task force after having conducted a lot of expert workshops

Table 16.11 Indexes of environmental control indicators and rankings of 30 provinces in China in 2009

Indicator	Environmental control indicators		Indicator	Environmental control indicators	
Province	Score	Ranking	Province	Score	Ranking
Beijing	0.089	1	Ningxia	−0.006	16
Inner Mongolia	0.058	2	Shaanxi	−0.009	17
Yunnan	0.057	3	Heilongjiang	−0.009	18
Guizhou	0.048	4	Henan	−0.010	19
Anhui	0.037	5	Liaoning	−0.010	20
Jiangsu	0.029	6	Hubei	−0.011	21
Shandong	0.027	7	Sichuan	−0.018	22
Guangdong	0.024	8	Hunan	−0.020	23
Shanxi	0.020	9	Guangxi	−0.021	24
Fujian	0.019	10	Jilin	−0.042	25
Hainan	0.013	11	Shanghai	−0.043	26
Hebei	0.007	12	Gansu	−0.044	27
Tianjin	0.002	13	Chongqing	−0.044	28
Jiangxi	0.001	14	Xinjiang	−0.062	29
Zhejiang	0.000	15	Qinghai	−0.082	30

Sources China Statistical Yearbook 2010, *Annual Statistical Report on Environment in China* 2009, *China Environmental Statistical Yearbook* 2010, *China Industrial Economic Statistical Yearbook* 2010, and *China City Statistical Yearbook* 2010

All eastern cities except Shanghai were above the national average, among which Beijing, Jiangsu, Shandong, Guangdong, and Fujian ranked among the top ten. The six central provinces fluctuated around the national average level, with Anhui,

Jiangxi, and Hainan slightly higher than the national average and the other three below it. Most western cities were below the national average, except Inner Mongolia, Yunnan and Guizhou that were among the top five; the last three were all western cities, namely Qinghai (No. 30), Xinjiang (No. 29) and Chongqing (No. 28). The eastern region had the best environmental control, for the region had taken the lead in developing economy and optimizing its industrial structure by lowering the share of industry, heavy industry in particular and improving that of service. As a result, the industrial structure was "lightened". Among the region, Shanghai had the lowest indicator value. Among the western provinces, however, Inner Mongolia and Guizhou had the highest value. On a comparison basis, we found a weak correlation between the Environmental Control Indicator (ECI) and SDGP, for in East China, the ECI values were higher than the SDGP values; in Central China, the two indicator values were basically the same; in West and Northeast China, the ECI values were higher than the SDGP values.

16.3 Comparison of City SDGP

In order to fully compare the inter-city SDGP, taking into account data availability, the city's representativeness, and regional differences, we measured 34 typical cities by analyzing three indicators, namely Green Investment, Infrastructure and Environmental Control. Based on the provincial measurement system, we designed a city measurement system where the evaluation and weight method were slightly adjusted. Weighing 33 % in GDI, the inter-city SDGP Index was composed of 13 positive Third-Level Indicators, each weighing between 2.48 and −2.75 %.

16.3.1 Measurement and Analysis of the Green Investment Indicator on the City Level

Weighing 25 % in urban SDGP, city Green Investment was made up of three indicators, namely ratio of environmental protection expenditures to government expenditures, ratio of investment in the control of industrial environmental pollution to GRP, and ratio of expenditures for science and technology, education, culture, and medical and health care to government expenditures. The three indicators were currently the main areas of green investment. Through the averaging weight method, we worked out their weight as 2.75 % identically (see Table 16.12). The Green Investment rankings of different cities were listed in Table 16.13.

Table 16.12 Third-class indicators, their weights and attributes of inter-city green investment

Sequence number	Indicator	Weight (%)	Attribute
1	Ratio of environmental protection expenditures to government expenditures correlated	2.75	Positively
2	Ratio of investment in the control of industrial environmental pollution to GRP correlated	2.75	Positively
3	Ratio of expenditures for science and technology, education, culture, and medical and health care to government expenditures correlated	2.75	Positively

Notes The contents in this table are determined by the task force after having conducted a lot of expert workshops

Table 16.13 Indexes of green investment indicators and rankings of 34 cities in China in 2009

Indicator	Green investment indicators		Indicator	Green investment indicators	
City	Score	Ranking	City	Score	Ranking
Shijiazhuang	0.072	1	Jinan	−0.007	18
Lanzhou	0.060	2	Guiyang	−0.009	19
Fuzhou	0.043	3	Changsha	−0.009	20
Beijing	0.035	4	Shanghai	−0.010	21
Harbin	0.029	5	Tianjin	−0.011	22
Hangzhou	0.029	6	Qingdao	−0.012	23
Yinchuan	0.028	7	Nanjing	−0.019	24
Guangzhou	0.027	8	Xi'an	−0.019	25
Shenzhen	0.015	9	Xiamen	−0.019	26
Zhengzhou	0.013	10	Xining	−0.020	27
Taiyuan	0.013	11	Wuhan	−0.025	28
Shenyang	0.007	12	Chongqing	−0.026	29
Haikou	0.004	13	Kunming	−0.026	30
Ningbo	0.002	14	Dalian	−0.026	31
Changchun	−0.005	15	Hefei	−0.038	32
Nanchang	−0.006	16	Chengdu	−0.039	33
Nanning	−0.007	17	Hohhot	−0.039	34

Sources China Statistical Yearbook 2010, *Annual Statistical Report on Environment in China* 2009, *China Environmental Statistical Yearbook* 2010, *China Industrial Economic Statistical Yearbook* 2010, and *China City Statistical Yearbook* 2010

Calculated results showed the values of city Green Investment ranged from 0.072 to −0.039, with a slight disparity. Shijiazhuang, capital of Hebei Province, ranked first, with a value 7.2 % higher than the national average; 14, or less than half of the cities, less than half of the cities were above the national average.

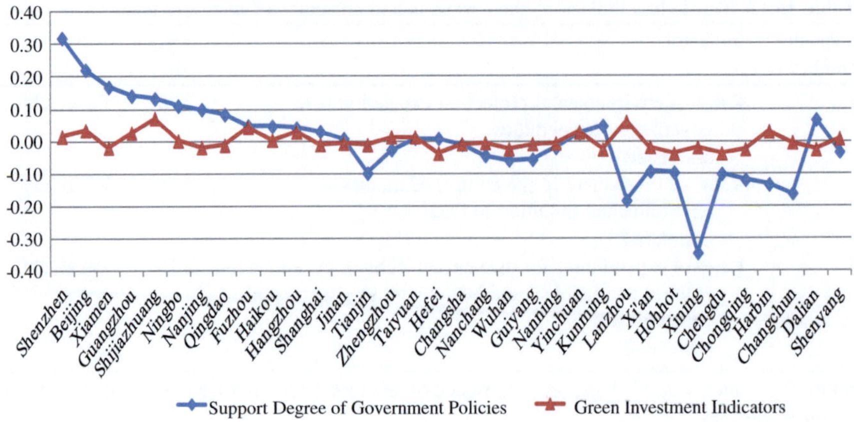

Fig. 16.8 Inter-city comparison of green investment index and SDGP index. *Notes* From the area division perspective of eastern, central, western and northeastern areas, this chart arranges these regions from *left–right* in terms of indexes of SDGP in descending order

In descending order they were Shijiazhuang, Lanzhou, Fuzhou, Beijing, Harbin, Hangzhou, Yinchuan, Guangzhou, Shenzhen, Zhengzhou, Taiyuan, Shenyang, Haikou and Ningbo. However, the other 20 such as Changchun and Hohhot were below the national average.

From the Fig. 16.8, the Green Investment Indicator (GII) values of all cities varied slightly. On a comparison basis, we found that the eastern cities had lower GII values compared with their SDGP values, which meant a small share of GII in SDGP; the central cities had basically the same GII and SDGP values; the western and northeastern cities had higher GII values than their SDGP values, which meant a large share of GII in SDGP.

16.3.2 Measurement and Analysis of the Infrastructure Indicator on the City Level

Given the current situation of policy making and implementation in infrastructure and urban management, we designated 45 % of weight to the Infrastructure Indicator in urban SDGP. Under the indicator were five positive Third-Level Indicators, namely area of green land per capita in urban areas, ratio of green covered area to completed urban area, coverage rate of urban population with access to tap water, treatment rate of urban waste water, ratio of urban consumption wastes treated, and public transportation vehicles per 10,000 urban population. Each of the indicators weighed 2.48 % Table 16.14.

Table 16.14 Third-class indicators, their weights and attributes of inter-city infrastructure

Sequence number	Indicator	Weight (%)	Attribute
4	Area of green land per capita in urban areas	2.48	Positively correlated
5	Ratio of green covered area to completed urban area	2.48	Positively correlated
6	Coverage rate of urban population with access to tap water	2.48	Positively correlated
7	Treatment rate of urban waste water	2.48	Positively correlated
8	Ratio of urban consumption wastes treated	2.48	Positively correlated
9	Public transportation vehicles per 10000 urban population	2.48	Positively correlated

Notes The contents in this table are determined by the task force after having conducted a lot of expert workshops

Table 16.15 Indexes of infrastructure indicators and rankings of 34 cities in China in 2009

Indicator	Infrastructure indicators		Indicator	Infrastructure indicators	
City	Score	Ranking	City	Score	Ranking
Shenzhen	0.287	1	Changsha	0.013	18
Beijing	0.077	2	Shanghai	0.010	19
Dalian	0.073	3	Chengdu	0.008	20
Xiamen	0.067	4	Zhengzhou	0.005	21
Qingdao	0.061	5	Xi'an	0.004	22
Nanjing	0.053	6	Jinan	−0.012	23
Hangzhou	0.049	7	Wuhan	−0.013	24
Nanchang	0.044	8	Yinchuan	−0.017	25
Shijiazhuang	0.042	9	Chongqing	−0.027	26
Nanning	0.036	10	Changchun	−0.031	27
Hefei	0.031	11	Taiyuan	−0.051	28
Shenyang	0.030	12	Tianjin	−0.052	29
Guangzhou	0.028	13	Hohhot	−0.090	30
Kunming	0.023	14	Xining	−0.119	31
Haikou	0.022	15	Guiyang	−0.119	32
Ningbo	0.020	16	Harbin	−0.194	33
Fuzhou	0.014	17	Lanzhou	−0.272	34

Sources China Statistical Yearbook 2010, *Annual Statistical Report on Environment in China* 2009, *China Environmental Statistical Yearbook* 2010, *China Industrial Economic Statistical Yearbook* 2010, and *China City Statistical Yearbook* 2010

Calculated results (Table 16.15) showed the values of urban Infrastructure ranged from 0.287 to −0.272, with a large disparity. There are 22 cities above the national average. In descending order they were Shenzhen, Beijing, Dalian, Xiamen, Qingdao, Nanjing, Hangzhou, Nanchang, Shijiazhuang, Nanning, Hefei,

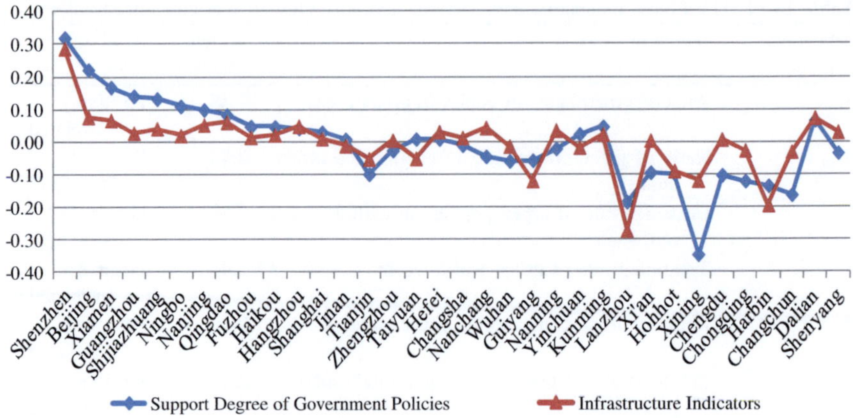

Fig. 16.9 Inter-city comparison of infrastructure index and SDGP index. *Notes* From the area division perspective of eastern, central, western and northeastern areas, this chart arranges these regions from *left–right* in terms of indexes of SDGP in descending order

Shenyang, Guangzhou, Kunming, Haikou, Ningbo, Fuzhou, Changsha, Shanghai, Chengdu, Zhengzhou and Xi'an. The No. 1 Shenzhen left Beijing far behind. The other 12 cities such as Jinan and Harbin were below the national average.

From a regional perspective, the eastern region had high Infrastructure value and ranked forward. The top ten included seven eastern cities and Dalian (No. 3), Nanchang (No. 8), and Nanning (No. 10) outside the region. The reason was that, compared with the other regions, eastern municipalities such as Beijing and Shanghai, provincial capital cities and cities specially designated in the state plan were highly urbanized and had sounder infrastructure, more scientific urban planning and more advanced urban management philosophy and levels. They had the following strengths: high green coverage, strong capacity in water supply and sewage Control, advanced roads and public transportation, fast-improving sanitation facilities and so on. The other cities outside the region should follow the example of the eastern counterparts to improve their infrastructure and urban management.

From the Fig. 16.9, the Infrastructure values varied greatly among the regions. The eastern cities, in general, had higher values than the rest. On a comparison basis, we found that a positive correlation existed between urban Infrastructure Indicator and urban SDGP and that SDGP mainly hinged on Infrastructure.

16.3.3 Measurement and Analysis of the Environmental Control Indicator on the City Level

Based on the goal and contents of the indicator, given the current situation of policy making and implementation in environmental protection and ecological

Table 16.16 Third-class indicators, their weights and attributes of inter-city environmental control

Sequence number	Indicator	Weight (%)	Attribute
10	Removal rate of industrial SO2 emissions	2.48	Positively correlated
11	Removal rate of COD in industrial waste water	2.48	Positively correlated
12	Removal rate of industrial NOx emissions	2.48	Positively correlated
13	Removal rate of ammonia nitrogen in industrial waste water	2.48	Positively correlated

Notes The contents in this table are determined by the task force after having conducted a lot of expert workshops

Table 16.17 Indexes of environmental control indicators and rankings of 34 cities in China in 2009

Indicator	Environmental control indicators		Indicator	Environmental control indicators	
City	Score	Ranking	City	Score	Ranking
Xiamen	0.120	1	Shenzhen	0.017	18
Beijing	0.109	2	Hefei	0.015	19
Ningbo	0.089	3	Yinchuan	0.014	20
Guangzhou	0.086	4	Fuzhou	−0.010	21
Guiyang	0.071	5	Changsha	−0.015	22
Nanjing	0.065	6	Wuhan	−0.021	23
Kunming	0.049	7	Tianjin	−0.035	24
Taiyuan	0.047	8	Hangzhou	−0.036	25
Qingdao	0.034	9	Zhengzhou	−0.045	26
Shanghais	0.031	10	Nanjing	−0.050	27
Hohhot	0.030	11	Chongqing	−0.068	28
Harbin	0.030	12	Shenyang	−0.071	29
Lanzhou	0.028	13	Chengdu	−0.072	30
Jinan	0.027	14	Xi'an	−0.080	31
Haikou	0.022	15	Nanchang	−0.084	32
Shijiazhuang	0.019	16	Changchun	−0.127	33
Dalian	0.018	17	Xining	−0.208	34

Sources China Statistical Yearbook 2010, *Annual Statistical Report on Environment in China* 2009, *China Environmental Statistical Yearbook* 2010, *China Industrial Economic Statistical Yearbook* 2010, and *China City Statistical Yearbook* 2010

control, we designated 30 % of weight to the urban Environmental Control Indicator in the SDGP. Under the indicator were four positive Third-Level Indicators, namely removal rate of industrial sulfur dioxide emissions, removal rate of chemical oxygen demand in industrial waste water, removal rate of industrial nitrogen oxide emissions, and removal rate of ammonia nitrogen in industrial waste water. Each indicator weighted 2.48 % Table 16.16.

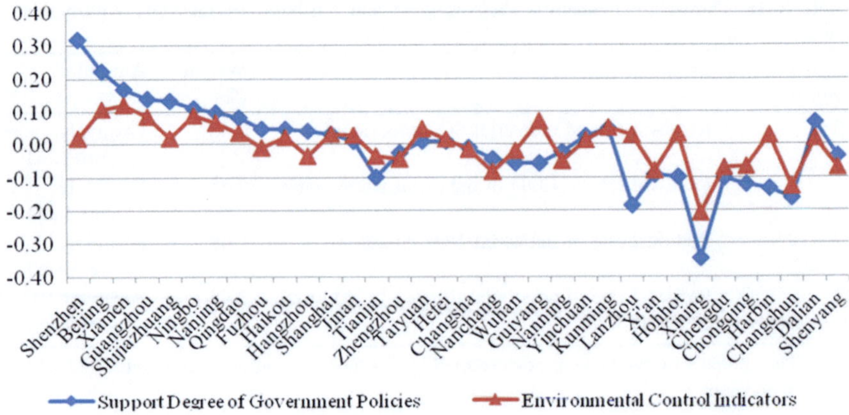

Fig. 16.10 Inter-city comparison of environmental control index and SDGP index. *Notes* From the area division perspective of eastern, central, western and northeastern areas, this chart arranges these regions from *left–right* in terms of indexes of SDGP in descending order

Calculated results (Table 16.17) indicated that the values of the urban Environmental Control Indicator ranged from 0.120 to −0.208. Xiamen had the highest value, which was 0.120 and about 12 % higher than the national average. 20 cities were above the national average, namely Xiamen, Beijing, Ningbo, Guangzhou, Guiyang, Nanjing, Kunming, Taiyuan, Qingdao, Shanghai, Hohhot, Harbin, Lanzhou, Jinan, Haikou, Shijiazhuang, Dalian, Shenzhen, Hefei and Yinchuan. The other 14 such as Fuzhou and Xining were below the national average.

From a regional perspective (see Fig. 16.10), the eastern cities had much higher indicator values and rankings. Seven of them ranked among the top ten and most of them were above the national average except Fuzhou (No. 21) and Tianjin (No. 24). Most of the central, western and northeastern cities were below the national average. The effective environmental control in eastern cities were mainly due to the early mover advantage in upgrading the industrial structure. For example, Beijing took the first place in this regard as a result of its sound and optimized industrial structure.

16.4 Conclusion

By measuring the SDGP Indicator values of 30 provinces (municipalities and autonomous regions) and 34 cities, we made the following conclusions:

All governments valued green development, but among them there was a huge disparity in terms of government policy support. The developed eastern region gave the strongest policy support, setting a good example to underdeveloped regions that economic development and green development were compatible.

To be specific, the disparity was mainly reflected in the disparity of infrastructure and urban management. The central and western regions fell far behind the eastern region. However, in terms of environmental control, different regions made similar progress.

We should coordinate infrastructure development with green development. Currently infrastructure was highly incompatible with environmental protection. The environmental infrastructure failed to catch up with other undertakings. Thus we should prioritize environmental protection and embed the ideal of green and low carbon development into the design, planning, construction and operation process, so that we could develop more environmental friendly facilities.

Urbanization should allow for green development. The Twelfth Five-Year Plan period would witness the acceleration of China's urbanization. Different regions had different carrying capacity of natural resources and environment, and foundation and potential for future development. Given their unique conditions, they should optimize the urban layout, intensify the use of urban land and solve the problems of resource constraints, urban functions, and urban management in a sustainable way.

Governments should play a dominant role in channeling investment. They should invest more in green development and spend more on environmental protection and pollutants control. The eastern provinces, in particular, should match investment with their economic strength. Meanwhile, governments should carry out incentives to prompt social investment, and improve investment and operational efficiency. In addition, science and technology played a critical role in greener economic development. Therefore, governments should, through investing more in this regard and encouraging technological innovation, speed up optimizing the industrial structure and transforming the economic development pattern so as to ease the conflict between economic development and resources and environment.

Session IV
Features on Green Development

Feature 1
Development and Reform of Environmental Statistics

Yuru Mao

The connotation of generalized environmental statistics includes statistical work, materials and science on the environment. Its broad study range determines a wide content for environmental statistics, for example, the surveys and analysis on the environmental condition and protection to provide statistical information and advice for supervision. Environmental statistics also helps environmental protection departments at all levels learn about the situation of environmental pollution, control and protection within their jurisdiction through legitimate collection, transmission, compilation, analysis and dissemination of statistical data. Although all of this work is just basic, it is an important part of the social economic statistics with difficulties.

1.1 Development of Environmental Statistics

China's statistics on land, water, meteorology, minerals conducted in the 1950s began to involve part of environmental statistical content. After the first national conference on environment protection in 1973, the pollution surveys on industrial sources were carried out in big cities such as Beijing, Shenyang and Nanjing. Since then, Environmental administrative organizations and monitor stations have been gradually established in all the provinces and cities (regions). China's environmental statistics began to be set up and a series of actions were carried out. In 1979, the leadership group of the state council on environmental protection organized a survey on the basic environmental conditions of more than 3,500 large and medium-size enterprises. In November 1980, in order to strengthen the environmental management and provide supports to draft environmental protection policies and plans, the leadership group convened the first national conference on environmental protection in Beijing and approved to establish a report system for environmental statistics. In 1981, the environmental statistics on discharges and management of industrial waste at or above county level were conducted;

X. Li and J. Pan (eds.), *China Green Development Index Report 2011*, Current Chinese Economic Report Series, DOI: 10.1007/978-3-642-31597-8_17,

meanwhile, the environmental protection work teams were formally constructed and developed. All of these marked the formal establishment of environment statistical system in China. Through 30 years' practice and development, China now has established relatively standard environmental statistical indicator system and work system via which statistical data can be submitted following the administrative levels. China also has formed a set of statistical methods that are base on regular general surveys, sample surveys and scientific estimations with special surveys as an effective complement. However, there still exist difficulties for china's environmental statistics such as weak work foundations, insufficient funds, backward technologies and equipments and lacks of basic researches. Furthermore, statistical objects are heterogeneous, fast-changing and difficult to be obtained, so some local environmental protection departments don't like to pay more attention to the statistical surveys and don't think they are able to figure out the statistical data. And even if they would have figured out the data, they don't have enough time and energy to manage them. It is clear that the environmental statistical work of China still can't completely meet the needs of environmental management and the requirements of scientificalness, pertinence, timeliness and accuracy to statistical data. So, it urgently needs to be strengthened and improved. The main difficulties to do so maybe include the following five aspects. First, the scope of environmental statistics doesn't cover all the areas that need to be investigated and the dynamic adjustment on pollution sources is not timely enough. Actually, the scope of environmental statistics now only covers industrial pollution, centralized pollution control facilities and urban life pollution. Agricultural pollution has not been involved yet. Although the statistical method of industrial pollution combines key-point surveys and non- key calculation, when the situation is unclear in the overall sample, this kind of survey method may not accurately reflect the conditions of industrial pollution and may frequently lead to underreporting. Second, the in-depth researches on environmental statistical method are not enough and should be further developed. For instance, the timely and liable basic data are still not enough to support the related analysis; the instrumental and institutionalized auditing is also needed to be greatly improved; the level of informatization and network is relatively low. Additionally, the needs of local government management have great impacts on the quality of data, which will lead to data distortions and ineffective supervision and management mechanisms. Third, the indicator system of environmental statistics is not rationally enough. It is too complicated and can't clearly and fully reflect the key situations of pollution and its management, so statistical indicators and the scope of statistics should be defined more clearly. For example, the environmental statistic indicator system of the 11th *Five-Year Plan* includes 990 indicators with 351 annual composite indicators, but some of them are not suitable to apply in the current environmental management. Moreover, due to the lack of some important indictors on environmental management, some annual indicators within the system can't professionally and correctly estimate the operability at the grass root levels. Fourth, the application and comprehensive analysis on environmental statistical data is rather weak and can't meet the needs of environmental management and policy analysis. Because of the time lag and the limitation of submission frequency, the analysis level of environmental statistical data is still low so that it can neither find the rules hidden in the statistical information nor satisfy the users' requirements on environment statistical data. Instead, the data are mainly provided to meet the requirements of related management departments rather than the needs of active and in-depth analysis, because most data describe and summarize the current conditions, less can be used to predict, warn or provide suggestions for the macro social economy. Naturally, it is difficult to use these data to make macro economic analysis and provide quantitative timely supports for environmental management. Fifth, the environmental statistical work doesn't get enough attention so that its basic functions are not fully developed. Lack of stable and professional personnel, knowledge and practical experience are all the difficulties faced by the environmental statistics of China and lead to the needs of practice work cannot be satisfied.

1.2 Explore and Advance

Environmental statistics is not only the main way to collect environmental information but also the foundation of environmental management. Since the 11th five-year plan, the state council has paid more attention to environmental protection and pollution reduction, emphasizing the importance of statistical accounting again. Premier Wen Jiabao requested to construct the auditing and monitoring systems for indicators on pollution reduction through fully discussing and encouraged them to achieve the first-class level by introducing the advanced international experience and innovating of our own. This is the clear guidance of the environmental statistics from the state council and the CCP and reflects the realistic needs of pollution reduction and environmental statistics. As environmental protection and pollution reduction advance, environmental statistics is becoming more and more important, however, it faces more heavy tasks and increasing contradictions including adjustments, cohesions, reforms and so forth. Thanks to environmental protection departments' efforts to boost the reform and the innovations on environmental statistics, environmental statistical work has been greatly strengthened particularly in management modes, statistical methods, quality guarantee and indicator systems.

1.2.1 Adjustment and Innovation on Concepts of Environmental Statistics

With the advancement of environmental protection and pollution reduction, the scientific decision making and quantitative management on environmental protection became more and more dependent on accurate and timely environment statistical data, and thus put forward the new higher requirements to environmental statistics. So, the ideas and concepts should be adjusted and innovated timely to meet the demands of environmental protection conditions. The following steps can help to achieve the commissions and requirements confronted by environmental statistics.

- Strengthen the linkage between environmental statistics and pollution reduction and improve the scientificalness and standardization of the statistical work;
- Strengthen the comparative analysis between environmental statistics and general surveys of pollution sources and research on how to speed up the reforms of management systems and technology methods on environmental statistics;
- Improve the support services of environmental statistics and the level of data quality by taking effective measures.

To achieve the above-mentioned goals and explore the new survey methods for environmental statistics in the future, the first national general survey on dynamic updating of pollution sources was conducted following the principle "concise indicators, real data and refined system". It combined with the 12th five-year plan, aimed to construct a statistical indicator system of environment resources and meet the needs of auditing and accounting on pollution reduction. The general idea involves:

- Explore a new way for environmental protection with Chinese characteristics;
- Strengthen support and assurance capacities of environmental statistics;
- Reform environmental statistical management and optimize the indicator system;
- Improve the scientificalness and authenticity of environment statistical data;
- Coordinate the relationship among environment statistic, pollution reduction and general surveys of pollution sources;
- Strengthen constructions of basic statistical abilities, updates of statistical concepts, innovations of statistical systems and normalization of statistical management;
- Enhance statistical services and try to build up a work system that can meet the requirements of environmental protection during the 12th *Five-Year Plan.*

1.2.2 Update Surveys to Support the Environmental Management During the 12th Five-Year Plan Period

In order to coordinate with the first national general survey of pollution sources, determine the level of pollution reduction, keep the continuity of environmental management and pollution reduction between the later 11th *Five-Year Plan* period and the 12th *Five-Year Plan* period and improve the level of environmental statistics, after carefully repeated studying, the environmental protection ministry determined that during the current environmental statistics, the dynamic updating of general surveys on pollution sources should be conducted based on the data from the first national general survey of pollution sources and follow the scheme "updating the general survey data, simplifying the survey methods, doing two-year skills training, making the baseline of pollution reduction and integrating the general and special surveys". Survey updates in 2009 and 2010 faced the shortage of time to finish data collection from more than 300,000 investigated objects covering industry, agriculture, life and centralized pollution control facilities. Through the phases of preparation, training, household surveys, scheduling guidance, audit and summarization, the latest data in 2009 and 2010 were obtained and provided supports to help further understand the discharge conditions of pollutants and its survey method and technology successfully linked with the first national general survey of pollution sources. All the practice provided trainings for work teams and helped them accumulate experience. At the same time, during the process, the ideas and methods on how to update surveys have been innovated; the survey scope has been expanded; the quantity of pollution sources, the technique and the working mechanism all meet the requirements of the environmental statistical reform of the 12th *Five-Year Plan*.

1.2.3 Develop Support Services of Environmental Statistics and Promote the Integration of Environmental Statistics and Practical Work

In recent years, China's environmental protection work has carried out in a deep-going way and the scope of work has been expanded, which also put forward new requirements to environmental statistics. Along with the development of environmental protection, the levels of environmental techniques and environmental monitoring have been improved, which provides the technical support for environmental statistics. In fact, environmental statistics can be greatly developed and paid more attention only if it is really integrated into the environmental management. So, there is some work needed to be done. First, in the practical work, agricultural sources and motor vehicles should be gradually integrated into environmental statistical system to meet the needs of development of pollution reduction. Second, as for the technical support, the auditing and accounting system of pollution reduction should be fully used to provide practical references for improving the methods of environmental statistics. Meanwhile, the analysis and applications of environmental statistics can, in turn, provide references to help determine the optimal level of pollution emission. Moreover, the auditing and accounting of pollution reduction data and environment statistical data can effectively guarantee the quality of environmental statistics and then make it possible for environmental statistics to play an important role in decision making.

1.2.4 Insist Data Quality is the Life of Environmental Statistics and Adopt Effective Ways to Strengthen Data Management

The system construction is the foundation to strengthen environmental statistics. In recent years, based on the a series of basic statistical laws and regulations such as Statistics Law, Implementation of Statistics Law and Management Regulation of Sector Statistics, China has gradually established and improved the relevant systems on environmental statistical data such as the management system, the quality control system, the audit system to meet the needs of deeper environmental protection management and standardized environmental statistics in the first national general survey of pollution sources. The concrete work conducted involved:

- Enacted a series of management regulations and technique specifications such as Data Regulation of The First National General Survey of Pollution Sources, Management Regulation of Environmental Statistical Data Use and Audit Method of Environmental Statistical Data to strengthen the management of environmental statistical data and guarantee the institutional building on data quality;
- Established the audit system of data quality which combined joint reviews and field surveys, following the principle "macro control, moderate comparison and micro audit"; Learned the practical experience from the auditing and accounting of pollution reduction and strengthened comparisons and verifications among data to guarantee their logic and authenticity;
- Required the establishment of environmental statistical accounts in main pollution enterprises and improve the management of original data;
- Established a dynamic update mechanism for production and emission and standardized the accounting methods for main pollutants;
- Conducted relevant basic researches and provided technical supports for establishing the control system of data quality.

The above-mentioned work not only enriches the content of environmental statistics, but also significantly enhances its quality. To further strengthen the functions of environmental statistics, environmental protection departments at all levels gradually set up the work mode that under the unique administrative management and guidance, monitoring stations at all levels take charge of providing technical supports for pollution reduction. This mode effectively facilitates the development of environmental statistics and survey updates.

1.2.5 Improve the Survey Method System of Environmental Statistic

The survey methods of environmental statistics can be classified by objects, purposes, ways, time dimensions and application areas of surveys. Surveys can be conducted via questionnaires, interviews, telephones, news media, diaries, documents, on-site observations and so on. The components of a survey method system of environmental statistics include the methodology, basic approaches and specific techniques. Methodology is the theoretical basis and the guiding principle, involving philosophical methodology, logic methodology and subject methodology. The basic approach has multiple classifications and can be classified as general surveys, sample surveys, key-point surveys, typical surveys and case surveys according to the different survey objects. The specific technique is a whole process including survey designing, information inputting, transmitting, auditing, analyzing, releasing and summary writing. All the three components are interrelated and mutually constrained, that is, methodology is the foundation and determines the direction and value of surveys and the choice among basic methods and practical technologies which the implementation of surveys depends on. In turn, the development of basic methods and practical technologies can promote the development of methodology. It is the interactions of the three components that make the method system of environmental statistics become a strict and scientific one. At last, it should be pointed out that the report system of environmental statistics composed of indicator systems, statement forms, term provisions, the filling guidance and the data auditing is the core of the environmental statistical survey.

Reference

Ministry of Environmental Protection, National Bureau of Statistics (2008) Environment statistical statement filling guidance. China Environmental Science Press, Beijing

Feature 2
The Dilemma of Food Safety in China and Perspectives

Fangfang Tang

A Chinese saying: "The foundation of any country is people, the fundamental issue of people is food." "People's food" safety matters fundamentally with the health and life safety of people and the economic development and social harmony. In recent years, with the frequent occurrence of food safety incidents, food safety has become a focus issue of current social discussion. The series work on food safety has illustrated the thinking of "people first" and social harmony. What is the dilemma of food safety in China and what is the way out?

2.1 The Development of Food Safety Issue in China

2.1.1 Definition of Food Safety

What is food safety? According to Wikipedie, food safety is a scientific discipline describing handling, preparation, and storage of food in ways that prevent food-borne illness. This includes a number of routines that should be followed to avoid potentially severe health hazards. Food can transmit disease from person to person as well as serve as a growth medium for bacteria that can cause food poisoning. WHO defines the food safety issue as a public health issue of hazardous substance in food on human health.

In a narrower sense, food safety includes mainly two categories of issues: one is the nutritious value and quality problem of food itself, such as the unsatisfactory nutritious value in food, expired food and so on; another aspect is the impact of artificially changing the food attributes during production, transportation, storage

Research assistance by ZHA Jianping of Southwestern Jiao Tong University and PAN Bo of Beijing Normal University is acknowledged.

X. Li and J. Pan (eds.), *China Green Development Index Report 2011*, Current Chinese Economic Report Series, DOI: 10.1007/978-3-642-31597-8_18,

and sale on the damage of health even death, such as excessive pesticide use, food additives etc.

More broadly speaking, food safety implies the following aspects: First, the quantity safety of food, that is, people have access and can afford to purchase the necessary basic food from a quantitative point of view; second, the quality safety of food, that is, people's health and life are not damaged on the basis of consumer demand satisfaction, requesting nutrition coverage, balanced structure and healthy contents; third, the sustainability safety of food, that is, the sustainability of obtaining food on the basis of environmental protection and ecological utilization of resources.

With the deepening reform in China and the rapid economic development, the food safety problem in China has gradually changed from the quantity safety aspect towards the quality safety aspect or sustainability safety perspective. In particular, the food quality issue has become the current focus of food safety problem.

2.1.2 The Development of Food Safety in China

Reform and development in more than 30 years in China has freed people from the "food quota" constraint in the central planning era. Living standard in China has jumped to a much higher level qualitatively. The history of food safety has shifted from grain safety period (1949–1984) and starting period of food safety (1984–2001) towards the development stage of food safety (from 2001 on). However, the fundamental nature of large population with a relative shortage of cultivation land has determined that the quantity safety of food issue will always be a potential risk of China's economic and social development, while the excessive imbalance of wealth distribution has worsened this problem. Tai Zong Emperor in Tang dynasty has mentioned that "food is fundamental for people and agriculture is the basis of governance". Thus, food safety issue in China must be based on the guarantee of sufficient supply of grain and then to improve the quality safety and sustainability safety of food, for the basic rights of people.

In recent years, there occurred a number of sudden food safety events, most of which were due to food quality, thus the current food safety issue in on food quality. For the current work on food safety, quantity safety aspect is still the foundation, while the sustainability safety is the future direction and the focus is on quality safety. With the gradual development of the food safety monitoring and prevention system, the quality safety aspect in China has risen to a much higher level overall. According to the data from health department, the overall food qualification rate has risen from 61.5 % in 1982 to 82.3 % in 1994 and 88.6 % in 2001, among which 13 categories of food products such as grain, alcohol and cans have risen to over 90 % from random samples. In the year 2010, the overall qualification rate of vegetables, cattle and poultry, and aquatic product are 96.8, 99.6 and 96.7 %, respectively. Further, more than 3,800 kinds' manufactured food of 23 categories has reached qualification rate of 94.6 % through random sample test and the export food has reached 99.8 % qualification rate. Random sampling test has shown that the pollutant, additives, heavy metal and pesticide leftover residuals in food have significantly fallen. Overall, food poison events and the number of people affected have been falling as well.

On the other hand, individual food safety incidents, sometimes at a large scale, have occurred from time to time. For example, the words on media associated with food safety have become familiar terms, such as melamine, hogwash oil, man-made red-date, clenobuterol hydrochloride, ampullaria gigas spix, red-hearted salt duck egg, colorful dumpling, and so on. The overall achievement of food safety progress in China is widely questioned and criticized. For example, health database have revealed 220 public health event reports in 2010, with 7,383 people being poisoned and 184 deaths. Among them, food poison events affecting more than 100 people account for 7. Compared to 2009 data, food poison report and affected people have actually fallen by 18.32 and 32.92 %, respectively, although the death toll has risen by 1.66 %. As the reasons of food poison events, microbiological food poison reports and affected people account most, for 36.82 and 62.1 %, respectively. This mainly includes two aspects, one is the traditional food pollution problem, such as salmonella, mycotoxin, agricultural pesticide and parasite pollution, another aspect is a series new pollution in food as occurred in the developed countries, such as Escherichia coli which has caused serious outbreaks in many provinces in China. Most death tolls are caused by poisonous animal and plants (especially mushrooms), accounting for 60.87 % in total. Compared to 2009, the number of reported microbiological poison events, affected people and the resulted death tolls have reduced by 31.36, 41.83 and 20 %, respectively. The numbers caused by chemical food poison aspect have reduced by 27.27, 38.17 and 27.27 %, respectively. Compared to 2009, the number of reported poisonous animal and plants (especially mushrooms) events, affected people and the resulted death tolls have reduced by 4.94, 9.3 and increased by 20.43 %, respectively, while the number of reported unknown-reason food poison events, affected people and the resulted death tolls have increased by 29.41 and 28.15 %, but reduced by 6, respectively.

In addition, although the application of transgenic farming technology may provide some grain supply safety, the discussion on its further production is debated. Some scholars point out that transgenic plants may threaten the environment, human health and biological diversity protection. Such a large number of negative press related to food safety at the information age have deeply hurt the already-sensitive nerves of people. According to the “2010–2011 Consumer Food Safety Confidence Report”, close to 70 % people who were surveyed do not feel a sense of safety concerning the food safety situation in China, among them over half (52.3 %) feel psychologically “rather uncomfortable”, while 15.6 % “especially unsafe”. It is worth noting that food safety legislation system, technology regulation and technical testing have made significant progress in China recent years. It shows the complexity of food supply channel and market in China, implying that it is not easy to accomplish food safety by a single jump in China.

Therefore, Chinese government must step by step and stage by stage improve the effectiveness of legislation system, perfection of technology regulations and the operationalization of testing technologies, to further build the food safety guarantee network and system.

2.2 The Profound Impact of Food Safety Events in China

The negative impact of frequent food safety events has exceeded industrial accidents on the life, health and financial assets in the aspect of direct harm. First, the frequent food safety events have shadowed the development of food industry in China. In 201, the food industry in China has accounted for 5.3 trillion RMB Yuan, about 9.1 % of total industry output in China. It is estimated that this figure will reach 12–14 trillion RMB Yuan by 2015, a pillar industry for stimulating domestic consumption, increasing employment and economic growth. However, food safety is the presumption of virtuous-cycle development of the food industry. Unfortunately, the sequence of food safety events makes such a picture of development worrisome. Melamine has not only destroyed the "San Lu (Three Deers)" company, but also badly hurt other domestic dairy product brands. Clenobuterol hydrochloride has trapped the "Shuang Hui" sausage company until now, while hogwash oil etc., has seriously bothered the public about what they can eat. The impact on Chinese food industry and restaurants industry and their development perspective is unknown yet, but less promising.

Second, the series of food safety events have gradually evolved into a serious crisis of social moral and credibility system. As Chinese Premier WEN Jiabao pointed out, "the poisonous milk powder, clenobuterol hydrochloride, hogwash oil, colorful dumplings and other incidents have sufficiently shown how badly the lack of honesty and lowering moral integrity have become". All the historical evidence has demonstrated that the rising-up of a country is not only in the creation and accumulation of material wealth but even more importantly the construction and improvement of an advanced culture system. Only through constructing a set of advanced and influential advanced culture systems, the soft power of China can be improved in the international community. From the viewpoint of new institutional economics, the lack of social credibility and the degrading moral integrity system can further increase social transaction costs and hurdle the deepening political and economic reforms, which is not beneficial for a sustainable development of healthy economy and hazardous for social stability and harmony.

Finally, frequent food safety events can provide an excuse for the developed economies to create "green barrier" in international trade, unfavorable for the Chinese trade development. In the recent years, some developed economies have gradually given up the so-called "hard barriers" (such as tariffs) but to use more subtle "soft barriers", including technical, green and safety barriers in environmental protection standards, animal and plant health quarantine and so on. Such barriers will disadvantage the export position of China in international and hurt the Chinese economic development.

2.3 Reasons of Frequent Food Safety Events in China

There are superficial reasons and also deep reasons among the inducing factors. Viewed from the surface, frequent food safety events can be attributed to the ineffectiveness of government regulation, unclear responsibility and duty system, incomplete legislation framework, unclear technology standards, low level of testing techniques, and the degrading morality of some businesspeople, etc. However, if pressed hard, one can notice that such frequency of food safety events comes from not only randomness but also causality. Briefly speaking, we summarize the inducing factors of underlining reasons as following:

First, the imbalance of economic development between cities and countryside and among various regions, and peasants and farmers of some regions have lower income and their income revenues and channels are limited with limited income increase. Since the reform and open-door policy in place, Chinese economy has grown by 9.8 % annually on average between 1979 and 2007, and 11.2 % during the 11th *Five-Year Plan* period. Income per capita in China has increased by 11 times during past 30 years. However, one must note the increasing imbalance between various regions as well, for example, the absolute gap of income per capita between the eastern areas and western areas has increased from 3308.45 RMB Yuan in 1994 to 16002.7 RMB Yuan in 2006, almost 4.8 times, while the absolute numbers and ratios are still widening. The ratio of average income between urban and rural residents has increased from 2.52:1 in 1998 to 3.31:1 in 2008. Some peasants and farmers can only depend on limited quantity of land or cattle to maintain low standard of living thus are easy to be lured by the illegal middlemen and producers in the downstream markets. For instance, according to an investigation by "March 15 Consumer Association", the cost of adding clenobuterol hydrochloride in the hog farming process may be only a few Yuan, but the hog sold can increase dozens of Yuan for them. That means, 100 hogs can sell for several thousand more Yuan, while ironically the hog without being fed with clenobuterol hydrochloride may have a difficulty in selling to the middlemen. Some farmers may have moral integrity problems, but quite some others may well be pressed to take such legal risk for their desire to improve their living standard, because this may well be the only reliable income channel, unfortunately.

Second, the food production industry is not well developed in China yet, with various participating enterprises at different quality levels. The raw material procurement, processing and sale channels within the food production industry are quite complex. Take the example of the grain processing enterprise, the grain procurement channels are complicated and chaotic, most from small-scaled and scattered farmers, grain processing chains are long and sale channels are messy with even some selling points without licenses in some regions. The resulting consequence is low quality in management and products. This is a typical feature of Chinese food production enterprises, most being small in scale and many in the regions between urban and rural. Such features have resulted enormous difficulties in the monitoring efforts by government regulatory bodies, thus the difficulty in food safety guarantee.

Third, consumer perception in food safety is weak and a biased consumption culture. Chinese "eating" culture has a long tradition and people enjoy talking about the art of eating. However, one question is whether the traditional Chinese way of eating is safe? Take the example of recent poisonous eggplants, poisonous ginger, "fitness" hog and so on, the illegal methods of adding chemicals to make the vegetables look tender and bright have exactly captured the misunderstanding of Chinese consumers' perception on food safety. Such poisonous food can make its way to the dinner tables in Chinese households from selling points without even licenses. Some Chinese consumers prefer lean pork to the extent of unreasonable demand, which certainly induces the middlemen to request leaner meat from farmers thus the farmers may well have every incentive to add clenobuterol hydrochloride in the hope of probably producing only lean meat. With the economic growth and wealth accumulation, some Chinese consumers blindly pursue "nutrition" in their perspective, resulting typically in exotic types of food demand and sometimes even strange combination of food sources. All these factors contribute to the potential danger of food safety events.

Forth, in the current Chinese governance structure with GDP-oriented and local taxation need, it is well too easy to form a local trade protection by local governments in favor of local enterprises. This factor is seriously hurdling the proper monitoring of large-scale food production enterprises' food safety issue. Because some large-scale food production enterprises contribute significantly to the local taxation for the local public finance, local governments can easily protect these companies for own interests of local economy. During some food safety events, local governments doubtfully hided certain truth and created certain frictions for proper and prompt handling of such issues. In China, there is not yet sufficient independence and authority by the food safety regulatory institutions for proper monitoring, thus some large-scale food production enterprises can operate with potential risks for a long time until they become the sources of sudden food safety events.

2.4 Suggestions for the Food Safety Issues in China

Based on the analysis of food safety problems in China, we suggest the following 6 points for the construction of an effective food safety network:

First, promote a more balanced development between cities and rural areas to increase the income levels of peasants and farmers and to widen their channels of generating revenues and incomes thus to enhance the resistance capability of peasants and farmers in sources for food safety incentives.

Second, enhance the monitoring and regulatory check on the food production industry in China, from raw material procurement, production and sale channels. The only way to lower monitoring and regulatory difficulty is to encourage scale-of-economy operation and market order in China, thus to enhance food safety monitoring effectiveness.

Third, enhance the publicity of food safety issues to improve the knowledge system of Chinese consumers. Consumers are the terminals of food chain, and also the victims of food safety events. We must actively promote the relevant knowledge of food safety to enhance the capability of consumers to distinguish and prevent hazardous food and their cooperation in food safety monitoring.

Forth, link food safety and local government officials' performance evaluation, and ensure the independence and authority of food monitoring and testing institutions. In this way, local governments can be motivated more effectively and local trade protection can be somehow mitigated.

Fifth, enhance the legislation system building and implementation, improve food safety standard and testing techniques. We must combine Chinese local conditions and learning from more advanced experiences in other countries for further capacity building in Chinese food safety legal system, safety standard and testing techniques.

Sixth, build early-warning mechanism in food safety systems to increase the handling capability in sudden food safety events. The most important mechanism is to prevent the problems, and the early-warning mechanism is inherently important. An effective mechanism of such kind can prevent food safety events from spreading thus to minimize the consequences. Meanwhile, a proactive approach and effective handling of such issues can mitigate the complex emotional reactions from people therefore to promote the public image of the relevant government bodies for a more harmonious society.

Feature 3
Research on the Environmental Information Disclosure System in Sudden Environmental Pollution Incidents: Taking Zijin Mining Pollution Incident as an Example

Luozhong Wang

In recent years, China suffers a variety of sudden environmental pollution accidents. Both the government and public attach great importance to the environmental information disclosure in the process of environmental emergency management, because the environment information publicity is not only an important mode according to which the public realize the right to know the environment truth, but also a premise that the public defend their rights of life health and property when the sudden environmental pollution takes place and actively take part in dealing with this event. In this paper, the author takes 2010 Zijin Mining pollution incident as an example, analyzing the issues and providing solutions to environmental information disclosure in sudden environmental pollution incident.

3.1 Analysis of Zijin Mining Pollution Incident

3.1.1 Introduction of Zijin Mining Pollution Incident

On July 12, 2010, the A-shares and H-shares of the listed company Zijin mining were suspended all of a sudden. In the afternoon of the same day, Fujian Provincial Department of Environmental Protection announced the environmental pollution accident, which resulted from a leakage of sewage reservoir of Zijin copper mine. The acidic wastewater leaked into Tingjiang River, containing copper ions and sulfate ions, causing large fish kills. Preliminary statistics showed that the amount of fish poisoned or killed within the basin of Tingjiang River of the Mianhuatan reservoir area only has reached 1.89 tons. The incident, happened on July 3rd, had been concealed for nine days. After then, Zijin Group announced on its official

X. Li and J. Pan (eds.), *China Green Development Index Report 2011*,
Current Chinese Economic Report Series, DOI: 10.1007/978-3-642-31597-8_19,

website a series of "Accident Briefings", without mentioning any information concerning about the way of discharging of the main pollutant or amount of the exceeded discharge.[1]

Before the pollution incident, Zijin mining had been blacklisted repeatedly in inspections of environmental protection departments. As early as May 2010, the Ministry of Environmental Protection criticized 11 listed companies for unscheduled rectification in serious environmental problems; Zijin Mining Group is among them. In the follow-up disposal of this pollution incident, another huge leak occurred again in the evening of July 16, five hundred cubic meters of wastewater leaked into the Tingjiang River.

As a result of the pollution incident, vice-president of Zijin mining was detained, and the head of Shanghang County was suspended from duties. On September 30, Zijin mining received a 9.56 million RMB fine, which was the highest one ever, issued by Fujian Provincial Environmental Protection Office. Meanwhile, Zijin Group was ordered to take actions to clear up all the pollutants.

3.1.2 China's Transparency Problem of Environmental Information Reflected from the Zijin Mining Pollution Incident

From the pollution incident of Zijin mining, it is not so optimistic for the transparency of environmental information of Chinese government and companies.

First of all, companies consider too much about their own interests, which makes the publicity of environmental information delayed, incomplete or untruthful.

Throughout the whole pollution incident of Zijin mining, the delay of environmental information publicity lies in: After the accident, Zijin concealed the accident up to nine days. On answering the question why Zijin mining had not made known to the public the information of the pollution accident until nine days after it happened, the general manager of the company's securities department said, "When things just happen, the reasons for the incident should have a judge as soon as possible. After the judge clearly convey to the masses, it will not cause panic." According to his speech, the reason why Zijin mining concealed the accident is to make clear the reasons and avoiding public panics. It seems that Zijin mining covered up the incident on behalf of public interests. If the company announced the incident at the first moment, fishermen would have taken actions of preventions and remedies, which could have reduced the damages caused by the incident. But on the contrary, the company was trying to block all the negative news until the incident became too big to be covered.

[1] Retrieved from http://green.sohu.com/s2010/zijinmining/, revised by the author.

The incompletion of environmental information publishing lies in: After the incident, Zijin Group announced on its official website a series of "Accident Briefings", without mentioning any information concerning about the way of discharging of the main pollutant or amount of the exceeded discharge. Meanwhile, the vice president declared that the leakage mainly related to natural disasters, continuous heavy rain resulted in the rapid uplift of the water level in the solution pool, the uplift exceeded the hight mark of sewage pool and formed a shear due to this imbalanced upper and lower water pressure, which led to several crackings of leakage-proof barrier, resulting in leakage of sewage pool. However, according to later investigations, the sewage drainage was opened by someone illegally. Zijin tried to cover up the "man-made" disasters by "natural calamities".

The false corporation environmental information disclosure lies in: Zijin Mining declared in July 8 that "the water quality has been improved, and the total copper concentration and pH value have been restored Surface Water Quality Standards III." However, according to the local people, fish were still dying.

Secondly, the delay of local government environmental information release affects the government's credibility and authority.

The government should release environmental information immediately after the pollution accident. However, the source of government's environmental information is often from the enterprises. When an environmental pollution accident occurs, the trouble-making company fails to report to relevant government authorities in time, but tends to delay or conceal contamination. Deeply entangled in vested interests, the local government seldom reveals environmental information in an accurate and timely manner, which directly undermines the government's credibility and authority.

Two days after the accident, dead fish were observed in large-scale waters of Tingjiang River. It was not until one week later did the Fujian Provincial Department of Environmental Protection officially reported Zijin Mining pollution accident. Government authorities' delay in information dissemination and their failure in informing the public of its cause, hazards, emergency response, recovery work progress and the need for public co-ordination led to all kinds of rumors which greatly affected the public mood, triggering a general panic and trust crisis. Even worse, if the government and officials deliberately conceal environmental information for their own interests or even make up untrue information, social conflicts will definitely be intensified and even develop into social panic.

In short, there are four major problems concerning environmental information release in China's environmental pollution accidents: first, delay or cover-up; second, release under public pressure; third, selective release of incomplete information; fourth, make up untrue information to fool the public.

3.2 Defects of China's Environmental Information Disclosure System

Since 2003, Chinese government successively promulgated a series of environmental laws, regulations and policies, including the "Cleaner Production Promotion Law", "Environmental Impact Assessment Law" and "Regulation for Public Participation in the Environmental Impact Assessment", which guarantee and regulate the environmental information disclosure on varying degrees. On May 1, 2008, the "Regulations on Open Government Information" promulgated by the State Council and the "Measures for the Disclosure of Environmental Information (for Trial Implementation)" (hereinafter referred to as the "Measures") promulgated by Department of Environmental Protection came into effect simultaneously, which presented in more detail the requirements for environmental departments and corporations for disclosing the environmental information. However, due to the flaw of environmental information disclosure system in China, local governments and enterprises would feel it's hard to disclose the relevant environmental information accurately and timely, when similar environmental pollution accidents occur as Zijin Mining pollution.

3.2.1 In Terms of Law Making, Low Legislation Level, Unscientific and Infeasible Relevant Regulations in Environmental Information Disclosure

Firstly, there is no specialized "Environmental Information Disclosure Act" in China. Laws and regulations about environmental information disclosure scattered in the "Environmental Protection Law", "Water Pollution Control Act", "Law on Prevention and Control of Pollution From Environmental Noise", "Environmental Impact Assessment Law,", "Clean Production Promotion Law" and other separate laws and regulations; even the "Measures" issued specially by Ministry of Environmental Protection, it only sets general requirements of environmental information disclosure for the government and enterprises under normal condition, without considering the distinctiveness of environmental information disclosure in the sudden environmental pollution incidents.[2]

Secondly, the level of normative documents for environmental information disclosure legislation ranks low. As the guiding draft of the "Measures", "Regulations on Open Government Information" issued by the State Council is subject to the unrevised "State Secrets Protection Act" and other laws. The stance

[2] Zhu Q (2007).

of original confidential laws views too much confidentiality, t he development of "Government Information Disclosure Regulations" and its following the "Measures" would lead to a relative lower legislation level. Whenever there is a similar Zijin Mining pollution incident occurs, local governments and enterprises tend to refuse to disclose the relevant environmental information on the pretext of "commercial secret", "inconvenience, likely to affect the social stability", "off the record, as it would easily lead to the media hype".

Thirdly, some of the provisions of environmental information disclosure are not scientific enough. For example, according to the current regulations, it is mandatory for "two-exceed" companies to disclose their environmental information—"enterprises with severe pollution and whose emission of pollutants exceeds the national or local emission standards, or whose total emission of pollutants is greater than the quota of total controlled emission determined by local government". Listed companies, which are controlled strictly by the government, are not included in the mandatory list. In fact, as public companies, listed companies have greater obligation to disclose various kinds of important information to the public. "Measures" requires companies to disclose information based on the industrial characteristics rather than corporation scales, making a lot of companies escape from disclosing environmental information,[3] which violates the original intention of the "Measures". Furthermore, the "Interim Measures for Information Report of Major Water Pollution Incident" and "National Emergency Plans for Handling Sudden Environmental Incident" and other documents take the economic losses caused by environmental pollution as one of the conditions to define sudden pollution incidents, such as "direct loss of water pollution over 10 million RMB" can be identified as a major water pollution incident. In practice, it takes time to count and calculate the economic loss which brings pretext for trouble-making corporations not to report the incident on time, becoming an institutional reason for delaying reporting.[4]

Finally, requirements on environmental information disclosure are not detailed enough for implementation. For example, the "Measures" does not distinguish businesses' voluntary and mandatory environmental information disclosure, and the content is decided by the business itself. Companies tend to brag about their achievements in environment protection and keep silent about those may undermine their image or performance. In such cases, the released information may be one-sided, useless or even false, which is against the legislative intent.

[3] Zhang X (2009).

[4] Gu M, Xu F (2011).

3.2.2 In Terms of Law Enforcement, Weak Supervision by Government on Businesses' Environmental Information Disclosure

The "Measures" stipulates that for mandatory disclosure, enterprises should release their environmental information on major local media and file for record to local environmental protection departments, and environmental protection departments shall have the right to verify the environmental information released by the enterprise; enterprises that voluntarily disclose environmental information may release to the public their environmental information on the media, internet or in the form of disclosure of their annual environmental reports. Environmental protection departments may grant certain rewards to the enterprises that voluntarily disclose information on their environmental behavior and comply with environmental protection laws and regulations. Therefore, according to the "Measures", the authorities can only supervise enterprises through weak approaches such as "record" and "verification", so it is difficult to ensure the authenticity and timeliness of the environmental information disclosure. Without unified standards and legal procedures, government's rewards for enterprises that do voluntary disclosure may be arbitrary and lack of supervision intensity.[5]

3.2.3 In Terms of Relief Right, Lacks of Effective Relief Approaches in Government Environmental Information Disclosure

If citizens believe that an environmental protection department has failed to fulfill, in accordance with the law, its obligations in respect of open government environment information, they may report to the higher level environmental protection department. This is the approach of protecting their right to know about environmental information provided in the "Measures". However, such behavior of the environmental protection department belongs to environmental administrative inaction, of which the public have right to initiate administrative litigation or administrative review, rather than simply report to the higher level. According to Article 26 of the "Measures", if citizens, corporations and other organizations believe a specific administrative action of an environmental protection department in its open government environmental information work has infringed their lawful rights and interests, they may, in accordance with the law, apply for administrative review or institute an administrative lawsuit. However, the current "Administrative Procedure Law" only provides administrative lawsuit on administrative body's infringement on personal and property rights, while access to

[5] Zhang X (2009).

environmental information is a right to know. Under the existing regulations, violation of right to information can not directly lead to an administrative lawsuit. Meanwhile, for environmental protection authorities' unjustified refusal to provide environmental information, or providing incomplete and untrue environmental information which leads to the damage of party's personal or property rights, there is no clear provision in law of letting environmental protection authorities to shoulder compensation liability.

3.3 Measures to Improve China's Environmental Information Disclosure System

Information disclosure is the first step to dispose environmental pollution accidents. Measures should be taken to improve China's environmental information disclosure system as follows:

3.3.1 Theoretically, We Should Focus on Public-Oriented Environmental Information Disclosure and Follow "Disclosure is Principle While Non-disclosure is Exception"

First of all, no matter "National Emergency Plans for Handling Sudden Environmental Incident", or "Environmental Information Disclosure Measure" and even the newly revised "Water Pollution Control Act", all focus on the environmental information reporting system between businesses and government authorities and between different government organizations, while the publicity of environmental information to the common citizens is neglected. Information tends to be weakened and blocked in hierarchy of reporting process, and eventually the blind spots in the law become the barriers hindering environmental information to the public. One thing that needs to be pointed out is that reporting to different authorities step by step is not equal to information disclosure. Disclosure of environmental information should be public-oriented, because protecting and realizing the public's right to know is not only a prerequisite of reducing personal and property damage in environmental pollution accidents, but also an important approach to adopt public opinion, seek public engagement and jointly overcome the crisis.

Second, in practice, environmental information disclosure is often subject to the interests of business and government. Whether to disclose information or not, and what kind of information to be disclosed all depend on the need of government management and business operation. In the future, government officials and business manager must change ideas in considering the content, quantity and mode of environmental information, taking firm stand not unilaterally from their own needs, but also protect the citizens' right to know and public interests. Under the principle that "disclosure is principle while non-disclosure is exception", information that can help protect citizens' right to know, limit businesses emissions, improve environmental management should be disclosed; environmental information which involving state secrets and commercial secrets can be kept secret only after administrative permission.

3.3.2 Legally, We Should Extend the Scope of Body in Compulsory Environmental Information Disclosure and Refine the Content and Methods of Disclosure

First, we should extend the scope of body in mandatory environmental information disclosure. According to the "Measures", mandatory disclosure is only limited to "two-exceed" enterprises. But the scope is too narrow in the long run. Listed companies must be included in the scope of compulsory disclosure to facilitate public access to more complete environmental information. To that end, Article 11 and Article 20 of the "Measures" must be modified accordingly to formulate the content and approach of listed companies in environmental information disclosure.

Second, we should substantiate the content of disclosed environmental information. According to the "Measures", the mandatory disclosed environmental information of "two exceed" enterprises is very limited, only including the name, address, corporate legal representative, the name of the major pollutants, mode, content and total volume of emission, information on emission that had surpassed the standards or total emission that had surpassed the prescribed limits, information on the construction and operation of their environmental protection facilities and emergency plans for sudden environmental pollution accidents, etc. From the public point of view, these four categories of information provided under the "Measures" can not satisfy the public's information needs. Many environmental accidents' impacts are long-term. Compared with the four categories of information, the public want to know whether the incident will bring potential, long-term harm, or threat to future personal and property safety. If the answer is definite, what measures should the public take to cope the incident, considered to be highly relevant to public's interests.

Finally, unreasonable provisions of existing regulations should be amended. As mentioned above in current laws, the economic loss caused by environmental pollution are considered as important conditions for defining an environmental pollution incident, which becomes an institutional barrier for timely reporting and disclosing. It is recommended that we should combine the sudden incident and data of excessive amount of pollutants as standards for defining environmental pollution incidents, since data of excessive pollutants can be available shortly after sampling.[6] Once the data exceed the stipulated standards, the trouble-making enterprise must immediately inform the public major environmental information in order to control and minimize the damage.

3.3.3 Practically, We Should Improve the Relief System of Environmental Information Disclosure in Pollution Incidents

No remedies no rights. For violation of obligations in environmental information disclosure, government authorities should improve procedures of receiving complaints and administrative review, strengthen the judicial remedies simultaneously.

After an environmental pollution incident, if the pollution enterprise delays, covers up or fabricates environmental information, leading to damage or increasing damage of the victim, the victim can sue the pollution enterprise. According to current laws, units or individuals directly harmed in environmental pollution can turn to the people's court to seek compensation. In practice, the victim often requires immediate compensation to reduce losses. Since China has no special litigation procedures for environmental pollution incidents, and general civil litigations take long time, which rarely help the victim claim compensation easily.[7] In the Zijin Mining pollution incident, the fishermen would rather dump dead fish in front of the county government office and factories to protest the incident and require compensation, than sue the trouble-making enterprise. Therefore, special procedures for environmental pollution incident litigation have far-reaching importance for protecting victims' interests.

3.3.4 Systemically, We Should Set up Coordination Mechanisms Between the Government and Enterprises, as well as the Government and NGOs in Environmental Information Disclosure

First, we should set up coordination mechanisms between the government and enterprises in environmental information disclosure. The main body of environmental information includes government departments and enterprises. Except for environmental pollution caused by natural factors, most environmental pollution incidents are caused by improper discharge of pollutants. Since enterprises know

---

[6] Gu M, Xu F (2011).

[7] Gu M, Xu F (2011).

environmental information better than any other type of organizations, they are important responsible parties in environmental information disclosure. The "Measures" underlines the importance of enterprises' environmental information disclosure. Under current laws, enterprises' environmental information disclosure can be mandatory and voluntary. "Two exceed" enterprises must conduct mandatory disclosure to the public in line with environmental information regulations. This mandatory disclosure is actually done by the government and enterprises together. The voluntary disclosure means that enterprises voluntarily disclose environmental information as an important approach of fulfilling corporate social responsibility and enhancing their green image. As a complement of the government's environmental information disclosure, voluntary disclosure is widely encouraged in foreign countries. EU, the US and other developed countries have established effective systems of enterprises' environmental information disclosure, while in China such system is still in its infancy. In the future, China should improve and perfect the system of enterprises' environmental information disclosure which is coordinated with that of the government. This is an important trend of China's development in environmental information disclosure system.

Second, we should set up coordination mechanisms between the government and NGO. Environmental protective NGO plays multiple roles in environmental information disclosure: NGO can supervise the information disclosure of government and businesses on behalf of the public, report violation of environmental information disclosure obligations or even file an administrative litigation for public interest. Besides, environmental protective NGO can provide part of environmental information to the public. Through further processing and sorting environmental information of government and businesses, NGO can present more understandable information to the public. Environmental protective NGO also plays a very important role in environmental education and training, raising public environmental awareness, helping and promoting public participation in environmental protection.[8]

In conclusion, the *12th Five-Year Plan* period will witness more environmental pollution accidents in China. To improve the environmental information system is only the first step to prevent environmental pollution incidents and establish an environmental emergency management system. Enterprise and government that refuse information disclosure and public supervision will become a natural breeding ground for the spread of contamination. Zijin Mining pollution incident reflects many problems in China's environmental information disclosure. Reflecting on these issues, we should take effective measures to improve China's environmental information system. Considering the special nature of sudden environmental pollution incidents, environmental information disclosure must be timely, accurate and complete in this process.

[8] Shen J (2010).

References

Gu M, Xu F (2011) The study of corporate environmental information disclosure issues in sudden environmental pollution incidents. J Changsha Railway Univ 1 35
Gu M, Xu F (2011) The study of corporate environmental information disclosure issues in sudden environmental pollution incidents. J Changsha Railway Univ 1 36
Shen J (2010) On the governmental disclosure system of environmental information. Nankaiy J 2 55
Zhang X (2009) On the corporate environmental information disclosure system. Rural Econ 2 125
Zhu Q (2007) On the environmental information publicity of sudden environmental pollution event. Sci Law 3 158

Feature 4
Basic Framework for Building China's Green Financial System

Yuanlong Wang and Ma Yun

To develop green finance will be the key issue to promote green economy development, the development of green economy needs adaptable financial system. Health and complete green financial system is the only way for China to transform the pattern of economic development and push green development and sustainable development strategy. So, building green financial system already is an urgent theoretical and practical issue in China finance development strategy. Green financial system is a financial system serving for green development and sustainable development, in brief, green financial system is the financial system from protecting the environment, climate change, green development and sustainable development in a variety of angles to make important financial institutional arrangements and mechanisms for innovation. Basic framework of China financial system including a sound financial system, green financial market development, green financial instruments innovation, financial institutions cultivation and improve the green financial regulation. From above five aspects, following by scientific concept of development, building China green finance system comply with green finance and sustainable development.

4.1 Improve Green Financial System

Green financial system is a green financial system architecture established by national law, related to all aspect and sectors. The system is composed by a series of regulations, reflecting country's statutory and non-statutory, governmental

This paper is part of the achievements of the topic research GúÈStrategic Research of Building ChinaGúøs Green Financial SystemGúË. Members of the team: WANG Yuanlong (Economist of Tianda ResearchInstitute, senior researcher), MA Jun, WANG Sicheng, LIU Yutin and so on.

X. Li and J. Pan (eds.), *China Green Development Index Report 2011*,
Current Chinese Economic Report Series, DOI: 10.1007/978-3-642-31597-8_20,

regulations, rules, regulations, and industry conventions, established practice and so on.

System reference and migration is an important way to establish and improve the system, however, in view of different national resource endowments and different stages of development, system building should not only rigidly adhere to international standards and institutions, but also should more focus on the country's regional and industrial characteristics; More localization and self-orientation should be concerned, rather than simply a mirror and learn. Building and improving China's green financial system should be based on the premise of respect international conventions, under existed legal system and reference developed countries' rules and standards. Above including following aspects: green financial basic legal system, green financial business development system and green financial supervision system.

First, building green financial basic legal system is the premise and bases for making green financial business implement rules and supervision regulation, mainly include total control plan and emission validity establishment. Total control limit the cap for environmental capacity, clear the scarcity of resource, so making the emission trading as a commodity possible. Currently, though under the framework of "United Nations Framework Convention on Climate Change" and the "Kyoto Protocol", China temporarily has no obligation to emission reduction, only needs to control the emission intensity. In the long run, China will face increasing emission pressure from international society. Considering developing domestic green financial market, establishing total capacity control implement rules and emission right legalization, put it into property range as other factors of production.

Second, the specific implementation of green financial system should include improving the market for carbon emissions trading system and development of green credit, green insurance, green securities business rules for the operation. China's carbon emissions trading market, positioned as a cover project construction market, the voluntary market and the quota trading market, multi-level market trading system, so the content of carbon emissions trading system means to gradually standardize the behavior of the three main market subject, the object and related trading activities. In the business of green credit, green insurance and green securities, and The People's Bank of China and Ministry of Environmental Protection have jointly issued "Guidance of Carrying Out Environmental Policies & Regulations and Avoiding Credit Risk", " Guidance on Environmental Pollution Liability Insurance for the Work "and" Guidance on the Strengthening of Supervision of Listed Companies Environmental Protection " in 2007 and 2008, the initial establishment of the green credit, green insurance and green securities field of the basic institutional framework, but at the operational level there are still some rules needed to be improved.

Third, the particularities in green financial area raise new challenge for China current supervision regime, the market mechanism to solve environmental problems largely depends on effective government oversight and the necessary administrative intervention, otherwise the specific trading system cannot play its proper role. Currently, National Development and Reform Commission associating with Ministry of Environmental Protection, The People's Bank of China issued a series of measures to carry out green financial system,which subject to administrative rules in supervision regime. As China green financial market stepping forward and financial business deepening, the following rules will be made one by one, the supervision system will improve gradually.

4.2 Development of Green Financial Market

Green financial markets is constructed to protect the environment (including water, air, forests, soil, etc.) or invest and finance considering environmental factors. It is green investment and financing in place. The current international financial market mainly refers to carbon credit instruments which base on the form of carbon emissions rights trading financial markets under "Kyoto mechanisms",[1] including the project market, the voluntary market and the quota trading market. In addition according to the functional classification, the market can be divided into primary and secondary markets; according to market forms of organization, can be divided into exchange-traded market and the OTC market.

The future development of China's financial markets will be the development of green carbon trading market as the guide. As China current is in the stage of industrialization, the carbon market is lack of market demand, unrealistic total control, so the future of carbon markets pilot will carry out in key regions and key industries. The United Nations and the World Bank predicts that by 2020 global carbon trading market capacity will be more than the oil market, to become the world's largest market. From the current pattern of international carbon trading development perspective, the quota-based market, supplemented by project transactions. The voluntary market share of global carbon trading market is very small, such as its trading volume accounted for only 2.9 % of international carbon trading, trading revenue accounted for only 0.6 %. But in the past few years, the market demands experience a solid upgrade, showing a rapid rise in the appeal, the outlook cannot be ignored, especially in Asia and North America.[2] Given China's current stage of development, the mainstream view is that by 2020 China would not make mandatory emission reduction commitments, so the short term is still the Clean Development Mechanism (CDM) project focus, while developing inside and outside the voluntary emission reduction trading market; in the long-term should be gradually developing the quota-based market, transition from one force to the double force, first pilot transactions, then the general promotion.

---

[1] Three mechanisms according to "Kyoto Protocol": Clean Development Mechanism (CDM), Joint Imple-mentation (JI), Emissions Trade (ET).

[2] State And Trend Of The Carbon Market 2010, http://www.siteresources.worldbank.org.

In addition to forms of organization, exploration of trading system is another important aspect for market establishment. In addition to "Kyoto mechanisms" which is to service the international carbon finance market, many countries have conducted a useful attempt on the condition of national circumstances. Up to now, the world's leading carbon trading platform is different such as the EU emissions trading system of the European Union (EU ETS), the UK's emissions trading system (ETG), the United States Chicago Climate Exchange (CCX) and the Australia New South Wales trading system (NSW) the trading mechanism, but the similarities are based on the needs of their domestic emission reduction and development strategies. In a sense, independent, localized trading mechanism is the key to open up their financial markets. To emphasis, green development of financial markets and market mechanisms do not have to rigidly adhere to international standards, it should be at a higher level of the establishment of autonomous, suitable for China's localization mechanism of green financial transactions.

4.3 Innovation of Green Financial Instrument

Green financial instruments, is financial assets which could be traded in financial markets, is the financial intermediation between surplus and deficiency by documentary evidence, the basic elements is the payment amount and payment conditions. According to the traditional native and derivative financial instruments classification, green financial instruments can be divided into primary and derivative financial instruments. In the current financial markets, native financial instruments includes green carbon credits and carbon stock, which derived green derivative financial instruments including forwards, futures, options, swaps and structured products. Derivative green financial instruments price depends on the value of the native green financial instruments, its main function is not to regulate funds or to promote the transformation of savings to investment, but the risk exposure management of native related green management and financial instruments.[3] Green innovation financial instruments in the international market are in the ascendant, after the World Bank in 1999 launched the first Prototype Carbon Fund (PCF) for the CDM, various types of fund emerged one by one. According to the distribution of different subjects, the funds can be divided into the World Bank funds, sovereign wealth funds, government multilateral cooperation fund, the profitability fund established by financial institutions, carbon funds and private carbon funds organized and managed by non-governmental organization. Response to huge capital requirements for the carbon finance projects and to avoid high uncertainty of future prospects, carbon guarantees and carbon factoring, etc. financial products emerged.

[3] WANG Zengwu, YUAN Zengting : "Pushing the innovation and development of carbon financial instruments", Financial Forum"2010(8).

China's green financial product innovation is still in its infancy basically, the type of green financial products is unitary. Since the necessary infrastructure and conditions of green financial products development vary, the development of green Chinese financial market trading tools should adhere to the principle, from foundation to innovation, from simple to complex, providing more risk management and arbitrage tools for the market step by step, improving the green financial market interest compensation mechanism. Based on the development of financial markets overall layout, first development of the native tools of carbon trading market, such as a certified emission reduction units (CER) trading in the original Clean Development Mechanism (CDM) market, and voluntary emission reduction units (VER) transactions in the voluntary market, set up trading platform, exchange-traded market and the OTC market.

With the domestic green financial market system becoming more and more mature and accumulating some experience of financial derivatives, China could learn from the European Union, the United States and other developed countries, developing conventional derivatives to meet the hedging needs of market participants. On this basis, further attempts to develop new green financial derivatives, such as carbon-based assets or carbon assets linked with the tradition of weather derivatives and other structured products.

4.4 Cultivating Green Financial Institution

As China green financial markets gradually develop in a full range, multi-level, market participants will be increasing, regardless of project owners and developers or buyers and sellers of carbon credits need skills and experience, to provide intermediary services market to relevant institution who familiar rules of relevant institution, including low-carbon project development and investment and financing services, consulting, asset management, brokerage, credit enhancements, guarantees and other services, while a number of professional organizations such as accounting firms and law firms opened up a new business services. China's green construction system of financial intermediation services should include at least the following aspects.

First, encourage existing financial intermediaries actively involved in green business. On the one hand, to achieve the green transformation of commercial banks, vigorously promote the concept and practice of green finance green financial services. In the process of credit allocation, to increase the environmental protection industry and energy saving technological innovation credit support. To provide a package of projects for CDM financing program similarity the idea of "development loans supporting the mortgage amount". The existing bank letter of cooperation mode explore the financial leasing business, to provide equipment for the energy saving project support; On the other hand, actively encourage professional organizations to provide technical support for green financial services, such as project negotiations,consulting, project evaluation, project financing, guarantees, legal and audit services. Professional financial intermediaries involved in the green finance operation will reduce transaction costs and project risks, and promote green financial business conduct.

Second, cultivate professional green finance-based services intermediaries. For example, to develop local Designated Operational Entity(DOE), green credit rating agencies, research institutions with professional approach, independent green financial services clearing and settlement platform.

Third, set up policy green financial institutions. To support the development of green finance, China could learn from international experience, establish particular policy green financial institution, such as “green bank” or “eco-bank”, specifically provide policy financial activities for green development and sustainable development projects; or set up a special green fund, such as “environmental special fund” and “ecological special fund”, mainly used to support environmental protection industry, besides, it can also be used for environmental pollution compensation of victims of major events.

4.5 Improve Green Financial Regulation

Green financial regulation means financial regulatory authorities implement a comprehensive, regular inspection and supervision on entire green financing (including green financial services institutions and green financial business) in accordance with national laws and regulations, so as to promote financial institutions to operation and development to conform to the requirements of the development of green finance.

At present, China’s financial market supervision system has been relatively well, basically forming a “One Bank, Three commissions” regulatory model, while the green development of China’s current financial regulatory system put forward new challenges, including the regulatory regime, regulatory cooperation and international regulatory and other aspects. Sound green financial regulatory framework, on the one hand, should learn from existing supervisory experience; on the other hand, according to the green financial specific risks and operation model, to make rules, adjust regulation ways, to innovate regulation concept. For example, for this new field of green finance, business boundaries becomes even more unclear, many transactions involve multiple participants, including banks, insurance companies, securities companies, asset evaluation agencies and certification bodies, etc., so coordinated monitoring is particularly important. In addition, as the specialty of green finance, which is closely related to international cooperation from initial stage of development and based on international financial markets, so in the regulatory process international coordination of regulation should be considered. To make China’s green finance regulation more comprehensive and effective, further improving the external monitor should be considered, such as setting up green financial associations and other self-regulatory organizations, and the establishment of market-based credit system.

Needless to say, building a green financial system is a large and complex systems engineering, will never be done overnight. Therefore, China's development of green finance, building financial system should be gradual and step by step, considering comprehensively of the international economic situation and China's overall economic development.

Feature 5
The Significance of "Humanism" in Green Development

Baoyuan Li and Pengzhou Kuai

Based on many theoretical points from the **Economics of Human Development,** which I have studied for many years, and also the empirical clues provided from a perspective of "macro history," and the recently released Research Report on Chinese **Human Development** Strategy, which says that we should be integrating today's "low-carbon green development" background and many of the newer future trends. In this thesis, I will bring forth the opinion that "humanism's low-carbon development" namely, human civilization and its progress is taking a course of freedom as the axis and is continuously improving the all-round free development of humankind. Also a low-carbon green development course is constantly getting rid of external material constraints and gradually returning to the spiritual free homeland. Thus, we can infer that China is a developing agricultural country, who faces major concerns in terms of human health and safety protection, as well as the educational issues faced in its growing population. How is China going to achieve the multiple social and economic transformations needed to establish a strong institutional security system based on the natural order and in line with all of the public interests, in order to get rid of the plight of black industry-oriented education system? This will be a difficult and complex historical and strategic task that will be confronted by all Chinese people as they move forward pursuing a new green low carbon development plan.

5.1 A Macro History: The vision of Green Development. Human-Orientation Returns to a More Nature Way of Life and Rational Thinking

From a purely liberalistic point of view, the essential historical process in human development is a means in which all human beings pursue freedom in a relatively restrained manner of reality. Rousseau said, "… (the) human is not only a

X. Li and J. Pan (eds.), *China Green Development Index Report 2011*,
Current Chinese Economic Report Series, DOI: 10.1007/978-3-642-31597-8_21,

particular species that has free will, but also the ability to help create a great nature. Since the birth of human civilization, humans have been in a type of 'bondage' for all time. The whole process of human development is essentially a course of human beings making an attempt to get rid of the shackles of nature." He further states, that the "currently popular thought of 'green (economic) development' still belongs to the historical background, in which a human's negation on the 'black development' of an industrialized society, and 'white development' in the pre-industrial society, are formed from each individuals own interests under the context of a post-industrial era. Under a macro horizon of the evolution in nature, green development has endured and developed within humanism connotation."

The human, as an animal species in the natural world, has evolved from a far lesser advanced mammal known apes. This evolution, which happened over thousands of years, developed the species into an animal with the ability to work, think, and understand the world, and is what we know today as the real human being. In the long history of primitive society, the human being's existence was mainly used for meeting the living and production requirements, based on that human's tribal units needs. They developed used simple stone tools for to collecting plants, flowers, and fruits from nature, even using these same simple tools for hunting animals for food. In the later periods of development, human's gradually developed a "slash and burn" technique in primitive agriculture and engaged in a small amount of sporadic animal domestication. These barbaric times can be characterized as "parasitic in the nature," where human being's were entirely dependent on the ecosystem. As a relatively end link of the food chain, the human being and other organisms, are "dependent on each other" in the race of life. Groups formed together conduct collective labor activity. Many of these groups, or living units, formed by taking blood lines as a unit, while still others were considering the idea of a higher power such as God, where beasts or totem worship, as spiritual bond were the valued standard to unify and constrain the harmonious relationship between the human being's and nature. Human being's would "instinctively" join in the natural cycle and order, that these groups offered, and learned that they could depend on the groups resources and the conditions provided by nature, to prosper and survive in the most "economic" way, thus maintaining the most desirable lifestyle.

For the period starting 5,000 years before the industrial revolution, human being's lived in the farming era for subsistence, where people were dependent on the cultivation of land. With the development in plant cultivation, animal husbandry and metal smelting technologies, a human being's living status gradually shifted from nomadic tribesmen to civilianized settlement. This development in living was followed by a rapid growth in population and a higher demand for consumable goods. This development further expanded the production scale and gradually increased the deepening social division of labor. On this basis, animal husbandry, fishery, handicrafts and business, were all, separated from agriculture. The social organization had changed from a patriarchal clan system into one that favored slave owners and feudal land ownership society. Therefore, an agricultural civilization whose basic characteristics and ideals lied within an agriculture-based, decentralized management system of self-sufficiency, with ample food, clothing, and shelter, has become the sign of the times of the human being's social development.

In the agrarian age, a human being gave way to the law of nature, also known as "natural selection." With their special intelligence, wisdom from understanding and transforming the nature around them, human being's once created a series of brilliant ancient agricultural civilization. This civilization is often characterized by its richness in natural products and the flourishing of men and livestock within nature. After a certain period of time for development, the agricultural civilization will move into its prosperous period and have to face the obvious contradiction between population and land use issues, as well as, the increasingly tense relationships between the civilizations food supply and a demand, or need, for that supply. Consequently, the human beings were forced to burn forests for land reclamation. This over-exploitation often resulted in soil erosion and a further decline in land productivity. In order to save the declining situation, human being's often resort to, what Arnold J. Toynbee once said was a "suicidal force" or predatory way of waging war on other civilizations. The resulting action may cause an accelerated decline in an aggressor civilization or another nations' civilization, or even extinction of one of the civilizations forever. However, with all the ups and downs of any civilization, with all the transformations and the many vicissitudes of the human being, the farming era would have limited utilization and impact on nature and its destructive effect is minimal.

In 1784, Watt invented the steam engine, marking mankind's farewell to the agrarian age and its step into the grand new era of industry known as the time of machinery. Some 200 years later, the development of human productivity successively experienced the three major industrial technological revolutions known as, mechanization, electrification and electronization. Driven by the technological revolution, human being's gradually broke away from the living state which featured a dependence on nature, the passive use of nature, and solely relying on plant fuel sources as powered energy source. This transformation into the new era, known for its social large-scale machinery production, was characterized by its conquering of nature by transforming and controlling the human's natural surroundings. The rapid development of nuclear energy, microelectronics, molecular biology, and space technology, the humans were relying on powerful supernatural material and technical strength. Doing so, the humans not only completely changed the natural appearance of the earth, but also broke through the restrictions of earth's biosphere. Human technologies broke through all the boundaries, reaching its tentacles into the boundless magic macro universe and microscopic worlds they discovered. Meanwhile, humans have also undergone a profound global change in its relationship to production, lifestyles, and social systems they established. Capitalism "roundabout" way of large-scale production, has not only created the wealth accumulation miracle that was once difficult for human society to achieve over the past thousands years, but it also created a production relationship characterized by the antagonistic relation between both "capital" and "labor." The widening gap between rich and poor, as well as the aggravated benefit conflict, depletion of resources, and extravagant consumerism lifestyle, were all realized by the new push towards a more capitalistic society. In a modern industrialized society, the direct interdependence relationship where people are relying on nature to provide their needs has been substituted by an indirect contact relationship where people are relying on the market as intermediaries. Individual labor and working significance are split and alienated by the specialization-based mass flow production system. Personal needs and the meaning of life are driven and overwhelmed by the constantly expanding "materialistic" market demands to meet these sane human needs. At the same time, the expanding size of the population, an increasingly larger labor force, the infinitely stretched production and marketing chain, and the increased urbanization development which facilitates the continued gathering and stacking of elements, brought about increasing pressures and challenges to social infrastructure, natural resources, and ecological environment.

In summary, by examining the macro-perspective of nature evolution, we draw a basic conclusion that the human being not only considers himself in the right, but also considers himself to always be in the "shackles" of nature. Human's think of themselves as the master of others, but far too often remain as greater slave than they desire. Amartya (2002) When a human being gives into his rationality of the extreme and believes in the realization of freedom, constraints imposed by nature will become more intense and bring new human disasters that could have been avoided. Therefore, if the currently advocated "green development" theory is the inevitable future for the development of human civilization, its essence lies in giving up and opposing the "Anthropocentric" view and returning to the more natural and rational "Humanism" philosophy with an understanding on the basis of the natural philosophical sense of values, a world view, and a universal outlook, all having a rich connotation of humanism.

5.2 Humanism Orientation of Low-carbon Green Development: Expanding Freedom and Achieving the Comprehensive Development of Human Being's

From a strictly liberal point of view, the so-called "development" is essentially a historical process in which human being's are rational in the pursuit of "freedom" in a "non-free" reality. According to the viewpoint of development as freedom, Amartya Sen said that the extension of "freedom is the primary aim of development, and also the main means for that development. What composes development is to dispel various constraints in which the human being will hardly have his own choice and is also faced with almost with no opportunity to apply the function of reason." Blaise (1985) Specifically speaking, the real meaning and process of development can be summarized in the following manner:

Under the "certain" frame of natural order, which is not free for all humans, and by the nature-gift of "reason," or free will, the human being can set new strategies in balancing, compromising, and inter-action, known as the means for freedom, in different situations via a series of tortuous struggles and efforts, known as the pathway to freedom, thereby continuously approaching an ideal state and realm of freedom, which is a human being's ultimate goal.

But as the main objective and means of development, freedom has multiple levels. It does not have the real means to generally speculate or explain "freedom" on a single abstract level. According to Marx's "all-round development" theory, we can start from three angles, i.e. natural attributes, social attributes and cultural attributes, and divide freedom into three levels, i.e. survival freedom, social freedom and spiritual freedom. Li (2006) The extension of the three levels of freedom constitutes the primary purpose of the main means for green development, and reflects the human orientation of green development.

First of all, looking from the natural attributes of the human being, freedom means:

To gradually reduce the absolute oppression, restrictions, and limitations on humans, from both natural resources and the ecological environment, in which struggling for existence in the nature world gets an ever-increasing level in material productivity. Amartya Sen says about "capability," in the example of a human's ability to get the basic means of subsistence from nature through the material production activities thus being able to live comfortably or have the feeling of "well-being," within a full rich life, with sufficient clothes and food. Since a human being is one of nature's "creations" with free will, the most direct opposite to a human's freedom is, of course, from his "creator," nature. Therefore, man is born free, but he will always remain in the shackles of nature, and the shackles are first imposed on the human being, by nature. In order to break away from the shackles naturally and externally imposed on a human by nature forces, the human understand its specific position in nature by virtue of his rationality, the first of these is "intellectuality." The human must then think and decide, using his rational intellectuality, what kind of state to adapt to. The human can use natural forces when the rational, or "free will," is given into play. To a certain extent, the human being will forget that nature is limited in its own rationality and walk towards an irrational state, making all attempts to replace the position of God. As the result, mankind's relationship with nature will become "deadlock" and even "out of control." Such relationships between contradictions and conflicts, do not only limit the development of the contemporary human being, but also affects the survival and continuity of future generations. Therefore, on a survival level, human beings learn to "rationally" adjust their relationship with nature. Humans do this for themselves and their future generations, in order to win a deserving share of "freedom" within the "permission" scope of nature, and this is the basic starting point for advocating more green development.

Secondly, viewing from the social attributes of a human, freedom means that a human will gradually surrender their natural state of ignorance, as featured by their survival competition among other social animals, in the course of social activities that are gradually defined, standardized, and coordinated. Relationships between people are established through the act of gradually resorting to a continuously improved social contract, in a particular legal system, so as to ensure that every member of that society can get a real and personal freedom of equality and fairness, in the social order, to their greatest possibility. Human beings are social animals, but different from the many general social animals. Human society is composed of individuals with free will and social order. New systems and rules are gradually formed and consummated, by people, in the pursuit of "freedom" after a long and complex struggle and compromise with each other. Freedom issues on this level mainly involve two aspects. The first aspect being the interpersonal freedom, of the human being, within social interaction, or how does a person treat and deal with the freedom of others in a social interaction, while still achieving his own freedom. Secondly, the public freedom in a national political life, for example, what procedures and rules should be adopted in social public activities to ensure the consistency between an individual's rational choice and the rational choice's of social groups? As the green economy has clear spillover effect, individual green consumption, green production of enterprises, and a national green development program, are faced with the inconsistency between an individual's rational choice and the collective's rational choice. Therefore, on the level of social freedom through institutional innovation and change, while building a natural order to provide a system guarantee for all individuals' to spontaneously practice green development, serves as the basic premise for achieving green development.

Thirdly, viewing this from the spiritual attributes of a human, freedom means that a human has gradually detached from meeting the limited social needs of life and social existences to transcend into the truly free realm of continuing to pursue an unlimited spiritual need centered on the full realization of self value. A human being is a thinking animal. As the 17th century French thinker Blaise Pascal said, "Man is to himself the most wonderful object in nature; for he cannot conceive what the body is, still less what the mind is, and least of all how a body should be united to a mind. This is the consummation of his difficulties, and yet it is his very being." Li (2011) Therefore, a human's greatest idea of freedom and also the dream of freedom, are capable spiritually to "fly freely" in the infinite universe. He struggles with how to "rescue" his soul from the non-free state that it is inexplicably associated with it. The human being is always under the "control" of his physical emotional needs. In an industrialized society, the main body of human significance has been alienated, and more levels, more important spiritual pursuits, have often been overlooked resulting in that human's interest is a limited "materialistic" whirlpool where he is unable to free himself. Therefore, one the level of the spiritual freedom, through the dominant role of spiritual production and reproduction, especially using education and cultural enterprises and development, mankind can thereby obtain unlimited enjoyment and spiritual culture, on the basis of a limited physical need that can be met. This is the ultimate goal of green development.

In short, the expansion of the above three levels of "freedom" is the three realms of a human being's pursuit of rationality, together with the three basic objectives to achieve green development and the means of its implementation. However, in the process of emphasizing development, we should not just focus on the "survival freedom" field. The humanism orientation to green development is the general trend for achieving human freedom and comprehensive development. The human being must use a continuous improvement process when developing from one form of freedom to another, starting with the survival freedom, moving on to social freedom, and then ending with the spiritual freedom.

5.3 Tough Challenges for Green Development in China: Weak Economic Fundamentals, Lack of Institution and Culture

Today, people can really feel the "non-green" issues in the development of China. The so-called "China miracle" or "China rise" is just a rapid economic growth based in a small economic base. Without "humanism" and "green" development concept as a guide, such a rise would become "weak" and cannot be sustainable. This can even lead to increased conflicts of interests, a depletion of China's natural resources for energy, environmental degradation, and other life issues. In other words, the green development task of China still remains arduous in the future. Integrating China's green development with and from a human being's orientation of green development lies mainly in the difficulty of the following three aspects.

First, the economic base of green development is still very fragile, from an angle of the survival freedom theory. China's large population with excessive numbers, low quality, and aging structure, are basic contradictions confronted by the country today. Also, since China's agriculture is characterized by typical farming production, with vast rural areas, and some large-scale farms, how China walks out of the traditional agricultural economy with its rural society. China is rapidly realizing industrialization, urbanization, and marketization, the real subject of the current modernization and development in China. Basic contradictions in large population with low quality, and the triple transition of industrialization, urbanization and marketization, will influence how China development economically for quite a long time. It may even determine the speed and path of China's economic development changes to an intensive low-carbon greener development.

Secondly, from the perspective of the social freedom theory, the system protections for green development are not perfect. Although all levels of the government have clearly felt the resources and environmental constraints confronted in a new greener economic development, and these same government agencies have developed the energy saving task and goals in the relevant planning report, the enactment of these new laws and regulations is still not perfect. Under the background of imbalance in the these three areas, economic development, technological alienation, and globalization of social interaction, the building of regional, national, and international laws and a new regulation system to coordinate the conduct of individuals, enterprises, and countries, has undoubtedly become more important and complex. This requires that the government, and even all social sectors, to have far higher strategic vision and more resilient "greener" scientific attitude, as well as a more democratic and equal public selection process to integrate all countermeasures.

Finally, from the perspective of spiritual freedom, cultural cause development is far from compatibility of green development. Education as one of the most important means of cultural and its transmission plays a fundamental role in the dissemination of the green culture by promoting green development. Contemporary China's national education system came into being under the macro background of trying to break the old imperial examination system while learning the west's modern education experience and implementing a planned economy system. As the government is rigidly and uniformly monopolizing China's schools and education system across the country, the national education system has less relevance to social needs, and is placing more emphasis on receiving an academic certificate and examination performance. Placing less emphasis on the quality of practice has a serious orientation bias, structural distortions, and institutional block in the long term, and has resulted in a tremendous negative impact on the cultural foundation upon which green development is built. Therefore, getting rid of the exam-oriented education crisis is the biggest challenge faced by China in its green development.

In short, we draw this conclusion. From the perspective of "macro history," the so-called "green development" in the post-industrial era, is the elimination of "black development" within industrialized society and "white development" in the pre-industrial society, having a rich historical significance of "human development". The basic orientation of green development takes freedoms expansion as the axis to achieve the overall development of a human being on life, social, and spiritual aspects. China has a long way to go in the low-carbon green development. How can China achieve multiple social and economic transformations, establish a sound security system for low-carbon green development, and get rid of exam-oriented education crisis to build up cultural roots supporting the green development? These things become the most arduous and complex historical mission and strategic tasks confronted by Chinese people in their mission to reach a low-carbon green development program.

References

Amartya S (2002) Development as freedom [M]. China Renmin University Press, Beijing

Blaise P (1985) Thought record [M]. Commercial Press, Beijing

Li B (2006) Economics of Human Development [M]. Economic Science Press, Beijing

Li B (2011) Human resources road to a power nation: research report on China human-oriented development strategy [M]. Economic Science Press, Beijing

Feature 6
International Challenges Confronted by China's Environmental Problems

Xiaoming Yang

The development of China faces fundamental challenges during the first half of 21st century, including: (1) the challenge of resources and energy shortage; (2) the challenge of bottleneck of ecological development; (3) the challenge of a series of issues during economic and social development. How to treat the challenges and pressures of ecological environment for China from international point of view will play an important role in China's integration into the world and strengthening international environmental cooperation.

6.1 Major Global and Regional Environmental Issues Faced by China

In the mean time of China's economic development, the global environmental pressure for China is also intensifying with the advance of industrialization, urbanization and agricultural modernization. The global environmental pressure China needs to respond to is "double pressure": natural pressure of pure "environmental threat" and "threating the environment" caused or directly generated by global environmental issues, which is linear or unidirectional; and social pressure of "the environment threat theory" and "Theory of Big Power's Responsibility" led by or extended from global environmental issues, which is normally a surrounding type of pressure.

X. Li and J. Pan (eds.), *China Green Development Index Report 2011*,
Current Chinese Economic Report Series, DOI: 10.1007/978-3-642-31597-8_22,

6.1.1 Various Regional Environmental Issues and Complex Relationship of Trans-boundary Impact

The regional environmental issues of China can be divided into the following three types. (1) Trans-boundary environmental issues with environmental medium as carrier. The key issues of this type include trans-boundary water pollution of China–Russia, China–Kazakhstan, China–North Korea and Mekong River basin, long-distance air pollutants transfer such as acid rain and sand storm in Northeast Asia, marine pollution such as the Huanghai Sea and the Bohai Sea. (2) Environmental issues caused by product/waste trade. The issues of this type include ecological problems caused by China-Russia timber trade, ecological damage caused by road transportation of China-ASEAN Free Trade Area, environmental impact caused by export of waste from US, Japan and South Korea to China, biological invasion caused by marine transportation, etc. and (3) Environmental issues with capital flow as carrier. The key issues of this type include environmental problems caused by foreign direct investment in China and China's investment in foreign countries, the latter one has become the excuse of "China threat theory- environment" and "China threat theory-resource".

6.1.2 Increase of Resource Consumption Pressure Caused by International Trade

Although China has been improving its import and export structure, China is still at the lower end of international trade system and international product supply chain. Compared with the products imported by developed countries, the products China exports are low in added value but high in resource and energy consumption. Research shows that the embodied energy[1] behind import and export is large in terms of both absolute value and growth rate. In the meantime of trade surplus, China loses in "ecological deficit". According to calculation, from 2001 to 2006, the embodied energy of China's net export increased from 210 million tons of standard coals to 630 million tons standard coals, showing the trends of steady and rapid growth. In 2002, the total amount of embodied energy of China's net export was approximately 240 million tons standard coal, accounting for 16 % of the primary energy consumption in 2002, which mainly run to developed countries such as US and Japan with the embodied energy of net export of 75.24 million tons of standard coal and 48.94 million tons of standard coal respectively, the sum of the above two countries accounts for more than 50 % of the embodied energy of

[1] Embodied energy or embedded energy is referred to the sum of energy inputs consumed during the whole process of upstream processing, production and transportation.

China's net export. However, despite the large trade surplus, China is experiencing large biological deficit.

6.1.3 Illegal Import of Waste from US Influences China's Environment

With the development of China's economy, the demand of China for solid waste as raw materials such as waste steel is increasing gradually. China imported 40 million tons of waste steel in 2006, 3.6 times more than that of 1999. However, there are cases such as importing waste do not meet the requirements of environmental standards or illegal import of solid waste. US is the largest exporter to China and the largest waste raw material exporter to China whose overseas suppliers accounts for 20 % of total suppliers. China's inspection and quarantine departments have discovered many lots of unqualified waste during the inspection of waste raw material exported from US to China. The lots number of unqualified waste reached 60 in 2006, and ten US enterprises were disqualified to export waste raw materials to China. Two US environmental protection organizations—BAN and SVTC jointly published a long survey report named "Exporting harm: the high-tech trashing of Asia" which shows that between 50 and 80 % of the electronic wastes (E-waste) collected in US for recycling are not recycled domestically at all, but very quickly placed on container ships bound for destinations of Asia and 90 % of those waste are exported to China. Since US is not a contracting state of "Basel Convention on the Control of Trans-boundary Movements of Hazardous Wastes and Their Disposal" and is not limited by multilateral hazardous waste import and export management mechanism. Though the waste raw material containing hazardous waste (for example E-waste) from US is found, it is hard or even impossible to ship them back.

6.1.4 China-EU Environment-Cooperation in Competition

China and EU share many common interests in a number of global environmental issues. China is an important member of developing countries and EU is an important member of developed country groups. The relationship between China and EU will definitely contains the competition and cooperation of developing and developed countries. In addition, there is also competition among developing countries as well as among developed countries. Therefore, China-EU relationship in global environmental issues shows not only the cooperation and competition between developing countries and developed countries, but the cooperation between China and EU to win competition advantage in their respective interest group. The focus of the issues between China and EU is climate change. EU needs to dispute with China on undertaking the obligation of emission-reduction, and it also needs to cooperate with China to force US to undertake its obligation of emission-reduction, or cooperate with China on energy technologies to win competition advantage over other developed countries. In terms of biodiversity, EU needs to dispute with China on plunder and anti-plunder, limitation and breaking limitation concerning species resources and biotechnology, and it also needs to cooperate with China on bio-technology to win competition advantage over other developed countries, because China is one of the countries with the richest species resources. In terms of POPs and ozone layer conservation, EU needs to force China to phase out relevant products in advance in order to make its new technologies and new products win competition advantage in international market, and it also assists China to phase out relevant products, thus to accelerate the phase-out of similar products and technology in the world and win the competition advantage over other developed countries without related technologies and products. More over, EU is an important trade partner of China, so there are also environmental issues caused by product trade, i.e. a large amount of products are exported to EU, but the pollution is left in China.

6.1.5 China–Africa and China–Arabia Continuously Strengthen Cooperation on Environment

China and Africa, and China and Arabia are all developing countries who share many common interests and with main similar interests in a number of global environmental issues but this kind of relationship lies in multilateral relationship at global level, i.e. the common position and interests as developing countries. As to bilateral relationship, there are no unresolved issues on global environmental issues between China and Africa as well as Arabia. Thus we can strengthen the coordination in international position and cooperation in reciprocal programmes.

On the other hand, the ecological damage caused by China's investment in foreign countries is highly concerned by China–Africa and China–Arabia relation. China's direct investment in foreign countries has been developing rapidly in recent years, and the investment in foreign resource development is also developing quickly. For example, by the end of 2006, China had invested USD 6.27 billion in Africa and nearly a thousand enterprises set their offices in 49 countries of Africa. However, there is a problem that can not be ignored, which is the significant impact of China-invested overseas enterprises on local environment as well as the fund, technologies and development opportunities to host countries and local citizens they bring. Because of the incomplete legal system in host countries and lack of social responsibility of some enterprises, some China-invested overseas enterprises do not pay enough attention to environmental and resource protection, which invites much blame on China-invested overseas enterprises. This will be utilized, exaggerated or even distorted by a few competitors with ulterior motives to create "China Threat Theory", leading to negative impact on China's image and harmful to the long-term and smooth implementation of China's Going Global Strategy. China investment in Arab countries does not cause environmental problems or concerns in that field, but the experience and lessons of China's investment in Africa shall be learned to avoid environmental issues and adverse impact on China's image and enterprises' interest.

6.1.6 The Development in the Future will Intensify the Complexity and Toughness of International Environmental Issues

In addition to the complex international environmental issues, China set new national development strategy in the 12th *Five-Year Plan* to accelerate the transformation of economic development mode, to determine the scientific development and expanding domestic demand as the development direction for a certain period in the future. However, there is huge international environmental pressure behind these development goals. Improvement of China's living standard and consumption standard means more import and consumption of energy and resource. China's industrialization enters a new stage which may bring unpredictable impact on environment. Can the new road to industrialization be really realized? Both China's foreign trade and international development strategy in the future will bring much pressure to the future. Under this situation, we must positively respond to these pressures, take effective measures, create good international cooperation atmosphere, promote cooperation with nearby countries in Asia, US, EU, countries in Africa and Arab areas, and take all types of supporting measures to advance environmental cooperation.

6.2 Positively Respond to Global Environmental Issues

We need to accomplish the following to respond to global and regional environmental issues.

First, we need to fully understand and focus on the importance and particularity of global environmental issues, to review the global and national environmental issues from the global point of view, to recognize that China's environmental issues are also global environmental issues, the relationship between global environmental problems and China's environmental issues are bidirectional who are depend on each other. The solution of global environmental issues is helpful to resolving China's environmental problems and reducing China's environmental pressure; meanwhile, the improvement of China's environmental problems is a large contribution to the resolve of global environmental problems. Therefore, we shall take into consideration of pressure from domestic environmental issues and the pressure from global environmental issues, and our emphasis shall be changed from responding to pressure only from domestic environmental issues to paying equal attention to pressure from domestic environmental issues as well as pressure from global environmental issues.

Second, we shall strengthen our action. Responding to both natural pressure and social pressure of global environment, we shall first of all strengthen our own compressive resistance. In face of natural pressure, ecological quality shall be improved to strengthen adaptability and resilience of ecological environment. Responding to social pressure, we shall do well in putting forward measures, taking measures, conducting proper publicity and leaving no disputes. Secondly, we need to change external pressure into internal driving force. In face of natural pressure, we need to "take" and "utilize". The trans-boundary transfer of waste may lead to environmental pollution and soil damage, but the reasonable planning and utilization of waste as resource will turn waste into wealth and reduce domestic resource and energy consumption. In face of social pressure, we shall advance domestic development with outside factors, borrow force to make force and turn resistance into motives. For example, we shall accelerate domestic industrial structure adjustment and develop green economy to reduce global emission of greenhouse gas. We shall accelerate the phase-out of domestic POPs to implement environmental conventions on reduction of POPs and substance consuming ozone layer. We shall classify ecological function region, set up natural reserves and protect biological environment so as to reduce the loss of biodiversity.

Third, we shall strengthen supporting measures of advancing global environmental cooperation. (1) We shall strengthen publicity to improve China's right to speak on protecting global environment. For example, we shall hold high level and large scale international seminars, print and distribute relevant prospectus, the Ministry of Environmental Protection shall establish close and periodical information exchange mechanism to publicize the efforts and achievement of China on global environmental protection and to fight for right to speak on international environment. (2) We shall deepen researches. In face of natural pressure, we shall strengthen resistance through hi-tech. In face of to social pressure, we shall deepen the analysis on the impacts of global environmental issues on China so as to study the interest of policies and conduct strategic research. (3) We shall continuously improve the ability of people conducting global environmental issues research and decision-makers through training and seminars, thus to enable these people to transit from participation in negotiation of international environmental rules to actively stipulate green rules, to respond to external pressure with internal motive, such as drawing up convention articles and submit negotiation proposal. (4) We shall continuously complete mechanisms such as research mechanism, negotiation mechanism, convention implementation mechanism, funding mechanism, international cooperation mechanism and related mechanisms concerning responding to global environmental problems. (5) We shall establish teams. Western developed countries including US all have a strong team studying global environmental issues to provide powerful technical support to negotiation. China is still weak in research base with research people dispersed in some research institutes and universities and no joint force formed. It is suggested that a special research organization be set up under Ministry of Environmental Protection to establish a strong team specialized in research of international environmental issues.

Appendix A
The Ideas, Structure and Indicators of China Green Development Index System and Measurement Methods

Based on extensive comparison and analysis, we built a China green development index system composed of three first-grade indicators, nine second-grade indicators and 55 third-grade indicators in line with China's reality. In the following part, we will also introduce the measurement methods briefly.

A.1 Ideas of China Green Development Index System

The China Green Development Index System report focuses on the following aspects. First, the report emphasizes on environment-friendly development. It shows how far we have gone in green industrial development, environment and resource protection, and the government support in this field. This is also why we have so many interdisciplinary experts in our study. Second, the report highlights the comparison of provinces, cities, autonomous regions in green development. It compares 30 provinces, cities and autonomous regions (except for Tibetan autonomous regions for data deficiency) according to their performance in striking a balance between economic expansion and environmental protection. This comparison method encourages different areas to learn from each other and improve their conditions based on realities. For some reasons, green development in Hong Kong Special Administrative Region, Macao Special Administrative Region and China Taiwan are not included in this report, but we're looking forward to cooperate with experts in these areas in the future. Third, the report underlines government guidance in green development. Government guidance, scientific support and public involvement are three driving force in green development, among which the first one is the most important. In China,

This appendix is from the pandect of China Green Development Index Annual Report 2010, whose essence is in line with China Green Development Index Annual Report 2011.

X. Li and J. Pan (eds.), *China Green Development Index Report 2011*,
Current Chinese Economic Report Series, DOI: 10.1007/978-3-642-31597-8,

government plays an important role in social and economic development. As a result, we want to highlight the evaluations on government performance in green development to urge their progress in this field. We're also looking forward to see more government guidance in the collaboration between green industry and green products study institutions. Fourth, this report accents the importance of green production. In America, the National Geographic Society pays more attention to the environmental influences caused by consumer consumptions, while based on China's reality, we concerned more on production, especially industrial production. Views on green consumption are also shared in this report. Fifth, we strongly valued the openness and authority of data sources. Data in this report all come from the published yearbooks and released data from authorities. The data sources are mainly from China Statistical Yearbook 2009, China Statistical Yearbook on Environment 2009, China City Statistical Yearbook 2009, China Statistical Yearbook on Energy 2009, China Industry Economy Statistical Yearbook 2009, China Statistical Yearbook on Environment 2008, China Statistical Abstract 2010 and China's 60 Years Statistical Collection.

A.2 The Structures of First and Second-Class Indicators of Green Development Index System

The following paragraphs elaborate on the structure and content of the system.

First, let's reveal how the first-class indicators are reached.

As shown in the Fig. A.1, we examine green development in three aspects—green degree of economic growth, carrying capacity potential of natural resources and environment and support degree of government policies—which reflect how the economic growth influences resources as well as environment, the potential carrying capacity of the natural resources and how well the governments balance the resources, environment and economic growth at the same time respectively.

Then why we categorize the indicators in this way? The reasons are obvious. First, our goal to prepare this report is to underline the situation in these three areas in China. Second, the three categories reflect three aspects, namely, the current situation of green development, pressures we faced with and responses made by governments respectively. Third, they are generated from third-class indicators following the three aspects we try to emphasize in green development. Last but not the least, the category we made has undergone heated discussions of experts from different fields. We tried to classify the indicators into four, five and six categories at first; however, we decided to make it three to stress our characteristics.

To sum up, the category, embodied the principal of "one body with two forces", which means green economic growth as a body with resource and environment as driving force and government policy as guiding force, has laid a solid foundation for green development.

Second, let's take a look at the second-class indicators (Fig. A.2).

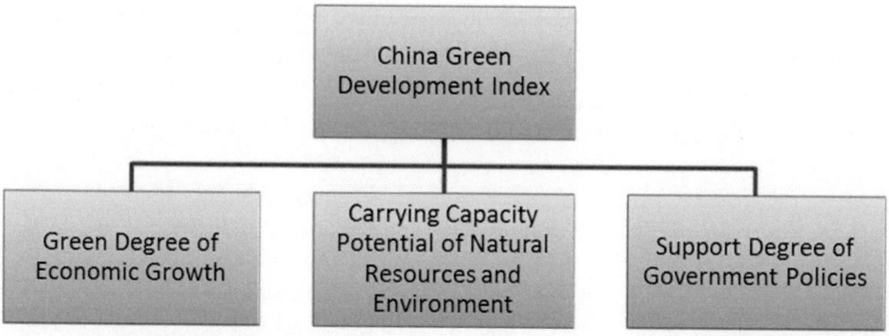

Fig. A.1 The structure of first-class indicators of China green development index system

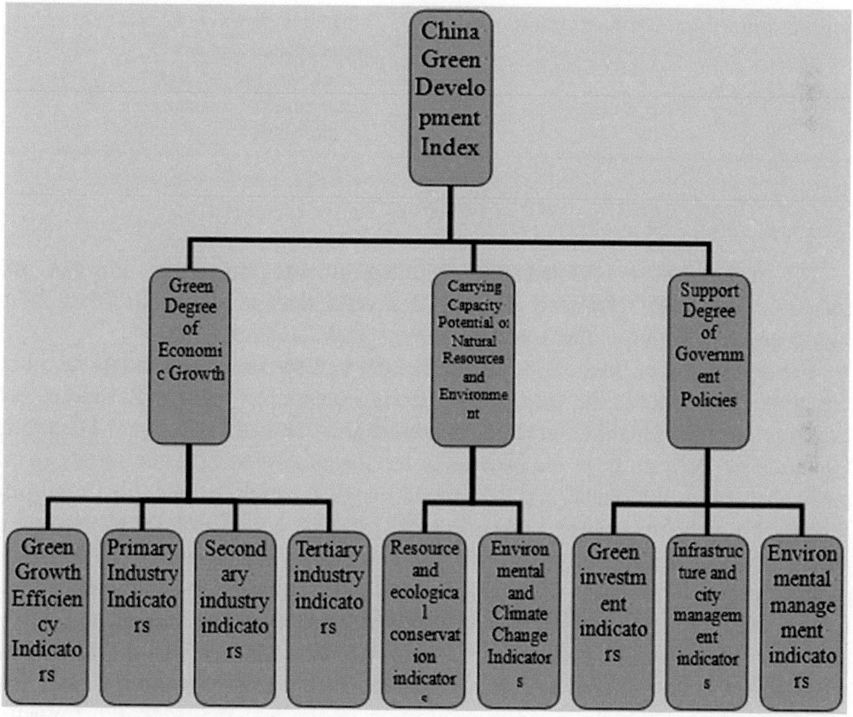

Fig. A.2 The structure of second-class indicators of China green development index system

Under the first-class structure, there are nine second-class indicators which are determined on the basis of both first-class and third-class indicators. Specifically, we first classified third-class indicators following the first-class structure and then further generated them into nine second-class ones by their characteristics.

Now, let's take a closer look at the first and second-class indicators and their proportions (Table A.1).

Table A.1 The first and second-class indicators and their proportions (Unit: %)

First-class indicators	Proportion (%)	Second-class indicators	Proportion (%)
Green degree of economic growth	30	Green growth efficiency indicators	40
		Primary industry indicators	10
		Secondary industry indicators	35
		Tertiary industry indicators	15
Carrying capacity potential of natural resources and environment	45	Resource and ecological conservation indicators	20
		Environmental and climate change indicators	80
Support degree of government policies	25	Green investment indicators	40
		Infrastructure and city management indicators	30
		Environmental management indicators	30

The proportion is determined according to the indicators' impacts and functions, and also following advices of several dozens of experts from fields like economy, resource, environment, energy and statistics.

The proportion of three first-class indicators is made on four foundations. First, the proportion order is the Carrying Capacity Potential of Natural Resources and Environment, the Green Degree of Economic Growth and the Support Degree of Government Policies from the highest to the lowest, as the first one serves as the basis of green development and the second one as its goal. Second, the proportion is relative so it's fine to take 5 or 10 % as its smallest unit. Third, the proportion is made due to that of the third-class ones, that is to say, the proportion of first-class indicators can be roughly estimated by calculating that of the third-class ones. Fourth, the latter two indicators are highly related to the development level of provinces; however, the indicator that weighted the most is related to the natural resource endowment, which is relatively high in the less developed central and western parts of the country. The proportion of the first indicator, the Carrying Capacity Potential of Natural Resources and Environment, is relatively high, as it emphasizes the balance between economic growth and natural resources and environment.

The proportion of second-class indicators is made in the same way as that of first-class ones. The total proportion of second-class indicators under one first-class indicator is 100 %. We can see that under the Green Degree of Economic Growth indicator, the Green growth efficiency indicators and Secondary industry indicators weight 40 and 35 % respectively, showing the key of green development is to use natural resources and environment efficiently in the process of industrialization and the total social input and output. Under the

Carrying Capacity Potential of Natural Resources and Environment, the environmental and climate change indicators weight 80 %, reflecting the public concern about waste, waste water and waste gas emission problems. It is quite encouraging that under the Support Degree of Government Policies, the green investment indicators weight a little higher than the other two, revealing the governments' efforts in striving the balance between economic growth and resources and environment. The rational weight of the indicators manifests the experts' consensus on China's reality.

A.3 Selection and Assortment of Third-Class Indicators

Third-class indicators of green development index constitute the foundation of our research and this section will elaborate on how these indicators are selected and what the selection standards are. Table A.2 shows the green development index system.

To start with, we need to explain how the third-class indicators, 55 altogether, are selected. The Green Index Information Group (GIG), specially formed for this research, is designated to perform the task. GIG initially collected 1,458 indicators after having a thorough study of relevant literatures home and abroad. GIG then cooperated with experts in this field and selected the final 55 third-class indicators that fit the reality in China the most.

Then we need to make sure what the selection standards are. Our researchers take the following six standards into consideration. First, indicators to be selected must be closely connected with one of the three first-class indicators (green degree of economic growth, carrying capacity potential of natural resources and environment, support degree of government policies) of green development index at least, which would be helpful for the selection of second-class indicators. Second, data needed to evaluate the indicators must be available. Our research finds that not all provinces record the data needed to measure relevant indicators. Furthermore, the data used must have continuous availability. Third, it should be clear-cut if an indicator is direct or inverse. The reason why we use the indicator of "non-fossil fuels" rather than that of water power, wind power or nuclear power is that each type of electric power has its advantages and disadvantages, so it's rather vague to determine a specific indicator of electric power is direct or inverse. Indicators like new energy and area of water and soil conservation are not included in the 55 third-class indicators because of noticeable regional differences. Fourth, our indicator evaluation depends on relevant data of current year rather than the rates of change from 2007 to 2008, for the latter are instable and the calculation is complicated. Fifth, the selected indicator should be typical in a series of similar indicators. For example, among various forms of electricity generation, we just evaluate the coal consumption of thermal power for which supplies the largest proportion of power and its coal consumption is a universal problems in the provinces. Sixth, we'd be cautious that the same indicator might bear different

Table A.2 Green development index system

First-class indicators	Second-class indicators	Third-class indicators	
Green degree of economic growth	Green growth efficiency indicators	1. GDP per capita	5. SO_2 emissions per unit of GRP
		2. Energy Consumption per unit of GDP	6. COD emissions per unit of GRP
		3. Ratio of non-fossil energy consumption to total energy consumption	7. NOx emissions per unit of GRP
		4. CO_2 emissions per unit of GRP	8. Ammonia nitrogen emissions per unit of GRP
			9. Industrial solid wastes discharged per unit of GRP
	Primary industry indicators	10. Labour productivity of primary sector	11. Output ratio of land
	Secondary industry indicators	12. Labour productivity of secondary sector	15. Ratio of industrial solid wastes utilized
		13. Water consumption per unit of industrial value added	16. Reuse rate of industrial water
		14. Energy consumption per unit of industrial value added at a cut-off level	17. Ratio of output value of high energy-bearing industrial sectors to gross industrial output value
			18. Coal consumption of thermal power generating units
	Tertiary industry indicators	19. Labour productivity of tertiary sector	21. Proportion of employees of tertiary sector in total employees
		20. Proportion of value added of tertiary sector in GRP	
Carrying capacity potential of natural resources and environment	Resource and ecological indicators	22. Water resources per capita	24. Forest coverage rate
		23. Forest area per capita	25. Proportion of area of natural reserves in total area of a region
	Environmental and climate change indicators	26. CO_2 emissions per unit of land area	34. Ammonia nitrogen emissions per unit of land area
		27. CO_2 emissions per capita	35. Ammonia nitrogen emissions per capita

(continued)

Table A.2 (continued)

First-class indicators	Second-class indicators	Third-class indicators	
		28. SO_2 emissions per unit of land area	36. Industrial solid wastes discharged per unit of land area
		29. SO_2 emissions per capita	37. Industrial solid wastes discharged per capita
		30. COD emissions per unit of land area	38. Consumption of chemical fertilizers per unit of area of cultivated land
		31. COD emissions per capita	39. Consumption of pesticides per unit of area of cultivated land
		32. Nitrogen oxides emissions per unit of land area	
		33. Nitrogen oxides emissions per capita	
Support degree of government policies	Green investment indicators	40. Ratio of environmental protection expenditures to government expenditures	43. Investment in converting cultivated land into forests and grassland per unit of area of cultivated land
		41. Ratio of investment in the treatment of environmental pollution to GRP	44. Ratio of expenditures for science and technology, education, culture, and medical and health care to government expenditures
		42. Per capita investment of water sanitation and toilet improvement in rural areas	
	Infrastructure and city management indicators	45. Area of green land per capita in urban areas	48. Ratio of urban consumption wastes treated
		46. Coverage rate of urban population with access to tap water	49. Public transportation vehicles per 10000 urban population
		47. Treatment rate of urban waste water	
	Environmental management indicators	50. Ecological restoration rate of the mineral area	53. Removing rate of industrial COD
		51. Area of afforestation per capita	54.Removing rate of industrial NOx emissions
		52. Removing rate of industrial SO_2 emissions	55. Removing rate of industrial NOx emissions

implications to various provinces. Take the indicator of "emission" as an example. When evaluating per capita emission of something, the results might be quite different between provinces with large population and those with small population. The same situation applies to the evaluation of per unit area emission and proportion of emission in per capita GDP. Therefore, to make sure the reasonability of this evaluation system, we evaluate the indicator of "emission" from three aspects: per capita emission, per unit area emission, and the proportion of emission in per capita GDP.

GIG has spared no efforts in indicators selection, but there is still room for improvement resulting from the four limitations. First, some indicators, though important for assessing green development, can not be evaluated for the lack of relevant data. Second, some significant indicators are evaluated by using surrogate data. Third, GDP adjustment poses a challenge for our work, but we have strived to resolve the problem with the help of statistics experts. Fourth, most of the data for evaluating indicators of green development in Tibet are not available, so this research just covers 30 provinces in China.

It should be pointed out that GIG also selected about 30 "adjusting indicator" to correct singular value. Some adjusting indicators can give bonus point to a province, but others may take points away. For the objectiveness of this study, we didn't use adjusting indicators.

For more detailed information about second-class indicators and third-class indicators, please refer to Chaps. 5, 10, and 14 and appendix of this report.

4 Measurement of the Green Development Index

The green development index is the core criterion in our measurement. This report has all the index indicators classified into three classes. As the first-class indicator, the combined green development index is a relative number derived from the evaluation of all the other indicators and its value is calculated in a scientific way based on the indicators' standardized values. In this report, we compare the green development levels of areas with the average, and then work out the green development index to measure their overall green economic growth.

The consistence of the indicators is a focus of our work. As a combined evaluation index, the green development index demands effective combining of different quantized indicators. Therefore, the related data is firstly collected and identified, and then positively-managed and standardized. In the 55 indicators, 30 are positively correlated with the green development index. These positive indicators need no standardization. The other 25 are negatively-correlated, called negative indicators. To combine the positive and negative indicators, we have to positively manage the negative ones through the reciprocal and complement methods. In the 25 negative indicators, the ratio of output value of high energy-bearing industrial sectors to gross industrial output value is obtained in the complement method; the industrial solid wastes discharge for per-unit acreage and the per-capita industrial solid wastes discharge are processed in the maximum method; the rest are positively-managed in the reciprocal method.

As the measurement units of the indicators differ, the dimension influences have to be eliminated before combining. Currently, the main standardization methods are the maximum and minimum method and the standard deviation method. Since great regional disparities exist in development level and resources, especially in per capita water resource, a maximum and minimum measurement may resulted to zero contribution for some regions and thus harms the scientific of the evaluation. Besides, results obtained in the maximum and minimum method are usually above normal values, that is to say, they are not reliable enough. This is why we have chosen the standard deviation method to mitigate the differences and enhance the stability of the measurement. The standard deviation method has been designed to evaluate the gap between the development level of a certain region and the average level. If the regional level surpasses the average, the value will be positive; if the regional level is below the average, the value will be negative. The larger the gap is, the greater the absolute value will be.

The indicators in this report are relative indicators. Among them, the 55 third-class indicators are classified into two categories: indicators showing ratios and indicators showing constructions. The 36 indicators showing ratios compare specific statistic to a related common value (such as the GDP, population and acreage), through which the influences from regional disparities can be eliminated. They mainly cover the evaluation of green development efficiency, environment and climate changes. On the other hand, the other 19 reflect the constructions of certain fields. They are mainly used for three purposes: to show the balance of industry and energy composition, such as the ratio of non-fossil energy consumption to total energy consumption; to reflect the environment and resource states and influences to the environment by industries, such as the forest coverage; to evaluate the government efforts in environment protection, such as the ratio of environmental protection expenditures to government expenditures.

Gross indicators and speed indicators are not adopted in consideration of the fairness and stability of the evaluation. To make the measurement impartial and objective, we have adopted practical ways to evaluate the green development level of those provinces lack of related data. To be prudent enough, we have kept very detailed record and illustration. With all the work mentioned above, we have come out with the final green development index of various areas and the other three sub-indexes.

To ensure the fairness and objectiveness of the measurement, all the indicators are consistent with the systems of the National Bureau of Statistics of China.

Appendix B
Explanatory Notes and Data Sources on Indicators Used to Calculate the Province Green Development Index in 2011

1. **GDP per capita**

Gross Domestic Product (GDP) refers to the final products at market prices produced by all resident units in a country (or a region) during a certain period of time. For a region, it is called as Gross Regional Product (GRP) or regional GDP. Its calculation formula is as follows:

$$\text{GDP per capita} = 2 \times \text{GRP}/$$
$$(\text{Population at the end of this year} + \text{ Population at the end of preceding year})$$

Data sources: *China City Statistical Yearbook 2010*, NBS, Beijing, China Statistical Publishing House, March 2011.

2. **Energy consumption per unit of GDP**

Total energy consumption refers to the total consumption of energy of various kinds by the production sectors and the households in the country in a given period. It can be divided into three parts: end-use energy consumption; loss during the process of energy conversion; and energy loss.

Energy consumption per unit of GRP refers to the energy consumption per unit of Gross Regional Product in a region in the same reference period, reflecting the increased energy consumption with an additional unit of GRP. The formula is:

$$\text{Energy consumption per unit of GDP} = \text{Total energy consumption}/\text{GRP}$$

Data sources: *China City Statistical Yearbook 2010*, NBS, Beijing, China Statistical Publishing House, March 2011.

3. **Ratio of non-fossil energy consumption to total energy consumption**

Non-fossil energy refers to other energy sources other than coal, oil and natural gas. The ratio of non-fossil energy consumption to total energy consumption refers

X. Li and J. Pan (eds.), *China Green Development Index Report 2011*, Current Chinese Economic Report Series, DOI: 10.1007/978-3-642-31597-8,

to the percentage of non-fossil energy consumption in total energy consumption. The formula is:

Ratio of non-fossil energy consumption to total energy consumption = 100 % × Non-fossil energy consumption/Total energy consumption

Data sources: *China Statistical Yearbook*, NBS, Beijing, China Statistical Publishing House.

4. **CO_2 emissions per unit of GRP**

CO_2 emissions per unit of GRP refer to the ratio of carbon dioxide emissions to GRP in a region in the same reference period. The formula is:

$$CO_2\text{ emissions per unit of GRP} = CO_2\text{ emissions}/\text{GRP}$$

5. **SO_2 emissions per unit of GRP**

Sulphur dioxide (SO_2) emissions cover SO_2 emissions through industrial activities, non-industrial and other activities, of which emissions through industrial activities refer to volume of sulphur dioxide discharged from fuel burning and production process by enterprises during a given period. The formula is:

$$\text{Industrial } SO_2\text{ emissions} = SO_2\text{ discharged from the process of fuel burning} + SO_2\text{ discharged from the process of production}$$

Emission through non-industrial and other activities are calculated on the basis of consumption of coal by households and other activities and the sulphur content of coal with the following formula:

$$\text{Emission through non-industrial and other activities} = \text{coals consumed by household and other activities} \times \text{Sulphur content} \times 0.8 \times 2$$

SO_2 emissions per unit of GRP refer to the ratio of SO_2 emissions to GRP in a region in a given period. The formula is:

$$SO_2\text{ emissions per unit of GRP} = SO_2\text{ emissions}/\text{GRP}$$

Data sources: *China Statistical Yearbook 2010*, NBS, Beijing, China Statistical Publishing House, September 2010; *China Environmental Statistical Yearbook 2009*, the Ministry of Environmental Protection, Beijing, China Environmental Science Publishing House, December 2010.

6. **COD emissions per unit of GRP**

Chemical Oxygen Demand (COD)refers to the amount of oxygen required when chemical oxidants are used to oxidize organic pollutants in water. A higher value of COD corresponds to more serious pollution. COD emissions mainly come from waste water discharged by industry and household, of which COD emissions of the latter refer to the annual amount of COD in the waste water discharged by urban households, which can be calculated by per capita coefficient as follows:

COD emissions in waste water discharged by urban households = Coefficient of COD generated through urban non-industrial waste water × Urban non-agricultural population × 365

$$\text{COD emissions per unit of GRP} = \text{COD}/\text{GRP}$$

Data sources: *China Statistical Yearbook 2010*, NBS, Beijing, China Statistical Publishing House, September 2010; *China Environmental Statistical Yearbook 2009*, the Ministry of Environmental Protection, Beijing, China Environmental Science Press, December 2010.

7. **NOx emissions per unit of GRP**

Nitrogen oxides (NO_x) emissions per unit of GRP refer to the ratio of NO_x emissions to GRP in a region in a given period. The formula is:

$$NO_x \text{ emissions per unit of GRP} = NO_x \text{ emissions} / \text{GRP}$$

Data sources: *China Statistical Yearbook 2010*, NBS, Beijing, China Statistical Publishing House,September 2010; *China Environmental Statistical Yearbook 2009*, the Ministry of Environmental Protection, Beijing, China Environmental Science Press, December 2010.

8. **Ammonia nitrogen emissions per unit of GRP**

Ammonia nitrogen emissions per unit of GRP refer to the ratio of ammonia nitrogen emissions to GRP in a region in a given period. The formula is:

$$\text{Ammonia nitrogen emissions per unit of GRP} \\ = \text{Ammonia nitrogen emissions} / \text{GRP}$$

Data sources: *China Statistical Yearbook 2010*, NBS, Beijing, China Statistical Publishing House, September 2010; *China Environmental Statistical Yearbook 2009*, the Ministry of Environmental Protection, Beijing, China Environmental Science Press, December 2010.

9. **Electricity consumption per capita in urban areas**

The formula is:

$$\text{Electricity consumption per capita in urban areas} = 2 \times \text{The amount of electricity consumption in urban areas} / (\text{Urban population at the end of this year} + \text{Urban population at the end of preceding year})$$

Data sources: *China City Statistical Yearbook 2010*, NBS, Beijing, China Statistical Publishing House, March 2011; *China Statistical Yearbook 2010*, NBS, Beijing, China Statistical Publishing House,September 2010.

10. **Labour productivity of primary sector**

Labour productivity of primary sector refers to the ratio of value added of primary sector to annual average of employees in primary sector. The formula is:

$$\text{Labour productivity of primary sector} = 2 \times \text{Value added of primary sector}/(\text{Employees in primary sector at the end of this year} + \text{Employees in primary sector at the end of preceding year})$$

Data sources: *China Statistical Yearbook 2010*, NBS, Beijing, China Statistical Publishing House, September 2010.

11. **Output ratio of land**

Output ratio of land refers to the output value of crop farming to the sown area of grain crops. The formula is:

$$\text{Output ratio of land} = \text{Output value of agriculture}/\text{Sown area of grain crops}$$

Data sources: *China Statistical Yearbook 2010*, NBS, Beijing, China Statistical Publishing House, September 2010.

12. **Proportion of water-saving irrigated area in effectively irrigated area**

Irrigated area refers to area of land that are effectively irrigated, i.e. relatively level land, where there are water sources or complete sets of irrigation facilities to lift and move adequate water for irrigation purpose under normal conditions. Under normal situations, irrigated area is the sum of watered fields and irrigated fields where irrigation systems or equipment have been installed for regular irrigation purpose. This important indicator reflects drought resistance capacity of the cultivated land in China. The formula is:

$$\text{Proportion of water-saving irrigated area in effectively irrigated area} = 100\,\% \times \text{Water-saving area}/\text{Effectively irrigated area}$$

Data sources: *China Rural Statistical Yearbook 2010*, NBS, Beijing, China Statistical Publishing House, November 2010; *China City Statistical Yearbook 2010*, NBS, Beijing, China Statistical Publishing House, March 2011.

13. **Proportion of effectively irrigated area in area of cultivated land**

Irrigated area refers to area of land that are effectively irrigated, i.e. relatively level land, where there are water sources or complete sets of irrigation facilities to lift and move adequate water for irrigation purpose under normal conditions. Under normal situations, irrigated area is the sum of watered fields and irrigated fields where irrigation systems or equipment have been installed for regular irrigation purpose. This important indicator reflects drought resistance capacity of the cultivated land in China.

Area of cultivated land refers to refers to area of land reclaimed for the regular cultivation of various farm crops, including crop-cover land, fallow, newly reclaimed land and land laid idle for less than 3 years.

Data sources: *China Statistical Yearbook 2010*, NBS, Beijing, China Statistical Publishing House, September 2010.

14. **Labour productivity of secondary sector**

Labour productivity of secondary sector refers to the ratio of value added of secondary sector to annual average of employees in secondary sector. The formula is:

$$\text{Labour productivity of secondary sector} = 2 \times \text{Value added of secondary sector}/(\text{Employees in secondary sector at the end of this year} + \text{Employees in secondary sector at the end of preceding year})$$

Data sources: *China Statistical Yearbook 2010*, NBS, Beijing, China Statistical Publishing House, September 2010.

15. **Water consumption per unit of industrial value added**

Value added of industry refers to the final results of industrial production of industrial enterprises in money terms during the reference period.

Water consumed by industry refers to water consumed by industrial enterprises for manufacturing, processing, cooling, air-conditioning, purifying, rinsing and so on in process of production, which is measured in terms of new withdrawals of water, excluding reuse of water within the same enterprises.

Water consumption per unit of industrial value added refers to the ratio of water consumed by industry to its value added in a given period. The formula is:

$$\text{Water consumption per unit of industrial value added} = \text{Water consumed by industry}/\text{Its value added}$$

Data sources: *China Statistical Yearbook 2010*, NBS, Beijing, China Statistical Publishing House, September 2010.

16. **Energy consumption per unit of industrial value added at a cut-off level**

Energy consumption per unit of industrial value added at a cut-off level refers to the ratio of energy consumed by industrial enterprises at a cut-off level to their value added in a region in a given period. The formula is:

$$\text{Energy consumption per unit of industrial value added at a cut-off level} = \text{Energy consumed by industrial enterprises at a cut-off level}/\text{Their value added}$$

Data sources: *China Statistical Yearbook 2010*, NBS, Beijing, China Statistical Publishing House, September 2010.

17. **Ratio of industrial solid wastes utilized**

Ratio of industrial solid wastes utilized refers to the percentage of industrial solid wastes utilized over industrial solid wastes produced (including stocks of the previous years). Its calculation formula is:

$$\text{Ratio of industrial solid wastes utilized} = 100\,\% \times \text{Industrial solid wastes utilized}/(\text{Industrial solid wastes produced} + \text{Stocks of previous utilized})$$

Of which, industrial solid wastes produced refers to total volume of solid, semi-solid and high concentration liquid residues produced by industrial enterprises from the process of production in a given period, including hazardous wastes, slag and ash, gangue, tailings, radioactive residues and other wastes, but excluding stones stripped or dug out in mining (gangue and acid or alkaline stones not covered). A stone is acid or alkaline if the pH value of the water is below 4 or above 10.5, when it is in, or soaked by, the water; Industrial solid wastes utilized refers to volume of solid wastes from which useful materials may be extracted or which can be converted into usable resources, energy or other materials by means of reclamation, processing, recycling and exchange (including utilizing the stocks of industrial solid wastes of the previous years in the years in question, e.g. fertilizers, building materials and road materials. The information concerned shall be collected by the producing units of wastes.)

Data sources: *China Environmental Statistical Yearbook 2010*, the NBS and the Ministry of Environmental Protection, Beijing, China Statistical Publishing House, November 2011.

18. **Reuse rate of industrial water**

Reuse rate of industrial water refers to the ratio of the amount of water re-used in industrial production to total amount of water used in a given period. The formula is:

$$\text{Reuse rate of industrial water} = 100\,\% \times \text{Reuse of water}/(\text{New withdrawals of water} + \text{reuse of water})$$

Data sources: *China Environmental Statistical Yearbook 2010*, the NBS and the Ministry of Environmental Protection, Beijing, China Statistical Publishing House, November 2011.

19. **Ratio of output value of six high energy-bearing industrial sectors to gross industrial output value**

Ratio of output value of six high energy-bearing industrial sectors to gross industrial output value refers to the percentage of output value of six high energy-bearing industries over gross industrial output value.

Gross industrial output value is the total volume of final industrial products produced and industrial services provided in a given period. It reflects the total achievements and overall scale of industrial production in a given period.

Output value of six high energy-bearing industrial sectors refers to the sum of output value of Processing of Petroleum, Coking, Processing of Nuclear Fuel, Manufacture of Raw Chemical Materials and Chemical Products, Manufacture of Non-metallic Mineral Products, Smelting and Pressing of Ferrous Metals, Smelting and Pressing of Non-ferrous Metals, Production and Supply of Electric Power and Heat Power.

Data sources: *China Industry Economy Statistical Yearbook 2009*, NBS, Beijing, China Statistical Publishing House, May 2010.

20. **Labour productivity of tertiary sector**

Labour productivity of tertiary sector refers to the ratio of value added of tertiary sector to annual average of employees in secondary sector. The formula is:

$$\text{Labour productivity of tertiary sector} = 2 \times \text{Value added of tertiary sector}/$$
$$(\text{Employees in tertiary sector at the end of this year} +$$
$$\text{Employees in tertiary sector at the end of preceding year})$$

Data sources: *China Statistical Yearbook 2010*, NBS, Beijing, China Statistical Publishing House, September 2010.

21. **Proportion of value added of tertiary sector in GRP**

Proportion of value added of tertiary sector in GRP refers to the percentage of value added of tertiary sector over GRP in a reporting period. The formula is:

$$\begin{aligned}&\text{Proportion of value added of tertiary sector}\\&\quad = 100\,\% \times \text{Value added of tertiary sector}/\ \text{GRP}\end{aligned}$$

Data sources: *China Statistical Yearbook 2010*, NBS, Beijing, China Statistical Publishing House, September 2010.

22. **Proportion of employees of tertiary sector in total employees**

Proportion of employees of tertiary sector in total employees refers to the percentage of employees of tertiary sector over total employees in a reporting period. The formula is:

$$\begin{aligned}&\text{Proportion of employees of tertiary sector} = 100\,\% \times\\&\quad \text{Employees of tertiary sector}/\ \text{Total employees}\end{aligned}$$

Data sources: *China Statistical Yearbook 2010*, NBS, Beijing, China Statistical Publishing House, September 2010.

23. **Water resources per capita**

Total Water Resources refers to total volume of water resources measured as run-off for surface water from rainfall and recharge for groundwater in a given area, excluding transit water.

Water resources per capita refer to run-off for surface water from rainfall and recharge for groundwater shared by each person in a region in a given period. The formula is:

$$\text{Water resources per capita} = \text{Total water resources}/\text{Year-end population}$$

Data sources: *China Statistical Yearbook 2010*, NBS, Beijing, China Statistical Publishing House, September 2010.

24. **Forest area per capita**

Forest Area refers to the area of forest where trees and bamboo grow with canopy density above 0.2, including land of natural woods and planted woods, but excluding bush land and thin forest land.

Forest area per capita refer to forest area shared by each person in a region in a given period. The formula is:

$$\text{Forest area per capita} = \text{Total area of forest}/\text{Year-end population}$$

Data sources: *China Statistical Yearbook 2010*, NBS, Beijing, China Statistical Publishing House, September 2010.

25. **Forest coverage rate**

Forest coverage rate refers to the ratio of area of afforested land to total land area. It is a very important indicator that reflects the status of abundance of forest resource and balance of the ecosystem. Forest area includes the area of trees and bamboo grow with canopy density above 0.2, the area of shrubby tree according to regulations of the government, the area of forest land inside farm land and the area of trees planted by the side of villages, farm houses and along roads and rivers. It is a very important indicator that reflects the status of abundance of forest resource and ecosystem balance. The formula for calculating forest coverage rate is as follows:

$$\text{Forest coverage rate} = 100\,\% \times \text{Area of afforested land}/\text{Area of total land} + 100\,\% \times \text{Area of shrubby tree/Area of total land} + 100\,\% \times \text{Area of forest land inside farm land/Area of total land} + 100\,\% \times \text{Area of tree planted by the side of villages, farm houses and along roads and rivers/Area of total land}$$

Data sources: *China Statistical Yearbook 2010*, NBS, Beijing, China Statistical Publishing House, September 2010.

26. **Proportion of area of natural reserves in total area of a region**

Natural reserves refer to certain areas of land, waters or sea that are representative in natural ecological systems or are natural habitants for rare or endangered wild animals or plants, or water conservation zones, or the location of important natural or historical relics, which are demarked by law and put under special protection and management. Nature reserves are designated by the formal approval of governments at and above county level (covering those still existing ones approved by related ministries or Revolutionary Commissions prior to the Sixth Five-Year Plan). Scenic spots and cultural preservation zones are not covered.

Proportion of area of natural reserves in total area of a region can be calculated as follows:

$$\text{Proportion of area of natural reserves in total area of a region} = 100\,\% \times \text{Area of natural reserves/Total area of a region}$$

Data sources: *China Statistical Yearbook 2010*, NBS, Beijing, China Statistical Publishing House, September 2010.

27. **Proportion of area of wetlands in total area of a region**

Wetlands refer to marshland and peat bog, whether natural or man-made, permanent or temporary; water covered areas, whether stagnant or flowing, with fresh or semi-fresh or salty water that is less than 6 m deep at low tide; as well as coral beach, weed beach, mud beach, mangrove, river outlet, rivers, fresh-water marshland, marshland forests, lakes, salty bog and salt lakes along the coastal areas. Its calculating formula is:

$$\text{Proportion of area of wetlands in total area of a region} = 100\,\% \times \text{Area of wetlands /Total area of a region}$$

Data sources: *China Statistical Yearbook 2010*, NBS, Beijing, China Statistical Publishing House, September 2010.

28. **Total standing stock volume per capita**

Total standing stock volume refers to the total stock volume of trees growing in land, including trees in forest, trees in sparse forest, scattered trees and trees planted by the side of villages, farm houses and along roads and rivers. Its calculating formula is:

$$\text{Total standing stock volume per capita} = \text{Total standing stock volume/Year-end population}$$

Data sources: *China Statistical Yearbook 2010*, NBS, Beijing, China Statistical Publishing House, September 2010.

29. **CO_2 emissions per unit of land area**

Area under land survey refers to the total area of land, under the land survey, within the jurisdiction of the administrative region, including land for agriculture use, land for construction and unused land.

CO_2 emissions per unit of land area in a region can be calculated as follows:

$$CO_2 \text{ emissions per unit of land area} = CO_2 \text{ emissions} / \text{Area under land survey}$$

30. **CO_2 emissions per capita**

CO_2 emissions per capita can be calculated as follows:

$$CO_2 \text{ emissions per capita} = \text{Total } CO_2 \text{emissions} / \text{Mid-year population}$$

31. **SO_2 emissions per unit of land area**

SO_2 emissions per unit of land area in a region can be calculated as follows:

$$SO_2 \text{ emissions per unit of land area} = CO_2 \text{ emissions} / (\text{Area under land survey-Area of desert and gobi})$$

Data sources: *China Environmental Statistical Yearbook 2010*, the NBS and the Ministry of Environmental Protection, Beijing, China Statistical Publishing House, November 2011; Wuzheng, *Desertification and Its Control in China,* Beijing, The Commercial Press, January 1995.

32. **SO_2 emissions per capita**

SO_2 emissions per capita in a region can be calculated as follows:

$$SO_2 \text{ emissions per capita} = \text{Total } SO_2 \text{ emissions} / \text{Mid-year population}$$

Data sources: *China Statistical Yearbook 2010*, NBS, Beijing, China Statistical Publishing House, September 2010.

33. **COD emissions per unit of land area**

COD emissions per unit of land area in a region can be calculated as follows:

$$\text{COD emissions per unit of land area} = \text{COD emissions} / (\text{Area under land survey})$$

Data sources: *China Statistical Yearbook 2010*, NBS, Beijing, China Statistical Publishing House, September 2010.

34. **COD emissions per capita**

COD emissions per capita in a region can be calculated as follows:

COD emissions per capita = Total COD emissions /Mid-year population

Data sources: *China Statistical Yearbook 2010*, NBS, Beijing, China Statistical Publishing House, September 2010.

35. **Nitrogen oxides emissions per unit of land area**

Nitrogen oxides emissions per unit of land area in a region can be calculated as follows:

Nitrogen oxides emissions per unit of land area = Nitrogen oxides emissions / (Area under land survey)

Data sources: *China Environmental Statistical Yearbook 2009*, the Ministry of Environmental Protection, Beijing, China Environmental Science Press, December 2010; *China City Statistical Yearbook 2010*, NBS, Beijing, China Statistical Publishing House, March 2011.

36. **Nitrogen oxides emissions per capita**

Nitrogen oxides emissions per capita in a region can be calculated as follows:

Nitrogen oxides emissions per capita = Nitrogen oxides emissions / Mid -year population

Data sources: *China Environmental Statistical Yearbook 2009*, the Ministry of Environmental Protection, Beijing, China Environmental Science Press, December 2010; *China City Statistical Yearbook 2010*, NBS, Beijing, China Statistical Publishing House, March 2011.

37. **Ammonia nitrogen emissions per unit of land area**

Ammonia nitrogen emissions per unit of land area in a region can be calculated as follows:

Ammonia nitrogen emissions per unit of land area = Ammonia nitrogen emissions/ (Area under land survey)

Data sources: *China Environmental Statistical Yearbook 2009*, the Ministry of Environmental Protection, Beijing, China Environmental Science Press, December 2010; *China City Statistical Yearbook 2010*, NBS, Beijing, China Statistical Publishing House, March 2011.

38. **Ammonia nitrogen emissions per capita**

Ammonia nitrogen emissions per capita in a region can be calculated as follows:

Ammonia nitrogen emissions per capita
= Ammonia nitrogen emissions / Mid-year population

Data sources: *China Environmental Statistical Yearbook 2009*, the Ministry of Environmental Protection, Beijing, China Environmental Science Publishing House, December 2010; *China City Statistical Yearbook 2010*, NBS, Beijing, China Statistical Publishing House, March 2011.

39. **Consumption of chemical fertilizers per unit of area of cultivated land**

Consumption of chemical fertilizers in agriculture refers to the quantity of chemical fertilizers applied in agriculture in the year, including nitrogenous fertilizer, phosphate fertilizer, potash fertilizer, and compound fertilizer. The consumption of chemical fertilizers is calculated in terms of volume of effective components by means of converting the gross weight of the respective fertilizers into weight containing effective component (e.g. nitrogen content in nitrogenous fertilizer, phosphorous pentoxide contents in phosphate fertilizer, and potassium oxide contents in potash fertilizer). Compound fertilizer is converted in regard to its major components. The formula is:

$$\text{Volume of effective component} = \text{physical quantity} \times \text{effective component of certain chemical fertilizer} \ (\%)$$

Area of cultivated land refers to refers to area of land reclaimed for the regular cultivation of various farm crops, including crop-cover land, fallow, newly reclaimed land and land laid idle for less than 3 years. Consumption of chemical fertilizers per unit of area of cultivated land can be calculated as follows:

$$\text{Consumption of chemical fertilizers per unit of area of cultivated land} = \text{Chemical fertilizers consumed} / \text{Area of cultivated land}$$

Data sources: *China Statistical Yearbook 2010*, NBS, Beijing, China Statistical Publishing House, September 2010.

40. **Consumption of pesticides per unit of area of cultivated land**

Consumption of pesticides per unit of area of cultivated land refers to pesticides consumed by each unit of area of cultivated land in a region in a given period, which can be calculated as follows:

$$\text{Consumption of pesticides per unit of area of cultivated land} = \text{Pesticides consumed} / \text{Area of cultivated land}$$

Data sources: *China Environmental Statistical Yearbook 2009*, the Ministry of Environmental Protection, Beijing, China Environmental Science Press, December 2010; *China City Statistical Yearbook 2010*, NBS, Beijing, China Statistical Publishing House, March 2011.

41. **Nitrogen oxides emissions from road traffic per capita**

Nitrogen oxides emissions from road traffic per capita can be calculated as follows:

$$\text{Nitrogen oxides emissions from road traffic per capita} = \text{Nitrogen oxides emissions from road traffic / Mid-year population}$$

Data sources: *China Environmental Statistical Yearbook 2009*, the Ministry of Environmental Protection, Beijing, China Environmental Science Press, December 2010; *China City Statistical Yearbook 2010*, NBS, Beijing, China Statistical Publishing House, March 2011.

42. **Ratio of environmental protection expenditures to government expenditures**

Expenditures for environment protection refer to the spending of government on environment protection, including the expense on administration of environment protection, environment monitoring and supervision, pollution control, natural ecology protection, project of virgin forests protection, reforesting farmland, controlling the sources of dust storms, returning pastureland to grassland, returning pastureland to grassland, returning cultivated land to grassland, energy conservation, emissions reduction, comprehensive utilization of renewable energy and resources, etc. Ratio of environmental protection expenditures to government expenditures refers to the percentage of expenditures for environment protection over government expenditures. Its formula is:

$$\text{Ratio of environmental protection expenditures to government expenditures} = 100\,\% \times \text{Expenditures for environmental protection / Government expenditures}$$

Data sources: *China Statistical Yearbook 2010*, NBS, Beijing, China Statistical Publishing House, September 2010.

43. **Ratio of investment in the control of environmental pollution to GRP**

Investment in the control of environmental pollution refers to the formation of investment in fixed assets in the total investment in harnessing pollution and in the construction of urban environment infrastructure facilities. It includes investment in harnessing sources of industrial pollution, investment in environment protection facilities designed concurrently with construction projects, and Investment in urban environment infrastructure facilities.

Ratio of Investment in the control of environmental pollution to GRP refers to the percentage of Investment in the control of environmental pollution over GRP in a region in a given period. Its formula is:

$$\text{Ratio of investment in the control of environmental pollution to GRP} = 100\,\% \times \text{Investment in the control of environmental pollution / GRP}$$

Data sources: *China Environmental Statistical Yearbook 2009*, the Ministry of Environmental Protection, Beijing, China Environmental Science Press, December

2010; *China City Statistical Yearbook 2010*, NBS, Beijing, China Statistical Publishing House, March 2011.

44. **Per capita investment of water sanitation and toilet improvement in rural areas**

Rural population refer to population living in towns and villages under the jurisdiction of counties. Per capita investment of water sanitation and toilet improvement in rural areas can be calculated as follows:

$$\begin{aligned}&\text{Per capita investment of water sanitation and toilet improvement in rural areas}\\&= \text{Investment of water sanitation and toilet improvement in rural areas} /\\&\quad \text{Mid-year rural population}\end{aligned}$$

Data sources: *China Environmental Statistical Yearbook 2009*, the Ministry of Environmental Protection, Beijing, China Environmental Science Press, December 2010; *China City Statistical Yearbook 2010*, NBS, Beijing, China Statistical Publishing House, March 2011.

45. **Investment in converting cultivated land into forests and grassland per unit of area of cultivated land**

Investment in converting cultivated land into forests and grassland per unit of area of cultivated land can be estimated as follows:

$$\begin{aligned}&\text{Investment in converting cultivated land into forests and grassland per unit of area}\\&\quad\text{of cultivated land} = \text{Investment in forestry/ Area of cultivated land}\end{aligned}$$

Data sources: *China Environmental Statistical Yearbook 2009*, the Ministry of Environmental Protection, Beijing, China Environmental Science Press, December 2010; *China City Statistical Yearbook 2010*, NBS, Beijing, China Statistical Publishing House, March 2011.

46. **Ratio of expenditures for science and technology, education, culture, and medical and health care to government expenditures**

Expenditure for science and technology refers to the spending of government on science and technology (S&T), including the expense on the administration of S&T, basic research, applied research, research and development, conditions and services of S&T, popularization of social science, science and technology, exchanges and cooperation of S&T, etc.

Expenditure for education refers to the spending of government on education, including the expense on the administration of education, pre-primary education, primary education, secondary education, high school education, regular higher education, primary vocational education, secondary vocational education, technical school education, vocational high school education and higher vocational education, radio and television education, student abroad education, special education, on the job training of cadres, education authorities services, etc.

Expenditure for culture, sport and media refers to the spending of government on culture, cultural heritage, sports, radio, film, television, press and publication, etc.

Expenditure for medical and health care refers to the spending of government on medical and health care, including the expense on administration of medical and health care, medical services, health care, disease prevention and control, health inspection and supervision, women and children's health, rural health care, etc.

Ratio of expenditures for science and technology, education, culture, and medical and health care to government expenditures can be estimated as follows:

$$\begin{aligned}&\text{Ratio of expenditures for science and technology, education, culture, and}\\&\text{medical and health care to government expenditures}\\&= 100\,\% \times \big(\text{Expenditures for science and technology}\\&+ \text{Expenditures for education} + \text{Expenditures for culture} + \text{Expenditures}\\&\text{for medical and health care}\big) / \text{Local government general budgetary expenditures}\end{aligned}$$

Data sources: *China Statistical Yearbook 2010*, NBS, Beijing, China Statistical Publishing House, September 2010.

47. **Area of green land per capita in urban areas**

Area of green land refers to total area occupied for green projects at the end of the reference period, including park green land, green land attached to institutions, residential quarter green land, production green land, protection green land, and scenic forest land. Area of green land per capita in urban areas can be calculated as follows:

$$\begin{aligned}&\text{Area of green land per capita in urban area}\\&= 100\,\% \times \text{Area of green land} \;/\; \text{Year-end urban non-agricultural population}\end{aligned}$$

Data sources: *China Municipal Construction Statistical Yearbook 2009*, the Ministry of Housing, Urban and Rural Development, Beijing, China Planning Press, November 2010.

48. **Coverage rate of urban population with access to tap water**

Coverage rate of urban population with access to tap water refers to ratio of urban population with access to tap water to total urban population. The formula is:

$$\begin{aligned}&\text{Coverage rate of urban population with access to tap water} =\\&100\,\% \times \text{Urban population with access to tap water} \;/\; \text{Urban population}\end{aligned}$$

Data sources: *China Statistical Yearbook 2010*, NBS, Beijing, China Statistical Publishing House, September 2010.

49. **Treatment rate of urban waste water**

Treatment rate of urban waste water refers to the percentage of waste water treated over waste water discharged in urban areas.Its formula is:

$$\text{Treatment rate of urban waste water} = 100\,\% \times \text{Waste water treated} / \text{Waste Water discharged}$$

Data sources: *China Environmental Statistical Yearbook 2010*, the NBS and the Ministry of Environmental Protection, Beijing, China Statistical Publishing House, November 2011.

50. **Ratio of urban consumption wastes treated**

Ratio of urban consumption wastes treated refers to consumption wastes treated over those produced in urban areas. In practical statistics, as it is difficult to estimate the volume of consumption wastes produced, which is replaced with the volume of consumption wastes transported. It is calculated as follows:

$$\text{Ratio of urban consumption wastes treated} = 100\,\% \times \text{Consumption wastes treated/Consumption wastes produced}$$

Data sources: *China Statistical Yearbook 2010*, NBS, Beijing, China Statistical Publishing House, September 2010.

51. **Public transportation vehicles per 10000 urban population**

Public transportation vehicles per 10000 urban population refers to the number of public transportation vehicles converted in terms of common standards, at the end of the reference period, per 10000 population in the city district. Its calculation formula is:

$$\text{Public transportation vehicles per 10000 urban population} = \text{Number of public transportation vehicles} / \text{City district population}$$

Data sources: *China City Statistical Yearbook 2010*, NBS, Beijing, China Statistical Publishing House, March 2011.

52. **Per capita length of urban public transit operating routes**

Per capita length of urban public transit operating routes refers to the length of public transit operating routes shared by each person in urban areas, which can be estimated as follows:

$$\text{Per capita length of urban public transit operating routes} = \text{Length of urban public transit operating routes} / \text{Mid-year urban population}$$

Data sources: *China Statistical Yearbook 2010*, NBS, Beijing, China Statistical Publishing House, September 2010.

53. **Ratio of rural population benefiting from water improvement projects to total rural population**

Population benefiting from water improvement projects refer to population who have benefited from various forms of water improvement projects. Rural population refer to population living in towns and villages under the jurisdiction of counties.

Ratio of rural population benefiting from water improvement projects to total rural population= 100 % × Rural population benefiting from water improvement projects Rural population

Data sources: *China Environmental Statistical Yearbook 2009*, the Ministry of Environmental Protection, Beijing, China Environmental Science Press, December 2010; *China City Statistical Yearbook 2010*, NBS, Beijing, China Statistical Publishing House, March 2011.

54. **Ratio of green covered area to completed urban area**

Completed urban area of green land refers to the total area occupied for green projects at the end of the reference period, including park green land, production green land, protection green land, green land attached to institutions, and other green areas.

Ratio of green covered area to completed urban area refers to the percentage of green covered area over completed urban area. Its calculation formula is:

$$\text{Ratio of green covered area to completed urban area} = 100\,\% \times \text{Green covered area} / \text{Completed urban area}$$

Data sources: *China Statistical Yearbook 2010*, NBS, Beijing, China Statistical Publishing House, September 2010.

55. **Area of Afforestation per capita**

Total area of afforestation refers to the total area of land suitable for afforestation, including barren hills, idle land, sand dunes, non-timber forest land, woodland and "grain for green" land, on which acres of forests, trees and shrubs are planted through manual planting. Area of afforestation per capita can be calculated as follows:

$$\text{Area of Afforestation per capita} = \text{Total area of afforestation} / \text{Mid-year population}$$

Data sources: *China Environmental Statistical Yearbook 2009*, the Ministry of Environmental Protection, Beijing, China Environmental Science Press, December 2010; *China City Statistical Yearbook 2010*, NBS, Beijing, China Statistical Publishing House, March 2011.

56. **Removal rate of industrial SO_2 emissions**

SO_2 emissions through industrial activities refer to volume of sulphur dioxide emission from fuel burning and production process by enterprises during a given period. Volume of industrial $\mathbf{SO_2}$ **e**missions removed refers to volume of $\mathbf{SO_2}$ removed in waste gas from fuel burning and production process after it being treated by various facilities for control of waste gas.

Removal rate of industrial SO_2 emissions refers to the percentage of volume of industrial $\mathbf{SO_2}$ **e**missions removed over the sum of SO_2 emissions through industrial activities and volume of industrial $\mathbf{SO_2}$ **e**missions removed. Its calculation formula is:

$$\text{Removal rate of industrial } SO_2 \text{ emissions} = 100\,\% \times \text{Volume of industrial } SO_2 \text{ emissions removed}/(SO_2 \text{ emissions through industrial activities} + \text{Volume of industrial SO 2 emissions removed})$$

Data sources: *China Statistical Yearbook 2010*, NBS, Beijing, China Statistical Publishing House, September 2010.

57. **Removal rate of COD in industrial waste water**

Removal rate of COD in industrial waste water refers to the percentage of amount of COD removed in industrial waste water over the sum of amount of COD discharged in industrial waste and amount of COD removed in industrial waste water. Its calculation formula is:

$$\text{Removal rate of COD in industrial waste water} = 100\,\% \times \text{Amount of COD removed in industrial waste water}/\left(\text{Amount of COD discharged in industrial waste} + \text{Amount of COD removed in industrial waste water}\right)$$

Data sources: *China Environmental Statistical Yearbook 2009*, the Ministry of Environmental Protection, Beijing, China Environmental Science Press, December 2010; *China City Statistical Yearbook 2010*, NBS, Beijing, China Statistical Publishing House, March 2011.

58. **Removal rate of industrial NOx emissions**

Volume of industrial $\mathbf{NO_x}$ emissions refers to volume of $\mathbf{NO_x}$ discharged into the atmosphere generated from production processes by enterprises during a given period. Volume of industrial $\mathbf{NO_x}$ **e**missions removed refers to volume of $\mathbf{NO_x}$ removed in waste gas from production process after it being treated by various facilities for control of waste gas.

Removal rate of industrial $\mathbf{NO_x}$ emissions refers to the percentage of volume of $\mathbf{NO_x}$ emissions removed over the sum of volume of $\mathbf{NO_x}$ emissions and volume of $\mathbf{NO_x}$ emissions removed. Its calculation formula is:

$$\text{Removal rate of } NO_x \text{ emissions} = 100\,\% \times \text{Volume of } NO_x \text{ emissions removed} / (\text{Volume of } NO_x \text{ emissions} + \text{Volume of } NO_x \text{ emissions removed})$$

Data sources: *China Environmental Statistical Yearbook 2009*, the Ministry of Environmental Protection, Beijing, China Environmental Science Press, December 2010.

59. **Removal rate of ammonia nitrogen in industrial waste water**

Removal rate of ammonia nitrogen in industrial waste water refers to the percentage of amount of ammonia nitrogen removed in industrial waste water over the sum of amount of ammonia nitrogen discharged in industrial waste and amount of ammonia nitrogen removed in industrial waste water. Its calculation formula is:

$$\text{Removal rate of ammonia nitrogen in industrial waste water} = 100\,\% \times \text{Amount of ammonia nitrogenremoved in industrial waste water}/ (\text{Amount of ammonia nitrogen discharged inindustrial waste} + \text{Amount of ammonia nitrogenremoved in industrial waste water})$$

Data sources: *China Environmental Statistical Yearbook 2009*, the Ministry of Environmental Protection, Beijing, China Environmental Science Press, December 2010; *China City Statistical Yearbook 2010*, NBS, Beijing, China Statistical Publishing House, March 2011.

60. **Sudden accidents effecting environment**

Sudden accidents effecting environment refer to sudden accidents, due to economic or social activities that are contrary to environment protection laws or due to unforeseen factors or natural disasters, that lead to environment pollution, destruction of protected wild animals, plants or nature reserves, damage to human health, economic and property losses, and other negative impacts on the society.

Data sources: *China Statistical Yearbook 2010*, NBS, Beijing, China Statistical Publishing House, September 2010.

Appendix C
Explanatory Notes and Data Sources on Indicators Used to Calculate the City Green Development Index in 2011

1. **GDP per capita**

Gross Domestic Product (GDP) refers to the final products at market prices produced by all resident units in a country (or a region) during a certain period of time. For a region, it is called as Gross Regional Product (GRP) or regional GDP. Its calculation formula is as follows:

$$\text{GDP per capita} = 2 \times \text{GRP}/(\text{Population at the end of this year} \\ + \text{Population at the end of preceding year})$$

Data sources: *China City Statistical Yearbook 2010*, NBS, Beijing, China Statistical Publishing House, 2011.

2. **Energy consumption per unit of GDP**

Total energy consumption refers to the total consumption of energy of various kinds by the production sectors and the households in the country in a given period. Energy consumption per unit of GRP refers to the energy consumption per unit of Gross Regional Product in a region in the same reference period, reflecting the increased energy consumption with an additional unit of GRP. The formula is:

$$\text{Energy consumption per unit of GDP} = \text{Total energy consumption}/\text{GRP}$$

Data sources: *China City Statistical Yearbook 2010*, NBS, Beijing, China Statistical Publishing House, 2011.

3. **Per capita electricity consumption by urban households**

Per capita electricity consumption by urban households refers to the ratio of electricity consumption by urban households to the mid-year urban population in a region in a given period. The formula is:

$$\text{Per capita electricity consumption by urban households} = \\ \text{Electricity consumption by urban households}/\text{Mid-year urban population}$$

X. Li and J. Pan (eds.), *China Green Development Index Report 2011*,
Current Chinese Economic Report Series, DOI: 10.1007/978-3-642-31597-8,

Data sources: *China City Statistical Yearbook 2010*, NBS, Beijing, China Statistical Publishing House, 2011.

4. **CO_2 emissions per unit of GRP**

CO_2 emissions per unit of GRP refer to the ratio of carbon dioxide emissions to GRP in a region in the same reference period. The formula is:

$$CO_2 \text{ emissions per unit of GRP} = CO_2 \text{ emissions} / \text{GRP}$$

5. **SO_2 emissions per unit of GRP**

Sulphur dioxide (SO_2) emissions cover SO_2 emissions through industrial activities, non-industrial and other activities, of which emissions through industrial activities refer to volume of sulphur dioxide discharged from fuel burning and production process by enterprises during a given period. The formula is:

$$\text{Industrial } SO_2 \text{ emissions} = SO_2 \text{ discharged from the process of fuel burning} + SO_2 \text{ discharged from the process of production}$$

Emission through non-industrial and other activities are calculated on the basis of consumption of coal by households and other activities and the sulphur content of coal with the following formula:

$$\text{Emission through non-industrial and other activities} = \text{coals consumed by household and other activities} \times \text{Sulphur content} \times 0.8 \times 2$$

SO_2 emissions per unit of GRP refer to the ratio of SO_2 emissions to GRP in a region in a given period. The formula is:

$$SO_2 \text{ emissions per unit of GRP} = SO_2 \text{ emissions} / \text{GRP}$$

Data sources: *China Statistical Yearbook 2010*, NBS, Beijing, China Statistical Publishing House, 2010; *China Environmental Statistical Yearbook 2009*, the Ministry of Environmental Protection, Beijing, China Environmental Science Press, 2010.

6. **COD emissions per unit of GRP**

Chemical Oxygen Demand (COD)refers to the amount of oxygen required when chemical oxidants are used to oxidize organic pollutants in water. A higher value of COD corresponds to more serious pollution COD emissions mainly come from waste water discharged by industry and household, of which COD emissions of the latter refer to the annual amount of COD in the waste water discharged by urban households, which can be calculated by per capita coefficient as follows:

COD emissions in waste water discharged by urban households = Coefficient of COD generated through urban non-industrial waste water × Urban non-agricultural population × 365

COD emissions per unit of GRP **refer to the ratio of** COD emissions to GRP in a region in a given period.

$$\text{COD emissions per unit of GRP} = \text{COD}/\text{GRP}$$

Data sources: *China Statistical Yearbook 2010*, NBS, Beijing, China Statistical Publishing House, 2010; *China Environmental Statistical Yearbook 2009*, the Ministry of Environmental Protection, Beijing, China Environmental Science Press, 2010.

7. **NO_x emissions per unit of GRP**

Volume of industrial **NO_x** emissions refers to volume of **NO_x** discharged into the atmosphere during a given period.

Nitrogen oxides (NO_x) emissions per unit of GRP refer to the ratio of NO_x emissions to GRP in a region in a given period. The formula is:

$$NO_x \text{ emissions per unit of GRP} = NO_x \text{ emissions} / \text{GRP}$$

Data sources: *China Statistical Yearbook 2010*, NBS, Beijing, China Statistical Publishing House,2010; *China Environmental Statistical Yearbook 2009*, the Ministry of Environmental Protection, Beijing, China Environmental Science Press, 2010.

8. **Ammonia nitrogen emissions per unit of GRP**

Ammonia nitrogen emissions refer to the amount of ammonia nitrogen in industrial waste water discharged by enterprises and urban households.

Ammonia nitrogen emissions per unit of GRP refer to the ratio of ammonia nitrogen emissions to GRP in a region in a given period. The formula is:

$$\text{Ammonia nitrogen emissions per unit of GRP} \\ = \text{Ammonia nitrogen emissions} / \text{GRP}$$

Data sources: *China Statistical Yearbook 2010*, NBS, Beijing, China Statistical Publishing House, 2010; *China Environmental Statistical Yearbook 2009*, the Ministry of Environmental Protection, Beijing, China Environmental Science Press, 2010.

9. **Labour productivity of primary sector**

Labour productivity of primary sector refers to the ratio of value added of primary sector to annual average of employees in primary sector. The formula is:

$$\text{Labour productivity of primary sector} = 2 \times \text{Value added of primary sector}/(\text{Employees in primary sector at the end of this year} + \text{Employees in primary sector at the end ofpreceding year})$$

Data sources: *China Statistical Yearbook 2010*, NBS, Beijing, China Statistical Publishing House, 2010.

10. **Labour productivity of secondary sector**

Labour productivity of secondary sector refers to the ratio of value added of secondary sector to annual average of employees in secondary sector. The formula is:

$$\text{Labour productivity of primary sector} = 2 \times \text{Value added of primary sector}/(\text{Employees in primary sector at the end of this year} + \text{Employees in primary sector at the end ofpreceding year})$$

Data sources: *China Statistical Yearbook 2010*, NBS, Beijing, China Statistical Publishing House, 2010.

11. **Water consumption per unit of industrial value added**

Water consumption per unit of industrial value added refers to the ratio of water consumed by industry to its value added in a given period. Of which value added of industry refers to the final results of industrial production of industrial enterprises in money terms during the reference period; Water consumed by industry refers to water consumed by industrial enterprises and non-industrial activities within them, which is measured in terms of new withdrawals of water and reuse of water. The formula is:

$$\text{Water consumption per unit of industrial value added} = \text{Water consumed by industry}/\text{Its value added}$$

Data sources: *China Statistical Yearbook 2010*, NBS, Beijing, China Statistical Publishing House, 2010; *China Environmental Statistical Yearbook 2009*, the Ministry of Environmental Protection, Beijing, China Environmental Science Press, 2010.

12. **Energy consumption per unit of industrial value added**

Energy consumption per unit of industrial value added refers to the ratio of energy consumed by industrial enterprises to their value added in a region in a given period. The formula is:

$$\text{Energy consumption per unit of industrial value added} = \text{Energy consumed by industrial enterprises/Their value added}$$

Data sources: *China Statistical Yearbook 2010*, NBS, Beijing, China Statistical Publishing House, 2010.

13. **Ratio of industrial solid wastes utilized**

Ratio of industrial solid wastes utilized refers to the percentage of industrial solid wastes utilized over industrial solid wastes produced (including stocks of the previous years). Its calculation formula is:

$$\text{Ratio of industrial solid wastes utilized} = 100\,\% \times \text{Industrial solid wastes utilized}/(\text{Industrial solid wastes produced} + \text{Stocks of previous utilized})$$

Of which, industrial solid wastes produced refers to total volume of solid, semi-solid and high concentration liquid residues produced by industrial enterprises from the process of production in a given period, including hazardous wastes, slag and ash, gangue, tailings, radioactive residues and other wastes, but excluding stones stripped or dug out in mining (gangue and acid or alkaline stones not covered). A stone is acid or alkaline if the pH value of the water is below 4 or above 10.5, when it is in, or soaked by, the water; Industrial solid wastes utilized refers to volume of solid wastes from which useful materials may be extracted or which can be converted into usable resources, energy or other materials by means of reclamation, processing, recycling and exchange (including utilizing the stocks of industrial solid wastes of the previous years in the years in question, e.g. fertilizers, building materials and road materials. The information concerned shall be collected by the producing units of wastes.)

Data sources: *China Environmental Statistical Yearbook 2010*, the NBS and the Ministry of Environmental Protection, Beijing, China Statistical Publishing House, November 2011.

14. **Reuse rate of industrial water**

The amount of water used in industrial production refers to the amount of water used in industrial and non-industrial activities within enterprises, which covers new withdrawals of water and reuse of water. New withdrawals of water refer to the amount of fresh water used in industrial and non-industrial activities within enterprises in a given period (the amount of fresh water used in non-industrial activities is measured separately, excluding houschold sewage that is not discharged altogether with household waste water), which cover their new withdrawals of water from urban running water and their self-supplied water; Reuse of water refers to the amount of water reused by enterprises in a given period, including recycling water, multi-purpose water, and cascade of water (covering water reused after control).

Reuse rate of industrial water refers to the ratio of the amount of water re-used in industrial production to total amount of water used in a given period. The formula is:

$$\text{Reuse rate of industrial water} = 100\,\% \times \text{Reuse of water}/$$
$$(\text{New withdrawals of water} + \text{ reuse of water})$$

Data sources: *Annual Statistical Report on Environment in China 2009*, the Ministry of Environmental Protection, Beijing, China Environment Science Press, 2010.

15. **Labour productivity of tertiary sector**

Labour productivity of tertiary sector refers to the ratio of value added of tertiary sector to annual average of employees in secondary sector. The formula is:

$$\text{Labour productivity of tertiary sector } = 2\times$$
$$\text{Value added of tertiary sector}/(\text{Employees in tertiary sector at the end of this year} +$$
$$\text{Employees in tertiary sector at the end of preceding year})$$

Data sources: *China City Statistical Yearbook 2010*, NBS, Beijing, China Statistical Publishing House, 2011.

16. **Proportion of value added of tertiary sector in GRP**

Proportion of value added of tertiary sector in GRP refers to the percentage of value added of tertiary sector over GRP in a reporting period. The formula is:

$$\text{Proportion of value added of tertiary sector} = 100\,\%$$
$$\times \text{ Value added of tertiary sector}/\text{ GRPProportion of value added of tertiary}$$
$$\text{sector} = 100\,\% \times \text{Value added of tertiary sector}/\text{ GRP}$$

Data sources: *China City Statistical Yearbook 2010*, NBS, Beijing, China Statistical Publishing House, 2010.

17. **Proportion of employees of tertiary sector in total employees**

Proportion of employees of tertiary sector in total employees refers to the percentage of employees of tertiary sector over total employees in a reporting period. The formula is:

$$\text{Proportion of employees of tertiary sector}$$
$$= 100\,\% \times \text{Employees of tertiary sector}/\text{ Total employees}$$

Data sources: *China City Statistical Yearbook 2010*, NBS, Beijing, China Statistical Publishing House, 2010.

18. **Water resources per capita**

Total Water Resources refers to total volume of water resources measured as run-off for surface water from rainfall and recharge for groundwater in a given area, excluding transit water.

Water resources per capita refer to run-off for surface water from rainfall and recharge for groundwater shared by each person in a region in a given period. The formula is:

$$\text{Water resources per capita} = \text{Total water resources}/\text{Year-end population}$$

Data sources: *Annual Statistical Report on Environment in China 2009*, the Ministry of Environmental Protection, Beijing, China Environment Science Press, 2010.

19. **CO_2 emissions per unit of land area**

CO_2 emissions per unit of land area in a region can be calculated as follows:

$$CO_2 \text{ emissions per unit of land area} = CO_2 \text{ emissions} / \text{Land area of an administrative region}$$

20. **CO_2 emissions per capita**

CO_2 emissions per capita can be calculated as follows:

$$CO_2\text{emissions per capita} = \text{Total } CO_2\text{emissions} /\text{Mid-year population}$$

21. **SO_2 emissions per unit of land area**

Land area of an administrative region refers to all land area (including water area) within it, which is measure in terms of administrative divisions.

SO_2 emissions per unit of land area in a region can be calculated as follows:

$$SO_2 \text{ emissions per unit of land area} = CO_2 \text{ emissions} / (\text{Land area of an administrative region})$$

Data sources: *Annual Statistical Report on Environment in China 2009*, the Ministry of Environmental Protection, Beijing, China Environment Science Press, 2010; *China Urban Living and Price Statistical Yearbook 2010*, NBS, Beijing, China Statistical Publishing House, 2011.

22. **SO_2 emissions per capita**

SO_2 emissions per capita in a region can be calculated as follows:

$$SO_2 \text{ emissions per capita} = \text{Total } SO_2 \text{ emissions} /\text{Mid-year population}$$

Data sources: *China Environmental Statistical Yearbook 2009*, the Ministry of Environmental Protection, Beijing, China Environmental Science Press, 2010; *China Urban Living and Price Statistical Yearbook 2010*, NBS, Beijing, China Statistical Publishing House, 2011.

23. **COD emissions per unit of land area**

COD emissions per unit of land area in a region can be calculated as follows:

$$\text{COD emissions per unit of land area} = \text{COD emissions} / (\text{Land area of an administrative region})$$

Data sources: *China Environmental Statistical Yearbook 2009*, the Ministry of Environmental Protection, Beijing, China Environmental Science Press, 2010; *China Urban Living and Price Statistical Yearbook 2010*, NBS, Beijing, China Statistical Publishing House, 2011.

24. **COD emissions per capita**

COD emissions per capita in a region can be calculated as follows:

$$\text{COD emissions per capita} = \text{Total COD emissions} / \text{Mid-year population}$$

Data sources: *China Environmental Statistical Yearbook 2009*, the Ministry of Environmental Protection, Beijing, China Environmental Science Press, 2010; *China Urban Living and Price Statistical Yearbook 2010*, NBS, Beijing, China Statistical Publishing House, 2011.

25. **Nitrogen oxides emissions per unit of land area**

Nitrogen oxides emissions per unit of land area in a region can be calculated as follows:

$$\text{Nitrogen oxides emissions per unit of land area} = \text{Nitrogen oxides emissions} / (\text{Land area of an administrative region})$$

Data sources: *China Environmental Statistical Yearbook 2009*, the Ministry of Environmental Protection, Beijing, China Environmental Science Press, 2010; *China Urban Living and Price Statistical Yearbook 2010*, NBS, Beijing, China Statistical Publishing House, 2011.

26. **Nitrogen oxides emissions per capita**

Nitrogen oxides emissions per capita in a region can be calculated as follows:

$$\text{Nitrogen oxides emissions per capita} = \text{Nitrogen oxides emissions} / \text{Mid-year population}$$

Data sources: *China Environmental Statistical Yearbook 2009*, the Ministry of Environmental Protection, Beijing, China Environmental Science Press, 2010;

China Urban Living and Price Statistical Yearbook 2010, NBS, Beijing, China Statistical Publishing House, 2011.

27. **Ammonia nitrogen emissions per unit of land area**

Ammonia nitrogen emissions per unit of land area in a region can be calculated as follows:

$$\text{Ammonia nitrogen emissions per unit of land area} = \text{Ammonia nitrogen emissions} / (\text{Land area of an administrative region})$$

Data sources: *China Environmental Statistical Yearbook 2009*, the Ministry of Environmental Protection, Beijing, China Environmental Science Press, 2010; *China Urban Living and Price Statistical Yearbook 2010*, NBS, Beijing, China Statistical Publishing House, 2011.

28. **Ammonia nitrogen emissions per capita**

Ammonia nitrogen emissions per capita in a region can be calculated as follows:

$$\text{Ammonia nitrogen emissions per capita} = \text{Ammonia nitrogen emissions} / \text{Mid-year population}$$

Data sources: *China Environmental Statistical Yearbook 2009*, the Ministry of Environmental Protection, Beijing, China Environmental Science Press, 2010; *China Urban Living and Price Statistical Yearbook 2010*, NBS, Beijing, China Statistical Publishing House, 2011.

29. **Ratio of days with air quality at and above level 2 to the whole year**

Air Pollution Index (API) is defined as a measure of pollution index ranges and corresponding pollutant concentrations in terms of air quality standards and each pollutant's effects on human health and ecological environment. Up to now API adopted in China may be divided into 5 levels, of which level 1 means that the air quality is excellent and meets the requirements for air quality in natural reserves, scenic spots and other regions needing special protection, if API is at or below 50; Level 2 means that the air quality is good, if API is above 50 but less than 100; Level 3 means that the air quality is lightly polluted, if API is above 100 but less than 200; Level 4 means that the air quality is moderately polluted, if API is above 200 but less than 300; Level 5 means that the air quality is heavily polluted, if API is above 200.

Ratio of days with air quality at and above level 2 to the whole year refers to the percentage of days with air quality at and above level 2 over all days in a year at an administrative region.

Data sources: The Data Centre of the Ministry of Environmental Protection, http://datacenter. Mep.gov.cn.

30. Ratio of days with principal pollutants as respirable suspended particulate to the whole year

Principal pollutants refer to the most polluted ones, including sulphur dioxide, nitrogen dioxide and respirable suspended particulate to be measured. Respirable suspended particulate refers to solid matter in the air with diameter from 0.1 to 100 μm that can be suspended in the air, instead of falling onto the ground under the gravity force for long periods of time.

Ratio of days with principal pollutants as respirable suspended particulate to the whole year refers to the percentage of days with principal pollutants as respirable suspended particulate over all days in a year at an administrative region.

Data sources: The Data Centre of the Ministry of Environmental Protection, http://datacenter. Mep.gov.cn.

31. Ratio of environmental protection expenditures to government expenditures

Expenditures for environment protection refer to the spending of government on environment protection, including the expense on administration of environment protection, environment monitoring and supervision, pollution control, natural ecology protection, project of virgin forests protection, reforesting farmland, controlling the sources of dust storms, returning pastureland to grassland, returning pastureland to grassland, returning cultivated land to grassland, energy conservation, emissions reduction, comprehensive utilization of renewable energy and resources, etc. Ratio of environmental protection expenditures to government expenditures refers to the percentage of expenditures for environment protection over government expenditures. Its formula is:

$$\begin{aligned}&\text{Ratio of environmental protection expenditures to government expenditures}\\&= 100\,\% \times \text{Expenditures for environmental protection} /\\&\text{Local government budgetary expenditures}\end{aligned}$$

Data sources: *China Statistical Yearbook 2010*, NBS, Beijing, China Statistical Publishing House, 2010.

32. Ratio of investment in the control of industrial environmental pollution to GRP

Investment in the control of industrial environmental pollution refers to investment in fixed assets in harnessing industrial waste water, waste gas, solid wastes, noise and other environmental pollution. Ratio of Investment in the control of industrial environmental pollution to GRP refers to the percentage of Investment in the control of industrial environmental pollution over GRP in a region in a given period. Its formula is:

$$\text{Ratio of investment in the control of industrial environmental pollution to GRP} = 100\,\% \times \text{Investment in the control of industrial environmental pollution} / \text{GRP}$$

Data sources: *Annual Statistical Report on Environment in China 2009*, the Ministry of Environmental Protection, Beijing, China Environment Science Press, 2010.

33. **Ratio of expenditures for science and technology, education, culture, and medical and health care to government expenditures**

Expenditure for science and technology refers to the spending of government on science and technology (S&T), including the expense on the administration of S&T, basic research, applied research, research and development, conditions and services of S&T, popularization of social science, science and technology, exchanges and cooperation of S&T, etc.

Expenditure for education refers to the spending of government on education, including the expense on the administration of education, pre-primary education, primary education, secondary education, high school education, regular higher education, primary vocational education, secondary vocational education, technical school education, vocational high school education and higher vocational education, radio and television education, student abroad education, special education, on the job training of cadres, education authorities services, etc.

Expenditure for culture, sport and media refers to the spending of government on culture, cultural heritage, sports, radio, film, television, press and publication, etc.

Expenditure for medical and health care refers to the spending of government on medical and health care, including the expense on administration of medical and health care, medical services, health care, disease prevention and control, health inspection and supervision, women and children's health, rural health care, etc.

Ratio of expenditures for science and technology, education, culture, and medical and health care to government expenditures can be estimated as follows:

$$\begin{aligned}&\text{Ratio of expenditures for science and technology, education, culture,}\\&\text{and medical and health care to government expenditures} = 100\,\% \times\\&(\text{Expenditures for science andtechnology} + \text{Expenditures for education} +\\&\text{Expenditures for culture} + \text{Expenditures for medical andhealth care})\\&/\text{Local governmentgeneralbudgetary expenditures}\end{aligned}$$

Data sources: *China Statistical Yearbook 2010*, NBS, Beijing, China Statistical Publishing House, 2010.

34. **Area of green land per capita in urban areas**

Area of green land refers to total area occupied for green projects at the end of the reference period, including park green land, green land attached to institutions, residential quarter green land, production green land, protection green land, and scenic forest land. Area of green land per capita in urban areas can be calculated as follows:

$$\begin{aligned}&\text{Area of green land per capita in urban area}\\&\quad = 100\,\% \times \text{Area of green land} \,/\, \text{Year-end urban non}\\&\qquad - \text{agricultural population}\end{aligned}$$

Data sources: *China City Statistical Yearbook 2010*, NBS, Beijing, China Statistical Publishing House, 2010.

35. **Ratio of green covered area to completed urban area**

Ratio of green covered area to completed urban area refers to the percentage of green covered area over completed urban area at the end of the reference period. Its calculation formula is:

$$\begin{aligned}&\text{Ratio of green covered area to completed urban area}\\&\quad = 100\,\% \times \text{Green covered area} \,/\, \text{Completed urban area}\end{aligned}$$

Data sources: *China Municipal Construction Statistical Yearbook 2009*, the Ministry of Housing, Urban and Rural Development, Beijing, China Planning Press, 2010.

36. **Coverage rate of urban population with access to tap water**

Coverage rate of urban population with access to tap water refers to ratio of urban population with access to tap water to total urban population. The formula is:

$$\begin{aligned}&\text{Coverage rate of urban population with access to tap water}\\&\quad = 100\,\% \times \text{Urban population with access to tap water} \,/\, \text{Urban population}\end{aligned}$$

Data sources: *China Statistical Yearbook 2010*, NBS, Beijing, China Statistical Publishing House, 2010.

37. **Treatment rate of urban waste water**

Treatment rate of urban waste water refers to the percentage of waste water treated over waste water discharged in urban areas in a given period.Its formula is:

$$\begin{aligned}&\text{Treatment rate of urban waste water } =\\&100\,\% \times \text{Waste water treated} \,/\, \text{Waste Water discharged}\end{aligned}$$

Data sources: *Annual Statistical Report on Environment in China 2009*, the Ministry of Environmental Protection, Beijing, China Environment Science Press, 2010.

38. **Ratio of urban consumption wastes treated**

Ratio of urban consumption wastes treated refers to consumption wastes treated over those produced in urban areas in a given period. In practical statistics, as it is difficult to estimate the volume of consumption wastes produced, which is replaced with the volume of consumption wastes transported. It is calculated as follows:

$$\text{Ratio of urban consumption wastes treated} \\ = 100\,\% \times \text{Consumption wastes treated} \;/\; \text{Consumption wastes produced}$$

Data sources: *China Statistical Yearbook 2010*, NBS, Beijing, China Statistical Publishing House, 2010.

39. **Public transportation vehicles per 10000 urban population**

Public transportation vehicles per 10000 urban population refers to the number of public transportation vehicles converted in terms of common standards, at the end of the reference period, per 10000 population in the city district. Its calculation formula is:

$$\text{Public transportation vehicles per 10000 urban population} \\ = \text{Number of public transportation vehicles} \;/\; \text{City district population}$$

Data sources: *China City Statistical Yearbook 2010*, NBS, Beijing, China Statistical Publishing House, March 2011.

40. **Removal rate of industrial SO_2 emissions**

SO_2 emissions through industrial activities refer to volume of sulphur dioxide emission from fuel burning and production process by enterprises during a given period.

Volume of industrial **SO_2** emissions removed refers to volume of **SO_2** removed in waste gas from fuel burning and production process after it being treated by various facilities for control of waste gas.

Removal rate of industrial SO_2 emissions refers to the percentage of volume of industrial **SO_2** emissions removed over the sum of SO_2 emissions through industrial activities and volume of industrial **SO_2** emissions removed. Its calculation formula is:

$$\text{Removal rate of industrial } SO_2 \text{emissions} = 100\,\% \times \\ \text{Volume of industrial } SO_2 \text{emissions removed} / \\ (SO_2 \text{emissions through industrial activities} + \\ \text{Volume of industrial } SO_2 \text{emissions removed})$$

Data sources: *Annual Statistical Report on Environment in China 2009*, the Ministry of Environmental Protection, Beijing, China Environment Science Press, 2010.

41. **Removal rate of COD in industrial waste water**

COD emissions through industrial activities refer to pure weight of COD in industrial waste water discharged by enterprises during a given period.

Volume of industrial **COD** emissions removed refers to pure weight of **COD** removed in industrial waste water after it being treated by various facilities for control of waste water.

Removal rate of COD in industrial waste water refers to the percentage of amount of COD removed in industrial waste water over the sum of amount of COD discharged in industrial waste and amount of COD removed in industrial waste water. Its calculation formula is:

$$\text{Removal rate of COD in industrial waste water} = 100\,\% \times \text{Amount of COD removed in industrial waste water}/ \text{Amount of COD discharged in industrial waste} + \text{Amount of COD removed in industrial waste water})$$

Data sources: *Annual Statistical Report on Environment in China 2009*, the Ministry of Environmental Protection, Beijing, China Environment Science Press, 2010.

42. **Removal rate of industrial NO_x emissions**

Volume of industrial **NO_x** emissions refers to volume of **NO_x** discharged into the atmosphere generated from production processes by enterprises during a given period.

Volume of industrial **NO_x** emissions removed refers to volume of **NO_x** removed in waste gas from production process after it being treated by various facilities for control of waste gas.

Removal rate of industrial **NO_x** emissions refers to the percentage of volume of **NO_x** emissions removed over the sum of volume of **NO_x** emissions and volume of **NO_x** emissions removed. Its calculation formula is:

$$\text{Removal rate of } NO_x \text{ emissions} = 100\,\% \times \text{Volume of } NO_x \text{ emissions removed} / (\text{Volume of } NO_x \text{ emissions} + \text{Volume of } NO_x \text{ emissions removed})$$

Data sources: *Annual Statistical Report on Environment in China 2009*, the Ministry of Environmental Protection, Beijing, China Environment Science Press, 2010.

43. **Removal rate of ammonia nitrogen in industrial waste water**

Ammonia nitrogen emissions through industrial activities refer to pure weight of ammonia nitrogen in industrial waste water discharged by enterprises during a given period.

Volume of industrial ammonia nitrogen emissions removed refers to pure weight of ammonia nitrogen removed in industrial waste water after it being treated by various facilities for control of waste water.

Removal rate of ammonia nitrogen in industrial waste water refers to the percentage of amount of ammonia nitrogen removed in industrial waste water over the sum of amount of ammonia nitrogen discharged in industrial waste and amount of ammonia nitrogen removed in industrial waste water. Its calculation formula is:

$$\begin{aligned}&\text{Removal rate of ammonia nitrogen in industrial waste water} =\\&100\,\% \times \text{Amount of ammonia nitrogen removed in industrial waste water}/\\&\quad(\text{Amount of ammonia nitrogen discharged in industrial waste} +\\&\text{Amount of ammonia nitrogen removed in industrial waste water})\end{aligned}$$

Data sources: *Annual Statistical Report on Environment in China 2009*, the Ministry of Environmental Protection, Beijing, China Environment Science Press, 2010.

MIX
Papier aus verantwortungsvollen Quellen
Paper from responsible sources
FSC® C105338

FSC
www.fsc.org

If you have any concerns about our products,
you can contact us on
ProductSafety@springernature.com

In case Publisher is established outside the EU,
the EU authorized representative is:
Springer Nature Customer Service Center GmbH
Europaplatz 3, 69115 Heidelberg, Germany

Printed by Libri Plureos GmbH
in Hamburg, Germany